Stefan Felsner

Geometric Graphs and Arrangements

W0225838

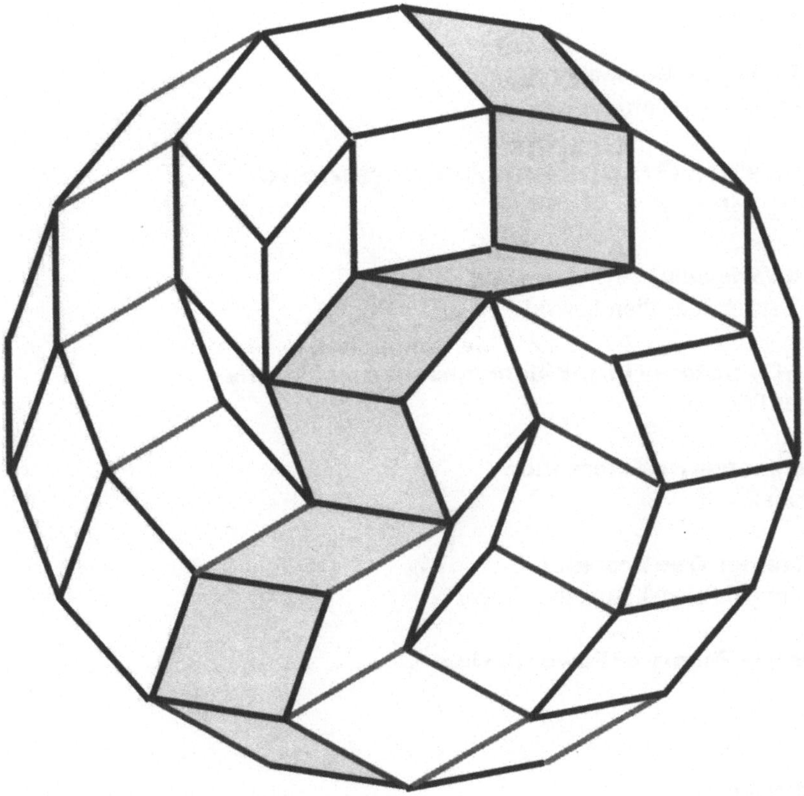

Advanced Lectures in Mathematics

Editorial Board
Prof. Dr. Martin Aigner, Freie Universität Berlin, Germany
Prof. Dr. Peter Gritzmann, Technische Universität München, Germany
Prof. Dr. Volker Mehrmann, Technische Universität Berlin, Germany
Prof. Dr. Gisbert Wüstholz, ETH Zürich, Switzerland

Introduction to Markov Chains
Ehrhard Behrends

Einführung in die Symplektische Geometrie
Rolf Berndt

Wavelets – Eine Einführung
Christian Blatter

Lattices and Codes
Wolfgang Ebeling

Local Analytic Geometry
Theo de Jong, Gerhard Pfister

Geometric Graphs and Arrangements
Stefan Felsner

Ruled Varieties
Gerd Fischer, Jens Piontkowski

Dirac-Operatoren in der Riemannschen Geometrie
Thomas Friedrich

Hypergeometric Summation
Wolfram Koepf

The Steiner Tree Problem
Hans Jürgen Prömel, Angelika Steger

The Basic Theory of Power Series
Jesús M. Ruiz

vieweg

Stefan Felsner

Geometric Graphs and Arrangements

Some Chapters from Combinatorial Geometry

vieweg

Bibliographic information published by Die Deutsche Bibliothek
Die Deutsche Bibliothek lists this publication in the Deutsche Nationalbibliografie; detailed bibliographic data is available in the Internet at <http://dnb.ddb.de>.

Prof. Dr. Stefan Felsner
Technische Universität Berlin
Institut für Mathematik (MA 6-1)
Straße des 17. Juni 136
10623 Berlin, Germany
felsner@math.tu-berlin.de

Mathematics Subject Classification
MSC 2000: 05-01, 05C10, 05C62, 52-01, 52C10, 52C30, 52C42

First edition, February 2004

All rights reserved
© Friedr. Vieweg & Sohn Verlag / GWV Fachverlage GmbH, Wiesbaden 2004

Vieweg is a company in the specialist publishing group Springer Science+Business Media.
www.vieweg.de

No part of this publication may be reproduced, stored in a retrieval system or transmitted, mechanical, photocopying or otherwise without prior permission of the copyright holder.

Cover design: Ulrike Weigel, www.CorporateDesignGroup.de

Printed on acid-free paper

ISBN 978-3-528-06972-8 ISBN 978-3-322-80303-0 (eBook)
DOI 10.1007/978-3-322-80303-0

Preface

The body of this text is written. It remains to find some words to explain what to expect in this book. A first attempt of characterizing the content could be:

MSC 2000: 05-01, 05C10, 05C62, 52-01, 52C10, 52C30, 52C42.

In words: The questions posed and partly answered in this book are from the intersection of graph theory and discrete geometry. The reader will meet some graph theory with a geometric flavor and some combinatorial geometry of the plane. Though, the investigations always start in the geometry of the plane it is sometimes appropriate to pass on to higher dimensions to get a more global understanding of the structures under investigation. This is the in Chapter 7, for example, when the study of triangulations of a point configuration leads to the definition of secondary polytopes.

I like to think of the book as a collection which makes up a kind of bouquet. A bouquet of problems, ideas and results, each of a special character and beauty, put together with the intention that they supplement each other to form an interesting and appealing whole.

The main mathematical part of the text contains only few citations and references to related material. These additional bits of information are provided in the last section of each chapter, 'Notes and References'. On average the bibliography of a chapter contains about thirty items. This is far from being a complete list of the relevant literature. The intention is to just indicate the most valuable literature so that these sections can serve as entry points for further studies. The text is supplemented by many figures to make the material more attractive and help the reader get a sensual impression of the objects. In some cases, I have confined the presentation to results which fall behind today's state of the art. I wanted to emphasize the main ideas and stop before technical complexity starts taking over. This strategy should make the mathematics accessibility to a relatively broad audience including students of computer science, students of mathematics, instructors and researchers.

The book can serve different purposes. It may be used as textbook for a course or as a collection of material for a seminar. It should also be helpful to people who want to learn something about specific themes. They may concentrate on single chapters because all the chapters are self-contained and can be read as stand alone surveys.

Topics

Chapter 1. We introduce basic notion graph theory and explain what geometric and topological graphs are. Planar graphs and some important theorems about them are reviewed. The main results of this chapter are bounds for some extremal problems for geometric graphs.

Chapter 2. We show that a 3-connected planar graph with f faces admits a convex drawing on the $(f-1) \times (f-1)$ grid. The result is based on Schnyder woods, a special cover of the edges of a 3-connected planar graph with three trees. Schnyder woods bring along connections to geodesic embeddings of planar graphs and to the order dimension of planar graphs and 3-polytopes.

Chapter 3. This is about non-planar graphs. How many crossing pairs of edges do we need in any drawing of a given graph in the plane? The Crossing Lemma provides a bound and has beautiful applications to deep extremal problems. We explain some of them.

Chapter 4. Let $\mathcal{P}$ be a configuration of n points in the plane. A k-set of $\mathcal{P}$ is a subset $\mathcal{S}$ of k points of $\mathcal{P}$ which can be separated from the complement $\bar{\mathcal{S}} = \mathcal{P} \setminus \mathcal{S}$ by a line. The notorious k-set problem of discrete geometry asks for asymptotic bounds of this number as a function of n. We present bounds, Welzl's generalization of the Lovász Lemma to higher dimensions and close with the surprisingly related problem of bounding the rectilinear crossing number of complete graphs from below.

Chapter 5. This chapter contains selected results from the extremal theory for configurations of points and arrangements of lines. The main results are bounds for the number of ordinary lines of a point configuration and for the number of triangles of an arrangement.

Chapter 6. Compared to arrangements of lines, arrangements of pseudolines have the advantage that they can be nicely encoded by combinatorial data. We introduce several combinatorial representations and prove relations between them. For each representation we give an applications which makes use of specific properties. The encoding by triangle orientations has a natural generalization which leads to higher Bruhat orders.

Chapter 7. In this chapter we study triangulations of a point configuration. The flip operation allows to move between different triangulations. The Delaunay triangulation is investigated as a special element in the graph of triangulations. This graph is shown to be related to the skeleton graph of the secondary polytope. In the special case of a point configuration in convex position they coincide. In this case, we make use of hyperbolic geometry to get a lower bound for the diameter of the graph of triangulations.

Chapter 8. Rigidity allows a different view to geometric graphs. We introduce rigidity theory and prove three characterizations of minimal generically rigid graphs in the plane. Pseudotriangulations are shown to be the planar minimal generically rigid graphs in the plane. The set of pseudotriangulations with vertices embedded in a fixed point configuration $\mathcal{P}$ has a nice structure. There is a notion of flip that allows to move between different pseudotriangulations. The flip-graph is a connected graph and it turns out that it is the skeleton graph of a polytope. The beautiful theory finds a surprising application in the Carpenter's Rule Problem.

The selection of topics is clearly governed by my personal taste. This is a drawback, because your taste is likely to differ from mine, at least in some details. The advantage is that though there are several books on related topics each of these books is clearly distinguishable by its style and content. In the spirit of 'Customers who bought this book also bought' I recommend the following four books:

- J. Matoušek
  Lectures on Discrete Geometry
  Graduate Texts in Math. 212
  Springer-Verlag, 2002.
- J. Pach and P. K. Agarwal
  Combinatorial Geometry
  John Wiley & Sons, 1995.
- G. M. Ziegler
  Lectures on Polytopes
  Graduate Texts in Math. 152
  Springer-Verlag, 1994.
- H. Edelsbrunner
  Geometry and Topology for Mesh Generation,
  Cambridge University Press, 2001.

Feedback

There will be something wrong. You may find errors of different nature. Inadvertently, I may not have given proper credit for certain contribution. You may know of relevant work that I have overlooked or you may have additional comments. In all these cases: Please let me know.

- `felsner@math.tu-berlin.de`

You can find a list of errata and a collection of comments and pointers related to the book at the following web-location:

- `http://www.math.tu-berlin.de/~felsner/gga-book.html`

Acknowledgments

There are *sine qua non* conditions for the existence of this book: The work of a vast number of mathematicians who laid out the ground and produced what I am retelling. Friends and family of mine, in particular Diana and my parents. without their persistent trust -blind or not- I would not have been able to finish this task. I am truly grateful to all of them.

In Berlin we have a wonderful environment for discrete mathematics. While working on this book, I had the privilege of working in teams of discrete mathematicians at the Freie Universität and at the Technische Unversität. I want to thank the nice people in these groups for providing me with such a friendly and supportive 'ambiente'.

I have been involved in the European graduate college 'Combinatorics, Geometry and Computing' and in the European network COMBSTRU and I am glad to acknowledge support from their side.

I'm grateful to Martin Aigner, he furthered this project with advice and the periodically renewed question: "What about your book Stefan?". For valuable discussions and for help by way of comments and corrections I want to thank Patrick Baier, Peter Braß, Frank Hoffmann, Klaus Kriegel, Ezra Miller, Walter Morris, János Pach, Günter Rote, Sarah Renkl, Volker Schulz, Ileana Streinu, Tom Trotter, Pavel Valtr, Uli Wagner, Helmut Weil, Emo Welzl and Günter Ziegler.

Berlin, December 2003
Stefan Felsner

Contents

1 Geometric Graphs: Turán Problems

The first part of this chapter collects rather elementary and well known material: We start with definitions of geometric and topological graphs, rush through basic notions from graph theory and report on facts about planar graphs. Beginning with Section 1.4 we discuss problems from the extremal theory for geometric graphs. That is, we deal with questions of Turán type: How many edges can a geometric graph avoiding a specified configuration of edges have?

1.1 What is a Geometric Graph?

As usual in texts dealing with graph theory, we shall begin by defining a *graph* G to be a pair (V, E) where V is the set of *vertices*, E is the set of *edges* $\{v, w\}$ each joining two vertices $v, w \in V$. To illustrate the definition, let us look at some pictures.

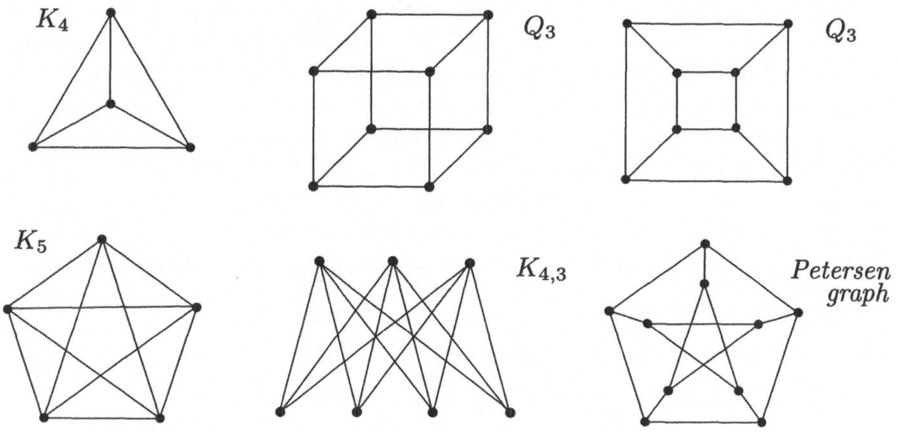

Figure 1.1 A little gallery of graphs, the cube graph Q_3 is shown with two drawings.

A *geometric graph* is a graph $G = (V, E)$ drawn in the plane with straight edges. That is, V is a set of points in the plane and E is a set of line segments with endpoints in V. For convenience, we often assume that the points of V are in general position, i.e., no three points of V are collinear. Pictures of graphs, as Figure 1.1, display geometric graphs. Hence, the study of geometric graphs can be viewed as the study of straight line drawings of a graph.

A *topological graph* is a graph drawn in the plane such that edges are represented by Jordan curves with the property that any two of these curves share at most one point. Obviously, geometric graphs are a subclass of topological graphs

In general, a drawing of a graph G will display *crossings* of edges that are disjoint as edges of G. Many interesting questions about geometric and topological graphs are concerned with crossing patterns in one way or the other. Among the simplest and most natural questions of this type is the quest of characterizing those graphs admitting a drawing without crossing edges. These *planar graphs* have been intensely studied for more than hundred years. Before initiating our studies on geometric graphs with a section on planar graphs, we continue with a brief overview of basic concepts of graph theory.

1.2 Fundamental Concepts in Graph Theory

Two graphs $G = (V, E)$ and $G' = (V', E')$ are *isomorphic* if there is a bijection $V \to V'$ such that two vertices are joined by an edge in G iff the corresponding vertices are joined by an edge in G'. When discussing the structure of a graph G the concrete set of vertices is, usually, of no relevance and G is used as an arbitrary representative of its isomorphism class. If this is not the case, then we emphasize the fact that the vertex set of G is distinguished by calling G a *labeled graph.* Two vertices joined by an edge are called *adjacent.* If vertex v is a vertex of edge e, then v and e are said to be *incident.* The *neighborhood* $N(v)$ of vertex v is the set of adjacent vertices. The size of the neighborhood is the degree $d(v) = |N(v)|$. A *complete graph* is a graph such that every pair of its vertices is an edge. With K_n we denote a complete graph (the isomorphism class of complete graphs) on n vertices.

A *path* in G is a sequence $v_1, v_2, \ldots, v_k$ of vertices of G such that $\{v_i, v_{i+1}\}$ is an edge for $1 \leq i < k$. We call $v_1, v_2, \ldots, v_k$ *a path from* v_1 *to* v_k and vertices v_1 and v_k the *endpoints* of the path. A path is a *simple path* if all its vertices are distinct. A *cycle* is a path with identical endpoints, i.e., $v_1 = v_k$. A graph without cycles is a *forest.*

A graph G is *connected* if any two vertices can be joined by a path, otherwise G is *disconnected.* The maximal connected pieces of a graph G are the *components* of G. A connected forest is a *tree.* Trees have many characterizations.

Proposition 1.1 *For a graph G with n vertices the following are equivalent:*

(a) *G is connected and has no cycles (i.e., G is a tree).*

(b) *G is connected and has $n - 1$ edges.*

(c) *G has no cycles and $n - 1$ edges.*

A *leaf* is a vertex of degree 1, every tree with at least two vertices has at least two leaves. If $G = (V, E)$ is connected and W is a set of vertices, such that removing W and all edges incident to vertices in W from E leaves a disconnected graph, then we call W a *separating set* of G. For $k \geq 2$ we say that a graph G is *k-connected* if every separating set of G has cardinality at least k. For complete graphs we make the exceptional agreement that K_{k+1} is k-connected.

A *subgraph* of $G = (V, E)$ is a graph $H = (W, F)$ with $W \subseteq V$ and $F \subseteq E$. The subgraph H of G is *spanning* if H has the full vertex set of G, i.e., $W = V$. A particularly important class of spanning subgraphs are the *spanning trees* An *induced subgraph* of G is a subgraph H containing all edges of G joining two vertices in W. The subgraph of G induced by W is denoted $G[W]$. A *clique* in $G = (V, E)$ is a set W of vertices such that $G[W]$ is a complete graph. If $G[W]$ has no edges, then we call W an *independent set.* A graph $G = (V, E)$ is a *bipartite* graph if V can be partitioned into two independent sets V_1, V_2 such that $E \subseteq V_1 \times V_2$. The graphs Q_3 and $K_{4,3}$ from Figure 1.1 are examples of bipartite graphs. Similarly, G is *k-partite* if there is a partition $V_1, V_2, \ldots, V_k$ of V such that each V_i is independent. With $K_{n_1,n_2,\ldots,n_k}$ we denote a *complete k-partite* graph: It has a k-partition $V_1, V_2, \ldots, V_k$ such that V_i contains n_i vertices and the edges are all pairs of vertices from distinct classes.

Many problems in *extremal graph theory* can be stated in the form: How many edges can a graph on n vertices satisfying a certain property P have? An important special case of this question is when P is the property of avoiding a subgraph isomorphic to some fixed graph H, graphs avoiding such a subgraph are called *H-free*. Turán initiated extremal graph theory by solving a problem of this type.

Theorem 1.2 (Turán 1941)
The number of edges of a K_{k+1}-free graph on n vertices is at most $(1-\frac{1}{k})\frac{n^2}{2}$.

Turán also provided a complete characterization of the extremal examples. There is -up to isomorphism- a unique K_{k+1}-free graph with a maximum number of edges, this graph, the *Turán graph* $T_{k+1}(n)$ is the complete k-partite graph with $|V_i| = \lceil n/k \rceil$ or $|V_i| = \lfloor n/k \rfloor$ for all $i = 1, \ldots, k$.

Before studying extremal problems for geometric graphs, we have a section on planar graphs.

1.3 Planar Graphs

A graph is *planar* if it can be drawn in the plane without crossing edges. A *plane graph* is a planar graph together with such a drawing. In this context we do not insist that the edges are straight, that is, a plane drawing of a planar graph shows a topological graph but not necessarily a geometric graph. The existence of a straight line drawing for every planar graph is a nontrivial fact (the next chapter contains a proof). A planar

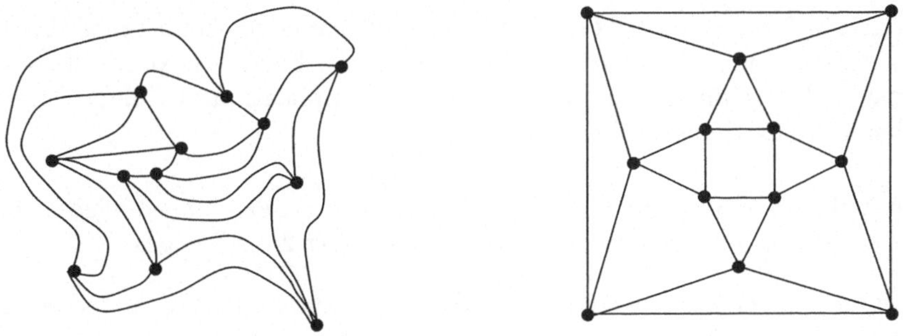

Figure 1.2 A topological and a straight plane drawing of the same graph.

drawing decomposes the plane into connected regions, one of them is the unbounded (outer) region, these regions are referred to as *faces*. To circumvent the special role taken by the outer face it is sometimes convenient to think of a planar drawing as a drawing on the sphere. A graph has a non-crossing drawing in the plane iff it has such a drawing on the sphere. This can be shown by stereographic projections: Place a sphere S on the plane and identify a point x in the plane with the point x' on the sphere where the ray from x to the north-pole N intersects S, see Figure 1.3.

One of the most fundamental and useful facts about planar graphs is Euler's Formula.

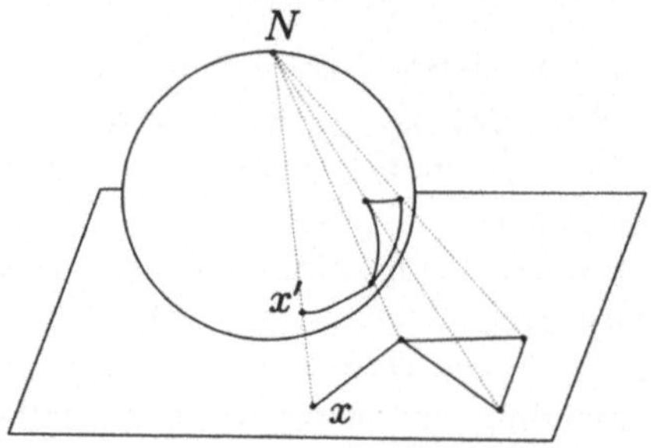

Figure 1.3 A stereographic projection.

Theorem 1.3 (Euler's Formula)
If G is a connected plane graph with v vertices, e edges and f faces, then

$$v - e + f = 2.$$

Proof. Remove the edges of G in any order from the drawing. Upon removal of any given edge a there are two possibilities:

(1) Edge a separates two different faces. Removing a reduces the number of faces by one and leaves the number of components invariant.

(2) Edge a is incident to the same face on both sides. Removing a increases the number of components by one and leaves the number of faces invariant*.

In the original graph we have one component and f faces. The final graph consisting of isolated vertices has v components and one face. Therefore, the number of edge removals of type (1) and (2) was $f - 1$ and $v - 1$ respectively. The total number of edge removals was e, hence, $e = (f - 1) + (v - 1)$. □

Note that the formula even holds if G has multiple edges connecting the same pair of vertices or loops, i.e, edges with a double endpoint. For the next result, however, we have to stay in the realm of (simple) graphs. This application of Euler's Formula gives the solution of an extremal problem.

Theorem 1.4 *A planar graph G with n vertices has at most $3n - 6$ edges. Moreover, if G is planar and contains no 3-cycle then it has at most $2n - 4$ edges.*

Proof. Assume that G is connected, otherwise some edges could be added such that the graph remains planar. Let f_k be the number of faces with k bounding edges, clearly, $\sum f_k = f$. Counting the bounding edges of all the faces we count every edge exactly twice, i.e., $2e = \sum k f_k$. Since every face has at least three edges on its boundary, this yields $2e \geq 3f$. Inserting this into Euler's Formula we obtain $3n - 6 = 3e - 3f \geq e$.

If G contains no 3-cycle every face has at least four edges on its boundary and $2e \geq 4f$. In this case we obtain $2n - 4 = 2e - 2f \geq e$. □

The proof shows that $e = 3n - 6$ and $f = f_3$ are equivalent. Hence, the extremal graphs are exactly those having only triangular faces, they are known with the name *planar triangulation*.

---

* This innocent statement is based on the Jordan Curve Theorem.

The count of edges alone allows the conclusion that K_5 and $K_{3,3}$ are not planar. In fact, K_5 has ten edges, one to much for a planar graph with five vertices. For $K_{3,3}$ it can be used that this graph contains no 3-cycle and $e = 9 > 8 = 2n - 4$.

Since every subgraph of a planar graph is planar we may ask for a characterization of planarity in terms of forbidden subgraphs. In a classical theorem Kuratowski has shown that the two graphs K_5 and $K_{3,3}$ are the only minimal obstructions against planarity. Let a *subdivision* of a graph G be a graph obtained from G by inserting new vertices of degree two into the edges of G. Planarity is invariant under taking subdivisions, in particular, subdivisions of K_5 or $K_{3,3}$ are not planar.

Theorem 1.5 (Kuratowski 1930)
A graph G is planar if and only if G contains no subgraph which is a subdivision of K_5 or $K_{3,3}$.

A drawing of a planar graph is called *convex* if every face boundary (including the unbounded face) is a convex polygon. Note that the 2-connected planar graph $K_{2,4}$ has no convex drawing.

Theorem 1.6 (Tutte 1960)
Every 3-connected planar graph admits a convex drawing.

Tutte's theorem can also be derived from a much older result from geometry, the theorem of Steinitz.

Theorem 1.7 (Steinitz 1922)
Every 3-connected planar graph is the skeleton graph, i.e., the graph of 0-dimensional faces (vertices) and 1-dimensional faces (edges), of a 3-dimensional polytope.

Given a 3-connected planar graph G, let P be a 3-dimensional polytope with skeleton G. Place P on the plane and choose a point a below the face of contact of P and the plane. Project each point x from P to the point x' in the plane where the ray from x to a intersects the plane. If a was chosen close enough to the plane this yields a convex drawing of G.

Planar graphs have many special properties. One of the most fundamental is the existence of a dual graph. A *dual* G^* of a plane graph G is a plane graph having a vertex in each face of G. Every edge e of G has a corresponding dual edge e^* in G^*: If F and F' are the faces on the two sides of e then e^* connects the vertices of G^* in F and F'.

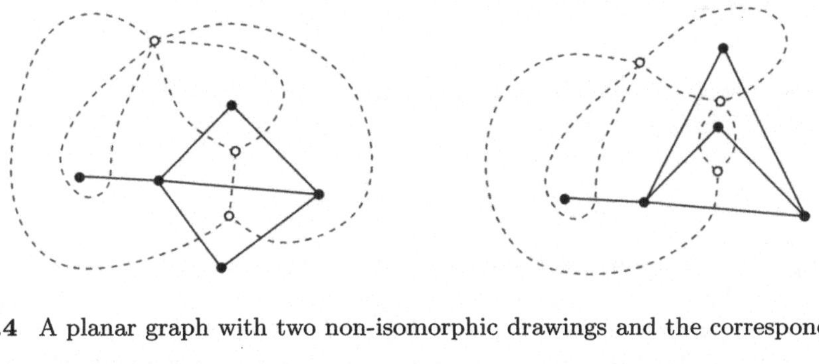

Figure 1.4 A planar graph with two non-isomorphic drawings and the corresponding duals.

Figure 1.4 illustrates that a dual may have some features which make it necessary to extend our concept of graphs. The dual G^* may have loops and multiple edges even though G is a simple graph. Having extended the definition of a graph accordingly it is almost evident from the drawings that $(G^*)^* = G$ for every connected plane graph G.

The figure also shows that a planar graph may have different drawings which have non-isomorphic duals. Whitney has characterized the well behaving planar graphs.

Theorem 1.8 (Whitney 1933)
A planar graph G has a unique dual iff G is 3-connected. Moreover, the dual is a simple graph in this case.

Let G be a plane graph and C be the edge set of a cycle in G. This cycle C splits the plane into at least two connected regions and each of these regions contains at least one face of G. If v^* and w^* are dual vertices from different regions, then every path connecting v^* and w^* has to cross some edge of C and, hence, contain an edge from the dual set C^*. Therefore, C^* is a *cut*, i.e., a disconnecting set of edges for G^*. In fact there is a bijection between simple cycles of G and minimal cuts of G^* and vice versa.

1.4 Outerplanar Graphs and Convex Geometric Graphs

A planar graph is *outerplanar* if it has a plane drawing such that all vertices lie on the boundary of the exterior face. Equivalently, G is outerplanar if the graph obtained by adding a new vertex x joined to all vertices of G is still planar.

Using Kuratowski's Theorem and the second definition of outerplanar graphs we conclude that a graph is outerplanar if and only if it contains no subgraph which is a subdivision of K_4 or $K_{2,3}$. The maximum number of edges of an n vertex outerplanar graph is $2n-3$. The extremal graphs can be drawn as convex n-gons with a triangulation of the interior by chords.

A *convex geometric graph* is a geometric graph with the property that the vertices are the vertices of a convex polygon.

A convex geometric graph without crossings is just a maximal outerplanar, therefore, a convex geometric graph with n vertices and without crossings has at most $2n-3$ edges. The next theorem deals with a less trivial extremal problem for convex geometric graphs. A pair of edges of a geometric graph is called *disjoint* if they do not cross in the geometric representation and share no vertex.

Theorem 1.9 *Let $dc_k(n)$ be the maximum number of edges that a convex geometric graph can have without containing $k+1$ pairwise disjoint edges. If $n \geq 2k+1$ then*

$$dc_k(n) = kn.$$

Proof. Let $x_0, x_1, \ldots, x_{n-1}$ be the cyclic order of the vertices of a convex geometric graph. Consider the set of all interior segments x_ix_j. Two segments x_ix_j and x_kx_l, with i, j, k, l distinct are called *parallel*, if there is a t such that $k = i+t$ and $l = j-t$, where indices are taken modulo n. This equivalence relation partitions the segments into n classes of pairwise parallel segments. If G has no $k+1$ pairwise disjoint edges, then G can have at most k edges from each class. This implies $dc_k(n) \leq kn$.

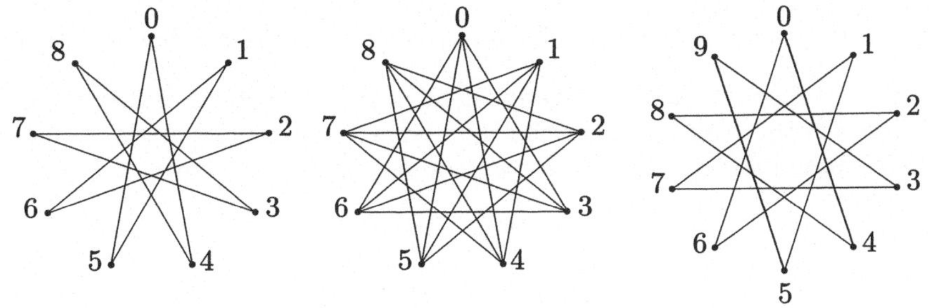

Figure 1.5 The graphs $G_1(9)$, $G_2(9)$ and $G_1(10)$.

To see that the bound can be attained we need appropriate graphs. The idea is to build a graph whose edges are the long segments: Let $G_k(n)$ be the graph whose edges are the segments $x_i x_{i+\lfloor \frac{n}{2} \rfloor + j}$ for $0 \le i \le n-1$ and $1 \le j \le k$. Figure 1.5 shows three examples.

Suppose that the vertices $x_0, x_1, \ldots, x_{n-1}$ are in clockwise order. Think of the edge $x_i x_{i+\lfloor \frac{n}{2} \rfloor + j}$ as being oriented $x_i \to x_{i+\lfloor \frac{n}{2} \rfloor + j}$. The oriented edge has $\lceil \frac{n}{2} \rceil - j - 1$ vertices on its right side and $\lfloor \frac{n}{2} \rfloor + j - 1$ vertices on its left side. Hence, the right side is always smaller than the left side. Therefore, all the edges in the definition of are different and $G_k(n)$ has nk edges.

Let P be a set of pairwise disjoint edges of $G_k(n)$, we want to show $|P| \le k$. Consider the plane graph G_P consisting of the cycle $x_0, x_1, \ldots, x_{n-1}, x_0$ together with the edges of P, these edges are disjoint chords in the cycle. Therefore, G_P is an outerplanar graph. Consider the dual graph of G_P and remove the vertex corresponding to the outer face of G_P. The remaining truncated dual is a tree, hence, has at least two leaves. They correspond to faces F_1 and F_2 of G_P which have only one edge e_1 resp. e_2 of P on the boundary. The vertices of F_i, $i = 1, 2$, are the two vertices of e_i together with all vertices on one side of e_i. From the above we know that each side of each edge of $G_k(n)$ contains at least $\lceil \frac{n}{2} \rceil - k - 1$ vertices. Vertices of F_i which are not on e_i cannot be incident to edges of P. We conclude, that there are at most $n - 2(\lceil \frac{n}{2} \rceil - k - 1)$ vertices which are incident to edges of P. In general this number is $\le 2k+2$ and if n is odd it is $\le 2k+1$. Edges of P are disjoint, therefore, the same bound holds for the number of incidences between edges of P and vertices. This later number, however, is twice the number of edges of P. Hence $|P| \le k$ for n odd.

For n odd and $n \ge 2k+1$ we have shown that the graph $G_k(n)$ proves $dc_k(n) \ge kn$. This is also true if n and k are both even and $n \ge 2k$. For n even and k odd however, $G_k(n)$ contains sets of $k+1$ pairwise disjoint edges. For an example look at the graph $G_1(10)$ from Figure 1.5. It is possible to modify $G_k(n)$ to show that $dc_k(n) \ge kn$ also holds for the pairs n even and k odd. □

1.5 Geometric Graphs without $(k+1)$-Pairwise Disjoint Edges

We consider the question: How many edges a geometric graph on n points can have without having $k+1$ pairwise disjoint edges? This is an old problem with a long history, see the notes at the end of the chapter. We begin with the easiest special case.

Theorem 1.10 *A geometric graph with no two disjoint edges can have at most n edges.*

Proof. Every vertex may mark an edge incident to it. For vertices of degree one there is no choice. For all other vertices mark the right edge at the largest angle at the vertex. If there remains an unmarked edge $e = uv$ we find a situation as shown in Figure 1.6. The

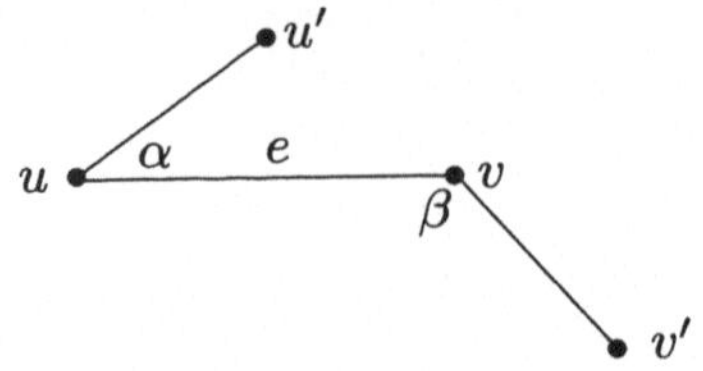

Figure 1.6 An unmarked edge e and its two disjoint adjacent edges.

angles α and β cannot be the largest angles at u and v. Hence, they are both less than π and the edges uu' and vv' must be disjoint because they reach into different halfplanes of the line supporting e. We conclude: In the absence of disjoint edges there is no unmarked edge and the number of edges is at most as large as the number of vertices. □

More complex marking arguments have been used to show that $3n+1$ edges force three disjoint edges and $10n+1$ edges force four disjoint edges. The following theorem deals with the general case.

Theorem 1.11 *A geometric graph with no $k+1$ disjoint edges cannot have more then $256\,k^2n$ edges.*

Proof. Let G be a geometric graph on n vertices containing no $k+1$ disjoint edges. Let $x(v)$ and $y(v)$ denote the x- and y-coordinate of point v. If there are two vertices with the same x-coordinate we slightly rotate the plane to make all x-coordinates different. Edge e is said to *lie above* edge e' if every vertical line intersecting both intersects e above e'. Define four relations $\prec_1, \prec_2, \prec_3$ and $\prec_4$ on disjoint pairs of edges. Let $e = v_1w_1$ and $e' = v_2w_2$ be two edges such that e is above e' and $x(v_1) < x(w_1)$ and $x(v_2) < x(w_2)$, we define:

$$
\begin{aligned}
e \prec_1 e' &\quad \text{iff} \quad x(v_1) < x(v_2) \text{ and } x(w_1) < x(w_2),\\
e \prec_2 e' &\quad \text{iff} \quad x(v_1) > x(v_2) \text{ and } x(w_1) > x(w_2),\\
e \prec_3 e' &\quad \text{iff} \quad x(v_1) < x(v_2) \text{ and } x(w_1) > x(w_2),\\
e \prec_4 e' &\quad \text{iff} \quad x(v_1) > x(v_2) \text{ and } x(w_1) < x(w_2).
\end{aligned}
$$

The definitions are illustrated in Figure 1.7. It is important to note that each of these relations is transitive, i.e., an order relation.

For any vertex v_i we partition the edges incident to v_i into *left edges*, those edges v_i, v_j with $x(v_j) < x(v_i)$, and *right edges*, those edges v_i, v_j with $x(v_i) < x(v_j)$. The *left degree* l_i is the number of left edges at v_i and the *right degree* r_i is the number of right edges at vertex v_i.

For every vertex v define two linear orders on the right edges of v. $R_s(v)$ is the order by decreasing slope and $R_x(v)$ is the order by increasing x-coordinate. The intersection $R_s(v) \cap R_x(v)$ is a partial order, actually a two dimensional order.

The following theorem is a corollary from the Greene-Kleitman duality theory:

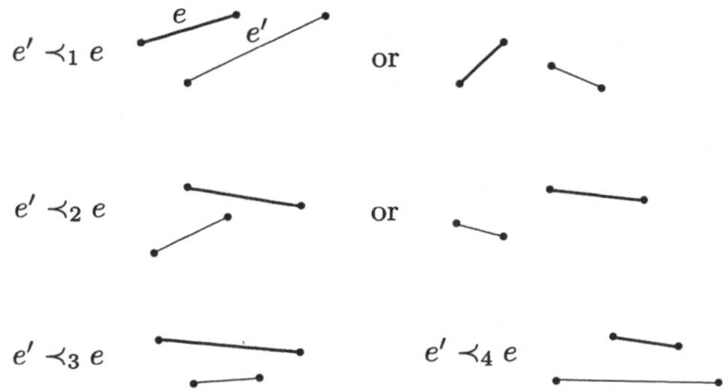

Figure 1.7 The four orders defined on disjoint edges.

Theorem 1.12 *Let $P = (X, <)$ be a partial order on n elements, then there is a family $\mathcal{C}$ of at most $\sqrt{n}$ chains and a family $\mathcal{A}$ of at most $\sqrt{n}$ antichains such that the chains and antichains in $\mathcal{C} \cup \mathcal{A}$ cover all elements of P.*

Hence, the right edges of a vertex v can be covered with at most $\sqrt{r_v}$ chains and $\sqrt{r_v}$ antichains. Let $C^r(v)$ be the set of edges covered by chains and $A^r(v)$ be the set of edges covered by antichains.

One of $\bigcup_v C^r(v)$ and $\bigcup_v A^r(v)$ contains at least one half of all edges of G. Let G' be the graph restricted to the edges of the larger class, clearly $e(G') > e(G)/2$. The *right block* of an edge of G' is the set of edges of the chain (or antichain) it belongs to. The order of edges of a right block is the order by decreasing slope, in this order the edges of a right block form an x-increasing (chain) or x-decreasing (antichain) sequence.

Let l'_v be the left degree of vertex v in G', clearly $l'_v \leq l_v$. We now uniformize the left edges of every vertex v in G'. Let $L_s(v)$ be the order by increasing slope and $L_x(v)$ be the order by increasing x-coordinate. As before, the intersection order $L_s(v) \cap L_x(v)$ can be covered by at most $\sqrt{l'_v}$ chains and $\sqrt{l'_v}$ antichains. Let $C^l(v)$ be the set of edges covered by chains and $A^l(v)$ be the set of edges covered by antichains. Let G'' be the graph restricted to the edges of the larger class of $\bigcup_v C^l(v)$ and $\bigcup_v A^l(v)$, clearly $e(G'') > e(G)/4$. The *left block* of an edge of G'' is the set of edges of the chain (or antichain) it belongs to. The order of edges of a left block is the order by increasing slope, in this order the edges of a left block form an x-increasing (chain) or x-decreasing (antichain) sequence.

The restriction of a right block of G' to the edges of G'' is a right block of G''. For two edges with a common endpoint, $e = vu$ and $e' = vw$, we say that (e, e') is a *right-zag*, if e' follows immediately after e in the same right block at vertex v. Analogously, for two edges, $e = uv$ and $e' = wv$, we say that (e, e') is a *left-zag*, if e' follows immediately after e in the same left block at v.

A path $e_1, e_2, \ldots, e_m$ of G'' is a *zig-zag* if in every pair e_i, e_{i+1} of consecutive edges is either a right-zag or a left-zag with left and right alternating along the path.

Claim 1. Every zig-zag of G'' contains at most $2k$ edges.

There are four cases, the right/left blocks can be increasing/decreasing, these cases correspond to the four order relations $\prec_i$. We detail one case, the other cases can be handled analogously.

Assume that all right blocks of G' (and hence of G'') are x-increasing and all left blocks of G'' are x-decreasing. Let $e_1, e_2, \ldots, e_{2k+1}$ be a zig-zag of length $2k+1$. We claim that $e_{i+2} \prec_3 e_i$ for all $1 \leq i \leq 2k-1$. Consider e_i, e_{i+1}, e_{i+2} and let a, b, c, d be the sequence of vertices of this zig-zag. The pair (e_i, e_{i+1}) is a right-zag or a left-zag at vertex b. We distinguish these two cases, see Figure 1.8.

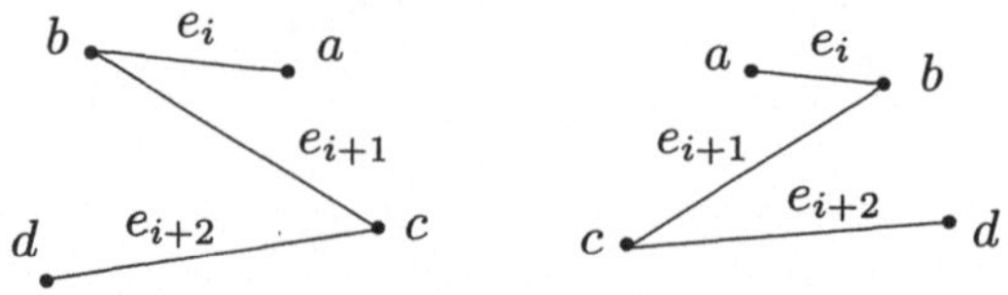

Figure 1.8 Edges e_i, e_{i+1} form a right-zag or a left-zag at b.

(1) If (e_i, e_{i+1}) is a right-zag then $x(a) < x(c)$. Since (e_{i+1}, e_{i+2}) is a left-zag at c it follows that $x(b) > x(d)$. Clearly, e_{i+2} is below e_i, so $e_{i+2} \prec_3 e_i$.

(2) If (e_i, e_{i+1}) is a left-zag then $x(a) > x(c)$. Since (e_{i+1}, e_{i+2}) is a right-zag at c it follows that $x(b) < x(d)$. Again, e_{i+2} is below e_i and $e_{i+2} \prec_3 e_i$.

Consequently, a zig-zag of length $2k+1$ contains a chain $e_{2k+1} \prec_3 e_{2k-1} \prec_3 \ldots \prec_3 e_1$ of length $k+1$. This is a contradiction, since the edges of a chain are pairwise disjoint. $\triangle$

Claim 2. If Z is the number of maximal zig-zags in G'', then $Z \leq 2\sqrt{e(G)n}$.

Let $e_1, e_2, \ldots, e_m$ be a maximal zig-zag if v is the vertex of e_1 which is not incident to e_2 then e_1 is the first element of its block at v. Therefore, the number of maximal zig-zags starting at a vertex v is at most the number of blocks of edges at v. By construction v has at most $\sqrt{r_v}$ right blocks and at most $\sqrt{l'_v} \leq \sqrt{l_v}$ left blocks. This can be turned into an estimate for Z. Apply a variant of the Cauchy-Schwarz inequality, namely, $(\sum_{i=1}^n a_i)^2 \leq n(\sum_{i=1}^n a_i^2)$ twice and use the obvious equation $\sum_{i=1}^n (r_i + l_i) = 2e(G)$, to obtain:

$$Z \leq \sum_{i=1}^n (\sqrt{r_i} + \sqrt{l_i}) \leq \sqrt{n \sum_{i=1}^n (\sqrt{r_i} + \sqrt{l_i})^2} \leq \sqrt{n \sum_{i=1}^n 2(r_i + l_i)} = \sqrt{4ne(G)}.$$

$\triangle$

Each edge of G'' is covered by at least one maximal zig-zag and each zig-zag contains at most $2k$ edges, hence $e(G'') \leq 2kZ$. By construction $e(G'') \geq e(G)/4$, together with Claim 2 this yields $e(G) \leq 16\,k\sqrt{e(G)n}$. Squaring the inequality and dividing by $e(G)$ we obtain $e(G) \leq 256\,k^2 n$. $\square$

1.6 Geometric Graphs without Parallel Edges

Assuming general position for the endpoints of two disjoint segments in the plane we can distinguish two cases. Either the convex hull of their four endpoints is a triangle or a quadrangle, in the second case we say that the two *edges are parallel*, see Figure 1.9.

A geometric graph on n points with no $k+1$ pairwise parallel edges can have more edges than a geometric graph with no $k+1$ pairwise disjoint edges. In the case of parallel edges, already the solution to the extremal problem for $k = 1$ requires the use of interesting techniques.

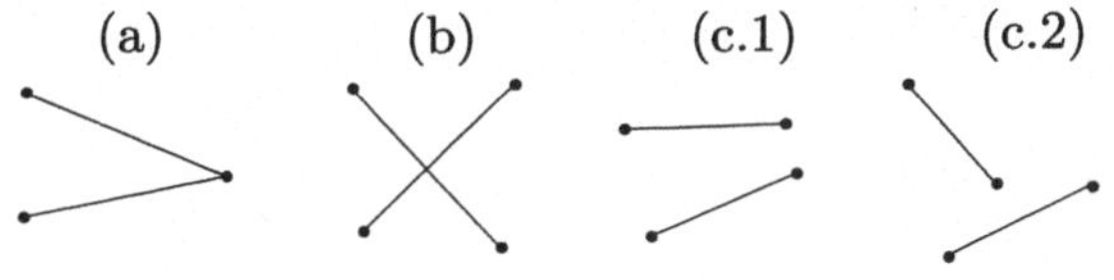

Figure 1.9 The four possible positions of two edges in a geometric graph, (a) adjacent, (b) crossing, (c.1) parallel, (c.2) stabbing.

Kupitz gave examples showing that in the absence of parallel edges a geometric graph can have as much as $2n - 2$ edges, see Figure 1.10, he conjectured that this is the maximum.

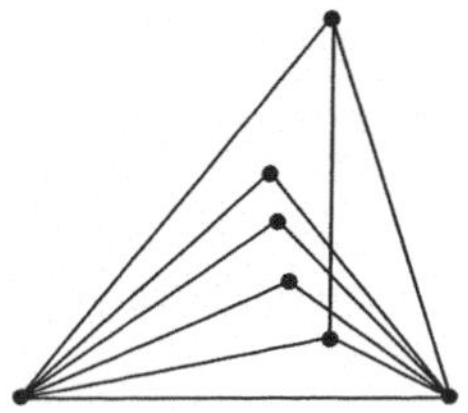

Figure 1.10 The construction of Kupitz for $n = 7$.

Theorem 1.13 *A geometric graph on n vertices with no two parallel edges can have at most $2n - 2$ edges.*

Let G be a geometric graph on n vertices and with $e(G)$ edges. For an edge e of G let l_e be the line supporting e. Choose a circle C such that all vertices $v_1, v_2, \ldots, v_n$ of G are in the interior of C. For every edge $e = v_i v_j$ consider the two points of intersection of line l_e with circle C and label them with the index of the closer vertex, i.e., traversing l_e we find point i, vertex v_i, vertex v_j and point j in this order. Let $S(G)$ be the circular sequence obtained from reading the labels of the intersection points along C in clockwise order. The length $|S(G)|$ of $S(G)$ is $2e(G)$.

From $S(G)$ we construct a reduced circular sequence $RS(G)$ by erasing all the labels which are equal to their predecessor labels. The reduced sequence for the graph in Figure 1.11 is $RS(G) = (4, 2, 3, 5, 3, 1, 3, 2, ($.

Lemma 1.14 *If G is a geometric graph without parallel edges, then $|RS(G)| \geq e(G)$.*

Proof. Consider an edge $e = v_i v_j$ and the labels $i = i_e$ and $j = j_e$ in $S(G)$ that come from e. Assuming that the successor of i_e in $S(G)$ is i and the successor of j_e is j we find parallel edges e_i and e_j as indicated in Figure 1.12. This shows that if G has no parallel edges, then every edge has at least one remaining label in $RS(G)$. In other words $|RS(G)| \geq e(G)$. □

Lemma 1.15 *If G is a geometric graph without parallel edges, then there is no circular subsequence i, j, i, j in $|RS(G)|$.*

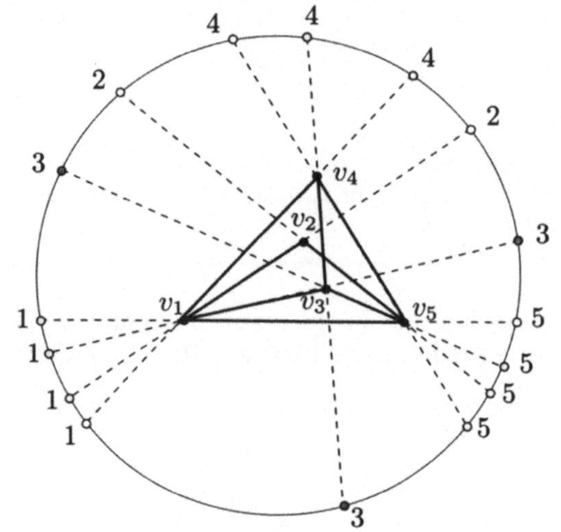

Figure 1.11 A graph G with circular sequence $S(G) = (4,4,4,2,3,5,5,5,5,3,1,1,1,1,3,2($.

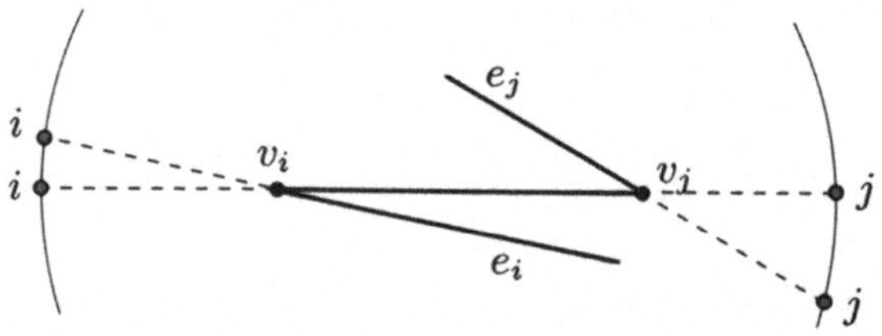

Figure 1.12 An edge with two erased successor labels implies a pair of parallel edges.

Proof. Suppose that there is a subsequence i, j, i, j in $RS(G)$. Let ℓ be the line through v_i and v_j. A rotation makes ℓ vertical, such that v_i is above v_j. One of the following two events is unavoidable: Reading clockwise there is a ji subsequence right of ℓ or a ij subsequence left of ℓ. In the second case exchange the labels i and j and rotate. Now we have the situation illustrated in Figure 1.13. The crossing right of ℓ of the supporting lines

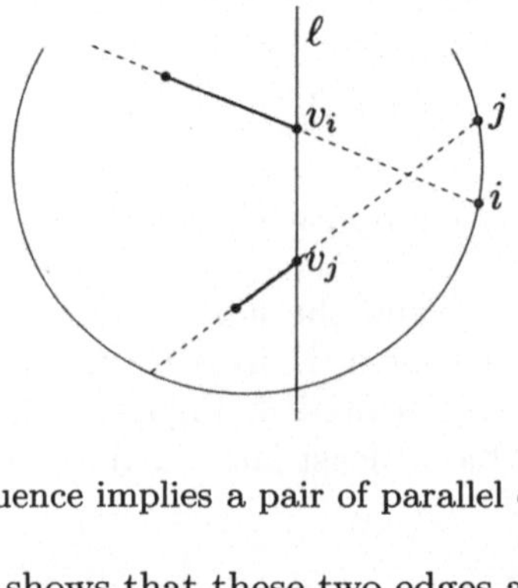

Figure 1.13 A i, j, i, j subsequence implies a pair of parallel edges.

of two edges situated left of ℓ shows that these two edges are parallel, a contradiction. □

The next lemma implies that $|RS(G)| \leq 2n - 2$. Together with Lemma 1.14 this completes the proof of Theorem 1.13.

Lemma 1.16 *If R is a circular sequence with entries from a set of $n > 1$ symbols such that no two adjacent entries are identical and R contains no circular subsequence of type abab, then the length of R is at most $2n - 2$.*

Proof. Let R be a circular sequence on n symbols such that no two adjacent entries are identical and R contains no circular subsequence of type *abab*. In the literature such a sequence is called a circular Davenport-Schinzel sequence of order 2. Let a be a symbol with $k \geq 2$ occurrences in R and decompose the sequence using this symbol, $R = aS_1aS_2a \dots aS_k$. The forbidden pattern enforces that for $i \neq j$ the subsequences S_i and S_j have no symbol in common. Let λ_i be the number of symbols of S_i. If we consider S_i as a circular sequence it contains no *abab* but the first and the last element may be equal. After removal of such a duplication the length of the sequence is at most $2\lambda_i - 2$, by induction. Therefore, $|S_i| \leq 2\lambda_i - 1$ and

$$|R| \leq \sum_{i=1}^{k}(2\lambda_i - 1) + k = 2\sum_{i=1}^{k}\lambda_i = 2(n-1).$$ □

1.7 Notes and References

A good introduction to many aspects of graph theory is the book of West [214], with its more then 500 references this book can serve as a starting point for deeper studies. Biggs et al. [26] give a nice account to the early history of graph theory. Extremal graph theory is the topic of a survey by Bollobás [29]. Four different proofs of Turán's theorem can be found in The Book by Aigner and Ziegler [8]. The existence of a straight line drawing for every planar graph is a theorem independently obtained by Wagner [208], Fáry [79] and Stein [187]. As part of his "Geometry Junkyard" Eppstein has collected fifteen proofs of Euler's Formula [72]. A chapter in The Book [8] is devoted to applications of the formula. The classification of regular polytopes is another application, see e.g. [214]. Extensions of Euler's formula in the theory of polytopes, in particular the Euler-Poincaré formula, are presented in the polytope book of Ziegler [219]. West [214] reproduces ideas of Thomassen to give a joint proof for the theorems of Kuratowski and Tutte. The theorem of Tutte is one of the main results in the next chapter of this book. The hard direction of Steinitz's theorem, to produce a polytope with a prescribed skeleton, can be proven in two ways. An inductive approach based on ΔY transformations is detailed by Ziegler [219]. Richter-Gebert [162] extends the approach used by Tutte to prove his theorem, this leads to a special plane drawing of the graph which has the property that the vertices can be lifted so that they form the desired polytope. More detailed comments about this approach can be found in the notes section of Chapter 2..

The book [142] of Mohar and Thomassen contains a rich chapter on planar graphs with many references. In particular they are careful about the use of the Jordan Curve Theorem in this theory.

Theorem 1.8 is a byproduct of Whitney's characterization of all plane embeddings of a 2-connected planar graph. He shows [217] that all these embeddings can be obtained from a given one by a series of *switchings* of two connected components. The construction of the dual graph introduced here is geometric. A multi-graph G^* is an *algebraic dual* of G if there is a bijection $E(G^*) \leftrightarrow E(G)$ between the edge-sets which maps simple cycles of G to minimal cuts of G^* and vice versa. Whitney characterized planar graphs as those graphs admitting an algebraic dual. The study of cuts and duals leads to the definition of a *matroid* and the rich theory thereof, West [214] gives an introduction and further references.

Outerplanar graphs and convex geometric graphs.

According to Pach [148] Theorem 1.9 was first obtained by Kupitz 1982. To show that the bound of the theorem is best possible the family $G_k(n)$ is described in [148]. The "dual" extremal problem was solved by Capoyleas and Pach [45]: The maximum number of edges that a convex geometric graph can have without containing $k+1$ pairwise crossing edges is $k(2n-2k-1)$ for all $n \geq 2k+1$.

Kupitz and Perles [127] study the maximum number of edges that a convex geometric graph can have without containing $k+1$ disjoint edges in convex position, that is all the edges are edges of the convex hull of their vertices. If $n \geq 2k+1$ then this number is $t_k(n)+n-k$, where $t_k(n)$ is the Turán number of Theorem 1.2.

Geometric graphs with no $k+1$ pairwise disjoint edges.

Besides the study of planar graphs, the first investigations on geometric graphs were in the context of repeated distances. For example the problem of determining the maximum number of diametral pairs of a set of n points in the plane is closely related to the maximum number of edges of a geometric graph with no disjoint edges. A related problem was posed by Hopf and Pannwitz in 1934. They ask for a proof that only for odd n it is possible to choose n points such that all the pairs x_i, x_{i+1}, $i = 1, .., n$ are diametral. Sutherland [191] and Fenchel [90] proposed solutions. Erdős [74] proved the bound od Theorem 1.10 in the context of distance problems.

The question about the maximum number $d_k(n)$ of edges of a geometric graph with no $k+1$ pairwise disjoint edges was raised by Kupitz and Perles [125] and again stated by Akiyama and Alon [10]. Alon and Erdős proved, $d_2(n) \leq 6n$. Goddard et al. [99] improved to $d_2(n) \leq 3n+1$, they also proved $d_3(n) \leq 10n+1$ and gave the first general upper bound $d_k(n) \in O(m(\log n)^{k-3})$. Pach and Törőcsik [154] introduced the order relations $\prec_i$, $i = 1, 2, 3, 4$, on disjoint edges. As an application of Dilworth's Theorem about chain decompositions of orders they could show that $d_k(n)$ is linear in n, more precisely $d_k(n) \leq k^4 n$. Tóth and Valtr [199] added the concept of a zig-zag and improved to $d_k(n) \leq k^3(n+1)$. They also construct examples to show the lower bound $d_k(n) > 1.5(k-1)n - 2k^2$. Tóth [197] obtained the bound on $d_k(n)$ of order ck^2n (Tóth says $c = 2^9$ but his argument only yields $c = 2^{13}$). Tóth is using a greedy approach based on the Erdős-Szekeres Theorem to define the blocks of edges at each vertex. Replacing this by an application of the more sophisticated Theorem 1.12 made the slightly improved constant in Theorem 1.11 possible.

Theorem 1.12 is an offspring of the Greene-Kleitman theory about chain and antichain families in orders. A nice proof based on min-cost flows is given by Frank [92]. The Greene-Kleitman Theory was surveyed by West [213] and more recently by Britz and Fomin [40].

It is widely believed that $d_k(n) \sim ckn$ for some very moderate c. It would already be very nice to know this for the restricted class of those geometric graphs admitting a line which intersects all the edges. Let $dl_k(n)$ be the maximum number of edges such a geometric graph can have if it has no $k+1$ pairwise disjoint edges. If $dl_k(n) \leq ckn$ the bound on $d_k(n)$ would drop to $d_k(n) \leq c'(k \log k)n$.

Conway defines a *thrackle* as a graph drawn in the plane such that any two distinct edges either have a common vertex or meet at exactly one point where they cross, i.e., a thrackle is a topological graph without disjoint edges. Every cycle is a thrackle (exercise).

Conway conjectured that a thrackle with n vertices has at most n edges, i.e., that the bound of Theorem 1.10 remains valid in the more general context of topological graphs.

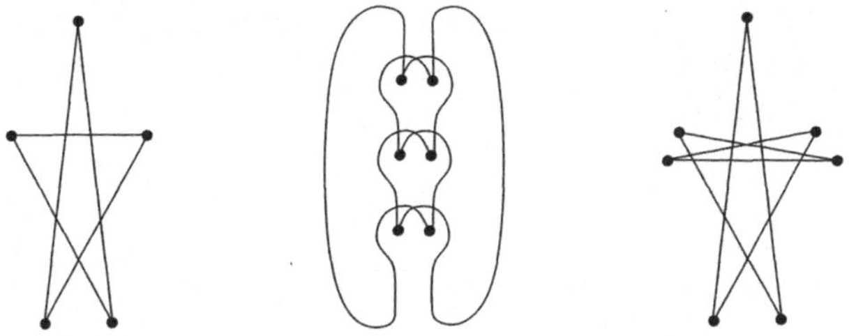

Figure 1.14 Thrackle-representations of the 5-, 6- and 7-cycle.

For about 40 years the best known upper bound on the number of edges of a thrackle was $O(n^{3/2})$. This follows from the fact that a subgraph of a thrackle is a thrackle and the four-cycle is not a thrackle. In a very nice paper Lovász, Pach and Szegedy [134] show that the number of edges of a thrackle on n vertices cannot exceed $2n-3$. Cairns and Nikolayevsky [43] improved the bound from $2n-3$ to $(3/2)(n-1)$.

Graphs with no $k+1$ pairwise parallel edges.

Kupitz [126] was the first who considered geometric graphs with no pair of parallel edges. He constructed examples of such graphs with n vertices and $2n-2$ edges and conjectured that this is the maximum. Katchalski and Last [119] proved an upper bound of $2n-1$. The argument was sharpened by Valtr [206] to yield the conjectured bound. The proof given here is a simplification of the original argument, it first appeared in Valtr [207]. Valtr also considers geometric graphs with no $k+1$ pairwise parallel edges, using Dilworth's theorem and generalized Davenport-Schinzel sequences he proves that such graphs have $O(n)$ edges.

Graphs with no $k+1$ pairwise crossing edges.

For $k=1$ these graphs are just the planar graphs. Graphs with no three pairwise crossing edges have been named quasi-planar. Agarwal et al. [2] have shown that quasi-planar graphs have a linear number of edges. A nice question stated in that paper is: Can the edges of a quasi-planar graph G be colored with a constant number of colors, such that the edges in each color class form a planar subgraph of G? For general k the currently best bound for the number of edges of a graph with no $k+1$ pairwise crossing edges is $O(n \log n)$, due to Valtr [206]. He makes use of the bound for the extremal problem for pairwise parallel edges. The key idea is to use the mapping $T : (x,y) \to (1/x, y/x)$, this sends a pair of crossing edges, both with one endpoint in the halfplane $x<0$ and one in the halfplane $x>0$, to a pair of parallel edges.

In the case of convex geometric graphs the problem has a complete solution. Capoyleas and Pach [45] have shown that for $n \geq 2k+1$ the maximum number of edges of a convex geometric graph without $k+1$ pairwise crossing edges is exactly $(2n-2k-1)k$.

Several authors have considered more complex extremal problems for convex geometric graphs. A convex geometric graph H with vertices $w_1, \ldots, w_r$ in cyclic order is a *convex subgraph* of a convex geometric graph G if there is a circular subsequence of vertices $v_1, \ldots, v_r$ of G, such that $w_i w_j \in E(H)$ implies $v_i v_j \in E(G)$. Kupitz and Perles [127] solve the extremal problem for the graph H on $2k$ vertices with k edges in convex position, i.e., $E(H) = \{w_1w_2, w_3w_4, \ldots, w_{2k-1}w_{2k}\}$.

Gritzmann et al. [106] study the following situation: H is a 2-connected outerplanar graph on $k+1$ vertices represented as a convex $(k+1)$-gon with some non-crossing chords. They show that a convex geometric graph not containing H as convex subgraph can have at most $t_k(n)$ edges, where $t_k(n)$ is the Turán number. In a recent paper [34] Braß, Károlyi and Valtr review the known results and work towards an extremal theory for convex geometric graphs.

2 Schnyder Woods or How to Draw a Planar Graph?

One of the most fundamental problems around a planar graph is the question: How should the graph be drawn? This, of course, is less a mathematical question and more a matter of taste. In the graph drawing literature many answers are offered. In this chapter we present some results about drawings and other representations of (3-connected) planar graphs. The results are based on the structure of Schnyder woods.

The main drawing result shown here is Tutte's Theorem (Theorem 1.6) it says that every 3-connected planar graph admits a convex drawing. This implies that every planar graph can be drawn with straight line segments. Early proofs for convex or straight line embeddings, like Tutte's proof, would produce drawings of little practical use, because, under realistic assumptions, they could not be perceived by a human eye. More formally, the ratio between the largest and the smallest distance of vertices would be unreasonably large. This motivated the problem of computing a straight line embedding placing the verices on a grid of small size. Schnyder proved the existence of an embedding on the $(n-2) \times (n-2)$ grid. Here we use extensions of Schnyder's ideas to produce embeddings of 3-connected planar graphs with f faces on the $(f-1) \times (f-1)$ grid.

In Section 2.3 we relate planar graphs and orthogonal surfaces in $\mathbb{R}^3$. Geodesic embeddings of planar graphs on orthogonal surfaces naturally lead to a notion of a dual Schnyder wood on an appropriately defined dual graph.

Section 2.5 deals with a concept of dimension for graphs and polytopes closely related to order dimension. Theorem 2.17 reproduces Schnyder's first application of Schnyder woods: A characterization of planar graphs as those graphs whose incidence order is of order dimension at most three. The Brightwell Trotter Theorem, (Theorem 2.19) is a generalization of Schnyder's theorem to polytopes: The dimension of a 3-polytope is four but removing any face makes the dimension drop to three.

2.1 Schnyder Labelings and Woods

A *planar map* M is a simple planar graph G together with a fixed planar embedding of G in the plane. A *suspension* M^σ of M is obtained by selecting three different vertices a_1, a_2, a_3 in clockwise order from the outer face of M and adding a half-edge that reaches into the outer face to each of these special vertices.

Let M^σ be the suspension of a 3-connected planar map. A *Schnyder labeling* with respect to a_1, a_2, a_3 is a labeling of the angles of M^σ with the labels $1, 2, 3$ (alternatively: red, green, blue) satisfying three rules*:

(A1) The two angles at the half-edge of the special vertex a_i have labels $i+1$ and $i-1$ in clockwise order.

(A2) *Rule of vertices:* The labels of the angles at each vertex form, in clockwise order, nonempty intervals of 1's 2's and 3's.

* We assume a cyclic structure on the labels so that $i+1$ and $i-1$ is always defined.

(A3) *Rule of faces:* The labels of the angles at each face form, in clockwise order, a nonempty interval of 1's, a nonempty interval of 2's and a nonempty interval of 3's. At the outer face the same is true in counterclockwise order.

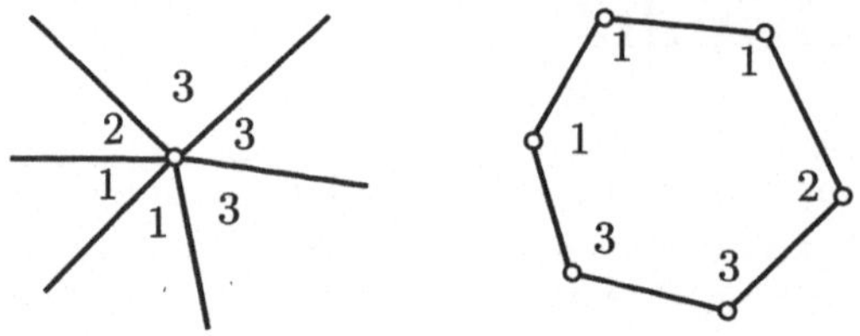

Figure 2.1 Rule of vertices and rule of faces

Let M^σ be the suspension of a 3-connected planar map. A *Schnyder wood* rooted at a_1, a_2, a_3 is an orientation and labeling of the edges of M^σ with the labels 1, 2, 3 satisfying the following rules.

(W1) Every edge e is oriented by one or two opposite directions. The directions of edges are labeled such that if e is bioriented the two directions have distinct labels.

(W2) The half-edge at a_i is directed outwards and labeled i.

(W3) Every vertex v has outdegree one in each label. The edges e_1, e_2, e_3 leaving v in labels 1,2,3 occur in clockwise order. Each edge entering v in label i enters v in the clockwise sector from e_{i+1} to e_{i-1}.

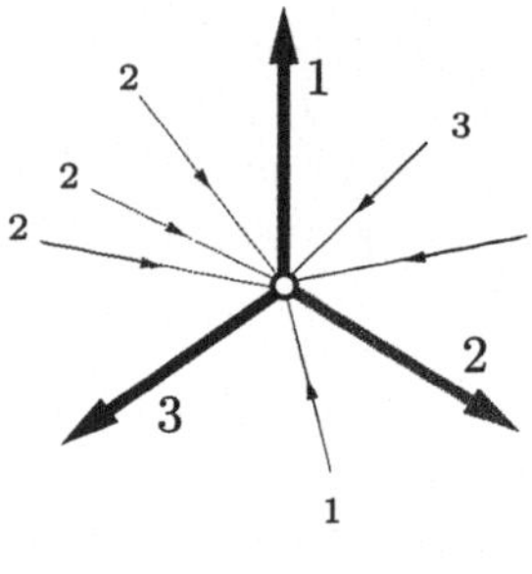

Figure 2.2 Edge orientations and edge labels at a vertex.

(W4) There is no interior face whose boundary is a directed cycle in one label.

The following lemma is central to the fact that there is a one-to-one correspondence between Schnyder labelings and Schnyder woods.

Lemma 2.1 *Let M^σ be a suspended planar map with a Schnyder labeling, then the four angles of each edge contain all three labels 1,2,3. Thus every edge has one of the two types shown in Figure 2.3.*

Proof. The proof is based on double counting and Euler's formula. Define the degree $d(v)$ of a vertex v as the number of edges incident with v whose angles at v have distinct labels. By the rule of vertices $d(v) = 3$ for every vertex v. Similarly, the degree $d(F)$ of a face F is the number of boundary edges of F whose angles in F have distinct labels. By

Figure 2.3 The two types of labeling for an edge.

the rules of faces $d(F) = 3$ for every face, in the case of the outer face the half-edges are counted as edges with distinct labels. The sum S of degrees of vertices and faces is:

$$S = \sum_v d(v) + \sum_F d(F) = 3n + 3f = 3|E| + 6.$$

The same number S can be obtained by counting the changes of label around the edges. Each of the half-edges contributes two. Hence, the average contribution of full-edges is 3. Consider the four angles $\alpha_1, \alpha_2, \alpha_3, \alpha_4$ of an edge in counterclockwise order. Define $\epsilon_1, \epsilon_2, \epsilon_3, \epsilon_4$ so that $\alpha_2 = \alpha_1 + \epsilon_1$, $\alpha_3 = \alpha_2 + \epsilon_2$, $\alpha_4 = \alpha_3 + \epsilon_3$ and $\alpha_1 = \alpha_4 + \epsilon_4$. Form the rules of vertices and faces $\epsilon_j \in \{0, 1\}$, for all j. The cyclic nature of the linear system implies that $\sum_{j=1}^{4} \epsilon_j = 0 \mod 3$, hence, either $\sum_j \epsilon_j = 0$ or $\sum_j \epsilon_j = 3$. Since the contribution of an edge e to the degree sum S is $\sum_j \epsilon_j(e)$ we must have $\sum_j \epsilon_j(e) = 3$ for every full-edge e. Up to rotational symmetry this only leaves the two cases shown in Figure 2.3. □

Note that from the exterior labels at special vertices (A1) and the rule of faces for the outer face we know all the edge labels at outer angles. Every outer angle on the clockwise outer path from a_i to a_{i+1} has label $i - 1$. With Lemma 2.1 we can deduce two labels i at angles incident to a_i. Together with rule (A2) applied to vertex a_i this implies:

Corollary 2.2 *In a Schnyder labeling all interior angles at the special vertex a_i are labeled i.*

Let a Schnyder labeling of the angles of a planar map be given, using the following rule the labeling induces a Schnyder wood.

($\star$) If edge $\{u, v\}$ has different angular labels i and j at vertex u, then we direct the edge from u to v in the third label k.

This rule can be reversed, so that a Schnyder wood induces a Schnyder labeling. See Figure 2.4.

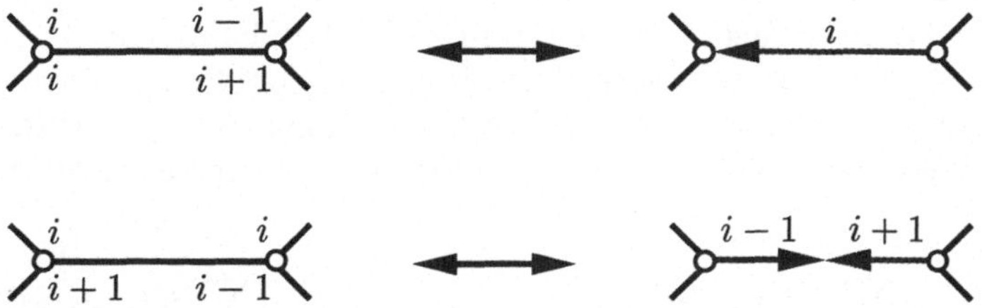

Figure 2.4 The correspondence between angle labels at an edge and the coloring and orientation of the edge.

Theorem 2.3 *Let M^σ be the suspension of a 3-connected planar map. The above correspondence is a bijection between the Schnyder labelings (axioms A1,A2,A3) and Schnyder woods (axioms W1,W2,W3,W4) of M^σ.*

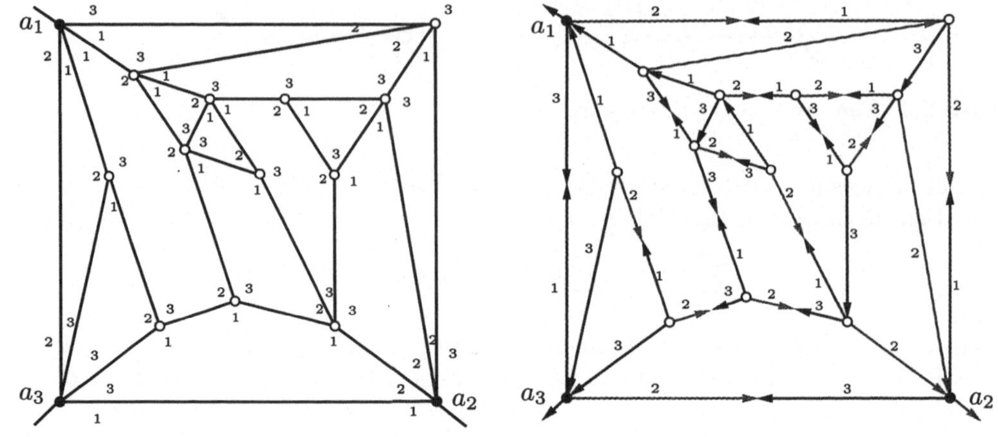

Figure 2.5 A suspended planar map with a Schnyder labeling (left) and the corresponding Schnyder wood (right).

Proof. Let a Schnyder angle labeling be given and use rule $\star$ to define orientations and labelings for the edges of the map. Lemma 2.1 shows that this orientation and labeling obeys (W1). There is an immediate correspondence of (A1) and (W2). Also the two rules of vertices (A2) and (W3) correspond to each other. If there is an interior face whose boundary is directed in one label, then we infer that all interior angles of this face have the same label. Hence, the rule of faces (A3) forces (W4). Together this shows that the construction yields a Schnyder wood.

Conversely, let a Schnyder wood be given. If the direction (u, v) of an edge is colored i then we color the angle at u to the left of edge uv with $i+1$ and the angle to the right of uv with $i-1$. If uv is unidirectional we also color the two adjacent angles at v with color i. From (W3) it follows that the two colors assigned to an angle from its two adjacent edges coincide, i.e., the coloring of angles is well-defined.

Again, the correspondence of (A1) and (W2) and of the two rules of vertices (A2) and (W3) is trivial. To show that (A3), i.e., the rule of faces, is valid is more subtle: As in the proof of Lemma 2.1 we count color changes of the angles at vertices, faces and edges. (W1) implies that at a full-edge e there are three changes, $d(e) = 3$, see Figure 2.4. For a half-edge e we let $d(e) = 2$. The contribution of vertices is $\sum_v d(v) = 3n$. The degree $d(F)$ of a bounded face F is the number of color changes at the angles when we cycle around F. If we cycle clockwise then an angle colored i is always followed by an angle colored i or $i+1$. Consequently, $d(F)$ must be a multiple of 3. Rule (W4) enforces $d(F) \neq 0$. For the unbounded face we count one color change at each half-edge, hence, $d(F) \geq 3$.

$$\sum_v d(v) + \sum_F d(F) = \sum_e d(e) \qquad \Longrightarrow \qquad 3n + \sum_F d(F) = 3|E| + 6.$$

With Euler's Formula $\sum_F d(F) = 3f$ which is only possible if $d(F) = 3$ for every face F.

We have already seen that the colors go clockwise at bounded faces and counterclockwise at the outer face. This proves the rule of faces (A3). □

Henceforth, when we have a given Schnyder wood or a Schnyder labeling we may be sloppy and refer to properties of the corresponding other structure.

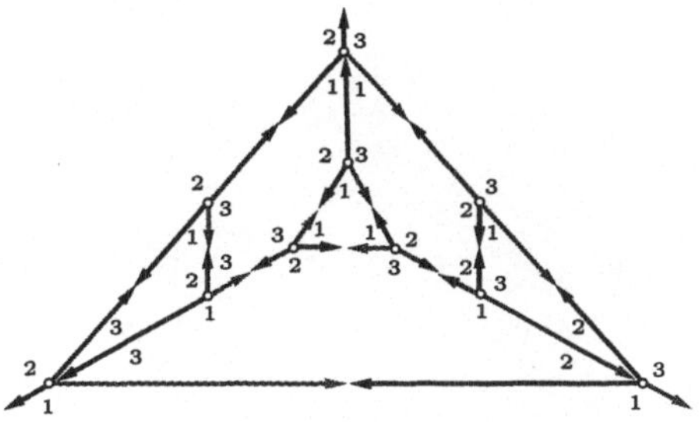

Figure 2.6 The shown orientation and coloring of edges obeys (W1), (W2) and (W3) but not (W4). The induced angle labeling is not a Schnyder labeling.

Let M be a planar map with a Schnyder wood. Let T_i denote the digraph induced by the directed edges of label i. Every inner vertex has outdegree one in T_i, therefore, every v is the starting vertex of a unique i-path $P_i(v)$ in T_i. The next lemma shows that each of the digraphs T_i is acyclic, actually we prove a bit more.

Lemma 2.4 *Let M be a planar map with a Schnyder wood (T_1, T_2, T_3). Let T_i^{-1} be obtained by reverting all edges from T_i. The digraph $D_i = T_i \cup T_{i-1}^{-1} \cup T_{i+1}^{-1}$ is acyclic for $i = 1, 2, 3$.*

Proof. Edges in $T_{i-1} \cap T_{i+1}$ remain bidirected in D_i, we do not consider bidirected edges and paths as directed cycles. Hence, a directed cycle in D_i will enclose a non-empty set of faces. Let Z be a directed cycle such that the number of faces enclosed by Z is minimum. Let F be the interior region of Z, we first show that F consists of a single face.

Suppose F contains a vertex x. Start at x and always use the outgoing edge of color i to leave a vertex. This defines the i-path $P_i(x)$ of vertex x. By minimality of Z there is a simple initial part $P_i'(x)$ of $P_i(x)$ connecting x to Z. Let $P_{i-1}'(x)$ be defined analogously. By the minimality of Z the paths $P_i'(x)$ and $P_{i-1}'(x)$ have no common vertex other than x. Together with one of the two segments they determine on Z these two paths form a directed cycle in $T_i \cup T_{i-1}^{-1} \cup T_{i+1}^{-1}$ which encloses fewer faces than Z. This contradiction shows that Z contains no vertex. An edge lying in F and joining two non-consecutive vertices of Z would similarly determine a cycle enclosing fewer faces than Z.

Therefore, F is a face and Z its boundary cycle. If the traversal of Z is clockwise no angle of F has label $i+1$ and if this traversal is counterclockwise no angle has label $i-1$. Both cases are excluded by the rule of faces. □

By the rule of vertices (W3) every vertex has out-degree one in T_i. Disregard the half-edges at special vertices. This makes the a_i a sink of T_i. Since T_i is acyclic and has $n-1$ edges we readily obtain:

Corollary 2.5 *T_i is a directed tree rooted at a_i, for $i = 1, 2, 3$.*

The i-path $P_i(v)$ is the unique path in T_i from v to the root a_i. Lemma 2.4 implies that for $i \neq j$ the paths $P_i(v)$ and $P_j(v)$ have v as the only common vertex. Therefore, $P_1(v), P_2(v), P_3(v)$ divide M into three regions $R_1(v)$, $R_2(v)$ and $R_3(v)$, where $R_i(v)$ denotes the region bounded by and including the two paths $P_{i-1}(v)$ and $P_{i+1}(v)$, see Fig. 2.7. The open interior of region $R_i(v)$, denoted $R_i^o(v)$, is $R_i(v) \setminus (P_{i-1}(v) \cup P_{i+1}(v))$.

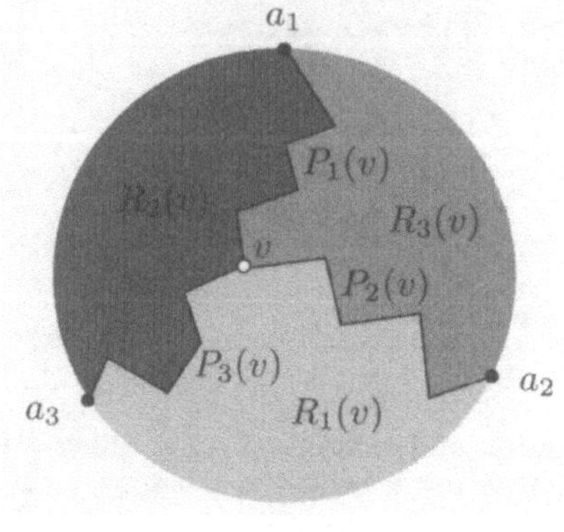

Figure 2.7 The three regions of a vertex

Lemma 2.6 *If u and v are vertices of a labeled graph with $u \in R_i(v)$, then $R_i(u) \subseteq R_i(v)$. If $u \in R_i^o(v)$, then the inclusion is proper: $R_i(u) \subset R_i(v)$.*

Proof. By symmetry it suffices to consider the case $i = 1$. Suppose $u \in R_1^o(v)$ and let x be the first vertex of $P_2(u)$ that belongs to $P_2(v) \cup P_3(v)$. From the edge orientations at x (Figure 2.2) it follows that $x \notin P_3(v)$. By the same reason $x \neq v$, hence, $x \in P_2(v)$. Similarly the first vertex y of $P_3(u)$ that belongs to $P_2(v) \cup P_3(v)$ is on $P_3(v)$ and $y \neq v$. Hence, $R_1(u) \subseteq R_1(v)$, see Figure 2.8, the inclusion is proper as $v \notin R_1(u)$.

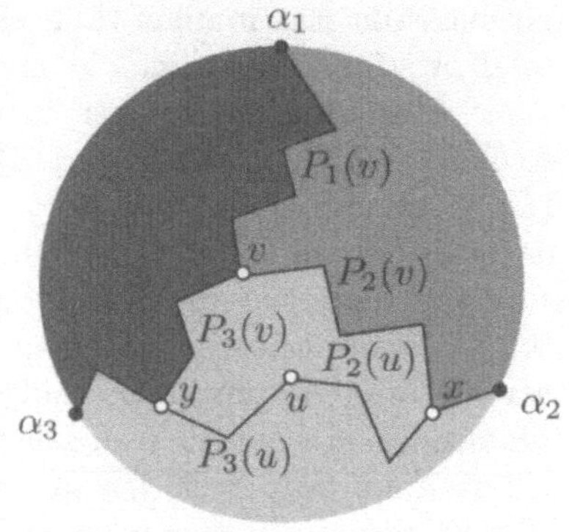

Figure 2.8 If $u \in R_1^o(v)$ then $R_1(u)$ is a proper subset of $R_1(v)$.

Now let $u \in R_1(v) \setminus R_1^o(v)$, by symmetry we only consider the case $u \in P_3(v)$. If at u the outgoing edge in label 2 is different from the incoming edge on $P_3(v)$ then a reasoning as in the previous case shows that the inclusion is proper, $R_1(u) \subset R_1(v)$. Otherwise, if u' is the other vertex of the bidirected edge leaving u in label 2 and entering in label 3, then $R_1(u') = R_1(u)$. However $R_1(u') \subseteq R_1(v)$ by induction on the number of vertices between u and v on $P_3(v)$. □

Let vertices u, v be neighbors such that the edge $e = (u, v)$ is directed from u to v in label i, see Figure 2.9. Since $v \in P_i(u)$ vertex v is contained in $R_{i-1}(u)$ and $R_{i+1}(u)$. The orientations of edges at v imply $u \in R_i(v)$. Therefore the following inclusions of regions hold:

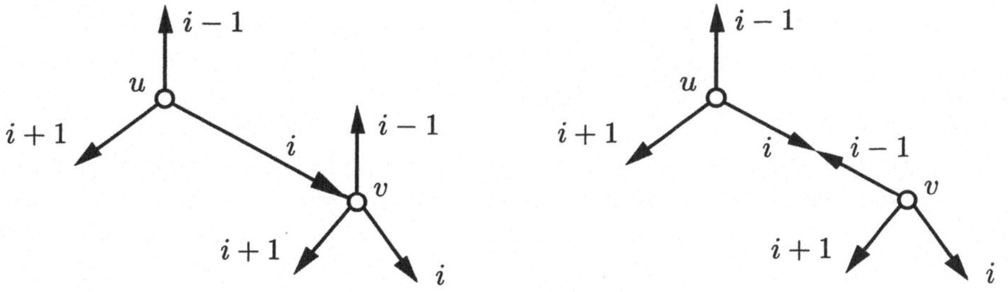

Figure 2.9 Two cases for an edge (u, v) with label i from u to v.

(E_1) If $e = (u, v)$ is an unidirectional edge with label i from u to v then $R_i(u) \subset R_i(v)$ and $R_{i-1}(u) \supset R_{i-1}(v)$ and $R_{i+1}(u) \supset R_{i+1}(v)$.

(E_2) If $e = (u, v)$ is bidirectional with label i from u to v and label $i - 1$ from v to u, then $R_{i+1}(u) = R_{i+1}(v)$ and $R_i(u) \subset R_i(v)$ and $R_{i-1}(u) \supset R_{i-1}(v)$.

2.2 Regions and Coordinates

Let M^σ be planar map with f faces and a Schnyder wood. With every vertex v of M we associate a *region vector* (v_1, v_2, v_3):

$$v_i = \#\text{ faces of } M \text{ in the region } R_i(v).$$

Note that the special vertices a_1, a_2, a_3 have region vectors $(f - 1, 0, 0)$, $(0, f - 1, 0)$ and $(0, 0, f - 1)$. Translating our knowledge about inclusion of regions, in particular (E_1) and (E_2), to the region vectors we obtain:

(1) $v_1 + v_2 + v_3 = f - 1$ for all vertices v.

(2) If $u \in R_i(v)$ then $u_i \leq v_i$ and if $u \in R_i^o(v)$ then $u_i < v_i$.

(3) If an edge of M is directed from u to v in label i then $u_i < v_i$, $u_{i+1} \geq v_{i+1}$ and $u_{i-1} \geq v_{i-1}$.

(4) For every edge (u, v) of a labeled graph there are indices i, j such that $u_i < v_i$ and $u_j > v_j$.

Given three non-collinear points α_1, α_2 and α_3 in the plane. These points and the region vectors of the vertices of M can be used to define an embedding of M in the plane. A vertex of M is mapped to the point

$$\mu : v \rightarrow \frac{1}{f-1}(v_1\alpha_1 + v_2\alpha_2 + v_3\alpha_3),$$

an edge (u, v) is represented by the line segment $(\mu(u), \mu(v))$. Note that any two drawings based on points α_1, α_2 and α_3 and β_1, β_2 and β_3 can be mapped onto each other by an affine map. Geometrically μ is a linear map from the affine plane $A_f \subset \mathbb{R}^3$ defined by $x_1 + x_2 + x_3 = f - 1$ to the standard plane $\mathbb{R}^2$.

Theorem 2.7 *If M is a planar map and the coordinate v_i of vertex v counts the number of faces in $R_i(v)$ with respect to a Schnyder wood of M, then the drawing $\mu(M)$ is a convex drawing of M.*

Modulo the existence of Schnyder woods for 3-connected planar graphs this theorem is Tutte's Theorem. The existence of Schnyder woods is shown in Section 2.6. With the special choice $\alpha_1 = (0, f-1)$, $\alpha_2 = (f-1, 0)$ and $\alpha_3 = (0,0)$ every vertex v of M is mapped to an integral point in the $(f-1) \times (f-1)$ grid. This yields the announced version of Tutte's Theorem.

Corollary 2.8 *If M is a 3-connected planar map with f faces, then there is a convex drawing of M on the $(f-1) \times (f-1)$ grid.*

Let v be a vertex of M with coordinates (v_1, v_2, v_3). The μ-images of the three lines in A_f given by $x_1 = v_1$, $x_2 = v_2$ and $x_3 = v_3$ cross in $\mu(v)$ and partition the triangle with vertices α_1, α_2 and α_3 into six regions, see Figure 2.10. Each of the three closed shaded parallelograms contains exactly one neighbor of v. This is because if (u, v) is directed towards v then by property (3) vertex u is contained in one of the three white triangles.

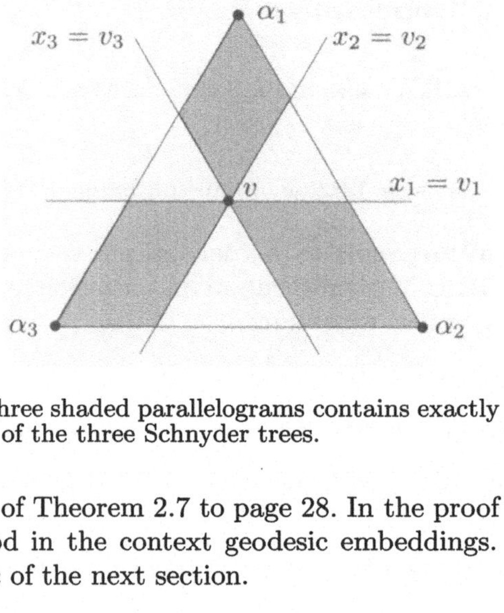

Figure 2.10 Each of the three shaded parallelograms contains exactly one neighbor of v, these are the outgoing edges at v of the three Schnyder trees.

We postpone the proof of Theorem 2.7 to page 28. In the proof we will use properties which are best understood in the context geodesic embeddings. These embeddings of planar maps are the topic of the next section.

2.3 Geodesic Embeddings of Planar Graphs

Consider $\mathbb{Z}^3$ as subsets of $\mathbb{R}^3$ and the *dominance order* on these sets, i.e, $(u_1, u_2, u_3) \leq (v_1, v_2, v_3)$ iff $u_i \leq v_i$ for $i = 1, 2, 3$. Use $u \vee v$ and $u \wedge v$ to denote the *join* (component-wise maximum) and *meet* (component-wise minimum) of $u, v \in \mathbb{R}^3$. Let $\mathcal{V} \subset \mathbb{Z}^3 \subset \mathbb{R}^3$ be an antichain, i.e., a set of pairwise incomparable elements. The *filter* generated by $\mathcal{V}$ in $\mathbb{R}^3$ is the set

$$\langle \mathcal{V} \rangle = \{\alpha \in \mathbb{R}^3 \mid \alpha \geq v \text{ for some } v \in \mathcal{V}\}.$$

The boundary $\mathcal{S}_\mathcal{V}$ of $\langle\mathcal{V}\rangle$ is the *orthogonal surface* generated by $\mathcal{V}$. Orthogonal projection onto the plane $x + y + z = 0$ yields a picture of $\mathcal{S}_\mathcal{V}$ in the plane. When the integral level curves are emphasized in the picture this results in a rhombic tiling of the plane, see Figure 2.11.

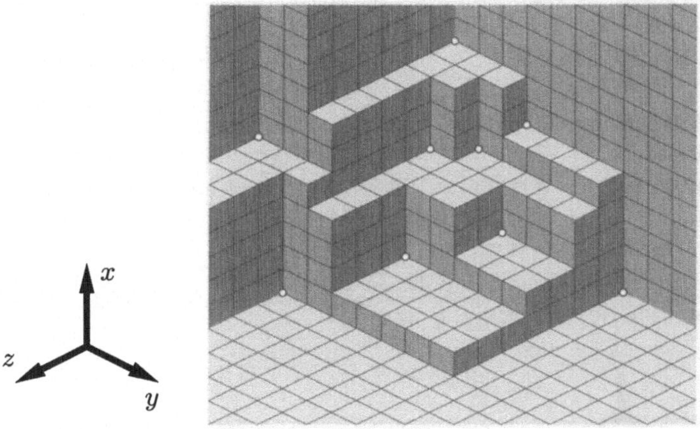

Figure 2.11 The orthogonal surface $\mathcal{S}_\mathcal{V}$ generated by $\mathcal{V} = \{(0,0,7),(0,7,0),(1,2,4),(2,4,2),$ $(4,1,2),(4,2,1),(5,-2,6),(5,3,0),(7,0,0)\}$.

If $u,v \in \mathcal{V}$ and $u \vee v \in \mathcal{S}_\mathcal{V}$ then $\mathcal{S}_\mathcal{V}$ contains the union of the two line segments joining u and v to $u \vee v$; we refer to such arcs as *elbow geodesics* in $\mathcal{S}_\mathcal{V}$. The *orthogonal arc* of $v \in \mathcal{V}$ in direction of the standard basis vector e_i is the intersection of the ray $v + \lambda e_i$, $\lambda \geq 0$, with $\mathcal{S}_\mathcal{V}$. Clearly every vector $v \in \mathcal{V}$ has exactly three orthogonal arcs, one parallel to each coordinate axis. Some orthogonal arcs are unbounded while others are bounded. Observe that $u \vee v$ must share two coordinates with at least one (and perhaps both) of u and v, so every elbow geodesic contains at least one bounded orthogonal arc.

Let M be a planar map, a drawing $M \hookrightarrow \mathcal{S}_\mathcal{V}$ is a *geodesic embedding* of M in $\mathcal{S}_\mathcal{V}$, if the following axioms are satisfied:

(G1) *Vertex axiom.* There is a bijection between the vertices of M and $\mathcal{V}$.

(G2) *Elbow geodesic axiom.* Every edge of M is an elbow geodesic in $\mathcal{S}_\mathcal{V}$, and every bounded orthogonal arc in $\mathcal{S}_\mathcal{V}$ is part of an edge of M.

(G3) There are no crossing edges in the embedding of M on $\mathcal{S}_\mathcal{V}$.

An antichain $\mathcal{V}$ in $\mathbb{Z}^3$ is called *axial* if it contains exactly three unbounded orthogonal arcs. The example from Figure 2.11 is not axial, however, removing the point $(5,-2,6)$ from the set $\mathcal{V}$ leads to an axial antichain, see Figure 2.12.

Theorem 2.9 *Let $\mathcal{V}$ be axial and $M \hookrightarrow \mathcal{S}_\mathcal{V}$ be a geodesic embedding, then the embedding induces a Schnyder wood of M^σ. Conversely, given a Schnyder wood of a planar graph M^σ define $\mathcal{V}$ as set of region vectors of vertices of M^σ. This yields a geodesic embedding of $M \hookrightarrow \mathcal{S}_\mathcal{V}$ with an axial $\mathcal{V}$.*

Proof. Let $M \hookrightarrow \mathcal{S}_\mathcal{V}$ be an axial geodesic embedding. The edges of M are colored with the direction of the orthogonal arc contained in the edge: Arcs parallel to the x_i-axis are colored i. The orientation of an edge is chosen in accordance with the axis used to color the edge, Figure 2.12 shows an example. The claim is that this coloring is a Schnyder wood for M. Since every elbow geodesic contains one or two orthogonal arcs, axioms (W1) and (W2) are obvious. A vertex $v \in \mathcal{V}$ has exactly three outgoing orthogonal arcs. In the standard projection of $\mathcal{S}_{\mathcal{V}}$ this yields a clockwise sequence of outgoing edges in colors 1,2,3. Consider an edge $\{u, v\}$ represented by an elbow geodesic that arrives at v through the sector $S_3(v)$ between or on the outgoing edges in colors 1 and 2. The sector $S_3(v)$ is contained in the plane $x_3 = v_3$. As an elbow geodesic the edge has to pass the join $u \vee v$ of its two vertices. The join has coordinates $u \vee v = (u_1, u_2, v_3)$. Therefore the geodesic representing $\{u, v\}$ contains the orthogonal arc leaving u in direction of the x_3-axis, whence $\{u, v\}$ is oriented as (u, v) in color 3. This together with symmetric arguments for the other sectors shows (W3). A path of edges colored i gives a sequence of vertices with increasing ith coordinate. Therefore, the directed graph T_i of i colored edges is acyclic, this implies (W4).

Note that the corresponding Schnyder labeling of the angles of M is the labeling by the three different shades in the tiling figure.

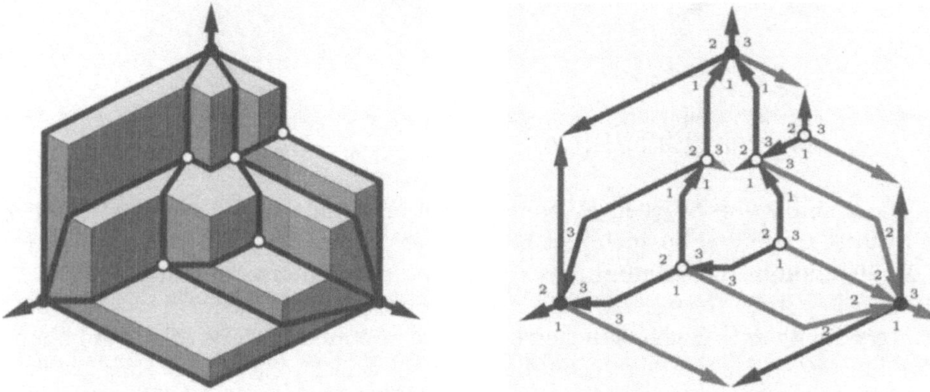

Figure 2.12 Schnyder labeling and wood induced by a geodesic embedding.

Given a Schnyder wood of M^σ embed every vertex v at its region vector $(v_1, v_2, v_3) \in \mathbb{N}^3 \subset \mathbb{Z}^3$, i.e., $\mathcal{V} = \{(v_1, v_2, v_3) : v \text{ is a vertex of } M\}$. Let f be the number of faces of M, then $v_1 + v_2 + v_3 = f - 1$ is independent of v. Hence, $\mathcal{V}$ is an antichain in $\mathbb{Z}^3$. Since the mapping is injective we have (G1). If $e = \{u, v\}$ is an edge of M and $x \notin e$ a vertex, then for some i edge e is contained in region $R_i(x)$. This implies $R_i(u) \subseteq R_i(x)$ and $R_i(v) \subseteq R_i(x)$ hence, $u_i \leq x_i$ and $v_i \leq x_i$. This shows that with $e = \{u, v\}$ the join $u \vee v$ and hence the elbow geodesic $[u, v]$ is on the surface $\mathcal{S}_{\mathcal{V}}$. If edge $e = (v, w)$ is directed in color i from v to w then $v_i < w_i$, $v_{i+1} \geq w_{i+1}$ and $v_{i-1} \geq w_{i-1}$ (property (3) on page 23). Therefore, the orthogonal arc of v in direction e_i is used by this edge. This yields (G2).

It remains to prove the non-crossing condition (G3). Every edge is represented by an elbow geodesic consisting of two straight legs. At least one of the legs is an orthogonal arc. An elbow geodesic cannot intersect another orthogonal arc. Suppose there is a pair $\{u, v\}$ and $\{y, z\}$ of crossing edges. The elbow geodesics representing these edges cross with their legs on a plane orthogonal to one of the coordinate axes. Up to symmetry the situation is as illustrated in Figure 2.13. We may thus assume that names of vertices and orientation are as in the figure, in particular $u_1 = y_1$, $u_3 > y_3$ and $z_3 > v_3$.

Between u and y there is a path consisting of orthogonal arcs only. With (G2) this

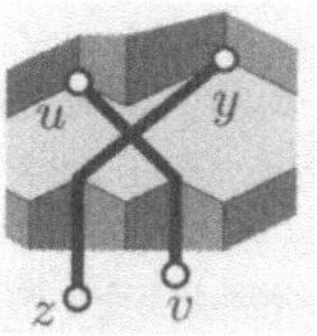

Figure 2.13 A pair of crossing elbow geodesics.

implies that in the Schnyder wood of M^σ there is a bidirected path P^* in colors 2 and 3 between u and y. P^* is directed from u to y in color 2. The crossing edges have color 1, they are unidirected as (v, u) and (z, y). Let s be the first common vertex of the color 2 paths $P_2(u)$ and $P_2(v)$. If $s \in P^*$ then there is a cycle in $T_1 \cup T_2^{-1} \cup T_3^{-1}$, hence $s \in P_2(y)$ and $s \neq y$. This proves $y \in R_3^\circ(v)$ and since $\{y, z\}$ is an edge also $z \in R_3(v)$. Expressed in terms of coordinates this yields $v_3 \geq z_3$, a contradiction. □

Let M be a planar map with a Schnyder wood. The corresponding surface $\mathcal{S_V}$ is above the plane X defined by $x_1 + x_2 + x_3 = f - 1$. The elements of $\mathcal{V}$ which are the minima of $\mathcal{S_V}$ are in this plane X. With a point $p \in \mathbb{R}^3$ above X consider the points of X dominated by p. This set is a triangle ∇_p. The border of ∇_p consists of those elements of X having a common coordinate with p.

Lemma 2.10 *Let $e = \{u, v\}$ be an edge of M and $\nabla_e = \nabla_{u \vee v}$. The triangle ∇_e has u and v on its border and the interior of ∇_e contains no vertex of M (see Figure 2.14).*

Proof. In the proof of property (G2) we have shown that $u \vee v \in \mathcal{S_V}$. □

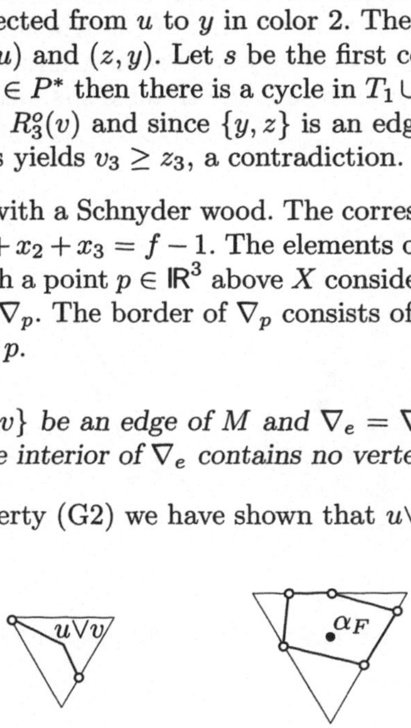

Figure 2.14 Triangles ∇_e of an edge and ∇_{α_F} of a face in the plane X.

Lemma 2.11 *Let F be a bounded face of M and $\alpha_F = \bigvee_{w \in F} w$ be the join of the vertices of F, then $\alpha_F \in \mathcal{S_V}$, moreover, α_F is a maximum of $\mathcal{S_V}$. All $v \in F$ are on the border of ∇_{α_F} and the interior of ∇_{α_F} contains no vertex of M (see Figure 2.14).*

Proof. Let w be any vertex of M and suppose that F is contained in region $R_i(w)$. For $v \in F$ let v^* be the last vertex of path $P_i(v)$ in region $R_i(w)$. From $v^* \in P_{i-1}(w) \cup P_{i+1}(w)$ it follows that $w_i \geq v_i^* \geq v_i$. Hence, for every w there is a coordinate i such that $w_i \geq [\alpha_F]_i$. This proves that α_F is on the surface $\mathcal{S_V}$.

For $v \in F$ we have $(v_1, v_2, v_3) \leq \alpha_F$ by definition and $v_i \geq [\alpha_F]_i$ if i is such that $F \subset R_i(v)$. Therefore, $v \in F$ is on the border of ∇_{α_F}.

Let v be a vertex with $v_i = [\alpha_F]_i$. For the clockwise neighbor u of v at F we find $R_{i+1}(u) \supset R_{i+1}(v)$, hence, $[\alpha_F]_{i+1} \geq u_{i+1} > v_{i+1}$. Considering the counterclockwise neighbor it follows that $[\alpha_F]_{i-1} > v_{i-1}$. Since there is a vertex $v \in F$ with $v_i = [\alpha_F]_i$ for $i = 1, 2, 3$ it can be concluded that α_F is a maximum of $\mathcal{S_V}$. □

Proof of Theorem 2.7. Projecting the geodesic embedding of M into the plane X gives a planar drawing of M. In this drawing every edge is composed by two straight segments. The claim is that the straight drawing, obtained by replacing each bend edge by the single segment connecting the vertices, is a convex drawing.

Claim 1: The straight drawing is crossing free. Suppose two edges e and e' cross in the straight drawing. These edges do not cross in the geodesic embedding. Therefore, one vertex of $e = \{u, v\}$ is contained in the triangle formed by the two representations of $e' = \{y, z\}$, or the other way round. This shows that a vertex is embedded in the interior of $\nabla_{e'}$ or of ∇_e. In either case a contradiction to Lemma 2.10.

Claim 2: The straight drawing is convex. Let F be a bounded face of M. By Lemma 2.11 all vertices of F are embedded on the border of the triangle ∇_{α_F}. The resulting shape of F in the straight drawing is a triangle with some truncated corners. In particular F is convex. The shape of the outer face is the triangle spanned by α_1, α_2 and α_3. □

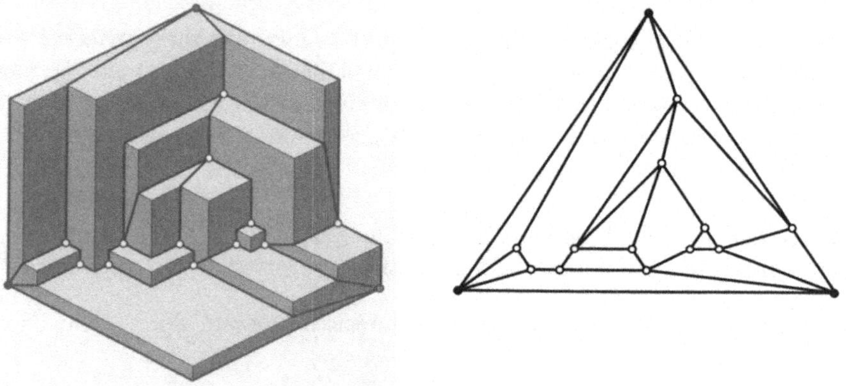

Figure 2.15 Geodesic and convex embedding for the graph of Figure 2.5

2.4 Dual Schnyder Woods

Let M be a planar map with a suspension M^σ and dual M^*. The *truncation* $M^{*\tau}$ of the dual of M is obtained by deleting the vertex corresponding to the unbounded face of M but leaving the edges incident to this vertex as half-edges. Furthermore these half-edges are partitioned into blocks B_1, B_2, B_3, where B_i contains the duals of the edges of M on the exterior path between the special vertices a_j and a_k, $\{i, j, k\} = \{1, 2, 3\}$.

Suppose a Schnyder wood of M^σ is given and $M^\sigma \hookrightarrow \mathcal{S}_\mathcal{V}$ is the corresponding geodesic embedding. With each bounded face F there is a maximum $\alpha_F \in \mathcal{S}_\mathcal{V}$ (Lemma 2.11). Actually, the maxima of $\mathcal{S}_\mathcal{V}$ are in bijection with bounded faces of M. With two faces F and F' sharing an edge $e = \{u, v\}$ we have $\alpha_F \wedge \alpha_{F'} = u \vee v \in \mathcal{S}_\mathcal{V}$. Let a dual elbow geodesic in a surface $\mathcal{S}_\mathcal{V}$ be the union of two line segments in $\mathcal{S}_\mathcal{V}$ connecting maxima α and α' to $\alpha \wedge \alpha'$. The dual of an edge $\{u, v\}$ separating the unbounded face from a bounded face F is a half-edge. It consists of the orthogonal arc reaching from α_F to $u \vee v$. A *dual geodesic embedding* is a set of dual elbow geodesics without crossings that uses all orthogonal arcs incident to maximal points.

Proposition 2.12 *With a geodesic embedding $M^\sigma \hookrightarrow \mathcal{S}_\mathcal{V}$ there is a dual geodesic embedding of the truncation $M^{*\tau}$ of the dual of M.*

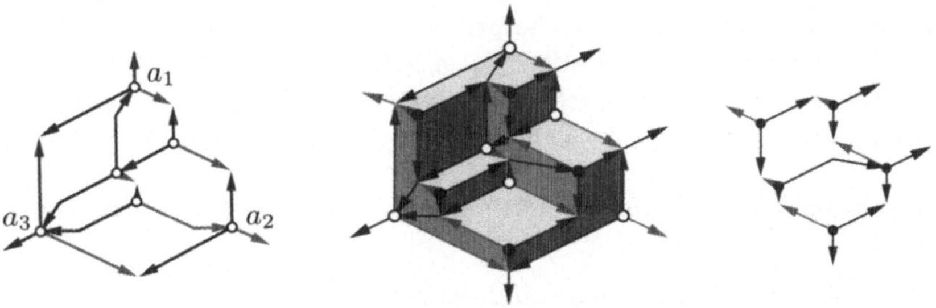

Figure 2.16 A suspended graph M^σ with a Schnyder wood, a corresponding embedding and the edge coloring and orientation of the truncation $M^{*\tau}$.

The dual geodesic embedding can be used to induce a coloring and orientation on the edges of $M^{*\tau}$ which is almost a Schnyder wood. The axioms (W1), (W3) and (W4) are fulfilled, instead of (W2) we have: all half-edges in B_i are colored i and directed outward. Based on $M^{*\tau}$ we construct the *suspension dual* M^{σ^*} of M^σ by connecting half edges in B_i to a new vertex b_i and adding the triangle edges $\{b_1, b_2\}, \{b_2, b_3\}, \{b_1, b_3\}$ and a half-edge for each b_i. Coloring and orientation of edges of $M^{*\tau}$ with the above properties can be extended in a unique way to a Schnyder wood of M^{σ^*}.

Proposition 2.13 *There is a bijection between the Schnyder woods of M^σ and the Schnyder woods of the suspension dual M^{σ^*}.*

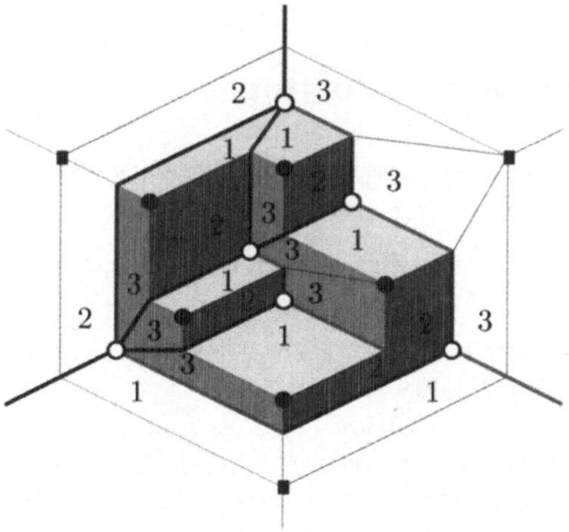

Figure 2.17 Bold edges show a suspended graph M^σ, light edges correspond to M^{σ^*}. The Schnyder angle labeling shown is valid for both graphs.

Proof. The proof becomes particularly simple in the terminology of Schnyder angle labelings. There is an obvious one-to-one correspondence between the angles of M^σ and the inner angles of M^{σ^*}. This correspondence yields an exchange between the rule of vertices

(A2) and the rule of faces (A3). Therefore, any Schnyder labeling of M^σ is a Schnyder labeling of M^{σ^*} and vice versa. This is exemplified in Figure 2.17. □

We now define the completion of a planar suspension M^σ and its dual M^{σ^*}. Superimpose M^σ and M^{σ^*} so that exactly the primal dual pairs of edges cross (the half edge at a_i has a crossing with the dual edge $\{b_j, b_k\}$, for $\{i,j,k\} = \{1,2,3\}$). The common subdivision of each crossing pair of edges by a new edge-vertex gives the *completion* $\widetilde{M^\sigma}$. The completion $\widetilde{M^\sigma}$ is planar and has six half-edges reaching into the unbounded face.

Proposition 2.14 *The Schnyder woods of a planar suspension M^σ are in bijection with orientations of $\widetilde{M^\sigma}$ such that*

- outdegree$(v) = 3$ *for all primal- and dual-vertices v,*
- outdegree$(v_e) = 1$ *for all edge-vertices v_e*

and all half-edges are oriented away from their incident edge-vertex.

This proposition leads to an easy technique for modifying Schnyder woods of M^σ. Given a Schnyder wood consider the corresponding orientation of $\widetilde{M^\sigma}$. If this orientation contains an oriented cycle C revert the orientation of all edges of C. This construction yields another orientation with the same outdegrees, hence, another Schnyder wood of M^σ. This observation is the starting point for the proof of the following theorem. It describes a global structure on all Schnyder woods of M^σ.

Theorem 2.15 *The set of Schnyder woods of a planar suspension M^σ form a distributive lattice.*

We abstain fom proving the theorem and pass on to another application of Schnyder woods.

2.5 Order Dimension of 3-Polytopes

Let $G = (V, E)$ be a finite simple graph. A nonempty family $\mathcal{R}$ of linear orders on the vertex set V of graph G is called a *realizer* of G provided:

(∗) For every edge $e \in E$ and every vertex $x \in V \setminus e$, there is some $L \in \mathcal{R}$ so that $x > y$ in L for every $y \in e$.

The *dimension* of G, denoted $\dim(G)$, is then defined as the least positive integer t for which G has a realizer of cardinality t.

An intuitive formulation for condition (∗) is as follows: For every vertex v and edge e with $v \notin e$ the vertex has to get over the edge in at least one of the orders of a realizer. From the defining condition it is obvious that dimension is monotone under

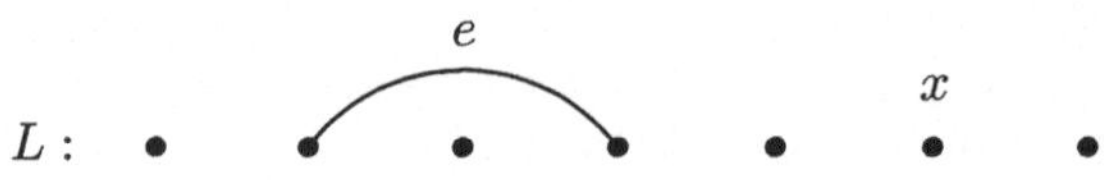

Figure 2.18 Vertex x is over edge e in L.

taking subgraphs, i.e.,

- if H is a subgraph of G then $\dim(H) \leq \dim(G)$.

For readers who are new to the concept of dimension for graphs, we first prove an elementary proposition.

Proposition 2.16 *If a graph G contains a cycle, i.e., if G is not a tree, then* $\dim(G) \geq 3$.

Proof. By the monotonicity of dimension we only have to show that $\dim(C_n) \geq 3$, where C_n is the cycle on $n \geq 3$ vertices. Assume that C_n has a realizer L_1, L_2. Suppose that two vertices u and v are in the same order $u < v$ in L_1 and L_2. Let $v' \neq u$ be a neighbor of v. One of the orders of a realizer has to bring u over (v, v'); this contradicts $u < v$ in L_1 and L_2. Therefore, L_2 has to be the reverse of L_1. In this case the two vertices of every edge have to be adjacent in L_1, otherwise, if (v, v') is an edge and $v < u < v'$ in L_1 then u does not get over (v, v') in L_1 and not in L_2. The cycle C_n has n edges but L_1 only has $n-1$ adjacent pairs, this shows that C_n has no realizer consisting of only two linear extensions. □

It is easy to construct a realizer consisting of 3 linear orders for C_n, $n \geq 3$. The dimension of the complete graph K_5 is 4, but the removal of any edge reduces the dimension to 3. Similarly, the dimension of the complete bipartite graph $K_{3,3}$ is 4 and again the removal of any edge reduces the dimension to 3. These examples are instances of the classical theorem of Schnyder.

Theorem 2.17 (Schnyder)
A graph G is planar if and only if its dimension is at most 3.

Proof. Let G be a non-planar graph and suppose $\dim(G) \leq 3$. Let $\{L_1, L_2, L_3\}$ be a realizer of G. For a vertex v of G let v_i be the position of v in L_i. Define an embedding ϕ of G in $\mathbb{R}^3$ by $v \to \phi(v) = (s_1v_1, s_2v_2, s_3v_3)$, where the s_i are scalars which will be fixed later. For an edge $e = \{u, v\}$ let $e_i = \max(u_i, v_i)$ and embed e by $e \to \phi(e) = (s_1(e_1 + \frac{1}{2}), s_2(e_2 + \frac{1}{2}), s_3(e_3 + \frac{1}{2}))$. Note that by the definition of a realizer we have

$(\star)$ $\phi(v)_i < \phi(e)_i$ for $i = 1, 2, 3$ if and only if $v \in e$.

Adjust the s_i such that under the orthogonal projection π to the plane $x_1 + x_2 + x_3 = 0$ all points in $\phi(V \cup E)$ project to distinct points and these points are in general position.

Now G is drawn in the plane by joining $\pi(\phi(v))$ and $\pi(\phi(e))$ with a straight line segment whenever $v \in e$. Assuming that G has no planar representation there are crossing segments $[\pi(\phi(u)), \pi(\phi(e))]$ and $[\pi(\phi(v)), \pi(\phi(f))]$ with $u \notin f$ and $v \notin e$. Let p be the crossing point and suppose that the ray starting in p and leaving the plane orthogonally meets the segment $[\phi(u), \phi(e)]$ in $\mathbb{R}^3$ at x no later than it meets $[\phi(v), \phi(f)]$ at y. Now consider the path formed by straight segments from $\phi(u)$ to x to y and $\phi(f)$. This path is increasing in each coordinate, hence $u \in f$ by property $\star$. The contradiction shows that G is planar.

It remains to show that every planar graph G admits a realizer $\{L_1, L_2, L_3\}$. By monotonicity we may assume that G is a maximal planar graph, i.e., a triangulation. Consider the trees of a Schnyder wood of G. Since each of the trees has $n-1$ edges and the graph has $3n-6$ edges the only bidirected edges are the three edges of the exterior triangle. Therefore, $R_i(u) \subset R_i(v)$ whenever $u \in R_i(v)$. For $i = 1, 2, 3$ let the inclusion order on the regions induce the order Q_i on the vertices, i.e., $u < v$ in Q_i iff $R_i(u) \subset R_i(v)$. For any edge (u, v) and vertex $w \neq u, v$ the edge is in one of the regions $R_i(w)$ of w, hence, $u < w$ and $v < w$ in Q_i. This shows that any choice of linear extensions L_i of Q_i, $i = 1, 2, 3$, will produce a realizer for G. □

In complete analogy to the definition of the dimension of a graph the dimension of a hypergraph can be defined. A particularly interesting instance is related to 3 dimensional polytopes and hence, by Steinitz's theorem also to planar graphs.

Let P be a polytope with vertex set $\mathcal{V}(P)$ and facets $\mathcal{F}(P)$. Given a subset $\mathcal{G}$ of $\mathcal{F}(P)$ a *realizer* for $(P, \mathcal{G})$ is a nonempty family $\mathcal{R}$ of linear orders on $\mathcal{V}(P)$ provided

(**) For every facet $F \in \mathcal{G}$ and every vertex $x \in \mathcal{V}(P) \setminus \mathcal{V}(F)$, there is some $L \in \mathcal{R}$ so that $x > y$ in L for every $y \in \mathcal{V}(F)$.

The *dimension* of $(P, \mathcal{G})$, denoted $\dim(P, \mathcal{G})$, is then defined as the least positive integer t for which $(P, \mathcal{G})$ has a realizer of cardinality t. In the case $\mathcal{G} = \mathcal{F}(P)$ we simply write $\dim(P)$ and call this the *dimension of the polytope P*.

Theorem 2.18 *If P is a d-polytope with $d \geq 2$, i.e., a polytope whose affine hull is d dimensional, then* $\dim(P) \geq d + 1$.

Proof. The proof is by induction on d. If $d = 2$ then the vertices and facets of P have the structure of the cycle C_n for $n = |\mathcal{V}(P)|$. It follows from Proposition 2.16 that $\dim(P) \geq 3$.

Let P be a d-polytope embedded in $\mathbb{R}^d$ for some $d > 2$ with realizer $L_1, L_2, \ldots, L_t$. Let v be the highest vertex in L_t and consider a hyperplane H which separates v from all the other vertices of P. The intersection $P \cap H$ is a $(d-1)$-polytope P/v, the so called

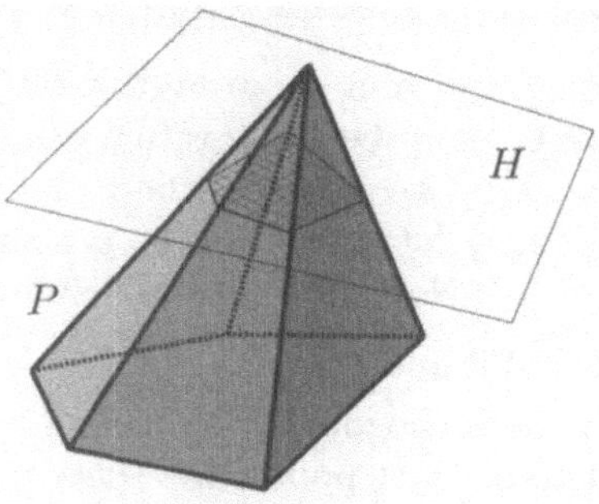

Figure 2.19 The *vertex figure* of the tip vertex of P.

vertex figure of P at v. The $(k-1)$-dimensional faces of P/v are in bijection with the k-dimensional faces of P that contain v. In particular an edge (u, v) of P corresponds to a vertex $u' = (u, v) \cap H$ of P/v and for every facet $\{u'_1, \ldots, u'_r\}$ of P/v there is a facet $\{v, u_1, \ldots, u_r, w_1, \ldots, w_s\}$ of P. Let $\mathcal{F}_v$ be the set of facets of P containing v. The correspondence $\mathcal{F}(P/v) \leftrightarrow \mathcal{F}_v$ shows that $\dim(P/v) \leq \dim(P, \mathcal{F}_v)$. Since P/v is $(d-1)$-dimensional $\dim(P/v) \geq d$ by induction. Now let $F \in \mathcal{F}_v$ and $w \notin \mathcal{V}(F)$, by the choice of v the order L_t cannot bring w over F. Therefore, $L_1, L_2, \ldots, L_{t-1}$ is a realizer for $(P, \mathcal{F}_v)$, i.e., $\dim(P, \mathcal{F}_v) \leq t - 1$. Combine the inequalities to deduce $t \geq d + 1$. □

It is known that for $d \geq 4$ a polytope in d-space can have arbitrarily high dimension. For $d = 3$, however, the situation is different. By Steinitz's theorem polytopes and 3-connected planar graphs are essentially the same. Making use of our knowledge about Schnyder woods we prove:

Theorem 2.19 (Brightwell–Trotter)
If P is a 3-polytope, then $\dim(P) = 4$. *Moreover, if $I \in \mathcal{F}(P)$ and $\mathcal{F}_I = \mathcal{F}(P) \setminus \{I\}$, then* $\dim(P, \mathcal{F}_I) = 3$.

Proof. Let $\mathcal{R}$ be a realizer of $(P, \mathcal{F}_I)$. To obtain a realizer for P we only have to add a single linear order with $v < w$ for all $v \in \mathcal{V}(I)$ and $w \in \mathcal{V}(P) \setminus \mathcal{V}(I)$ to $\mathcal{R}$. Combined with the lower bound from Theorem 2.18 this yields $4 \le \dim(P) \le \dim(P, \mathcal{F}_I) + 1$. To prove the theorem it remains to show $\dim(P, \mathcal{F}_I) \le 3$.

Let G be the graph of P. The graph G is planar and 3-connected by Steinitz Theorem, Theorem 1.7. Choose a planar embedding of the graph G of P with I as the exterior face. Specify vertices a_1, a_2, a_3 in clockwise order around I. As 3-connected planar graph G has a Schnyder woood for every choice of three special vertices at the exterior face, see Section 2.6. Consider a Schnyder wood of G with special vertices a_1, a_2, a_3. As in the proof of Theorem 2.17 we define linear extensions L_i of the inclusion order Q_i of regions $i = 1, 2, 3$, i.e., $u < v$ in Q_i iff $R_i(u) \subset R_i(v)$. To bring every vertex y over every face $F \in \mathcal{F}_I$ with $y \notin F$, however, more care in the choice of L_i is required.

Define Q_i^* such that $u < v$ in Q_i^* if either

(a) $u < v$ in Q_i or

(b) $u || v$ in Q_i and $u < v$ in Q_{i+1}.

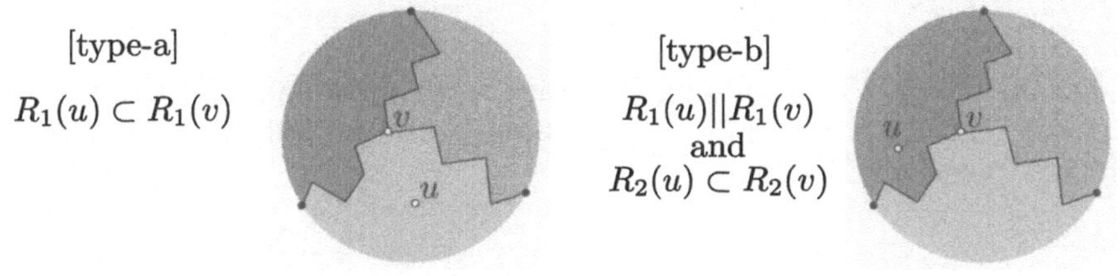

Figure 2.20 The two types of comparabilities $u < v$ in Q_1^*.

Lemma 2.20 Q_i^* *is acyclic for* $i = 1, 2, 3$.

Proof. Call (u, v) a type-a pair if $u < v$ in Q_i^* by part (a) of the definition and call it a type-b pair if $u < v$ in Q_i^* by (b). A cycle in Q_i^* has to contain both a type-a pair and a type-b pair. We claim that if $u < v$ is a type-a pair and $v < w$ is a type-b pair then $u < w$ is also in Q_i^*. Since $u < v$ and $v < u$ cannot be both in Q_i^* the claim yields a contradiction to the assumption that Q_i^* contains a cycle.

Claim. If $u < v$ is a type-a pair and $v < w$ is a type-b pair then $u < w$ is also in Q_i^*.

By symmetry we may assume that $i = 1$. If $R_1(v) = R_1(w)$ then with (u, v) the pair (u, w) also is type-a. Therefore, we assume $R_1(v) \not\subseteq R_1(w)$, since (v, w) is type-b this implies $w \notin R_1(v)$ and $w \notin R_2(v)$. Therefore, $w \in R_3^o(v)$ and $R_3(w) \subset R_3(v)$. Since $u \in R_1(v)$ we either find u in $R_1(w)$ or in $R_2(w)$. If u in $R_1(w)$ then $R_1(u) \subseteq R_1(w)$ but equality is impossible since $w \notin R_1(u)$, i.e., (u, w) is type-a pair in this case. Otherwise $u \in R_2^o(w)$, i.e., $R_2(u) \subset R_2(w)$, and the 1–regions of u and w are incomparable. This shows that (u, w) is a type-b pair in this case. □

Let L_i be a linear extension[†] of Q_i^*. We have to show that L_1, L_2, L_3 is a realizer for the incidence hypergraph of vertices and bounded faces of G, i.e., a realizer for $(P, \mathcal{F}_I)$. Consider a pair (F, y), where F is a face and y is a vertex not on F. Face F is contained in one of the regions of y, by symmetry we may assume that $F \in R_1(y)$. Hence, $R_1(x) \subseteq R_1(y)$ for all $x \in F$. If $R_1(x) \subset R_1(y)$ for all $x \in F$ then F is below y in L_1. Assume that there is an $x \in F$ with $R_1(x) = R_1(y)$.

Note that it is impossible that F contains vertices x and x' with $R_1(x) = R_1(y) = R_1(x')$ and $x \in P_3(y)$ while $x' \in P_2(y)$. This would lead to the placement of y on some edge bounding F in the drawing $\mu(G)$. This is a contradiction since $\mu(G)$ is a convex drawing.

Suppose that for all $x \in F$ either $R_1(x) \subset R_1(y)$ or $R_1(x) = R_1(y)$ and $x \in P_3(y)$. From (E_2), page 23, it follows that $R_2(x) \subset R_2(y)$ for all $x \in F$ with $R_1(x) = R_1(y)$. By the definition of Q_1^* this shows that F is below y in L_1.

Finally, consider the situation that for all $x \in F$ either $R_1(x) \subset R_1(y)$ or $R_1(x) = R_1(y)$ and $x \in P_2(y)$. We claim that F is below y in L_3 in this case. All x in F with $R_1(x) = R_1(y)$ have $R_3(x) \subset R_3(y)$ by (E_2), hence, they are below y in L_3. If $x \| y$ in Q_3 for all $x \in F$ with $R_1(x) \subset R_1(y)$ these vertices also go below y in L_3 and we are done. This is shown to be true in the lemma below which completes the proof of the theorem.

Lemma 2.21 *If $R_1(x) = R_1(y)$, $x \in P_2(y)$ and F is a face in $R_1(x)$ with $x \in F$ and $y \notin F$ then $R_3(y) \not\subseteq R_3(v)$ for all $v \in F$.*

Proof. Consider the triangle ∇_F enclosing F in the convex drawing $\mu(G)$ (Lemma 2.11). Vertex y is placed on the horizontal line ℓ_1 bounding ∇ and y is left of all vertices of F on ℓ_1, Figure 2.21 shows the situation.

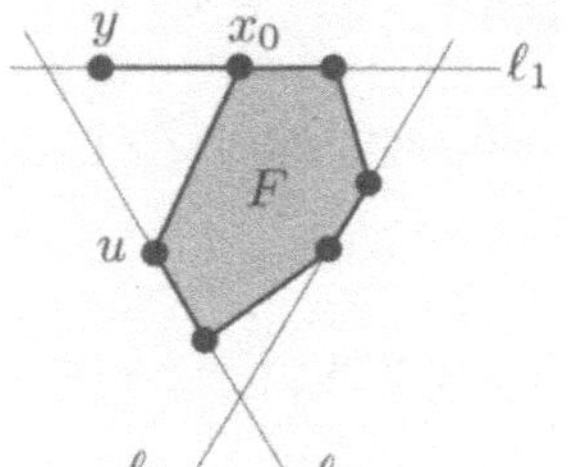

Figure 2.21 Crucial for this case is the orientation of u, x_0.

Let x_0 be the leftmost vertex of F on ℓ_1 and u be the uppermost vertex of F on ℓ_3. Let x_1 be the other neighbor of x_0 at F, i.e., $u \neq x_1$. Even so x_1 need not be on ℓ_1 the edge x_0, x_1 is the outgoing edge of x_0 in label 2, c.f. Figure 2.10. Also (x_0, y) is the outgoing edge of x_0 in label 3. The edge orientations at vertex x_0 imply that (u, x_0) is oriented from u to x_0 in label 1. This shows that x_0 is on $P_1(v)$ for all $v \in F \cap \ell_3$. The paths $P_1(v)$ for $v \in F \cap \ell_2$ clearly cross ℓ_1 to the right of x_0. This shows that $y \notin R_3(v)$ for all $v \in F$, hence $R_3(y) \not\subseteq R_3(v)$. □

[†] Actually Q_i^* is already a total order.

2.6 Existence of Schnyder Labelings

A route plan for the construction of a Schnyder labeling for a 3-connected planar graph could be the following:

(1) Choose an edge e of G and let G/e be the graph obtained by contraction of e.

(2) Recursively construct a Schnyder labeling of G/e.

(3) Expand the labeling of G/e to a Schnyder labeling of G.

A detail that deserves some cautiousness is the choice of edge e. For the induction it is required that G/e is again 3-connected. We call an edge e of a 3-connected graph G such that G/e is again 3-connected a *contractible edge.* The existence of a contractible edge is warranted by the following lemma of Thomassen [196]:

Lemma 2.22 *A 3-connected graph G with at least five vertices contains an edge e whose contraction leaves a 3-connected graph G/e.*

If we let e be an arbitrary contractible edge, however, the proof that the expansion of the labeling can be carried out may involve excessive case distinctions. To reduce the case analysis it would be desirable to have a contractible edge of a special form. But the existence of such an edge will likewise not come for free. Schnyder has taken this approach in his work about planar triangulations. Here we take a different inductive approach.

Let G be a 3-connected planar graph with three special vertices a_1, a_2, a_3 in clockwise order on the boundary cycle C of the outer face. Let $x \notin C$ be a neighbor of a_1.

Suppose that $e = (a_1, x)$ is contractible and let $a_1, x_1, x_2, \ldots x_k$ be the neighbors of x in clockwise order. Since e is contractible only x_1 and x_k may be neighbors of both a_1 and x. Figure 2.22 shows a generic contraction of the edge (a_1, x) into a_1.

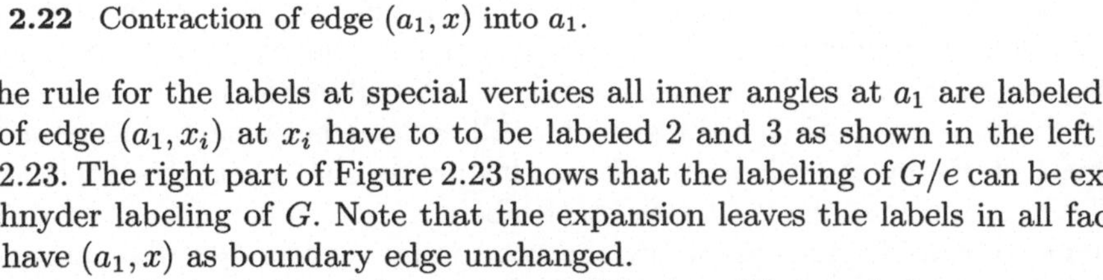

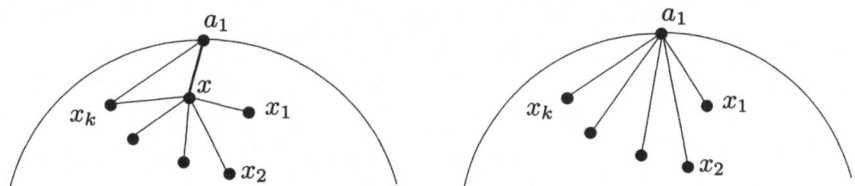

Figure 2.22 Contraction of edge (a_1, x) into a_1.

By the rule for the labels at special vertices all inner angles at a_1 are labeled 1. The angles of edge (a_1, x_i) at x_i have to to be labeled 2 and 3 as shown in the left part of Figure 2.23. The right part of Figure 2.23 shows that the labeling of G/e can be expanded to a Schnyder labeling of G. Note that the expansion leaves the labels in all faces that do not have (a_1, x) as boundary edge unchanged.

Next suppose that $e = (a_1, x)$ is not contractible, i.e., G/e is only 2-connected. Clearly, every cutset of size two in G/e has to contain a_1, let y be the second vertex of such a cutset. The set $S = \{a_1, x, y\}$ is a cutset of G, denote the components of $G \setminus S$ by H and K. Let H' be $G \setminus K$ and $K' = G \setminus H$. The idea is to take Schnyder labelings of the two smaller graphs H' and K' and to show that they can be pasted together.

The first problem is that H' and K' need not be 3-connected, resolve this by augmenting both graphs with the edges (a_1, y) and (x, y), provided these edges are not existent, this yields H'' and K''.

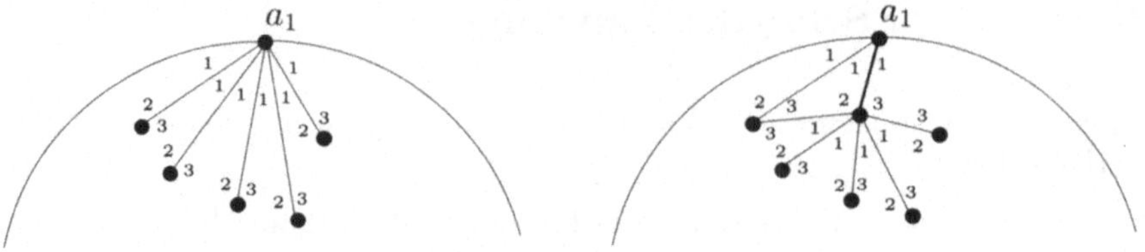

Figure 2.23 Extending the Schnyder labeling of G/e to G.

We have to consider two cases. First suppose that one of the graphs, say H'', contains all special vertices a_1, a_2 and a_3. A Schnyder labeling of H'' contains all three labels in the triangle $T = (a_1, x, y)$. The label at a_1 is 1, let the second special vertex for K'' be the vertex with label 2 in T and the third special vertex be the vertex with label 3 in T. Construct a Schnyder labeling for K'' with this assignment of special vertices. If all three edges of T have been present in G, then the pasting of the labelings makes no problem: the rules of vertices and faces can be verified in the labelings of H'' and K''.

It remains to consider a face that was cut by a new edge. We treat this case with the edge (x, y) with the assumption that the angle of x in T has label 2 in the labeling of H''. Let F be the face of G containing x and y and let F_H and F_K be the parts of this face after insertion of the edge (x, y) such that F_H belongs to H'' and F_K to K''. Figure 2.24 shows the situation.

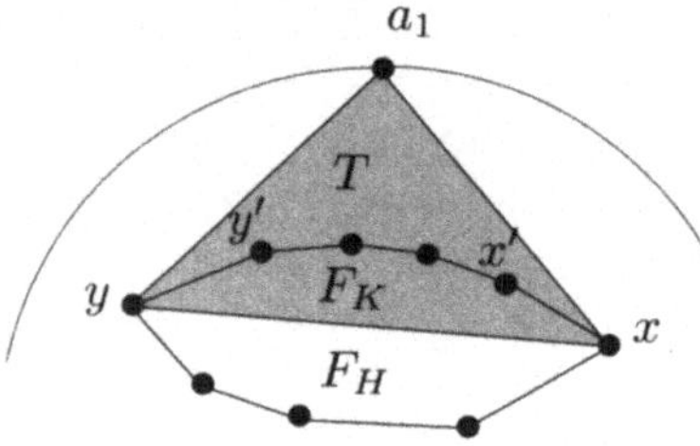

Figure 2.24

Since in labeling of H'' the angles of x and y at T are 2 and 3 the label of x at F_H is 1 or 2 and the label of y at F_H is 1 or 3. The claim is that we can use the same labels in G. Now consider the labeling of K''. Both labels of x at the edge (x, x') are 2 and both labels of y at the edge (y, y') are 3 by the rule for special vertices. Therefore, the labels of x' and y' at F_K are both 1, see Figure 2.3. All vertices between x' and y' in K'' also have label 1 at F_K by the rule for the face. This proves that using the labels of H'' for the angles at x and y in G gives a consistent labeling.

It remains to consider the case where the two special vertices a_2 and a_3 are separated by $\{a_1, x, y\}$. Assume $a_3 \in H''$ and $a_2 \in K''$ vertex y has to play the role of the missing special vertex in both graphs, i.e., the role of a_2 in H'' and the role of a_3 in K''. Figure 2.25 shows some of the labels in the Schnyder labelings of H'' and K'', we have to prove that they can be pasted together to yield a Schnyder labeling of G.

The edge a_1, y was not present in G, so we remove it from both graphs and identify the two copies of a_1, x and y. Since the edges (a_1, x) and (x, y) from the two graphs are also identified the labels in the triangles formed by a_1, x, y in H'' and K'' vanish. However, if we assign label 1 to the outer angle at y the rule of vertices is satisfied at x and y. If

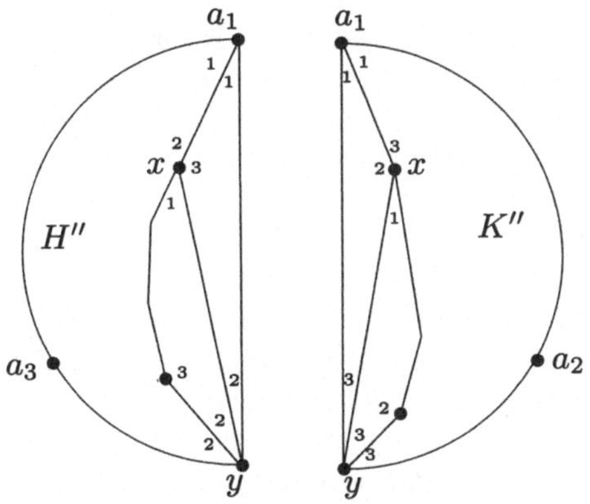

Figure 2.25

the edge (x, y) is in G then the rules of all the other vertices and faces can be verified in the labelings of H'' and K''. If (x, y) has to be removed assign label 1 to the angle at x and one of the labels 2 or 3 to the angle at y. Again all the conditions for a Schnyder labeling are easily verified.

2.7 Notes and References

The existence of straight line drawings for planar graphs was independently proven by Wagner [208], Fáry [79] and Stein [187]. The question whether every planar graph has a straight line embedding on a grid of polynomial size was raised by Rosenstiehl and Tarjan [165]. Unaware of the problem Schnyder [172] constructs a barycentric representation which immediately translates to an embedding on the $(2n-6) \times (2n-6)$ grid. The first explicit answer to the question was given by de Fraysseix, Pach and Pollack [56, 57]. They construct straight line embeddings on an $(2n-4) \times (n-2)$ grid and show that the embedding can be computed in $O(n \log n)$. De Fraysseix et al. also observed a lower bound of $(\frac{2}{3}n - 1) \times (\frac{2}{3}n - 1)$ for grid embeddings of the n vertex graph containing a nested sequence of $n/3$ triangles. Xin He [113] mentions the conjecture that every planar graph can be embedded on the $(\frac{2}{3}n - 1) \times (\frac{2}{3}n - 1)$ grid. In [173] Schnyder improved on his first result and shows the existence of an embedding on the $(n-2) \times (n-2)$ grid which can be computed in $O(n)$ time. The difference between the two algorithms of Schnyder is that in the first case coordinate v_i of vertex v is obtained by counting the faces in region $R_i(v)$. In the second algorithm v_i is computed using the vertices in this region. More compact representations can be found for 4-connected planar graphs. Xin He [113] shows that every such graph embeds on a $W \times H$ grid with $W + H \leq n$.

Tutte [202, 203] shows that every 3-connected planar graph G admits a strictly convex drawing. The idea for Tutte's proof is to nail down at least three vertices of the outer face of G and consider edges as springs. The equilibrium state of the self-stress of this framework is a convex drawing of G and can be computed by solving a system of linear equations. This basic idea of spring-embeddings has been modified and developed in many directions, confer the graph drawing book [61] for further references. Schnyder and Trotter [174] have worked on convex grid embeddings. Felsner [81] elaborated the idea of using Schnyder woods for convex drawings. Kant [118] has extended the approach of de Fraysseix et al. to construct convex drawings on the $(2n-4) \times (n-2)$ grid, the grid size was further reduced to $(n-2) \times (n-2)$ by Chrobak and Kant [49]. Straight line grid drawings with area $O(n^2)$ are not strictly convex, in fact, every strictly convex embedding of the n-cycle requires an area of $\Omega(n^3)$.

Every graph admits a straight line embedding in $\mathbb{R}^3$ without crossing edges. This can be achieved, e.g., by placing the vertices on different points of the moment curve $x \to (x, x^2, x^3)$. Three-dimensional grid drawings have been studied, among others, by Pach, Thiele and Tóth [153]. They prove that every r-colorable graph has a three-dimensional grid drawing in a box of volume cr^2n^2.

The self-stress approach of Tutte can be extended to prove Steinitz's Theorem. Details for the following outline are elaborated by Hopcroft and Kahn [115] and Richter-Gebert [162]. Let G be a planar graph with $V(G) = \{1, \dots, n\}$. A *stress* on G is a symmetric $n \times n$ matrix $W = (w_{ij})$ with $w_{ij} = 0$ for all non-edges ij of G. An embedding $i \to p_i$ of the vertices of G to $\mathbb{R}^2$ is said to be in equilibrium at i if $\sum_{j=1}^{n} w_{ij}(p_i - p_j) = 0$. Let W be a positive stress, i.e., $w_{ij} > 0$ for every edge ij of G and let G be a 3-connected planar graph. Suppose that a set C of at least three vertices of G is nailed down, i.e., has been assigned unchangeable positions in the plane. Then:

(1) There exists unique positions p_i for the vertices of $V \setminus C$ such that all these vertices are in equilibrium.

(2) If the set C is a face of G and C is nailed down as a convex polygon in the plane, then the equilibrium is a strictly convex drawing of G.

(3) Such a strictly convex equilibrium drawing of a 3-connected graph can be lifted to a 3-polytope P, see Figure 2.26.

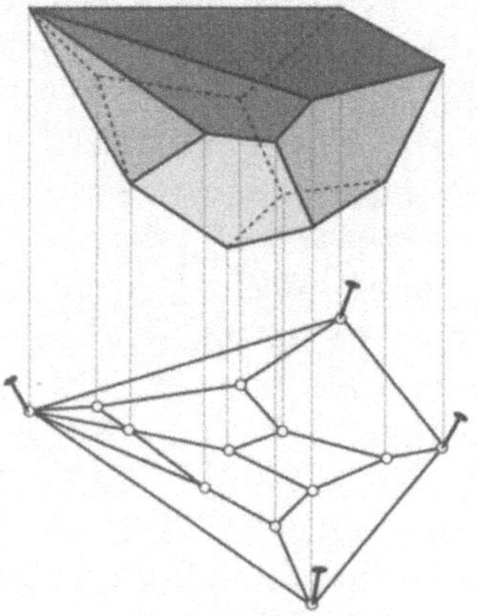

Figure 2.26 A stressed drawing lifted to space

A different and very beautiful approach to both, convex drawings and Steinitz's Theorem, is based on the Koebe Circle Packing Theorem.

Theorem 2.23 *Every planar graph G can be represented by a set of non-overlapping circles in the plane, a circle C_v for every vertex v, so that two vertices u, v are adjacent in G if and only if circles C_u and C_v touch in a point $t_{u,v}$, i.e., the two circles are tangent to each other.*

The result originated from complex analysis and was rediscovered at several times and places. Sachs [168] and Ziegler [219] comment on the history of the theorem. A decade ago the theorem and some generalizations where popularized in the graph theory community. Particularly appealing is the following generalization.

Theorem 2.24 *Let G be a 3-connected planar graph with dual G^* and let o be the vertex of G^* corresponding to the outer face of G. Then there are circle representations of G and $G^* \setminus o$ and a circle C_o so that:*

- *Circle C_o contains all face circles and touches C_f, in a point $t_{f,o}$, iff (f,o) is an edge of G^*.*
- *For every edge (u,v) of G and its dual edge (f,g) the touching points coincide, i.e., $t_{u,v} = t_{f,g}$, and the tangents of C_u, C_v in $t_{u,v}$ and of C_f, C_g in $t_{f,g}$ cross perpendicular.*

Furthermore, the representation is unique up to linear fractional transformations of the plane.

A proof of Theorem 2.23 can be found in Pach and Agarwal [148]. Theorem 2.24 is proved by Brightwell and Scheinerman [39]. We sketch the steps for a proof of the stronger of the two theorems.

From 3-connectivity it follows that either G or G^* contains a vertex of degree three, so we may assume that the outer face o of G is a triangle with vertices a, b, c. To each vertex $v \neq a, b, c$ and to each face $f \neq o$, assign a variable r_v, respectively r_f which is thought of as the radius of the corresponding circle. The radii of the circles corresponding to a, b, c are fixed at 1. With an assignment of radii there is a right angled triangle with side length r_v, r_f and $\sqrt{r_v^2 + r_f^2}$ corresponding to every incident pair v vertex, f face. Two such triangles are combined to form the kite k_{vf}, see Figure 2.27.

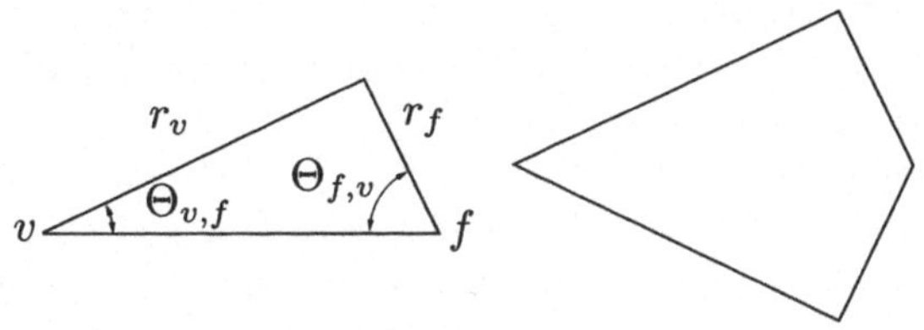

Figure 2.27 The triangle and the kite of v and f.

It has to be shown that some assignment of radii allows to lay out the kites nicely side by side as in Figure 2.28. A necessary local condition for this is that $\sum_{f:vIf} \Theta_{v,f} = \pi$ holds for every vertex v and of course a similar condition for faces. It can be shown that under the 3-connectivity assumption this set of conditions has a unique solution. The second main step in the proof is to show that the triangles obtained that way indeed fit together globally. This is shown using a discrete homotopy argument.

Lifting a primal-dual circle representation to the sphere and using the planes supported by the face cycles yields a polytope with skeleton graph G. This polytope can, in an essentially unique way, be adapted to have all edges tangent to the sphere, Schramm [175]. Higher dimensional sphere representations and connections to the Colin de Verdière number of a graph are discussed by Kotlov, Lovász and Vampala [124]. Another application

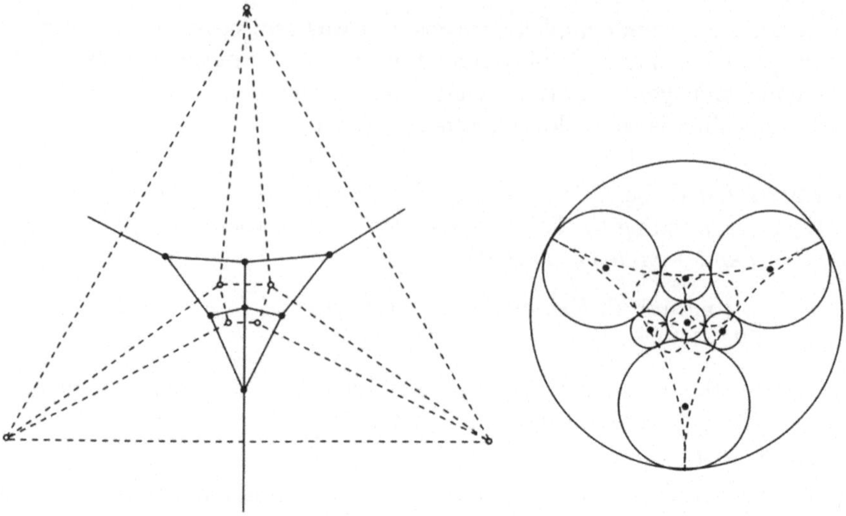

Figure 2.28 A planar graph layed out with kites (left) and with circles (right).

of circle representations is a new proof of the Lipton-Tarjan separator theorem for planar graphs, c.f. the book of Pach and Agarwal [148].

Schnyder woods and applications.

In his two fundamental papers [172, 173] Schnyder developed a theory of Schnyder labelings and Schnyder woods for planar triangulations.

In the first publication [172] Schnyder gave the characterization of planar graphs. He stated Theorem 2.17, however, he used the following slightly different concept of dimension. With a finite graph $G = (V, E)$, associate a height two order P_G whose ground set is $V \cup E$. The order relation is defined by setting $x < e$ in P_G if $x \in V$, $e \in E$ and $x \in e$. P_G is called the *incidence order* of G.

When $P = (X, <)$ is an order, and $\mathcal{R} = \{L_1, L_2, \ldots, L_t\}$ is a family of linear orders on X, we call $\mathcal{R}$ a realizer of P if $P = \cap\mathcal{R}$, i.e., $x < y$ in P if and only if $x < y$ in L_i for all $i = 1, 2, \ldots, t$. The *dimension* of an order is then defined as the minimum cardinality of a realizer.

With this notation at hand, here is the original form of Schnyder's theorem [172].

Theorem 2.25 *A graph is planar if and only if the dimension of its incidence order is at most 3.*

In the same paper Schnyder also shows that the dimension of the face lattice of a simplicial 3-polytope, i.e., of a polytope with only triangular faces, is 4 and drops to 3 upon removal of a face.

If G is a graph with minimum degree at least 2, then the dimension of G and the dimension of its incidence order P_G agree. Also the dimension of a polytope as defined in this chapter and the order dimension of the face lattice of the polytope agree. The two order theoretic facts that sit in the background of the phenomenon are: All critical pairs of face lattices are min-max pairs, and secondly, if all critical pairs of an order are min-max pairs then its interval dimension equals its order dimension. For additional information on the order theoretical background we recommend Trotter's monograph [200].

A particularly interesting problem about dimension of graphs is the dimension of the complete graph. The order theoretic counterpart to this problem is to determine the order dimension of the first two levels of the Boolean lattice. Spencer and Trotter determined the asymptotic growth of this function as:

$$\dim(K_n) \sim \log_2 \log_2 n + \big(\frac{1}{2} + o(1)\big) \log_2 \log_2 \log_2 n.$$

Hoşten and Morris [114] found a surprising connection to the number of intersecting antichains in the Boolean lattice. Building upon the Hoşten-Morris result Felsner and Trotter [86] found a variant of dimension for graphs such that this dimension of the complete graph relates to the Dedekind numbers, i.e., numbers of all antichains in Boolean lattices. Based on Schnyder's ideas there is also a dimension theoretic characterization of outerplanar graphs, this and further results and references about the dimension of graphs can be found in [86]. Studies of the order dimension of polytopes were carried out by Reuter [160], he proves the lower bound in lattice theoretic terms. Brightwell and Trotter [37] prove Theorem 2.19, however, their definition of the three linear orders for the realizer is more involved. The proof given here is adapted from [81]. Brightwell and Trotter [38] extend the approach to show that the inclusion order of vertices, edges and faces of any planar multi-graph is of dimension at most 4. The dimension of higher dimensional polytopes does not behave as nicely as the dimension of 3-polytopes. Four dimensional cyclic polytopes have a complete graph as skeleton graph, since the dimension of these graphs is unbounded the dimension of 4-polytopes is unbounded as well.

A connection between orthogonal surfaces and planar graphs came up in a series of papers by Sturmfels and others, e.g. [21, 140]. These authors were interested in the structure of minimal resolutions of monomial ideals in three variables. Miller [139] began studying geodesic embeddings from this perspective. He observes and exploits a connection with Schnyder labelings. Miller defines a *rigid geodesic* on a surface $\mathcal{S}_{\mathcal{V}}$ generated by $\mathcal{V}$ as an elbow geodesic connecting u and v via $u \vee v$ with the additional property that u and v are the only elements of $\mathcal{V}$ which are dominated by $u \vee v$. A rigid surface is an orthogonal surface such that all elbow geodesics on the surface are rigid. The main result in Miller's paper is that every 3-connected planar graph is induced by a rigid orthogonal surface. The Brightwell-Trotter Theorem (Theorem 2.19) is a corollary to the existence of rigid geodesic embeddings. Confirming a conjecture from Miller [139], Felsner [84] shows a bijection between planar graphs with a Schnyder wood and combinatorially different rigid orthogonal surfaces.

The set of Schnyder woods of a planar triangulation has the structure of a distributive lattice, this was shown by de Mendez [58] and Brehm [36]. The first step in the proof consists in showing that Schnyder woods of a suspended triangulation M^σ are in bijections with 3-orientations, i.e, orientations such that every vertex has outdegree 3. Given a 3-orientation of M^σ and a directed triangle we obtain another 3-orientation by reverting the orientation of the triangle. Let a flip be the operation from the counterclockwise orientation of the triangle to the clockwise, see Figure 2.29. The transitive hull of the flip operation is an order relation on the set of all 3-orientations and, hence, on the set of all Schnyder wood, of M^σ. This order is a distributive lattice.

As shown by Felsner [83] and de Mendez [58] a much more general result is true: If G is a plane map and $\alpha : V \to \mathbb{N}$ assigns a non-negative integer to every vertex of G, then the set of all orientations of G with $\mathsf{outdegree}(v) = \alpha(v)$ for all $v \in V$ has the

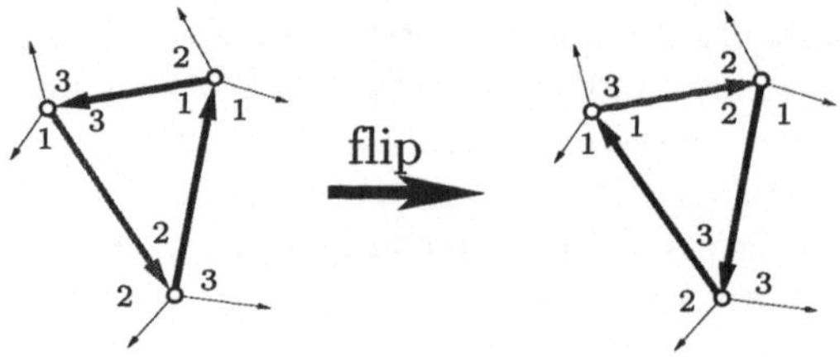

Figure 2.29 Flip of a triangle. These flips generate a distributive lattice on Schnyder woods of a planar triangulation.

structure of a distributive lattice. With this at hand Theorem 2.15 is a direct corollary of Proposition 2.14. The general theorem about α-orientations has several interesting applications (see [83]). To mention just one: The set of rooted spanning trees of a planar graph has the structure of a distributive lattice.

Besides the triangle-flip another type of local flip on Schnyder woods on planar triangulations was recently introduced by Bonichon and others, see [32]. The generic instance of this flip is shown in Figure 2.30. This type of flip makes a change in the underlying graph. Actually, the first application of this operation is a very nice proof of a theorem

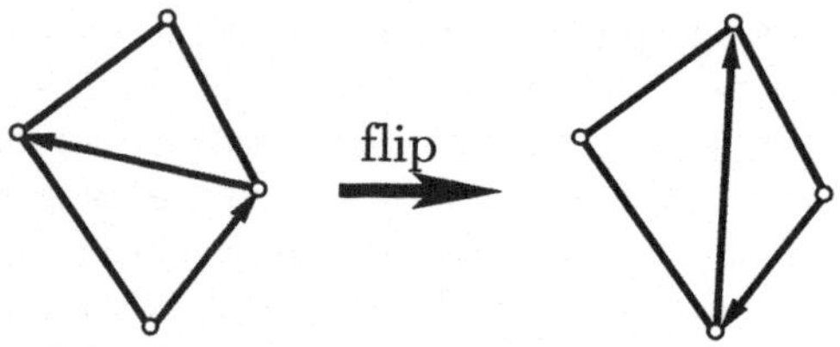

Figure 2.30 Flip of diagonals in a quadrangle.

of Wagner. This theorem says that any two planar triangulations on n vertices can be transformed into each other by a sequence of diagonal flips in quadrangles. The technical hard part of the usual proofs for this theorem is where it comes to show that multiple edges can be avoided in the course of the flipping. This assertion becomes very easy when transforming a given Schnyder wood of one triangulation into a Schnyder wood of the other triangulation via the flip of Figure 2.30. Other interesting applications of this flip lie in the area of counting various kinds of planar map, see [30, 31].

3 Topological Graphs: Crossing Lemma and Applications

Intuitively a handy drawing of a non-planar graph will be a drawing with few crossings. The *crossing number* of a graph G is the least possible number of pairs of crossing edges in a drawing of G. This measure for the non-planarity of a graph has been studied for more than thirty years now. The main result is the Crossing Lemma (Theorem 3.3) it provides a lower bound for the crossing number in terms of the numbers of vertices and edges of a graph. In Section 3.3 the constant in the Crossing Lemma is improved. This improvement is an application of bounds for the number of edges of topological graphs with the property that every edge participates at at most one or two crossings.

With the simple probabilistic proof the Crossing Lemma seems to be an innocent result. However, as first observed by Székely, it is the key to simplified proofs for some deep and important questions of Erdős type. Examples of this phenomenon are the subject of Section 3.4.

3.1 Crossing Numbers

A drawing of a graph is an embedding of the graph in the plane with vertices represented by points and edges represented by Jordan curves. Topological graphs are drawings with the following three additional restrictions:

- No edge has a crossing with itself.
- Two edges with a common endpoint do not cross.
- Two edges cross at most once.

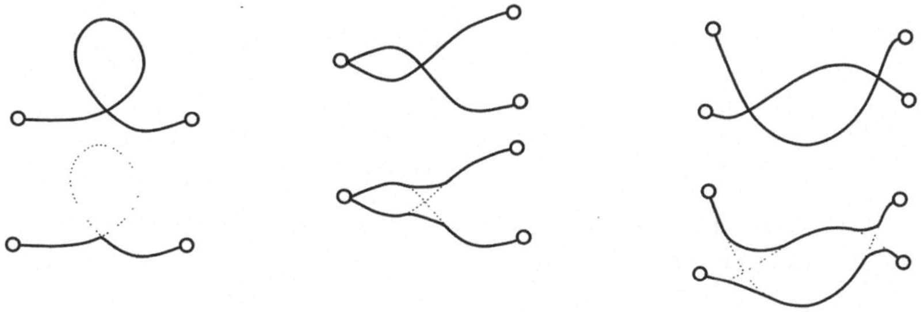

Figure 3.1 Simplifying and crossing reducing modifications of a topological graph.

Figure 3.1 indicates how to modify a drawing in order to obtain a topological drawing with fewer crossings. So, a drawing which achieves the minimal number $\mathrm{cr}(G)$ of crossings of G is a topological drawing. For a topological drawing it can also be assumed that there is no single point where three or more edges cross. In this chapter the drawings under discussion are topological drawings.

The attentive reader may already suspect that the reason for allowing curved edges in drawings is that they may help reduce the number of crossings. This is indeed true, $\mathrm{cr}(K_8) = 18$ but a straight-line drawing of K_8 contains at least 19 crossings. The least number of crossings in a straight-line drawing of a graph G is the *rectilinear crossing number* $\overline{\mathrm{cr}}(G)$. The ratio of $\overline{\mathrm{cr}}(G)$ and $\mathrm{cr}(G)$ can get arbitrarily large. For every m there is a graph with $\mathrm{cr}(G) = 4$ and $\overline{\mathrm{cr}}(G) \geq m$. The construction is based on the graph shown in Figure 3.2.

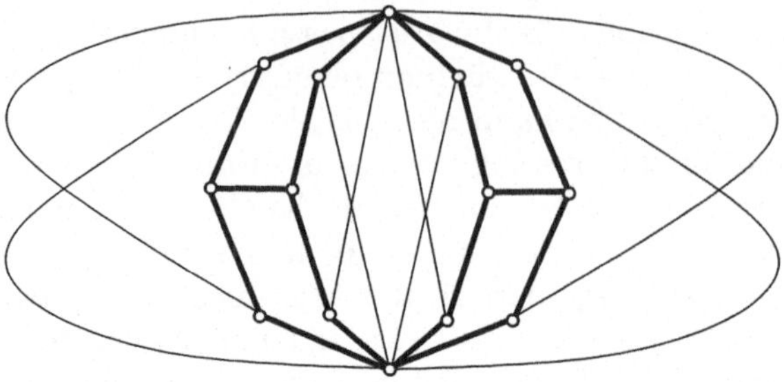

Figure 3.2 A graph constructed to separate $\mathrm{cr}(G)$ and $\overline{\mathrm{cr}}(G)$.

The key for the proof is the following: A straight line drawing of this graph either has a crossing between bold edges or between a bold and a skinny edge. Having shown this, replace each bold edge by a bundle of m edges and subdivide these edges by vertices of degree two. The resulting graph still has crossing number 4 but the rectilinear crossing number will be at least m.

3.2 Bounds for the Crossing Number

A planar graph with n vertices has at most $3n - 6$ edges (Theorem 1.4). This fact is the basis for estimates of the crossing number in terms of the number of vertices and edges of a graph.

Proposition 3.1 *A drawing of a graph G with n vertices and m edges has at least $m - 3n + 6$ crossings.*

Proof. Let H be a maximal planar subgraph of G. Every edge which is not in H has a crossing with some edge in H. Since H has at most $3n - 6$ edges the bound follows. □

With a slightly more subtle counting argument we can raise the lower bound on the crossing number to the order of $m^2/6n$.

Proposition 3.2 *Any drawing of a graph G with n vertices and m edges has at least $\sum_{i=1}^{\lfloor m/r \rfloor} i \cdot r$ crossings, where $r \leq 3n - 6$ is the number of edges of a maximal planar subgraph of G.*

Proof. Let $G_0 = G$ and H_0 be a maximal planar subgraph of G_0. Inductively, define $G_{i+1} = G_i \setminus H_i$ and let H_{i+1} be a maximal planar subgraph of G_{i+1}. Every edge of H_i has at least i crossings, one with each subgraph H_j, $j < i$. The observation $\sum_{i=1}^{\lfloor m/r \rfloor} i \cdot r \leq \sum_i i \cdot |E(H_i)|$ completes the proof. □

Erdős and Guy conjectured that the true order of magnitude of the crossing number is cm^3/n^2 for some constant c. The positive answer is known as the *Crossing Lemma*, it deserves the name lemma because it has very nice applications as we will see in subsequent sections.

Theorem 3.3 (Crossing Lemma)
If G is a graph with n vertices and $m \geq 4n$ edges, then

$$\operatorname{cr}(G) \geq \frac{1}{64}\frac{m^3}{n^2}.$$

Proof. Fix a drawing of G with a minimal number of crossings and construct a random subgraph H of G as follows. Take a p-biased coin, i.e., a coin showing head with probability p. The graph H is constructed by flipping the coin for every vertex v of G, if the coin shows head, then v is accepted for H, otherwise v is refused. The graph H is the graph induced by the accepted vertices.

Let $n(H)$ and $m(H)$ be the number of vertices and edges of H. Denote the number of crossings in the induced drawing of H by $\hat{\operatorname{cr}}(H)$, clearly $\hat{\operatorname{cr}}(H) \geq \operatorname{cr}(H)$. Since H is a random graph these quantities are random variables. The expected values are easily computed

$$\begin{aligned}
\operatorname{Ex}(n(H)) &= p \cdot n(G) \\
\operatorname{Ex}(m(H)) &= p^2 \cdot m(G) \\
\operatorname{Ex}(\hat{\operatorname{cr}}(H)) &= p^4 \cdot \operatorname{cr}(G).
\end{aligned}$$

The exponents of p on the right-hand side just tell how many coin-flips have to show head to make a single vertex, edge or crossing of G appear in H.

Proposition 3.1 implies $\hat{\operatorname{cr}}(H) - m(H) + 3n(H) \geq 0$. Taking expectations and making use of linearity of expectations:

$$\begin{aligned}
\operatorname{Ex}(\hat{\operatorname{cr}}(H) - m(H) + 3n(H)) &\geq 0, \\
\operatorname{Ex}(\hat{\operatorname{cr}}(H)) &\geq \operatorname{Ex}(m(H)) - \operatorname{Ex}(3n(H)), \\
p^4 \cdot \operatorname{cr}(G) &\geq p^2 \cdot m(G) - 3p \cdot n(G), \\
\operatorname{cr}(G) &\geq \frac{m(G)}{p^2} - \frac{3n(G)}{p^3}.
\end{aligned}$$

Set $p = 4n/m$, here we need the assumption that $m \geq 4n$, because as a probability $p \leq 1$. With this p we obtain:

$$\operatorname{cr}(G) \geq \frac{m^3}{16n^2} - \frac{3m^3}{64n^2} = \frac{1}{64}\frac{m^3}{n^2}.$$

□

The bound given by the Crossing Lemma is tight up to the constant: Consider a convex n-gon with vertices $x_0, x_1, \ldots, x_{n-1}$. Let G_k be the geometric graph on this set whose edges are the pairs $x_i x_j$ (indices modulo n) with $j \leq i + k$, i.e., edges are just the 'short' segments. This graph is regular of degree $2k$ and has $m = nk$ edges. An edge $x_i x_j$ with $j = i + (l+1)$ is involved in $l \cdot 2k - l(l+1)$ crossings. Summation over all edges yields:

$$\operatorname{cr}(G_k) = \frac{1}{3}nk^3 + O(nk^2) \approx \frac{1}{3}\frac{m^3}{n^2}.$$

3.3 Improving the Crossing Constant

The constant 1/64 resulting from the probabilistic proof of the crossing lemma is surprisingly good. However, with more effort an improvement by a factor of almost two is possible. The improvement is obtained by applying the probabilistic proof technique to an inequality of the form

$$\mathrm{cr}(G) \geq tm - (\binom{t}{2} + 3t)n. \tag{3.1}$$

This inequality is known to be valid for $t \leq 5$. The crossing constant 1/33.75 is obtained from the probabilistic argument using the above equation with $t = 5$ and $p = 7.5\,n/m$. Here we proof (3.1) only for $t = 3$ and obtain a slightly weaker constant. In the computation we use the probability $p = 6n/m$.

Theorem 3.4 *If G is a graph with n vertices and $m > 6n$ edges, then*

$$\mathrm{cr}(G) \geq \frac{1}{36}\frac{m^3}{n^2}.$$

The proof of (3.1) is based on another nice extremal problem. A drawing is said to be *k-restricted* if every edge is crossed by at most k other edges. The question is: How many edges can a graph G with n vertices have, if G admits a k-restricted drawing.

Theorem 3.5 *A graph with n vertices admitting a k-restricted drawing, $k = 0, 1$ or 2 has at most $(k+3)(n-2)$ edges.*

Before starting into the proof we show how to derive (3.1) for $t = 3$ from the theorem. Let $G_3 = G$ and for $i = 2, 1, 0$ let G_i be a maximal subgraph of G_{i+1} such that the induced drawing of G_i is i-restricted. An edge $e \in E(G_{i+1}) \setminus E(G_i)$ has at least $i+1$ crossings with edges in G_i, otherwise it could be included in G_i. Therefore, $\mathrm{cr}(G) \geq 3(|E(G_3)| - |E(G_2)|) + 2(|E(G_2)| - |E(G_1)|) + (|E(G_1)| - |E(G_0)|) \geq 3m - \sum_{k=0}^{2}(k+3)(n-2) > 3m - 12n$.

Proof. $\mathbf{k = 0}$: Since 0-restricted graphs are exactly the planar graphs they have at most $3n - 6 = 3(n-2)$ edges.

$\mathbf{k = 1}$: Consider a 1-restricted drawing of a graph G with n vertices maximizing the number of edges of such a graph. Let H be a maximal planar subgraph of G. Let e be an edge in $E(G) \setminus E(H)$. Since $e \notin E(H)$ there is an edge $xy \in E(H)$ crossed by e. Let v be a vertex of e and note that it is possible to include edges vx and vy in the drawing of G, so that they are drawn crossing free. The maximality of G implies that the edges vx and vy are in $E(G)$. Since they are crossing free they already belong to $E(H)$.

Consequently, every edge $e \notin E(H)$ traverses a triangular face of H at each of its vertices. Call such a triangular face a *final triangle* for e. Since the drawing of G is 1-restricted a triangular face of H can be the final triangle for only one edge $e \notin E(H)$. Since H has at most $2n - 4$ triangular faces and every $e \in E(G) \setminus E(H)$ requires two final triangles there are at most $n - 2$ edges in $E(G) \setminus E(H)$. Since H has at most $3n - 6$ edges we obtain $|E(G)| \leq 4(n-2)$.

k = 2 : Consider a 2-restricted drawing of a graph G with n vertices maximizing the number of edges of such a graph. Let $vw \in E$ be an edge with a crossing. Start traversing vw at vertex v and let xy be the first edge crossed by vw. Since xy has at most two crossings one of the endpoints of xy can be reached from v without crossing an edge, say, this vertex is x. By maximality of G the edge vx is in G and moreover, this edge is a crossing-free edge, see Figure 3.3.

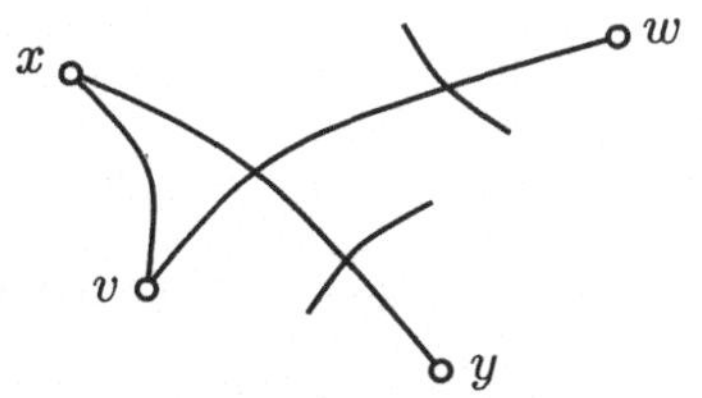

Figure 3.3 A crossing-free edge vx associated to vw.

Let H be the subgraph of G consisting of all edges of G which are crossing free in the drawing of G. We summarize the above argument as follows:

> If $e \in E(G) \setminus E(H)$ and v is a vertex of e, then one of the neighboring edges of e in the cyclic order of edges at v is in H. We say that this edge of H *takes care* of e at v.

An edge of H can take care of at most four crossed edges, i.e., edges of $E(G) \setminus E(H)$. Every crossed edge is taken care of at least twice. Hence, $2|E(G) \setminus E(H)| \leq 4|E(H)|$, or more simply $|E(G)| \leq 3|E(H)|$. To improve on this consider the faces of the plane graph H.

- If there is a four face F in H, then there are (at most) two edges interior to F and from the eight angles of edges of H in F only four are, actually, needed to take care of the interior edges, a loss of four.
- If there is a triangular face F in H, then from the six angles of edges of H in F non is needed to take care of interior edges, a loss of six.

Let f_4, f_3 be the number of 4-faces and 3-faces of H and let $m = |E(H)|$. We have shown $2|E(G) \setminus E(H)| \leq 4|E(H)| - 4f_4 - 6f_3$, or more simply $|E(G)| \leq 3m - 2f_4 - 3f_3$.

Let f be the number of faces H and $f^+ = f - f_4 - f_3$. Since H may be disconnected the Euler relation turns into an inequality: $n - m + (f^+ + f_4 + f_3) \geq 2$. Double counting the edge-face incidences we further obtain $2m \geq 5f^+ + 4f_4 + 3f_3$. Subject to these two inequalities together with $f_3, f_4 \geq 0$ the objective $3m - 2f_4 - 3f_3$ is maximized when $f_3 = f_4 = 0$, $f^+ = \frac{2}{3}(n-2)$ and $m = \frac{5}{3}(n-2)$. This completes the proof of $|E(G)| \leq 5(n-2)$. □

Remark. The bounds given in the theorem for the number of edges of k-restricted drawings, $k = 1, 2$ are best possible. For $k = 1$ let Q be a planar graph such that all faces of Q are 4-gons. Q has $n - 2$ faces and $2n - 4$ edges. Adding the two diagonals to each face of Q gives a 1-restricted graph with $4(n-2)$ edges. Candidates for Q are the cube-graph and graphs obtained by gluing cubes together, face by face, see Figure 3.4.

For $k = 2$ let D be a planar graph such that all faces of D are 5-gons. D has $\frac{2}{3}(n-2)$ faces and $\frac{5}{3}(n-2)$ edges. Adding the five diagonals to each face of D gives a 2-restricted graph with $5(n-2)$ edges. Candidates for D are the dodecahedron-graph and graphs obtained by gluing dodecahedra together, face by face, see Figure 3.4.

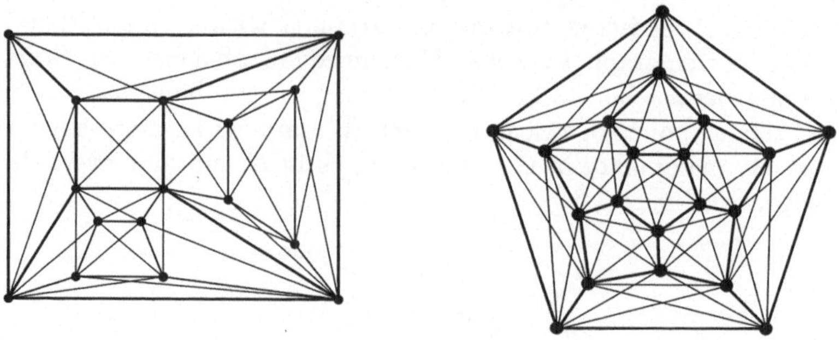

Figure 3.4 Examples of 1- and 2-restricted geometric graphs with maximal number of edges.

3.4 Crossing Numbers and Incidence Problems

Points, lines and incidences between them give raise to elementary and appealing problems. The celebrated Szemerédi–Trotter theorem gives a bound on the number of incidences in terms of the numbers of points and lines. The original proof of the theorem was a remarkable tour de force and would result in enormous constants. It came as a big surprise when László Székely observed that the crossing lemma could be used in an extremely short and elegant proof.

If $\mathcal{P}$ is a set of points and $\mathcal{L}$ is a set of lines, let $I(\mathcal{P}, \mathcal{L})$ be the number of incidences between points of $\mathcal{P}$ and lines of $\mathcal{L}$.

Theorem 3.6 (Szemerédi–Trotter)
Let $I(n, m)$ denote the maximum of $I(\mathcal{P}, \mathcal{L})$ taken over all sets $\mathcal{P}$ of n points and $\mathcal{L}$ of m lines, then

$$I(n, m) \leq 3.3\, n^{2/3} m^{2/3} + 6n + 2m.$$

Proof. Given $\mathcal{P}$ and $\mathcal{L}$ with I incidences we consider the arrangement as a geometric graph G. The vertices of G are the points in $\mathcal{P}$, i.e., $n(G) = n$. The edges of G are segments between consecutive points on a line. A line $\ell \in \mathcal{L}$ containing k points of $\mathcal{P}$ contributes $k - 1$ edges, hence, $m(G) = I - m$.

Since two lines have at most one intersection we have a bound on the crossing number: $\mathrm{cr}(G) \leq \binom{m}{2}$. With (Theorem 3.4) the Crossing Lemma

$$m^2 > \binom{m}{2} \geq \mathrm{cr}(G) \geq \frac{1}{36} \frac{(I - m)^3}{n^2}.$$

This converts to $(36\, m^2 n^2)^{1/3} > I - m$ and further to $I < 3.3\, n^{2/3} m^{2/3} + m$. But recall that $I - m > 6n$ was a precondition for the application of the Crossing Lemma. If this precondition fails $I < 6n + m$. Adding the two bounds on I yields the theorem. □

The following is a surprising application of the incidence bound to a problem from number theory. Let $A = \{a_1, \ldots, a_n\}$ be a set of distinct numbers. Let $\Sigma = |A + A|$ and $\Pi = |A \cdot A|$, i.e., Σ and Π are the number of different sums and products of pairs of elements of A. Clearly, $\Sigma, \Pi \leq n^2/2$. Considering only sums and products involving $\min(A)$ and $\max(A)$ it follows that $\Sigma, \Pi \geq 2n - 1$. Erdős and Szemerédi asked whether one of the two quantities is always large. Elekes made big progress at this problem by applying the Szemerédi–Trotter Theorem.

Let $\mathcal{P}_A$ be the set of all points with coordinates of the form $(a_i + a_j\ ,\ a_k \cdot a_j)$. The number of points in this set is upper bounded by $\Sigma \cdot \Pi$. Consider the set $\mathcal{L}_A$ of the n^2 lines with equations $y = a_k(x - a_i)$. Since point $(a_i + a_j\ ,\ a_k \cdot a_j)$ is on line $y = a_k(x - a_i)$ each line contributes at least n incidences making a total of at least n^3 incidences. From the Szemerédi–Trotter Theorem

$$n^3 \leq C \cdot (|\mathcal{P}_A|^{\frac{2}{3}} |\mathcal{L}_A|^{\frac{2}{3}} + |\mathcal{P}_A| + |\mathcal{L}_A|).$$

With $|\mathcal{L}_A| = n^2$ this yields $|\mathcal{P}_A| \geq cn^{5/2}$. Consequently, one of $A + A$ and $A \cdot A$ has to be of size $c^{1/2}n^{5/4}$.

Back to plane geometry. Let $\mathcal{P}$ be a set of n points and $k \in \mathbb{N}$. A line ℓ is called a *big line* if it contains at least k points from $\mathcal{P}$. How big can the number of big lines be?

Theorem 3.7 *If B_k denotes the maximal number of big lines of a set $\mathcal{P}$ of n points and $2 \leq k \leq \sqrt{n}$, then*

$$B_k \leq c\frac{n^2}{k^3}.$$

Proof. Let $\mathcal{L}$ be a maximal set of big lines for $\mathcal{P}$. As in the previous proof we have a geometric graph G with vertex set $\mathcal{P}$. Since all lines are big $m(G) \geq B_k(k-1)$. Assume $m(G) \geq 6n$ and apply the Crossing Lemma to obtain

$$B_k^2 \geq \mathrm{cr}(G) \geq \frac{1}{36} \frac{(B_k(k-1))^3}{n^2}.$$

Since $k/(k-1) \leq 2$ we can estimate the constant as $c \leq 288$. Now suppose that the graph has to few edges for the Crossing Lemma. From $B_k(k-1) < 6n$, i.e., $B_k < 12n/k$, the bound stated in the theorem follows from $n/k^2 \geq 1$ which is true since $k \leq \sqrt{n}$. □

The next question is about distances. Let $U(\mathcal{P})$ denote the number of unit distances among points in $\mathcal{P}$.

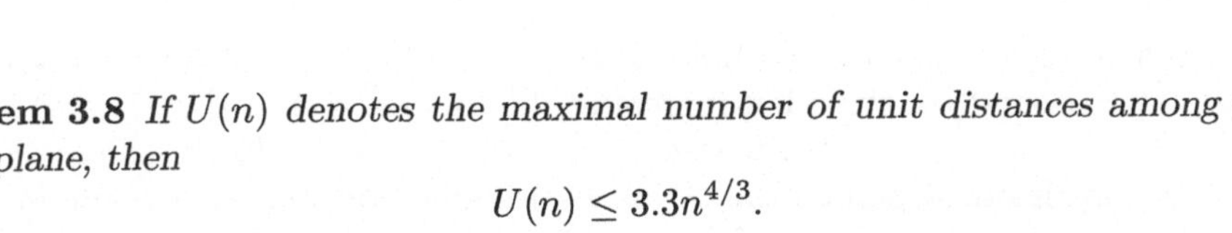

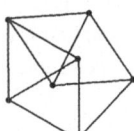

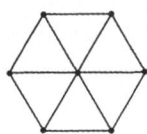

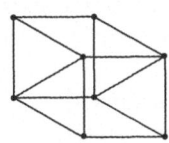

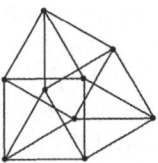

Figure 3.5 Four unit distance graphs, i.e., geometric graphs with edges of length one.

Theorem 3.8 *If $U(n)$ denotes the maximal number of unit distances among n points in the plane, then*

$$U(n) \leq 3.3n^{4/3}.$$

Proof. Draw a circle of radius one around each of the n points in a maximizing configuration of points $\mathcal{P}$. Clearly each of the n circles contains at least two points of $\mathcal{P}$. Consider the arcs between consecutive points on the circles as edges of a topological graph G. The circle centered at point p is subdivided into as many arcs as there are points at unit distance from p. Therefore, the number of arcs of G is $2U(\mathcal{P})$.

Some pairs of points may be connected by two different arcs of G, discard one of every such pair of arcs. For the number of edges of the resulting graph G' we have $m(G') \geq U(\mathcal{P})$. If $m(G') \leq 4n$, then $U(n)$ is only linear in n. Otherwise, note that any two circles cross at most twice and apply the Crossing Lemma

$$n^2 \geq 2\binom{n}{2} \geq \operatorname{cr}(G') \geq \frac{1}{36}\frac{m(G')^3}{n^2} \geq \frac{1}{36}\frac{U(\mathcal{P})^3}{n^2}.$$

□

3.5 Notes and References

The crossing number problems for complete graphs and complete bipartite graphs have a long history. The problem for complete bipartite graphs is known as Turán's brick-factory problem. Turán was prisoner in a German labor camp in 1944 when the problem occurred to him. In [201] he tells the following story:

We worked near Budapest, in a brick factory. There were some kilns where the bricks were made and some open storage yards where the bricks were stored. All the kilns were connected by rail with all the storage yards. The bricks were carried on small wheeled trucks to the storage yards. All we had to do was to put the bricks on the trucks at the kilns, push the trucks to the storage yards, and unload them there. We had a reasonable piece rate for the trucks, and the work itself was not difficult; the trouble was only at the crossings. The trucks generally jumped the rails there, and the bricks fell out of them, in short this caused a lot of trouble and loss of time which was precious to all of us. We were all sweating and cursing at such occasions, I too; but nolens volens the idea occurred to me that this loss of time could have been minimized if the number of crossings of the rails had been minimized. But what is the minimum number of crossings? I realized after several days that the actual situation could have been improved, but the exact solution of the general problem with m kilns and n storage yards seemed to be very difficult ... the problem occurred to me again ... at my first visit to Poland where I met Zarankiewicz. I mentioned to him my "brick-factory"-problem ... and Zarankiewicz thought to have solved (it). But Ringel found a gap in his published proof, which nobody has been able to fill so far-in spite of much effort. (Turán 1977)

The conjecture of Zarankiewicz asserts:

$$\operatorname{cr}(K_{m,n}) = \lfloor\frac{m}{2}\rfloor\lfloor\frac{m-1}{2}\rfloor\lfloor\frac{n}{2}\rfloor\lfloor\frac{n-1}{2}\rfloor.$$

This has been verified in some cases but the general problem remains unresolved. There is an easy upper bound construction matching this bound: Let $\lceil m\rceil = 2s$ and $\lceil n\rceil = 2t$ and define $X = \{(-a,0),(+a,0) : 0 < a \leq s\}$ and $Y = \{(0,-b),(0,+b) : 0 < b \leq t\}$ if m or n is odd delete a point from X or Y. Finally, connect each point in X by a straight segment with each point of Y. Due to this construction a proof of Zarankiewicz's conjecture would also imply $\operatorname{cr}(K_{m,n}) = \overline{\operatorname{cr}}(K_{m,n})$. Very readable accounts to the state of crossing number problems in the early 1970's are given by Guy [110] and by Erdős and Guy [75]. Guy [110] proved that rectilinear crossing number and crossing number can differ, he shows $\operatorname{cr}(K_8) = 18$ and $\overline{\operatorname{cr}}(K_8) = 19$. The example of Figure 3.2, which shows that $\overline{\operatorname{cr}}(G)/\operatorname{cr}(G)$ is unbounded, is due to Bienstock and Dean [25]. Garey and

Johnson [96] prove that computing the crossing number of a graph is a computationlly hard problems.

The Crossing Lemma (Theorem 3.3) was conjectured by Erdős and Guy [75]. The first proofs were given by Ajtai, Chvátal, Newborn and Szemerédi [9] and independently Leighton [131]. The probabilistic proof presented in the first section was communicated by Emo Welzl who attributed it to discussions with Chazelle and Sharir. Pach and Tóth [155] improved the constant from $1/64 \approx 0.015$ to 0.029. They prove the statement of Theorem 3.5 for k up to 4 and conclude inequality (3.1) for $t = 5$. The latest improvement of the crossing constant is due to Pach, Radoicic, Tardos and Tóth [150], the new constant is 0.31. The result is based on an improved bound for the number of edges in a 3-restricted drawing. They show that a graph admitting a 3-restricted drawing has at most $5.5(n-2)$ edges.

Pach and Tóth [155] also discuss upper bound constructions. The geometric graph consisting of all short edges (edges of length $\leq (2m/\pi n)^{1/2}$) in a slightly perturbed $n^{1/2} \times n^{1/2}$ grid is shown to have $\leq 0.06(m^3/n^2)$ crossings. Variants of crossing numbers and of the Crossing Lemma have been studied by Pach and Tóth [156] and Pach, Spencer and Tóth [151]. One of the variants is the *odd-crossing number*, this is the minimum number of pairs of edges which have an odd number of crossings. Planar graphs can be characterized as those graphs which have odd-crossing number zero. An easy proof for this remarkable result of Tutte [204] is still missing.

The geometric applications of the crossing lemma shown in Section 3.4 are taken from Székely's stupendous paper [193]. That paper also contains an improved bound on the number of different distances determined by some point from a set of n points based on the same technique. The number theoretic application of the Szemerédi–Trotter Theorem is from Elekes [69]. There are many more applications of the Crossing Lemma and the incidences bound than shown in this chapter. For further reading we recommend the books of Matoušek [138] and of Pach and Agarwal [148] and the survey of Erdős and Purdy [77].

The Szemerédi–Trotter theorem (Theorem 3.6) is one of the strong tools in combinatorial geometry. The original proof of the theorem [195] was extremely complicated and resulted in enormous constants. A proof using cuttings was found by Clarkson et al. [50], a simplified version of this proof is due to Aronov and Sharir [16]. Székely [193] gave the extremely short and elegant proof reproduced here.

The Szemerédi–Trotter paper [195] contains a first collection of geometric extremal problems which could be attacked using their result. One of these first applications is the big-line theorem (Theorem 3.7). At about the same time Beck [22] proved a slightly weaker result and used this for improvement in several other geometric extremal problems, among them Dirac's problem (cf. the notes section of Chapter 5).

Erdős [74] initated the study of the distribution of distances in planar configurations of points. In particular he asked for two quantities: The minimum number $D(n)$ of distinct distances among n points in the plane and the maximum number $U(n)$ of unit distances among n points in the plane. Erdős remarks that n points arranged on a square-grid of suitable step-size realize $n^{(1+c/\log\log n)}$ unit distances, the computation veifying this is detailed in the book [138] of Matoušek. As upper bound Erdős proved $U(n) \leq n^{3/2}$. The bound given in Theorem 3.8 was first obtained by Spencer, Szemerédi and Trotter [182]. It is conjectured that the actual growth of $U(n)$ is close to the lower bound. Erdős had offered \$500 for a proof or disproof of this conjecture.

Another fascinating problem related to unit distances is the *chromatic number of the plane*: What is the minimum number of colors needed to color the points of $\mathbb{R}^2$ such that any two points at unit distance receive different colors, this number is denoted $\chi(\mathbb{R}^2)$. The first of the small unit distance graphs of Figure 3.5 shows that $\chi(\mathbb{R}^2) \geq 4$. The upper bound $\chi(\mathbb{R}^2) \leq 7$ can be shown by an appropriate coloring of the cells of a regular honeycomb tiling with side-length slightly less than 1/2, see Figure 3.6. The

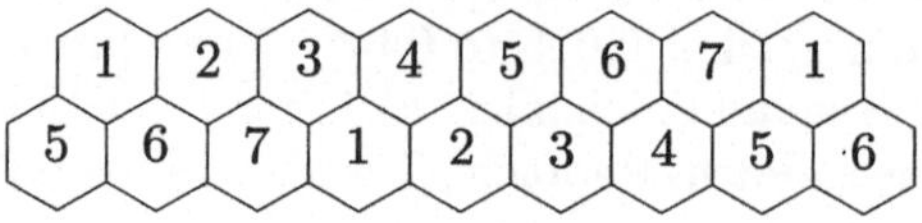

Figure 3.6 Part of a periodic 7 coloring of the plane.

bounds $4 \leq \chi(\mathbb{R}^2) \leq 7$ have stood for over 50 years now. A recent survey was given by Székely [194].

4 k-Sets and k-Facets

Let $\mathcal{P}$ be a set of n points in $\mathbb{R}^d$ in general position, this assumption means that no subset of $j \leq d+1$ points is affinely dependent.

A *k-set* of $\mathcal{P}$ is a k element subset $\mathcal{S}$ of $\mathcal{P}$ such that $\mathcal{S}$ and $\bar{\mathcal{S}} = \mathcal{P} \setminus \mathcal{S}$ can be separated by a hyperplane H. The problem of finding good upper and lower bounds for the number of k-sets is central to combinatorial geometry.

In this chapter we present some of the results that have been obtained. In Section 4.1 we deal with the k-set problem in the plane. We collect classical observations and bounds and continue with the new $O(n\,k^{1/3})$ upper bound (Theorem 4.5). One of the key ingredients is again the Crossing Lemma (Theorem 3.3).

With Section 4.2 we leave the plane. Welzl's Theorem (Theorem 4.7) gives a higher dimensional analog of the Lovász Lemma (Lemma 4.3) in the plane. The section ends with a bound for the 3-dimensional problem. Here the Crossing Lemma again plays a crucial role.

Section 4.3 shows a surprising application of a lower bound for the total number of t-sets with $t \leq k$ ($\leq$$k$-sets for short). The lower bound is used to prove that every straight line drawing of the complete graph K_n has at least $\frac{3}{8}\binom{n}{4}$ crossings. This result of Lovász, Vesztergombi, Wagner and Welzl is a very recent breakthrough at the old and notorious problem of rectilinear crossing numbers.

4.1 k-Sets in the Plane

Let $\mathcal{P}$ be a set of n points in the plane A k-edge of $\mathcal{P}$ is a directed edge with endpoints in $\mathcal{P}$ such that exactly k points of $\mathcal{P}$ are on the positive (left) side of the line supporting the edge.

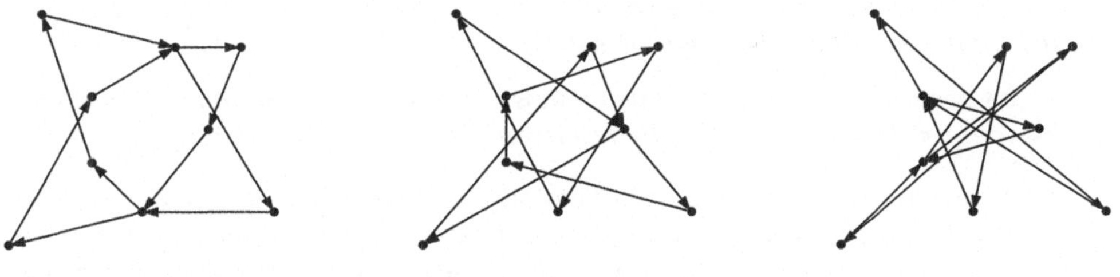

Figure 4.1 A set of 9 points with its k-edges for $k = 1, 2, 3$.

Most contributions in the area of k-sets of planar point sets actually work with k-edges. This is legitimized by the following simple observation.

Observation 1. Let $\mathcal{P}$ be a set of n points in the plane and $1 \leq k < n$. The number of k-sets of $\mathcal{P}$ equals the number of $(k-1)$-edges of $\mathcal{P}$.

Proof. Let $\mathcal{S}$ be a k-set of $\mathcal{P}$. There are exactly two lines which are tangent to $\mathcal{S}$ and $\bar{\mathcal{S}}$ and separate them. Note that being tangent these lines contain exactly one point from each of $\mathcal{S}$ and $\bar{\mathcal{S}}$. Orient these bi-tangents such that $\mathcal{S}$ is left and $\bar{\mathcal{S}}$ right. On exactly one of the lines the direction is from the point in $\mathcal{S}$ to the point in $\bar{\mathcal{S}}$, call this line $g(\mathcal{S})$. The segment between the two points on $g(\mathcal{S})$ is a $(k-1)$-edge $e(\mathcal{S})$ of $\mathcal{P}$. The mapping $\mathcal{S} \to e(\mathcal{S})$ is a bijection between k-sets and $(k-1)$-edges. △

The following lemma has an elementary proof but a wide range of applications (it really deserves to be called a lemma).

Lemma 4.1 (Alternation Lemma)
Let p be a fixed point from $\mathcal{P}$ and ℓ be a directed line which is split by p into a front-half and a back-half. Rotate ℓ with center p in clockwise direction and consider the out-events *where the front-half covers a k-edge which is outgoing at p and the* in-events *where the back-half covers a k-edge which is incoming at p. Out-events and in-events alternate during the rotation.*

Proof. Let $\mu(\ell)$ be the number of points on the positive (left) side of the rotating line ℓ. If the front-half of the line scans a point, then $\mu(\ell)$ is increased by one, $\mu(\ell) \to \mu(\ell)+1$. Symmetrically, $\mu(\ell) \to \mu(\ell)-1$ corresponds to the event that the back-half of the line scans a point. At an out-event $\mu(\ell)$ changes from k to $k+1$. Between two out-events the value of $\mu(\ell)$ must get back to k. This happens first when the value changes from $k+1$ to k and this corresponds to an in-event. Similarly, there is an out-event between any two in-events. Together this enforces the claimed alternation of out- and in-events. □

One immediate consequence of the lemma is that every point has the same number of incoming and outgoing k-edges. In particular, the number of k-edges incident to a point is even. The following lemma pinpoints another important fact.

Lemma 4.2 *Let ℓ be a line containing a unique point p from $\mathcal{P}$. If from the positive side of ℓ there are a incoming k-edges at p, then there are either $a-1$ or a or $a+1$ outgoing k-edges reaching from p into the negative side. If $k < \frac{n-1}{2}$ the precise statement is:*

- *If the positive side of ℓ contains at most k points, then the number of outgoing k-edges reaching from p into the negative side is $a+1$.*
- *If both sides of ℓ contain more than k points, then the number of outgoing k-edges reaching from p into the negative side is a.*
- *If the negative side of ℓ contains at most k points then the number of outgoing k-edges reaching from p into the negative side is $a-1$.*

Proof. As in the proof of the Alternation Lemma we consider rotations of a line centered at p. Starting in the position of ℓ consider a clockwise rotation. If left of ℓ there are at most k points, then the line first arrives at an out-event and if there are more than k points left of ℓ, then the line first arrives at an in-event.

Next we have to determine whether in counterclockwise direction the line first reaches an incoming k-edge from the left of ℓ or an outgoing k-edge to the right. Note that reverting the orientation of an k-edge, yields an $(n-k-2)$-edge and conversely. Therefore we consider $(n-k-2)$-edges, those which are outgoing left of ℓ and those which are incoming right of ℓ and adopt the terminology of out- and in-events for them. Rotating counterclockwise an out-event corresponds to a decrease from $n-k-1$ to $n-k-2$

points on the left side of the line. An in-event corresponds to an increase from $n-k-2$ to $n-k-1$ points on the left side. Therefore, starting with $n-k-1$ or more points on the left the line first arrives at an out-event while starting with $n-k-2$ or fewer points on the left side the line first arrives at an in-event.

The three cases of the lemma are shown schematically in Figure 4.2. The Alternation

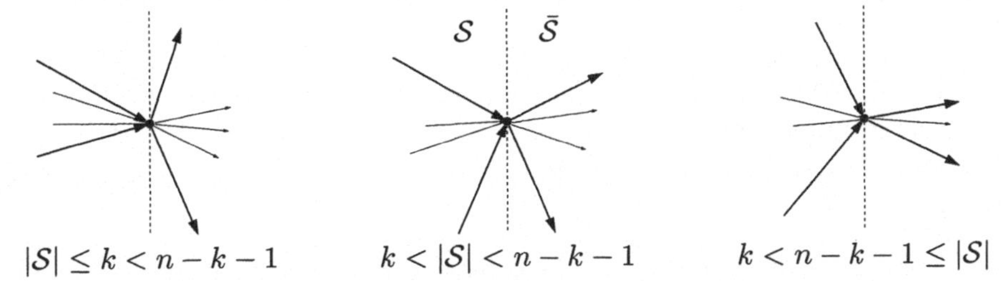

Figure 4.2 A small, medium and large set $\mathcal{S}$ left of the vertical through p.

Lemma allows to draw the claimed conclusions. □

An important special case is when the number n of points is even and $k = \frac{n-2}{2}$. In this case the negative and the positive side of a line supporting a k-edge contain the same number of points, hence, the reorientation of a k-edge is again a k-edge, i.e., k-edges come in bidirected pairs. Consider such pairs of k-edges as single undirected edges and call them *halving edges.* The lines supporting the halving edges are the *halving lines.* Note that the previous lemma implies that every point of an even point set $\mathcal{P}$ is incident to an odd number of halving edges.

Lemma 4.3 (Lovász Lemma)
Let ℓ be a line disjoint from $\mathcal{P}$ which separates $\mathcal{P}$ into a left set $\mathcal{S}$ and a right set $\bar{\mathcal{S}}$. If $|\mathcal{S}| = s$, then for $0 \leq k \leq \frac{n-2}{2}$, the number $e_k(\ell)$ of k-edges (p,q) which cross ℓ from left to right i.e., with $p \in \mathcal{S}$ and $q \in \bar{\mathcal{S}}$, is exactly $\min(k+1, s, n-s)$.

Proof. For simplicity we assume that ℓ is vertical and that no two points of $\mathcal{P}$ have the same x-coordinate. Consider a vertical line ℓ_0 which has $\mathcal{P}$ on its right side. Let ℓ_t be a continuously moving vertical line starting at ℓ_0 and moving across the plane such that ℓ_t, $t \in [0,1]$ are the intermediate positions of the line. Let $e_k(\ell_t)$ be the number of k-edges crossing ℓ_t from left to right. Starting with $e_k(\ell_0) = 0$ the value of $e_k(\ell_t)$ can only change when ℓ_t moves over a point of $\mathcal{P}$. At the first $k+1$ points met by the line the set of points left of ℓ_t has cardinality at most k. Therefore, the first case of Lemma 4.2 tells us that at each of these points the value of $e_k(\ell_t)$ is increasing by one. While the set of points left of ℓ_t has cardinality between $k+1$ and $n-k-2$ there is no change in the value $e_k(\ell)$ as shown by the second case of Lemma 4.2. Hence, $e_k(\ell_t) = k+1$ in this range. At each of the last $k+1$ points the value of $e_k(\ell_t)$ is decreased by one, this is implied by the thirds case of Lemma 4.2.

Since $\ell_t = \ell$ for some t the value of $e_k(\ell)$ only depends on the number of points left of ℓ and can be put into the closed form $e_k(\ell) = \min(k+1, s, n-s)$. □

The classical upper bound for the number of k-sets follows quite easily from the Lovász Lemma: With $n-1$ vertical lines we partition the plane into n stripes, each containing a single point of $\mathcal{P}$. Each vertical line is intersected by $2(k+1)$ or fewer k-edges. This gives a total of $\approx 2nk$ intersections between a k-edge and one of the vertical lines. It follows that the number of k-edges which cross $\sqrt{k}$ or more of the vertical lines is at most $2n\sqrt{k}$. A point of $\mathcal{P}$ is incident to at most $2\sqrt{k}$ of the remaining short edges. Therefore, the total number of short k-edges is again bounded by $2n\sqrt{k}$. Hence, the total number of k-edges is at most $4n\sqrt{k}$ and we have proven the following proposition.

Proposition 4.4 *A set of n points in the plane has at most $O(n\sqrt{k})$ k-edges.*

Proof. Here is another nice proof which makes use of the Crossing Lemma (Theorem 3.3): Lemma 4.3 implies that any given k-edge is crossed by at most $2k$ other k-edges. Consider the geometric graph G_k of all k-edges and let m_k be the number of edges. We have $\mathrm{cr}(G_k) \leq km_k$. In combination with the Crossing Lemma this yields, $km_k \geq c\frac{m_k^3}{n^2}$, i.e., $m_k \leq c'n\sqrt{k}$. □

Actually, the Crossing Lemma was the essential ingredient that led to the following improved bound for the number k-edges of planar point sets. We define the following quantities which depend on a point set $\mathcal{P}$ which is usually suppressed in the notation. With $e_j = e_j(\mathcal{P})$ we denote the number of j edges and $E_k = E_k(\mathcal{P})$ is the total number of j-edges for $0 \leq j \leq k$, i.e., $E_k = \sum_{j=0}^{k} e_j$.

Theorem 4.5 *A set of n points in the plane has at most $O(nk^{1/3})$ k-edges.*

Proof. The work in proving the theorem is to prove a sensible bound on the crossing number $\mathrm{cr}(G_k)$ of the graph of k-edges. Here this is done via the identity from Proposition 4.6, which implies $\mathrm{cr}(G_k) \leq E_{k-1}$. It is known* that $E_{k-1} \leq kn$, with the Crossing Lemma this implies $kn \geq c\frac{m_k^3}{n^2}$ and hence, $m_k \leq c'n\sqrt[3]{k}$. □

Proposition 4.6 *Let $\mathcal{P}$ be a set of n points in general position and $k \leq \frac{n-3}{2}$. If $\deg_k^+(p)$ denotes the out-degree of a point p in the graph G_k of k-edges of $\mathcal{P}$, then:*

$$\mathrm{cr}(G_k) + \sum_{p \in \mathcal{P}} \binom{\deg_k^+(p)}{2} = E_{k-1}.$$

Proof. The proof is based on a *continuous motion argument.* The idea is as follows: First, the identity is verified for a specific set $\mathcal{P}_o$. Then we consider a motion of the points starting with $\mathcal{P}_o$ and, eventually, stopping when the points are in the designated positions, i.e., when the point set is $\mathcal{P}$. The key observation is that changes in the graph of k-edges, which may affect the identity, correspond to situations where three points move through a collinearity. These situations, called *mutations*, are analyzed and it is shown that in each case the increase or decrease on both sides of the formula is the same so that the identity remains valid.

Convex position. Let $\mathcal{P}_o$ be a set of $n \geq 2k+3$ points in convex position. Each point $p \in \mathcal{P}_o$ has precisely one outgoing k-edge, $\deg_k^+(p) = 1$, hence, all binomial coefficients in the sum are of type $\binom{1}{2} = 0$. The outgoing k-edge at p is crossed from left to right by those k-edges which are outgoing at one of the k points left of the edge. In total this makes nk crossings and nk is the value of the left hand side. The number of j-edges is n for each $j = 0, \ldots, n-2$, therefore, the value of E_{k-1} is kn and the identity holds for $\mathcal{P}_o$.

* A proof for the slightly weaker bound $E_{k-1} \leq 2kn$ follows from by Proposition 6.15 via duality.

Moving points. For the movement of the points from $\mathcal{P}_0$ to $\mathcal{P}$ we make the following assumptions:

- The moving set is in general position, except for finitely many moments when collinearities of more than two points occur.
- The collinearities never involve more than three points and at any moment there is at most one collinearity.

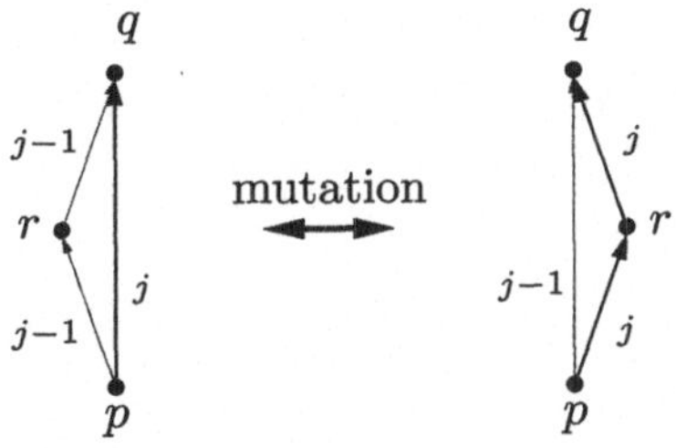

Figure 4.3 Point r crossing from the left through a j-edge or from the right through a $(j-1)$-edge.

Lets take a closer look at a mutation on three points $\{p, q, r\}$. We may assume that the mutation is such that in the moment of collinearity r is between p and q. We can imagine such a mutation as a movement of point r across the line spanned by p and q. Let (p, q) be a j-edge and let r cross the edge from left to right. It follows that immediately before the mutation the edges (p, r) and (r, q) both are $(j-1)$-edges. After the mutation (p, q) has lost the point r on the positive side, so (p, q) has become a $(j-1)$-edge and (p, r), (r, q) have become j-edges, see Figure 4.3. The reverse of the picture describes a mutation with r is crossing through a $(j-1)$-edge (p, q) from right to left.

From this analysis we conclude that a mutation involving j- and $(j-1)$-edges, keeps the value of both sides of the claimed identity unaffected unless $j = k$ or $j - 1 = k$.

Mutations involving a k-edge. We separately analyze the mutations involving k-edges and $(k-1)$-edges (type M_k^-) and those mutations involving k-edges and $(k+1)$-edges (type M_k^+), see Figure 4.4 for both types.

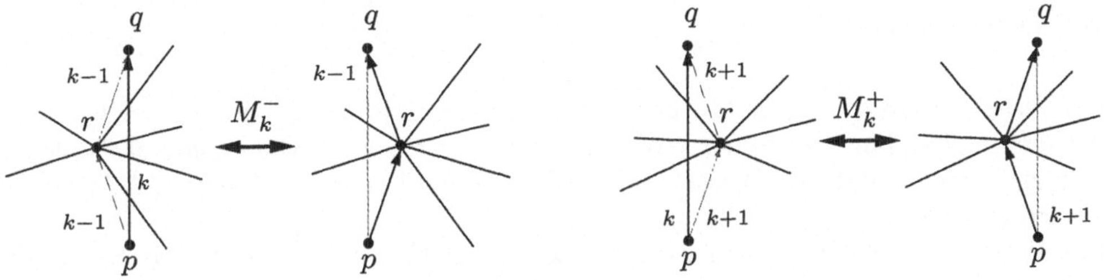

Figure 4.4 The two types M_k^- and M_k^+ of mutations involving k-edges.

Mutation M_k^-: Suppose M_k^- transforms the left configuration in the figure to the right configuration. The left configuration contains one $(k-1)$-edge more than the right one, therefore, E_{k-1} is decreased by one. Let $s = \deg_k^+(r)$ be the out-degree of r in the left configuration. The contribution of r in the sum changes from $\binom{s}{2}$ to $\binom{s+1}{2}$ which amounts to an increase by s.

It remains to consider the number b of crossings lost under the mutation. Let ℓ be a line through r parallel to (p, q). Note that b is just the number of k-edges incident to r and reaching into the right of ℓ. Let a count the number of k-edges incident to r and reaching into the left of ℓ. The sum $a + b$ is the degree of r in G_k which is $2\deg_k^+(r) = 2s$, (see Lemma 4.1). Refine the counts of a and b by separately counting incoming and outgoing edges as a^-, b^- and a^+, b^+. Since line ℓ has $k-1$ points on the left, Lemma 4.2 implies that $a^- = b^+ - 1$. Consider ℓ in the reverse direction, such that the line has $(n-k)$ points on the positive side. Again with Lemma 4.2 we obtain $b^- = a^+ + 1$. Together with the trivial equations $a = a^- + a^+$ and $b = b^- + b^+$ this yields $b = a + 2$ and, hence, $b = s + 1$.

Summarizing: A mutation of type M_k^- (from left to right) implies the following changes on the two rides of the identity:

- E_{k-1}: decrease by 1.
- $\mathrm{cr}(G_k) + \binom{\deg_k^+(r)}{2}$: (decrease by $\deg_k^+(r) + 1$) + (increase by $\deg_k^+(r)$).

Therefore, both sides of the identity are decreased by one if a mutation M_k^- is performed from left to right. If M_k^- is performed from right to left, then both sides of the identity are increased by one.

Mutation M_k^+. Again we assume that the mutation is from the left to the right configuration in Figure 4.4. No j-edges with $j < k$ are involved in the mutation, hence, the value E_{k-1} remains unaltered. Let $s = \deg_k^+(r)$ be the out-degree of r in the left configuration. The contribution of r in the sum changes from $\binom{s}{2}$ to $\binom{s+1}{2}$ which amounts to an increase by s.

It remains to consider the number b of crossings lost under the mutation. Let ℓ be a line through r parallel to (p, q) and let a and b be the number of k-edges incident to r from the left and right of ℓ. Again $a + b = 2\deg_k^+(r) = 2s$ as well as $a = a^- + a^+$ and $b = b^- + b^+$ where a^-, b^- count incoming and a^+, b^+ outgoing edges. Since line ℓ has $k+2$ points on the left Lemma 4.2 implies† that $a^- = b^+$. Consider ℓ in the reverse direction, such that the line has $n - (k+3)$ points on the positive side. Again with Lemma 4.2 we obtain $b^- = a^+$. Hence, $b = s$ and the value of the left hand side of the identity is also unaffected by a mutation of type M_k^+. □

4.2 Beyond the Plane

The Lovász Lemma (Lemma 4.3) is a key tool for bounding the number of k-edges of point sets in the plane. A similarly prominent role is taken by the analog of the lemma in higher dimensions. In this section we prove Welzl's Theorem (Theorem 4.7) which is a precise version.

Let $\mathcal{P}$ be a set of n points in general position in $\mathbb{R}^d$. A *k-facet* of $\mathcal{P}$ is an oriented $(d-1)$-simplex spanned by d points of $\mathcal{P}$, such that exactly k points of $\mathcal{P}$ are on the (strictly) positive side of the hyperplane affinely generated (spanned) by the d points of the simplex. The terminology is motivated by the fact that 0-facets are exactly the facets of the convex hull of $\mathcal{P}$, oriented such that the positive side contains no points from $\mathcal{P}$.

We say that *a directed line ℓ enters a k-facet F* if the line intersects F in a single point where it changes from the positive to the negative side of F.

† If $n = 2k + 3$ this is not quite true: In this case replace ℓ by a slightly tilted line through r which separates p and q. With the resulting $(k+1, n-(k-2))$ partition the argument works.

If F is a $(d-1)$-simplex and ℓ a directed line intersecting F, then we assign an orientation to F such that the facet F_ℓ corresponding to this orientation of F is entered by ℓ.

Let ℓ be a directed line which is disjoint from all convex hulls of $d-1$ points in $\mathcal{P}$ and define $\bar{h}_j = \bar{h}_j(\ell, \mathcal{P})$ as the number of j-facets entered by line ℓ.

The aim is to find upper bounds on the $\bar{h}_j$'s. The 0-facets are facets of the convex hull of $\mathcal{P}$. Since a line can enter the convex hull at most once, we have:

Fact 0. $\bar{h}_0 \leq 1$.

Define $s^0 = \sum_j \bar{h}_j$. Every $(d-1)$-simplex based on points of $\mathcal{P}$ that is intersected by ℓ is a k-facet for some k. Therefore, s^0 is just the total number of $(d-1)$-simplices intersected by ℓ.

Define $s^1 = \sum_j j\bar{h}_j$. This counts pairs (F, p), where F is a $(d-1)$-simplex and p is a point on the positive side of F_ℓ. Together F and p span a d-simplex intersected by ℓ and, indeed, every d-simplex Σ intersected by ℓ gives a unique pair (F, p): facet F of Σ is specified by the property that ℓ leaves Σ through F. The bijection shows that s^1 is the number of d-simplices intersected by ℓ.

More generally define $s^k = \sum_j \binom{j}{k}\bar{h}_j$. This counts pairs (F, B), where F is a $(d-1)$-simplex and B is a set of k points from the positive side of F_ℓ. A bijection as in the previous case shows that s^k counts the number of $(k+j)$-element subsets of $\mathcal{P}$ whose convex hull is intersected by ℓ.

Fact 1. The vector $(\bar{h}_0, \bar{h}_1, \ldots, \bar{h}_{n-d})$ and the vector $(s^0, s^1, \ldots, s^{n-d})$ completely determine each other.

Proof. By definition $s^k \sum_j \binom{j}{k}\bar{h}_j$ is expressed in terms of the $\bar{h}_j$. The explicit representation of $\bar{h}_j$ in terms of the s^k is $\bar{h}_j = \sum_k (-1)^{j+k}\binom{k}{j}s^k$. This inversion can be verified by elementary manipulations of binomial coefficients. △

Let ℓ' be the line parallel to ℓ through p. The interpretation of s^k directly implies: Replacing a point p by another point p' on ℓ' leaves the vector $(s^0, s^1, \ldots, s^{n-d})$ unaltered. Combined with Fact 1 this implies the same invariance for the $\bar{h}$-vector:

Fact 2. The vector $(\bar{h}_0, \bar{h}_1, \ldots, \bar{h}_{n-d})$ does not change when we replace a point p by another point p' on ℓ_p.

Let H be a hyperplane perpendicular to ℓ, such that all points of $\mathcal{P}$ are on one side of H. Let $p \to \hat{p}$ be the reflection at H, i.e., p and $\hat{p}$ determine a line parallel to ℓ and have the same distance from H. Let $\hat{\mathcal{P}}$ denote the reflected set and note that F is a j-facet of $\mathcal{P}$ iff $\hat{F}$ is a $(n-d-j)$–facet of $\hat{\mathcal{P}}$. With Fact 2 this implies the symmetry of the $\bar{h}$-vector:

Fact 3. $\bar{h}_j = \bar{h}_{n-d-j}$ for $0 \leq j \leq n-d$.

Fact 4. $\sum_p \bar{h}_j(\ell, \mathcal{P} \setminus \{p\}) = (n-d-j)\bar{h}_j + (j+1)\bar{h}_{j+1}$

Proof. The left hand side counts all pairs (p, F) where p is a point of $\mathcal{P}$ and F is a j-facet of $\mathcal{P} \setminus \{p\}$ entered by ℓ. These pairs are of two types. Either F is a j-facet of $\mathcal{P}$ and p a point on the negative side of F. Each j-facet entered by ℓ comes with $(n-d-j)$ points on the negative side. This gives a total of $\bar{h}_j(n-d-j)$ such pairs. The other possibility is that F is a $(j+1)$-facets of $\mathcal{P}$ and p is a point on the positive side of F. There are $\bar{h}_{j+1}(j+1)$ such pairs. △

Fact 5. $\bar{h}_j \geq \bar{h}_j(\ell, \mathcal{P} \setminus \{p\})$ for every $p \in \mathcal{P}$.

Proof. Fix $p \in \mathcal{P}$, above p in the direction of ℓ there is a point p' which is on the negative side of all $(j+1)$-facets generated by points in $\mathcal{P} \setminus \{p\}$. Replace p in $\mathcal{P}$ by p' to obtain $\mathcal{P}'$. Fact 2 implies $\bar{h}_j(\ell, \mathcal{P}) = \bar{h}_j(\ell, \mathcal{P}')$.

Let F be a j-facet of $\mathcal{P}' \setminus \{p'\}$. By construction p' is on the negative side of F. Therefore, F is also a j-facet of $\mathcal{P}'$. This shows $\bar{h}_j(\ell, \mathcal{P}' \setminus \{p'\}) \leq \bar{h}_j(\ell, \mathcal{P}')$. With $\mathcal{P}' \setminus \{p'\} = \mathcal{P} \setminus \{p\}$ and the identity $\bar{h}_j(\ell, \mathcal{P}) = \bar{h}_j(\ell, \mathcal{P}')$ we obtain $\bar{h}_j(\ell, \mathcal{P} \setminus \{p\}) \leq \bar{h}_j(\ell, \mathcal{P})$. △

Combining Fact 4 and 5 we obtain $n\bar{h}_j \geq (n-d-j)\bar{h}_j + (j+1)\bar{h}_{j+1}$ which can be rewritten as:

$$\bar{h}_{j+1} \leq \frac{j+d}{j+1}\bar{h}_j.$$

This allows to prove the next theorem with induction on j. The initial condition $\bar{h}_0 = 1$ for the induction is provided by Fact 0.

Theorem 4.7 (Welzl)
Let $\mathcal{P}$ be a set of n points in general position and ℓ be a directed line disjoint from all convex hulls of sets of $d-1$ points of $\mathcal{P}$. The number $\bar{h}_j$ of j-facets of $\mathcal{P}$ entered by ℓ satisfies

$$\bar{h}_j \leq \binom{j-1+d}{j} = \binom{j+d-1}{d-1}.$$

k-Facets in Three Dimensions

We apply previous findings to give an upper bound for the number of k-facets in three dimensions. The bound in the next theorem is weaker than the best known, still, the proof carries some ideas that are also used for the stronger results.

Theorem 4.8 *The number of k-facets of a set of n points in $\mathbb{R}^3$ is $O(n^2k^{2/3})$.*

Proof. In $\mathbb{R}^3$ a k-facet is an oriented triangle. Let T be the set of k-facets of a point set $\mathcal{P}$ of n points in general position. A pair of triangles Δ_1, Δ_2 is a *crossing pair* iff they share one vertex p and the edge of Δ_1 opposite of p intersects the interior of Δ_2, see Figure 4.5.

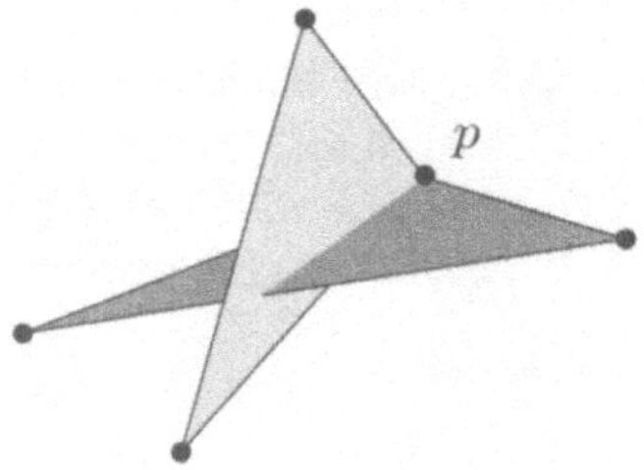

Figure 4.5 A pair of crossing triangles.

We give two estimates for the number X of crossing pairs. The first estimate is based on Theorem 4.7: A line spanned by two points x, y of $\mathcal{P}$ can enter at most $\binom{k+2}{2}$ triangles from T. Hence, a line can cross at most $2\binom{k+2}{2}$ triangles from T. Each crossed triangle Δ provides three corners p, such that $(\{p, x, y\}, \Delta)$ can be a crossing pair. Therefore $X \leq 3\binom{n}{2}2\binom{k+2}{2} \in O(n^2k^2)$.

For a second estimate of X we associate a geometric graph G_p with each $p \in \mathcal{P}$: Take a sphere S_p with center p and project all triangles of T which have p as a corner radially onto S_p. If there are t_p triangles with corner p we obtain a geometric graph G_p with t_p edges. Crossing pairs of edges in G_p correspond to pairs of crossing triangles with common vertex p. If x_p is the number of crossings of G_p, then $X = \sum_p x_p$. From the Crossing Lemma‡ $x_p \geq c\frac{t_p^3}{n^2}$. Using the inequality $\sum_p t_p^3 \geq n\left(\frac{1}{n}\sum_p t_p\right)^3$ this implies a lower bound on X:

$$X = \sum_p x_p \geq \sum_p c\frac{t_p^3}{n^2} = \frac{c}{n^2}\sum_p t_p^3 \geq \frac{c}{n^2}\frac{(\sum_p t_p)^3}{n^2} = \frac{c'|T|^3}{n^4}.$$

Combining the upper and the lower bound for X we obtain $|T|^3 \leq c'' n^6 k^2$, as claimed. □

4.3 The Rectilinear Crossing Number of K_n

We close the chapter with a problem about crossing numbers. Precisely, we ask for the rectilinear crossing number of the complete graph.

We begin with a simle observation: The number of crossings of a straight-line drawing of K_n with the vertices embedded in an n element point set $\mathcal{P}$ in general position is exactly the number $\square(\mathcal{P})$ of 4-element subsets of $\mathcal{P}$ which are in convex position, i.e., whose convex hull is a quadrilateral.

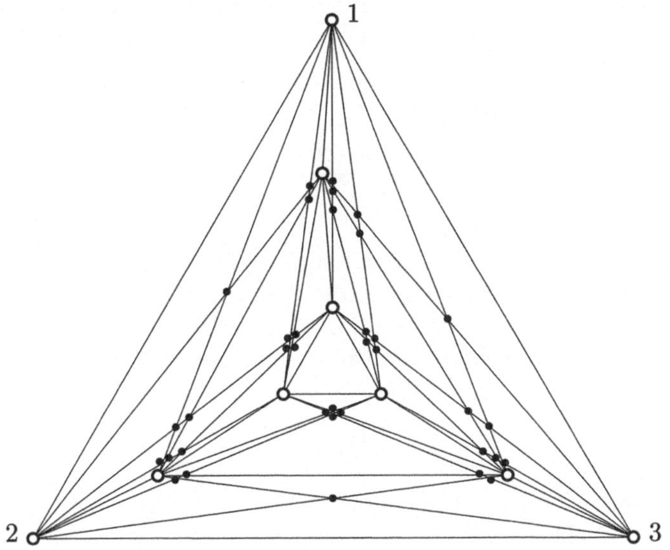

Figure 4.6 A drawing of K_9 with realizing the rectilinear crossing number $\overline{\text{cr}}(K_9) = 36$. Removing the vertices labeled 1, 2, 3 in this order yields optimal drawings realizing $\overline{\text{cr}}(K_8) = 19$, $\overline{\text{cr}}(K_7) = 9$, $\overline{\text{cr}}(K_6) = 3$.

Proposition 4.9 *A set of five points in general position in the plane always determines at least one convex quadrilateral.*

‡ Points with $t_p < 4n$ would help strengthening the bound.

Proof. If the convex hull of the five points contains four or five of the points we are done. Suppose, the convex hull is a triangle. One edge of the triangle is disjoint from the line which is spanned by the two points interior to the triangle. The two endpoints of this edge and the two points in the triangle form a quadrilateral. □

This gives a nice direct proof for the non-planarity of K_5 (with straight edges). The proposition implies $\square(\mathcal{P})(n-4) \geq \binom{n}{5}$ and, hence, $\square(\mathcal{P}) \geq \frac{1}{5}\binom{n}{4}$ for every set $\mathcal{P}$ of n points. The trivial upper bound is $\square(\mathcal{P}) \leq \binom{n}{4}$. The remaining question is to determine good bounds for the constant $C_\square$ in front of the binomial coefficient.

Very recently, the lower bound was strongly improved leading to a decrease in the gap between the bounds for $C_\square$ from ≈ 0.05 to ≈ 0.005.

Theorem 4.10 *For every point set $\mathcal{P}$ of n points in general position*

$$\square(\mathcal{P}) \geq \frac{3}{8}\binom{n}{4}.$$

The proof is based on a surprising connection between the number of convex quadrilaterals of a point set $\mathcal{P}$ and the numbers of k-edges. The connecting formula (Lemma 4.11) translates lower bounds for numbers of k-edges to lower bounds for the number of convex quadrilaterals.

From now on we suppress references to the underlying set of n points in general position on which all counting variables of the play are based. With $\square$ and $\triangle$ we denote the numbers of 4-element subsets which are in convex position and in non-convex position, in particular $\square + \triangle = \binom{n}{4}$. The number of k-edges of the point set is e_k and the number of $\leq k$-edges is $E_k = e_0 + \ldots + e_k$.

Lemma 4.11

$$\square = \sum_{k<\frac{n-2}{2}} e_k \left(\frac{n-2}{2} - k\right)^2 - \frac{3}{4}\binom{n}{3}.$$

Proof. We count the number Z of ordered 4-tuples $(u|v,w|x)$ such that the directed line v,w has u on its left and x on its right side. The contribution of a set $\{u,v,w,x\}$ in convex position to Z is four. The contribution of a set $\{u,v,w,x\}$ in non-convex position to Z is six. That is, $Z = 4\square + 6\triangle$ and after eliminating $\triangle$:

$$Z = 6\binom{n}{4} - 2\square \tag{4.1}$$

Now, consider a k-edge (v,w). This pair is in the central pair of $k(n-2-k)$ of the 4-tuples counted by Z. This gives another count for Z as:

$$Z = \sum_{k=0}^{n-2} k(n-2-k)e_k \tag{4.2}$$

Starting from equation 4.1 and using the obvious identity $\sum_k e_k = n(n-1)$ we get:

$$Z + 2\square + \frac{3}{2}\binom{n}{3} = 6\binom{n}{4} + \frac{3}{2}\binom{n}{3} = \left(\frac{n-2}{2}\right)^2 (n-1)n = \sum_{k=0}^{n-2}\left(\frac{n-2}{2}\right)^2 e_k \tag{4.3}$$

Below, we start from 4.3 and first replace Z with the expression from 4.2, then we use the symmetry of the e_j, namely $e_j = e_{n-j-2}$ together with the fact that in the even case the coefficient of $e_{\frac{n-2}{2}}$ vanishes:

$$\begin{aligned} 2\square &= \sum_{k=0}^{n-2} \left(\frac{n-2}{2}\right)^2 e_k - \sum_{k=0}^{n-2} k(n-k-2)e_k - \frac{3}{2}\binom{n}{3} \\ &= \sum_{k<\frac{n-2}{2}} 2\left[\left(\frac{n-2}{2}\right)^2 - k(n-k-2)\right] e_k - \frac{3}{2}\binom{n}{3} \\ &= 2\sum_{k<\frac{n-2}{2}} \left(\frac{n-2}{2} - k\right)^2 e_k - \frac{3}{2}\binom{n}{3} \end{aligned}$$

□

Corollary 4.12

$$\square = \sum_{k<\frac{n-2}{2}} (n-2k-3)E_k + O(n^3).$$

Proof. In Lemma 4.11 substitute $e_k = E_k - E_{k-1}$ for all k. Simplify the resulting formula with $\left(\frac{n-2}{2} - k\right)^2 - \left(\frac{n-2}{2} - (k+1)\right)^2 = n - 2k - 3$. Errors made at the boundary of the summation and by omitting the term $\frac{3}{2}\binom{n}{3}$ are subsumed in the big-O. □

In Theorem 4.16 we prove the lower bound $E_k \geq 3\binom{k+2}{2}$ for $k < n/2$. Substitute this into Corollary 4.12 and concentrate on the leading coefficient:

$$\square \geq \frac{3}{2}\left(\frac{n}{3}\left(\frac{n}{2}\right)^3 - 2\frac{1}{4}\left(\frac{n}{2}\right)^4\right) + O(n^3) = \frac{1}{64}n^4 + O(n^3).$$

This is the bound of Theorem 4.10.

A Lower Bound for $\leq k$-Edges

The lower bound on E_k is proved in the combinatorial disguise of allowable sequences. These objects will be central in Chapter 6 (pages 91ff). We briefly introduce the setting:

A sequence $\Sigma = \pi_0, \ldots, \pi_z$ of permutations of an n-element set is an *allowable sequence* if π_0 and π_z are reverse of each other, i.e., $\pi_0(t) = \pi_z(n-t+1)$ for all t, and each consecutive pair π_{i-1}, π_i differs by an adjacent transposition.

Let $\mathcal{P}$ be a set of n points. We require general position for $\mathcal{P}$ and assume in addition that no two lines spanned by the points are parallel. With $\mathcal{P}$ we associate an allowable sequence as follows: Let ℓ be a directed line such that the orthogonal projections of the points are all different. Let π_0 be the order of the projections of the points to ℓ. Start rotating ℓ keeping track of the order of the orthogonal projections of the points. This order changes whenever ℓ gets orthogonal to a line spanned by two points p_i, p_j of $\mathcal{P}$. The two orders before and after differ by the transposition exchanging these two adjacent points. The sequence of the different orderings, as permutations of labels, arising while ℓ rotates 180° is the allowable sequence associated with $\mathcal{P}$ (and ℓ). The following lemma emphasizes an easy observation, essential in our context.

Lemma 4.13 *Let Σ be an allowable sequence associated to a set $\mathcal{P}$ of n points by the above construction. The j-edges of $\mathcal{P}$ correspond bijectively to pairs π_{i-1}, π_i of permutations in Σ which are related by one of the two transpositions $(j+1, j+2)$ or $(n-j-1, n-j)$.*

To bound E_{k-1} we will prove a lower bound on the number of occurrences of transpositions from the set

$\{(1,2),(2,3),\dots,(k,k+1)\} \cup \{(n-k,n-k+1),(n-k+1,n-k+2),\dots,(n-1,n)\}$ these are the *critical transpositions.* To simplify the exposition let σ_i be the transposition $(k-i+1, k-i+2)$ and σ^i be the transposition $(n-k+i-1, n-k+i)$ so that the set of critical transpositions is $\{\sigma_i, \sigma^i : i = 1, \dots, k\}$. We let $m = n - 2k$ and assume that the first permutation of Σ is:

$$\pi_0 = (a_k, a_{k-1}, \dots, a_1, b_1, b_2, \dots, b_m, c_1, c_2, \dots, c_k)$$

and the last is:

$$\pi_z = \overline{\pi_0} = (c_k, c_{k-1}, \dots, c_1, b_m, \dots, b_2, b_1, a_1, a_2, \dots, a_k)$$

To get from π_0 to $\pi_z = \overline{\pi_0}$ the element a_j has to be moved to the right by critical transpositions of types $\sigma_j, \sigma_{j-1}, \dots, \sigma_1$ and $\sigma^1, \sigma^2, \dots, \sigma^j$. Symmetrically, element c_j is moved left by critical transpositions of types $\sigma^j, \sigma^{j-1}, \dots, \sigma^1$ and $\sigma_1, \sigma_2, \dots, \sigma_j$. Counting these critical transpositions used by the a_j's and c_j's we get a total of $2\sum_{i\le k} 2i = 4\binom{k+1}{2}$. The problem is that a critical transposition can simultaneously move an a right and a c left. Further analysis will allow us to bound the number of critical transpositions counted twice.

Let us think of the sequence Σ as a sequence $\tau_1, \tau_2, \dots, \tau_{z-1}$ of transpositions and of the order of the transpositions as time. We say that an element a_j (respectively, c_j) is *confined* until it is moved by a transposition of type σ_1 (respectively, σ^1), after that it becomes *free.* An element b_j is always free.

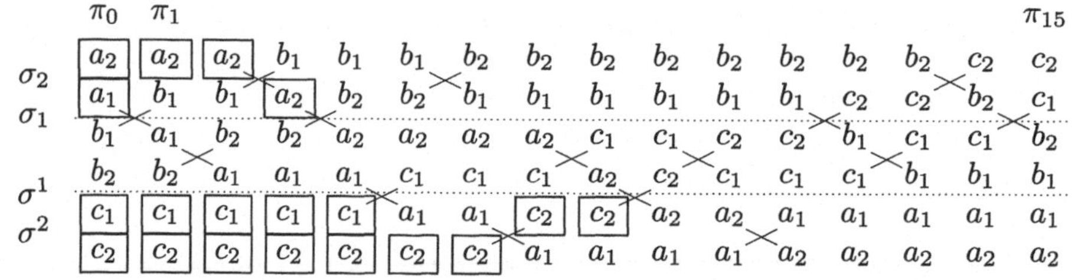

Figure 4.7 An example with $n = 6$ and $k = 2$, confined elements are boxed.

Lemma 4.14 *A sequence Σ' can be replaced by a sequence Σ which uses the same number of critical transpositions and without transpositions which move two confined elements.*

Proof. Suppose Σ' contains a transposition moving two confined elements, a *confined transposition.* Let τ_s be the first confined transposition. The elements moved by τ_s are either two a's or two c's, we assume they are a's. From the choice of τ_s it follows that the two elements exchanged by τ_s are adjacent in the starting permutation π_0. Otherwise, one of the elements exchanged by τ_s would have been involved in another confined transposition which precedes τ_s, contradiction. Let a_j and a_{j-1} be the two elements that are exchanged by τ_s. Define the sequence $\Sigma^<$ with transpositions $\tau_1, \dots, \tau_{s-1}, \tau_{s+1}, \dots, \tau_{z-1}, \sigma^j$, the effect of removing τ_s is that now a_{j-1} follows the path of a_j in Σ' and vice versa. Only at the end a_j and a_{j-1} are transposed by σ^j. The allowable sequence $\Sigma^<$ contains the same number of critical transpositions but one confined transposition less. Repeat this step until a sequence Σ without confined transpositions is reached. □

In the *liberation sequence* λ corresponding to Σ we list all a's and c's in the order in which they become free. We assume that Σ contains no confined transpositions (Lemma 4.14) which implies that the a's appear in λ in increasing order, i.e., a_i precedes a_j in λ iff $i < j$, and the same for the c's.

For $i \leq j$, let $[a_j, \sigma_i]$ be the first transposition of type σ_i that moves a_j in Σ. Define $[a_j, \sigma^i]$, $[c_j, \sigma^i]$ and $[c_j, \sigma_i]$ accordingly.

All the transpositions thus assigned to the a's move the defining element to the right, so they are pairwise distinct. The c's are moved left by the transpositions assigned to them. Therefore, only two cases are left for *duplication*:

$[a_j, \sigma_i] = [c_l, \sigma_i]$. This requires $i \leq \min(j, l)$, moreover, the transposition takes place while a_j is confined and c_l is free. Assign this duplication to c_l.

$[a_j, \sigma^i] = [c_l, \sigma^i]$. Again $i \leq \min(j, l)$, such a duplication takes place while c_l is confined and a_j is free. Assign this duplication to a_j.

Claim a. The number of duplications assigned to an a is upper bounded by each of the following quantities:

(i) One plus the number of a's preceding it in the liberation sequence.

(ii) The number of c's behind it in the liberation sequence.

Proof. The a's appear in λ in increasing order, therefore, the value from (i) just gives the index j for the a. The index is a bound since $i \leq j$ in $[a_j, \sigma^i]$. For (ii) recall that when we assign a duplication to a_j, then a_j is free and c_l confined, therefore, c_l is behind a_j in the liberation sequence. △

Claim c. The number of duplications assigned to a c is upper bounded by each of the following quantities:

(i) One plus the number of c's preceding it in the liberation sequence.

(ii) The number of a's behind it in the liberation sequence

The proof is as above. Given a liberation sequence we let $\mu_\lambda(a_j)$ and $\mu_\lambda(c_l)$ be the bounds from the above claims, i.e., if there are s of the c's behind a_j, then $\mu_\lambda(a_j) = \min(j, s)$.

Lemma 4.15 *If λ is a liberation sequence, then*

$$\sum_{j=1}^{k} \left(\mu_\lambda(a_j) + \mu_\lambda(c_j)\right) = \binom{k+1}{2}.$$

Proof. If $\lambda = \langle a_1, a_2, \ldots, a_k, c_1, c_2, \ldots, c_k \rangle$, then $\mu_\lambda(a_j) = j$ and $\mu_\lambda(c_j) = 0$ and the equation holds. Starting from this sequence every other sequence (with the same letters and the a's and c's in order) can be reached with a series of adjacent transpositions. We claim that the sum is invariant under adjacent transpositions. It is, obviously, enough to show this for an exchange of an a and a c which are adjacent. Consider $\lambda = \rho_1 * \langle a_j, c_l \rangle * \rho_2$ and $\lambda' = \rho_1 * \langle c_l, a_j \rangle * \rho_2$. Observe that $\mu_\lambda(x) = \mu_{\lambda'}(x)$ for $x \neq a_j, c_l$. We distinguish two cases:

$j+l \leq k$: From $j \leq k-l$ it is evident that $\mu_\lambda(a_j) = \min(j, k-l+1) = j = \min(j, k-l) = \mu_{\lambda'}(a_j)$. Similarly, $l \leq k-j$ implies $\mu_\lambda(c_l) = \min(l, k-j) = l = \min(l, k-j+1) = \mu_{\lambda'}(c_l)$.

$j+l > k$: From $j \geq k-l+1$ it is evident that $\mu_\lambda(a_j) = k-l+1$ and $\mu_{\lambda'}(a_j) = k-l$. Similarly, $l \geq k-j+1$ implies $\mu_\lambda(c_l) = k-j$ and $\mu_{\lambda'}(c_l) = k-j+1$.

In both cases $\mu_\lambda(a_j) + \mu_\lambda(c_l) = \mu_{\lambda'}(a_j) + \mu_{\lambda'}(c_l)$ □

From the lemma it follows that there are at most $\binom{k+1}{2}$ many duplicate transpositions in the family $\{[a_j, \sigma_i], [a_j, \sigma^i], [c_j, \sigma^i], [c_j, \sigma_i] : 1 \leq i \leq j \leq k\}$. Therefore, the number of critical transpositions is at least $3\binom{k+1}{2}$. Stating the result more explicitly:

Theorem 4.16 *If Σ is an allowable sequence on n elements and $k < n/2$, then Σ contains at least $3\binom{k+1}{2}$ transpositions from the following set:*
$\{(1,2), (2,3), \ldots, (k, k+1), (n-k, n-k+1), (n-k+1, n-k+2), \ldots, (n-1, n)\}$.

The next figure indicates a construction that can be used to show that the bound given of the theorem is tight for all $k \leq n/3$.

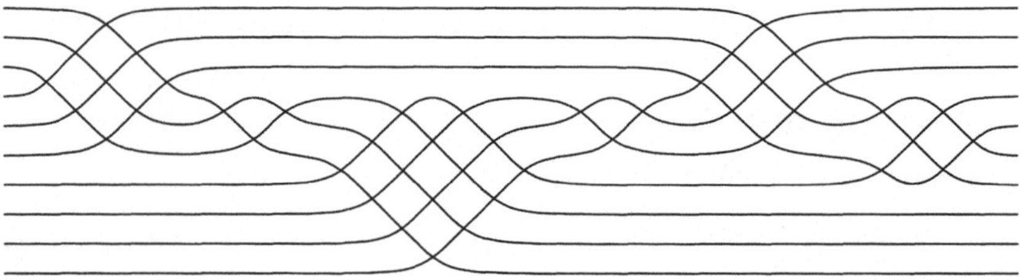

Figure 4.8 An allowable sequence corresponding to this arrangement shows that the theorem is tight for $n = 10$ and $k = 1, 2, 3$.

4.4 Notes and References

According to Lovász [133] it was Simmons who asked for the maximum number of halving lines. Lovász proves the upper bound $2n\sqrt{2n}$ and remarks that Straus constructed a set with $cn\log(n)$ halving lines. A paper by Erdős, Lovász, Simmons and Straus [76] contains upper and lower bounds for k-sets, respectively k-edges, which are of order $O(n\sqrt{k})$ and $\Omega(n\log(k))$. Up to the first proof of Proposition 4.4 this chapter presents the classical approach from [76].

The k-set bounds from [76] were later rediscovered by Edelsbrunner and Welzl [68]. A nice selection of proofs for the $O(n\sqrt{k})$ upper bound was published by Agarwal, Aronov and Sharir [3]. Pach, Steiger and Szemerédi [152] obtained a slightly improved bound of $O(\frac{n\sqrt{k}}{\log^*(k)})$. Significant progress was made by Dey [59] in 1997, he improved the upper bound to $O(nk^{1/3})$. Dey's proof combined the concept of convex chains, introduced by Agarwal et al [3], with an application of the Crossing Lemma. The proof of Theorem 4.5 given here follows ideas of Welzl, see Andrzejak et al. [13].

Tóth [198] gave a construction that improved the lower bound to $\Omega(n\exp(c\sqrt{\log(k)}))$. Tóth's examples can also be found in the book of Matoušek [138].

The dual question is to bound the complexity of the k-th level of an arrangement, i.e., what is the maximum number of vertices v of an arrangement of n lines such that exactly k lines pass below v. In this setting there is an obvious generalization to arrangements of pseudolines (c.f. Chapter 6), still the known bounds are the same as for k-sets. The proof of the lower bound for $\leq k$-sets (Theorem 4.16) actually is a proof in the generalized dual setting. Theorem 4.16 was obtained by Lovász, Vesztergombi, Wagner and Welzl [135]. The bound is best possible for the number of $\leq k$-sets in point configurations with $n \geq 3k$. The precise upper bound kn for $\leq k$-sets of n points was obtained by Alon and Györi [11] (cf. Proposition 6.15).

Welzl [211] proves the bound $O(n\sqrt{\sum_{k\in K} k})$ for the number of k-sets with $k \in K$. In [13] it is mentioned that the square-root in this bound can be improved to a third-root. Special classes of sets which allow smaller bounds on k-sets have been studied by Edelsbrunner, Valtr and Welzl [67] and Alt et al. [12]. The expected number of k-sets of a set drawn uniformly from a convex polygon was studied by Bárány and Steiger [20]. Edelman [63] investigates the expected size of the k-level of allowable sequences.

The complexity of the k-th level in arrangements of other objects, like segments or pseudocircles has been studied intensely over the past few years. These issues are covered in a survey of Agarwal and Sharir [4].

Bounds for k-sets have widespread applications. Among the most surprising ones is the connection with parametric matroid optimization first described by Eppstein [73]. As an example consider a graph with edge-weights $w(e) = a_e t + b_e$ which are time dependent and linear and let S_t be the minimum spanning tree at time t. The upper bound for k-sets implies that there are at most $O(mn^{1/3})$ many different trees S_t.

The study of k-sets in higher dimensions was initiated in a paper by Bárány, Füredi and Lovász [19]. The best known bound is of order $O(n^{\lfloor \frac{d}{2} \rfloor} k^{\lceil \frac{d}{2} \rceil - \varepsilon_d})$, where ε_d decreases exponentially with d, see Agarwal, Aronov, Chan and Sharir [1]. Nevertheless, this ε_d is the crucial point, since, for $\leq k$-facets there is an upper bound of order $O(n^{\lfloor \frac{d}{2} \rfloor} k^{\lceil \frac{d}{2} \rceil})$. Andrzejak and Welzl [14] give a beautiful proof for this result, it combines random sampling and the Upper Bound Theorem for polytopes.

The paper of Bárány et al. [19] already contains a higher dimensional Lovász Lemma. Several improvements were made before Welzl [212] proved the exact version (Theorem 4.7). Welzl describes mappings between pairs $(\mathcal{P}, \ell)$, where $\mathcal{P}$ is a set of n points in $\mathbb{R}^d$ and ℓ a directed line and simplicial polytopes Q with n vertices in $\mathbb{R}^{n-d-1}$, such that $\bar{h}_j(\mathcal{P}, \ell) = h_j(Q)$ where (h_j) is the h-vector of Q (for terminology of polytopes, see [219]). Via this mapping, which is essentially the Gale transform, Theorem 4.7 corresponds to the Upper Bound Theorem for convex polytopes, which is known to be best possible. Using the Generalized Lower Bound Theorem (GLBT) for polytopes in $\mathbb{R}^d$ with at most $d+4$ vertices Welzl can show that in $\mathbb{R}^3$ the number of $\leq k$-facets is minimized by point sets in convex position. More connections in particular with the GLBT are made by Sharir and Welzl in [178].

In $\mathbb{R}^3$ a bound close to the bound of Theorem 4.8, namely $O(n^{8/3} \log^{5/3} n)$ was first obtained by Aronov et al. [15]. Dey and Edelsbrunner [60] gave a more direct argument and got rid of the log-factors. Substantial progress was made by Sharir, Smorodinsky and Tardos [177], they prove a $O(nk^{3/2})$ bound. This improved bound in based on a more clever definition of graphs G_p. These graphs contain edges and rays to infinity, therefore, the Crossing Lemma is not directly applicable. The ideas are outlined in the book of Matoušek [138].

Crossing numbers of complete graphs. Related to the conjecture of Zarankiewicz for the crossing number of complete bipartite graphs there is a conjecture for the crossing number of complete graphs:

$$\mathrm{cr}(K_n) = \frac{1}{4}\lfloor\frac{n}{2}\rfloor\lfloor\frac{n-1}{2}\rfloor\lfloor\frac{n-2}{2}\rfloor\lfloor\frac{n-3}{2}\rfloor.$$

The conjecture and a matching upper bound construction was popularized by Guy [110]. Guy also describes a proof of the conjecture for $n \le 10$. Ringel comments on this: 'the proofs for $7 \le n \le 10$ are very uncomfortable'. Moon [143] suggested the following construction for drawings of K_n: Choose a set $\mathcal{P}$ of n points in general position from the unit sphere. For each of the points decide independently with probability $1/2$ to keep the point or replace it by its antipodal, let $\hat{\mathcal{P}}$ be this random point set. Connect every pair of points from $\hat{\mathcal{P}}$ by the shorter arc on the great cycle defined by them. Consider a four element subset A of $\mathcal{P}$ and the six great cycles defined by A. These cycles intersect in the points of A, their antipodals and six additional crossings. Each of these additional crossings requires a specific choice of $\hat{A}$ for its appearance in the drawing $D(\hat{\mathcal{P}})$. Hence, the probability that $\hat{A}$ determines a crossing is $6/16$. Linearity of expectation implies that the expected number of crossings in the drawing is $\frac{3}{8}\binom{n}{4}$. Accordingly, there must exist a drawing with at most that number of crossings. Note that this construction reaches the order of magnitude of Guy's conjecture.

The rectilinear crossing number $\overline{\mathrm{cr}}(K_n)$ is, in general, larger than $\mathrm{cr}(K_n)$. As shown by Guy [110] this is first true for $n = 8$ with $\mathrm{cr}(K_8) = 18$ and $\overline{\mathrm{cr}}(K_8) = 19$. Exact values for $\overline{\mathrm{cr}}(K_n)$ have been determined for n up to 12. The latest progress, by Aichholzer, Aurenhammer and Krasser [5], is a consequence of the generation of all order types of point sets with $n \le 10$. The known rectilinear crossing numbers have been used by Aichholzer et al. [5] to give the following asymptotic estimates:

$$0{,}3115\binom{n}{4} < \overline{\mathrm{cr}}(K_n) < 0{,}3807\binom{n}{4}$$

In 1865 Sylvester asked for the probability that four points chosen at random in a set $R \subset \mathbb{R}^2$ have a convex hull which is a quadrilateral. Depending on R and the probability distribution used to pick points from R, a number of different solutions are possible, cf. the web-page on Sylvester's Four-Point Problem [210]. Let R be open and of finite area Scheinerman and Wilf [171] consider $q_* = \inf_R \Box(R)$, where $\Box(R)$ is the probability that four points drawn independently from the uniform distribution on R are in convex position. They show $q_* = \lim_n \overline{\mathrm{cr}}(K_n)/\binom{n}{4}$.

Uli Wagner [209] improved the lower bound for q_* from $0{,}3115$ to $0{,}3288$. In this proof Wagner used an object called *staircase of encounter* which encodes information about the number of points on the left side of a rotating line through p. Further improvement from $0{,}3288$ to $3/8+\varepsilon = 0.375+\varepsilon$ was obtained by Lovász, Vesztergombi, Wagner and Welzl [135]. Section 4.3 is based on that paper. Note that we have not included the improvement from $3/8$ to $3/8+\varepsilon$. This ε, however, is required for the observation that the difference between $\overline{\mathrm{cr}}(K_n)$ and $\mathrm{cr}(K_n)$ is of order $\Omega(n^4)$. The ε improvement is made possible by using a second bound for E_k. The bound given in Theorem 4.16 is only tight for $k \le n/3$. Lovász et al. give a second bound which is better for $k > 0.495n$.

5 Combinatorial Problems for Sets of Points and Lines

In this chapter we study some fundamental questions of combinatorial geometry. The objects of this study are finite sets of points and finite arrangements, i.e., finite sets of lines or hyperplanes. As an introduction let us look at three classical contributions to this area.

- 1826 Steiner [188] enumerated the regions of Euclidean arrangements of lines and planes.
- 1893 Sylvester [192] asked for a proof of the following: If n points in the plane are not collinear then there is a line containing exactly two of the points.
- 1926 Levi [132] proved: A set of n lines in the projective plane determines at least n triangles.

Steiner's result (Theorem 5.1) is taken here as a warm-up in arrangements. Section 5.1 continues with an elementary discussion of planar geometry: We relate Euclidean and projective planes and explain the important concept of duality between point configurations and arrangements of lines. Sylvester's Problem and the stronger Kelly–Moser Theorem (Theorem 5.9) are the subject of Section 5.2. This is followed by a lower bound for the number of lines spanned by n points. Finally, in Section 5.4 we deal with Levi's Problem for the Euclidean plane. The result is that every simple Euclidean arrangement of n lines contains $n-2$ triangles. This is substantially harder than the question for the projective plane.

5.1 Arrangements, Planes, Duality

An *arrangement of lines* is a collection $\mathcal{A}$ of n lines $\ell_1, \ldots, \ell_n$ in the plane. The arrangement is called *trivial* if there exists a point p incident to all the lines ℓ_i of $\mathcal{A}$. If no point belongs to more than two lines of $\mathcal{A}$ the arrangement is called *simple*.

An arrangement partitions the plane into convex faces of dimensions 0, 1 or 2. The faces of dimension 0 are the intersection points of lines, these are the *vertices* of the arrangement. Maximal vertex free pieces of lines are the *edges* and *cells* are the connected pieces of plane after removal of the lines. Similar notions are used for arrangements of hyperplanes in d dimensions, in particular, the d-dimensional faces of such arrangements are again called cells. An arrangement of hyperplanes in d dimensions is *simple* if the intersection of any k hyperplanes is of dimension $d-k$, for $k = 1, \ldots, d$ and no point belongs to more than d hyperplanes. The following theorem gives Steiner's counting result.

Theorem 5.1 *The number of cells of a simple arrangement of n hyperplanes in $\mathbb{R}^d$ is*

$$\binom{n}{0} + \binom{n}{1} + \ldots + \binom{n}{d}.$$

Proof. We prove the formula by induction. The truth for $d = 1$ is easy, there is a line and n points on it. Removing the points leaves $n + 1$ connected pieces of line. Now let $\mathcal{H}$ be a simple arrangement in $\mathbb{R}^d$. Let H be an additional hyperplane, we will think of H as being horizontal and assume the following properties:

(a) H is intersected by every hyperplane in $\mathcal{H}$.

(b) All the vertices of $\mathcal{H}$ are strictly above H.

Place one marble in every cell of $\mathcal{H}$ and activate gravity so that a marble moves to the lowest vertex of its cell or is inhibited from disappearing vertically to $-\infty$ by the plane H (we assume that all marbles originally were placed above H). Property (b) and the assumption about $\mathcal{H}$ imply that the arrangement $\mathcal{H}_H$ induced by $\mathcal{H}$ on $H \equiv \mathbb{R}^{d-1}$ is simple. Hence, by induction, the number of marbles that hit H is $\binom{n}{0} + \binom{n}{1} + \ldots + \binom{n}{d-1}$. It remains to account for the marbles stopped by vertices of $\mathcal{H}$. A vertex of $\mathcal{H}$ is the intersection of d-hyperplanes and every set of d hyperplanes intersects in a vertex, hence, there are $\binom{n}{d}$ vertices in $\mathcal{H}$. Since there are no horizontal hyperplanes in $\mathcal{H}$ every vertex is the lowest point of exactly one cell. The number of marbles and hence the number of cells is $\binom{n}{0} + \binom{n}{1} + \ldots + \binom{n}{d}$. □

Before going ahead it seems appropriate to discuss different geometries hosting configurations of points and arrangements. The conception of the plane we have been working with has two main aspects. We want that a piece of paper represents a generic finite piece of plane, this is the basis for the extensive use of figures to illustrate and support conclusions. The second fundamental aspect is the coordinate system which allows to identify the plane with $\mathbb{R}^2$. Together these two aspects make a good intuitive model of the *Euclidean plane.*

Embed the Euclidean plane, i.e., $\mathbb{R}^2$, as an affine plane A not containing the origin into $\mathbb{R}^3$. This plane is described as $A = \{x \in \mathbb{R}^3 : \langle x - a, a\rangle = 0\}$ for some nonzero vector $a \in \mathbb{R}^3$ and the inner product $\langle a, b\rangle = \sum_i a_i b_i$. Every point in plane A spans a line through the origin of $\mathbb{R}^3$, intersecting this line with the unit sphere S^2 maps a point of A to a pair of antipodal points on S^2. The image of a line in the plane under this mapping is a great circle on S^2. The equator E_A of S^2 relative to A is the great circle of intersection between S^2 and the plane $A_0 = \{x \in \mathbb{R}^3 : \langle x, a\rangle = 0\}$, this is the plane parallel to A which contains the origin. Every pair of antipodal points corresponding to a point of A is separated in S^2 by E_A. Therefore, one of the open hemispheres of $S^2 \setminus E_A$ with half great circles as lines is a finite model of the Euclidean plane. Conceptually, the same thing can be obtained by identifying antipodal points of $S^2 \setminus E_A$.

The *projective plane* is obtained from S^2 by identifying all pairs of antipodal points, that is, by treating all pairs of antipodal points alike. Referring to our original plane A we can think of the projective plane as A enhanced by the pairs of antipodal points on E_A which form the line at infinity.

These considerations enable us to move a configuration of points and lines between different planes. The general ambient space for such configurations is S^2 or equivalently modulo identification of antipodal points the projective plane. By choosing an equator and a plane parallel to this equator the configuration can be mapped into Euclidean space. Figure 5.1 shows a configuration with two Euclidean representations. The important fact about these transformations is that they preserve incidences between points and lines as well as collinearity of points. Concurrency of lines is a bit more delicate as a set of concurrent lines can get parallel. In this case part of the configuration (the intersection point of the parallel lines) has to disappear with the line at infinity.

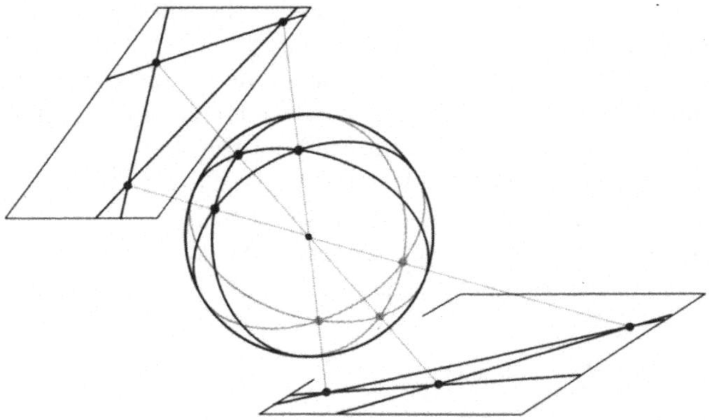

Figure 5.1 A spherical configuration and two Euclidean representations.

A very pleasing aspect of the geometry of antipodal points and great circles on S^2 is the natural notion of *duality*. With an antipodal pair $\{v, -v\}$ on S^2 associate the great circle $C_v = \{x \in S^2 : \langle x, v\rangle = 0\}$ and conversely with a great circle C associate $\{v \in S^2 : \langle x, v\rangle = 0 \text{ for all } x \in C\}$, this set is the pair $\{v, -v\}$ with $C = C_v$. Hence, duality $v \leftrightarrow C_v$ is a bijective mapping between points and great circles. Applying this duality mapping to each element of a configuration of points and lines gives a dual configuration. The importance of duality transformations originates from the fact that it preserves incidences and maps collinear points to concurrent lines and vice versa. By choosing affine planes A and B as primal and dual planes duality on the sphere can be used to map a configuration in plane A to a dual configuration on plane B. With Figure 5.2 we try to illustrate this construction of a dual configuration.

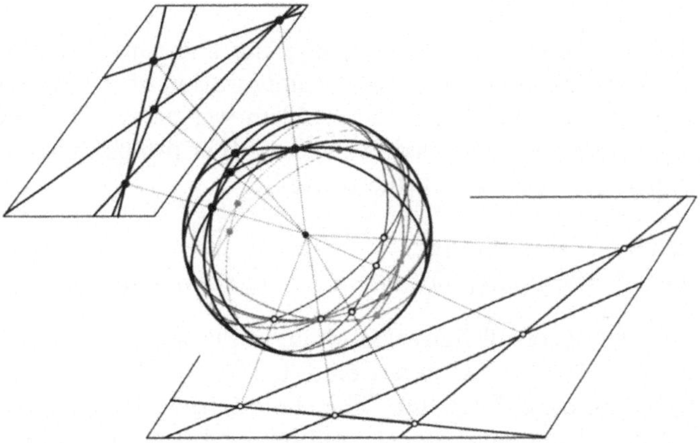

Figure 5.2 A configuration and its dual projected to different Euclidean planes.

We have kept these considerations intuitive and informal. It seems appropriate to conclude with a concrete example. Let A be the plane $z = 1$ in $\mathbb{R}^3$, this choice of A corresponds to the representation of points of $\mathbb{R}^2$ by homogeneous coordinates. Corresponding to the point $p = (p_x, p_y)$ in $\mathbb{R}^2$ we have $(p_x, p_y, 1) \in A$. Dual to this point in $\mathbb{R}^3$ is the plane $\{(x, y, z) : p_x x + p_y y + z = 0\}$. Intersecting with A we obtain the line $p_x x + p_y y = -1$ as the dual to p. Conversely, the dual of a line ℓ is obtained by bringing ℓ into the form $ax + by = -1$ and taking $p_\ell = (a, b, 1)$. Of course, the origin of A and all lines through it are exempt from this game, they map to the line at infinity. Things get even nicer if we take the plane $z = -1$ for B and identify points of A and B if they have the same x and y coordinates. Now the dual of $p = (p_x, p_y)$ is the line $\ell_p = \{(x, y) : p_x x + p_y y = 1\}$. With this duality $p \in \ell_p$ iff p is on the unit circle. This kind of dual mapping is called *polarity* with respect to the unit circle. In Figure 5.3 we indicate a plane geometric construction for this polarity. Similar constructions can be made with respect to other conics, e.g. the polarity at the parabola $y = x^2$ is frequently used.

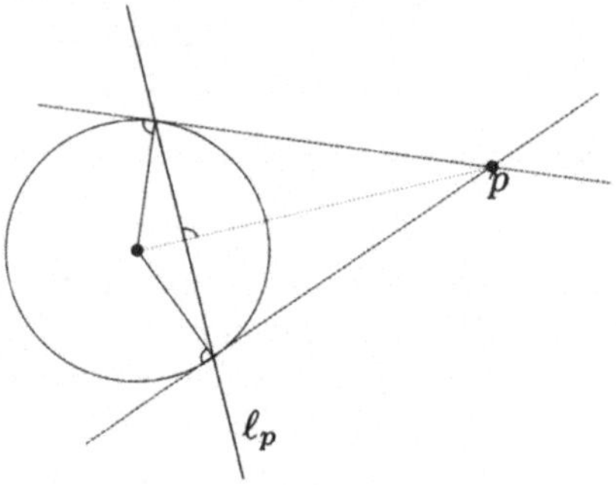

Figure 5.3 A plane construction for the polarity $p \leftrightarrow \ell_p$.

5.2 Sylvester's Problem

For a configuration $\mathcal{P}$ of points an *ordinary line* is a line containing exactly two of the points. Sylvester asked for a proof that every configuration of n points, not all on a line, admits an ordinary line. In the 1930's the problem was revived by Erdős and others. The first solution by Gallai was followed by several other proofs based on different ideas. We include the amazingly short proof due to Kelly.

Proposition 5.2 (Sylvester–Gallai)
Every configuration $\mathcal{P}$ of n points, not all on a line, admits an ordinary line.

Proof. Consider the set $\mathcal{L}$ of all lines containing at least two points of $\mathcal{P}$. Among all point-line pairs $(p, L) \in \mathcal{P} \times \mathcal{L}$ with p not on L let (p^*, L^*) be one which minimizes the distance between the point and the line. The claim is that L^* is ordinary. Let q be the closest point to p^* on L^* and suppose L^* contains at least three points, then there are two points p_1 and p_2 on the same side of q on L^*, Figure 5.4 shows the situation.

Let L be the line spanned by p^* and p_2. Clearly the distance of p_1 and L is smaller than the distance p^* and L^*, this contradicts the choice of the pair (p^*, L^*). □

By duality the role of points and lines in incidence statements may be interchanged. The dual version of Proposition 5.2 is: *Every arrangement $\mathcal{A}$ of n lines, not intersecting in a single point, and not all parallel, admits an ordinary point, i.e., a point contained in exactly two of the lines.*

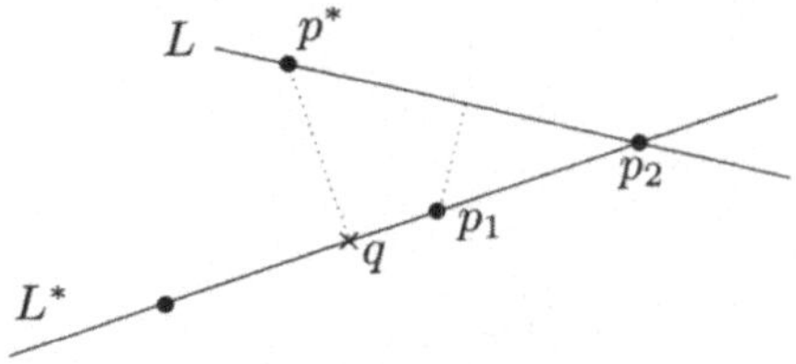

Figure 5.4 Illustration for Kelly's distance argument: $\mathsf{dist}(p^*, L^*) \geq \mathsf{dist}(p_1, L)$.

Below we give a nice proof for this dual version of Sylvester's problem whose main ingredient is Euler's formula for arrangements of lines living in the projective plane.

Proposition 5.3 (Euler)
If $\mathcal{A}$ is a projective arrangement with f_0 vertices, f_1 edges and f_2 faces, then

$$f_0 - f_1 + f_2 = 1.$$

Proof. The proof is by induction, if there are two lines, we have $f_0 = 1$, $f_1 = 2$ and $f_2 = 2$. Upon addition of a new line L to an arrangement of at least two lines some vertices and edges are created on L. Every edge on L splits an old face into two new faces and every vertex on L splits an old edge into two new edges. Therefore, the value of the alternating sum is not effected by the insertion of L. □

With an arrangement $\mathcal{A}$ and $j \geq 2$, $k \geq 3$ associate the following statistics:

$$\begin{aligned} t_j &= \#\text{vertices where } j \text{ lines cross}, \\ p_k &= \#\text{faces surrounded by } k \text{ edges } (k\text{-faces}), \end{aligned}$$

Following Melchior we cleverly define

$$Y = \sum_{j \geq 2} (3 - j)t_j + \sum_{k \geq 3} (3 - k)p_k.$$

Clearly $\sum_{j \geq 2} t_j = f_0$ and $\sum_{k \geq 3} p_k = f_2$ (this last equation only holds in the absence of 2-faces, i.e., if the arrangement has at least two vertices). Every edge is incident with two faces, therefore $\sum_{k \geq 3} k \cdot p_k = 2f_1$. A vertex where j lines cross is incident to $2j$ edges and every edge is incident to two vertices, therefore $\sum_{j \geq 2} j \cdot t_j = f_1$. Substituting the sums in the definition of Y by these formulas and applying Proposition 5.3 yields:

$$Y = (3f_0 - f_1) + (3f_2 - 2f_1) = 3(f_0 - f_1 + f_2) = 3.$$

Observe that in the definition of Y only the coefficient of t_2 is positive, it is 1. This gives a strengthening of Proposition 5.2 in its dual form.

Proposition 5.4 *If $\mathcal{A}$ is an arrangement of lines with at least two vertices and without parallel lines, then*

$$t_2(\mathcal{A}) \geq 3.$$

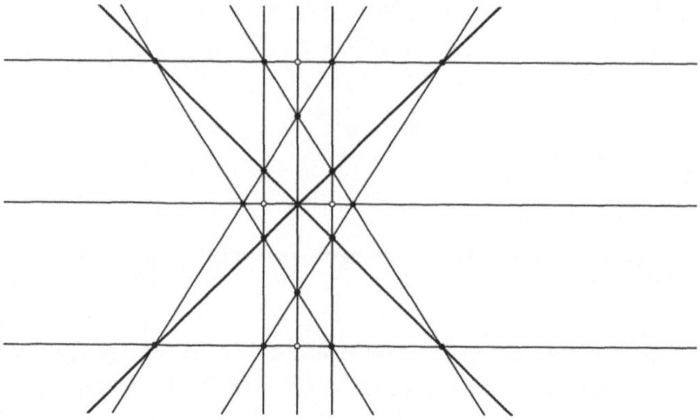

Figure 5.5 A projective arrangement of 13 lines (the 13th line is the line at infinity). The arrangement has 6 ordinary points: The 4 ordinary points shown in white and 2 more ordinary points where the bold lines intersect the line at infinity.

It is natural to ask for the minimum number of ordinary points in arrangements of n lines. A family of examples with few ordinary points is obtained from regular n-gons with n odd. Take the n supporting lines of the n-gon and add the n lines of symmetry. This arrangement of $2n$ lines only has n ordinary points, the midpoints of the edges of the n-gon. Two exceptional arrangements are known with an even smaller number of ordinary points, one with $n = 7$ and $t_2 = 3$ and a second one with $n = 13$ and $t_2 = 6$, see Figure 5.5.

Conjecture 5.5 *For $n \geq 14$ every non-trivial arrangement of n lines has at least $n/2$ ordinary points.*

The strongest result known today is $t_2 \geq 6n/13$. We proceed to show the weaker bound $t_2 \geq 3n/7$ which was first obtained by Kelly and Moser.

Let $\mathcal{A}$ be an arrangement of $n > 3$ lines in the projective plane. Further assume that every line of $\mathcal{A}$ contains at least three vertices. This assumption is legitimized by the observation that otherwise, either all lines intersect in a single point, or there are at least $n - 2$ ordinary points in $\mathcal{A}$. The general plan is to associate ordinary points to the lines and finally use a double counting argument. An ordinary point p is *attached* to a line L not containing p, if L together with the two lines crossing in p form a triangular face T of the arrangement. Sometimes it is useful to be more precise and say p is attached to L through T.

Lemma 5.6 *Let T be a triangle formed by three lines of $\mathcal{A}$. Let L be one of the defining lines of T and $[p, q]$ be the interval of intersection of L and T. If T is not a cell of $\mathcal{A}$ and every line intersecting T also intersects $[p, q]$ and there are no ordinary points on the interval $[p, q]$, then there exists an ordinary point x attached to L through some triangle T_x contained in T.*

Proof. We use a distance argument a la Kelly (c.f. Proposition 5.2). Let x be the vertex of $\mathcal{A}$ in T but not on L, which has the smallest distance to L. Suppose that there are three lines L_1, L_2, L_3 intersecting in x, let v_1, v_2, v_3 be their intersection points with L. By the assumptions all the v_i are in the interval $[p, q]$ we assume that v_2 is between v_1 and v_3.

Since v_2 is not ordinary there is a line $M \neq L_2$ entering T at v_2. Clearly, the intersection of M and L_1 or of M and L_2 is of smaller distance to L than x, a contradiction. □

Figure 5.6 shows a situation conforming with the assumptions of the lemma and two ordinary points attached to line L .

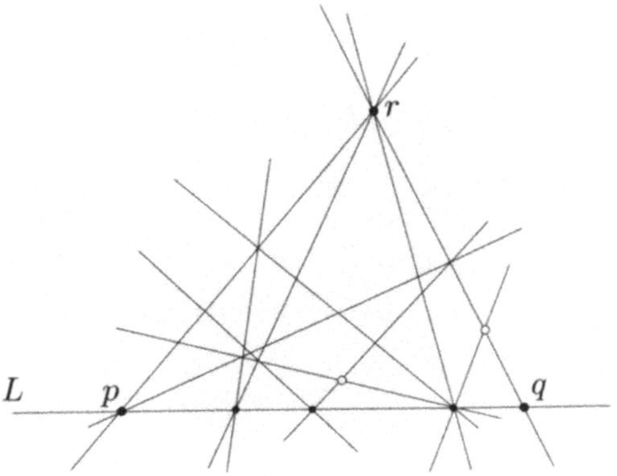

Figure 5.6 Two ordinary points attached to L in T.

Lemma 5.7 *If a line L of $\mathcal{A}$ contains no ordinary point, then there are at least three ordinary points attached to L.*

Proof. Since L contains no ordinary points a distance argument reveals that the vertex x of $\mathcal{A}$ with the smallest positive distance to L is an ordinary point attached to L. Let M and N be the lines crossing in x and let p and q be their intersection points with L. The three lines L, M and N partition the plane into four triangular cells. Let T_1 and T_2

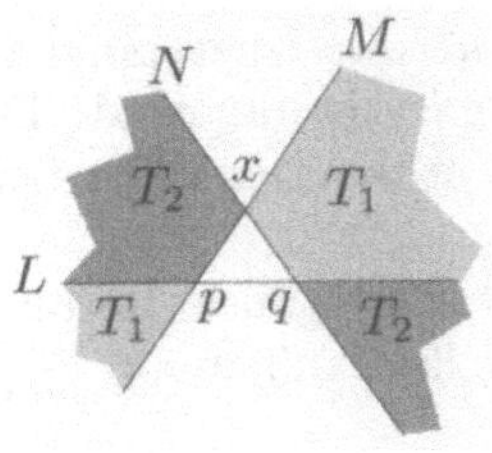

Figure 5.7 Illustration for Lemma 5.7.

be the two cells indicated in Figure 5.7. Both triangles have a side which is an edge of $\mathcal{A}$, px for T_1 and qx for T_2. Let v be a third vertex on L. All the lines crossing L at v intersect the interior of T_1 and T_2. Therefore, Lemma 5.6 implies that we find ordinary points attached to L in T_1 and T_2. Together with x this makes for at least three ordinary points attached to L. □

Lemma 5.8 *If a line L of $\mathcal{A}$ contains exactly one ordinary point, then there are at least two ordinary points attached to L.*

Proof. Let p be the ordinary point on L and let r be a neighbor of p on the other line M through p. Let N be a second line through r and q be the point of intersection of N and L. Line L is partitioned by p and q into the intervals $[p,q]$ and $[q,p]$. If both intervals contain further vertices of $\mathcal{A}$, then we consider the two triangular regions T_1 and T_2 as indicated in Figure 5.8. Both triangles have a side pr which is an edge of $\mathcal{A}$. With Lemma 5.6 we find ordinary points attached to L in T_1 and T_2. As second case we

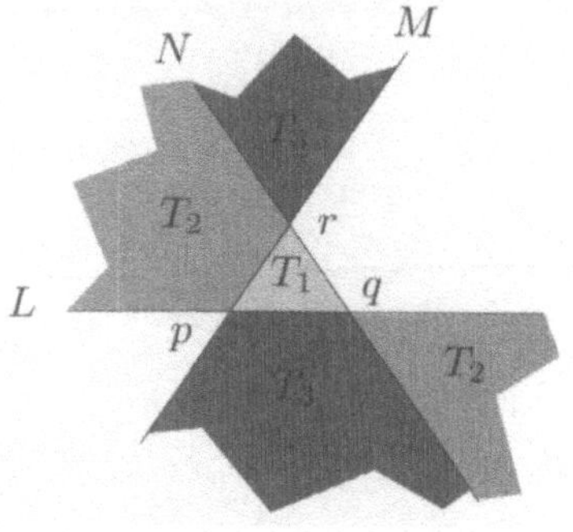

Figure 5.8 Illustration for Lemma 5.8.

have to consider the situation where all the vertices of L fall into one interval $[p,q]$. This time the triangular regions T_1 and T_3 as indicated in Figure 5.8 are of use. To verify the assumptions of Lemma 5.6 for T_3 note that a line intersecting T_3 but not the interval $[p,q]$ would have to intersect L in $[q,p]$, contradicting the assumptions for this case. Hence, again with Lemma 5.6 we find two ordinary points attached to L in T_1 and T_3. □

Theorem 5.9 (Kelly–Moser)
If $\mathcal{A}$ is a projective arrangement of $n \geq 4$ lines with at least two vertices, then

$$t_2(\mathcal{A}) \geq 3n/7.$$

Proof. Consider a matrix A with n rows correspond to the lines of $\mathcal{A}$ and $t = t_2(\mathcal{A})$ columns corresponding to the ordinary points of $\mathcal{A}$. The entry $A(L,p)$ is $3/2$ if $p \in L$, it is 1 if p is attached to L and 0 otherwise. With Lemma 5.7 and 5.8 it follows that $\sum_p A(L,p) \geq 3$ for every line L.

An ordinary point p is contained in two lines. Because p is incident to only four faces it can be attached to at most four lines. Therefore $\sum_L A(L,p) \leq 7$ for every ordinary point.

The combination of the two bounds on the entries of A completes the proof:

$$3n \leq \sum_L \sum_p A(L,p) = \sum_p \sum_L A(L,p) \leq 7t.$$

□

5.3 How many Lines are Spanned by n Points?

In this section we treat another question raised by Erdős in the early 1930's. Given a configuration $\mathcal{P}$ of n points, not all on a line, how many lines are determined by the points.

The *near-pencil* is the configuration with $n-1$ points on one line and one additional point. The near-pencil shows that we cannot hope for more than n lines. An easy inductive proof for the fact that there are at least n lines can be based on the existence of ordinary lines.

Theorem 5.10 *Every configuration $\mathcal{P}$ of $n \geq 3$ points in the plane, not all on a line, determines at least n lines each passing through at least 2 of the points.*

Proof. For $n = 3$ the result is obvious, assume that it is true for $n-1$ points. Let $\mathcal{P}$ consist of n points and let ℓ be an ordinary line spanned by two points of $\mathcal{P}$. Let p be one of the points on ℓ. Removing p from $\mathcal{P}$ there are two possible cases. Either all points of $\mathcal{P} \setminus p$ are collinear, in this case $\mathcal{P}$ is a near-pencil with n lines. Otherwise, the set $\mathcal{P} \setminus p$ determines at least $n-1$ lines, by induction, together with line ℓ which is not spanned by $\mathcal{P} \setminus p$ this makes for a total of at least n lines. □

The conclusion of Theorem 5.10 is valid in many more general situations where it is known under different names, e.g. de Bruijn-Erdős theorem or Fisher's inequality. Consequently there are proofs using only elementary facts, incidence and counting. Here we concentrate on a sharpening of the result for point configurations. Again it was Erdős who conjectured that, if we exclude the near-pencil and some small exceptional configurations, a set of n points will span $2n-4$ lines.

Theorem 5.11 *If $\mathcal{P}$ is a set of $n \geq 27$ points not all on a line and not forming a near-pencil, then $\mathcal{P}$ determines at least $2n-4$ lines.*

Lemma 5.12 *If $\mathcal{P}$ is a configuration of points such that there is a line ℓ containing exactly $n-s$ points of $\mathcal{P}$, then the number of lines determined by $\mathcal{P}$ is at least*

$$s\,(n-s) - \binom{s}{2} + 1.$$

Proof. Let A be the set of points on ℓ and B be the complementary set. For every pair (a, b) with $a \in A$ and $b \in B$ there is a connecting line. Not all of these $s\,(n-s)$ lines have to be different. However, for every pair (b, b') there is at most one a such that the three points a, b, b' are collinear. Therefore, there are at least $s\,(n-s) - \binom{s}{2}$ lines containing points from A and B. Adding one, to account for line ℓ, completes the proof. □

Fixing n the expression from the lemma is a polynomial $f(s)$ in s of degree two. The equation $f(s) = 2n-4$ has the solutions 2 and $(2n-5)/3$, for all values of s between these solutions $f(s) > 2n-4$. Since we have excluded the near-pencil, i.e., the configuration with $s = 1$, we obtain: If $\mathcal{P}$ is a configuration of n points conforming with the theorem and there is a line with at least $n - (2n-5)/3 = (n+5)/3$ points, then $\mathcal{P}$ determines at least $2n-4$ lines.

Now assume that $\mathcal{P}$ contains two points p and q each of degree at most 5. Let ℓ be the line connecting p and q and consider the set B of points not on ℓ. Every point $b \in B$ is the intersection point of a line through p and a line through q. The degree conditions for p and q imply $|B| \leq 16$. Therefore line ℓ contains at least $n-16$ points. But $n - 16 < (n+5)/3$ only if $n \leq 26$, hence, non-trivial configurations of $n \geq 27$ points with two points of degree at most 5 determine at least $2n-4$ lines.

Now let $\mathcal{P}$ contain at most one point incident to at most 5 lines. As a side product of the proof of Proposition 5.4 we had the following equation for arrangements

$$3 = \sum_{j \geq 2}(3-j)t_j + \sum_{k \geq 3}(3-k)p_k.$$

Disregarding the non-positive contribution of the second sum we obtain Melchior's inequality

$$3 \leq \sum_{j \geq 2}(3-j)t_j.$$

In the context of arrangements t_j was counting the number vertices where j lines cross. Dually the formula also holds for a point configuration with

$$t_j \quad = \quad \#\text{lines containing } j \text{ points of } \mathcal{P},\ j \geq 2.$$

Note that $t = \sum_j t_j$ is the number of lines determined by $\mathcal{P}$. For the points we have the counting coefficients

$$r_j \quad = \quad \#\text{points of } \mathcal{P} \text{ of degree } j,\ j \geq 2.$$

Double counting yields the equation $\sum_j jt_j = \sum_j jr_j$. Rewriting Melchior's inequality and using the assumption that there is at most one point of degree ≤ 5 we obtain:

$$3t - 3 \geq \sum_{j \geq 2} j \cdot t_j = \sum_{j \geq 2} j \cdot r_j \geq \sum_{j \geq 6} j \cdot r_j \geq 6(n-1).$$

Hence, the number t of lines spanned is at least $2n-1$ in this case. This completes the proof of the theorem.

□

The following infinite family of examples shows that Theorem 5.11 is best possible: Consider two lines, place one point at their intersection two additional points on the first line and $n-3$ points on the second line. This configuration of n points spans $2n-4$ lines.

5.4 Triangles in Arrangements

Ordinary points are vertices of an arrangement with minimal degree, i.e., where a minimal number of lines cross. In non-trivial arrangements the faces of minimal degree, i.e., faces with a minimal number of surrounding lines, are the triangles. In this section we discuss bounds on the number p_3 of triangles. An easy inductive argument shows that in the projective plane every arrangement which is not a star contains a triangle: Add a line L to an arrangement $\mathcal{A}$. If $\mathcal{A}$ is a star then it either remains a star or some triangles are created. Now suppose that $\mathcal{A}$ contains a triangle T. If T is not crossed by L it remains a triangle. If L is cutting through T, then T is either subdivided into a triangle and a quadrangle or into two triangles.

Using Euler's formula for projective arrangements a small improvement is possible. As for the proof of Proposition 5.4 we begin with a clever definition.

$$Z = \sum_{j \geq 2}(4-2j)t_j + \sum_{k \geq 3}(4-k)p_k.$$

Recall the four elementary equations $\sum_{j \geq 2} t_j = f_0$ and $\sum_{k \geq 3} p_k = f_2$ and $\sum_{j \geq 2} j \cdot t_j = f_1$ and $\sum_{k \geq 3} k \cdot p_k = 2f_1$. These formulas together with Euler's formula yield:

$$Z = (4f_0 - 2f_1) + (4f_2 - 2f_1) = 4(f_0 - f_1 + f_2) = 4.$$

In the definition of Z only the coefficient of p_3 is positive, it is 1. Hence, $p_3 \geq 4$ or more accurately:

$$p_3 \geq 4 + \sum_{k \geq 5} (k-4)p_k.$$

Levi [132] has obtained a sharp lower bound on p_3 in the projective case.

Proposition 5.13 *For every non-trivial arrangement $\mathcal{A}$ of n lines in the projective plane $p_3(\mathcal{A}) \geq n$.*

Proof. Let L be a line of $\mathcal{A}$ and consider L as the line at infinity. Denote by V_L the set of vertices of $\mathcal{A}$ which are not on L. If the convex hull of V_L is degenerate, i.e., a single point or all vertices of V_L are collinear, then there are $2n-2$ triangles in the arrangement. If the convex hull is not degenerate it consists of at least three vertices. Together with line L a vertex v of this convex hull with $t(v) = j$ determines $j-1$ triangular faces of $\mathcal{A}$. Therefore, there are at least three triangles incident to L.

Since the line L was chosen arbitrarily, every line is incident with at least three triangles. Every triangle is incident with exactly three lines. Therefore, $p_3 \geq n$. □

The result is best possible. To see this take the n supporting lines of the edges of a regular n-gon for $n \geq 4$. This arrangement of lines has $p_3 = n$.

Levi noted that his theorem does not make use of the straightness of the lines. He coined the term *pseudoline* to denote a curve whose intersection behavior with respect to some other pseudolines is as one would expect it from lines and proved his theorem in this more general context. We will frequently work with arrangements of pseudolines, here is a precise definition for these objects.

An *arrangement $\mathcal{B}$ of pseudolines* in the projective plane $\mathcal{P}$ is a family of simple closed curves, called *pseudolines*, in $\mathcal{P}$. Each pair of the pseudolines has exactly one point in common and they cross at this common point. With this notation we can state Levi's Theorem:

Theorem 5.14 *For every non-trivial arrangement $\mathcal{A}$ of n pseudolines in the projective plane $p_3(\mathcal{A}) \geq n$.*

Let $\mathcal{B} = \{P_0, P_1, \ldots, P_n\}$ be an arrangement of pseudolines in $\mathcal{P}$. Specifying a pseudoline P_0 in $\mathcal{B}$ as the line at infinity induces the arrangement $\mathcal{B}_{P_0}$ of pseudolines $\{P_1, \ldots, P_n\}$ in $\mathcal{P} \setminus P_0$. Since $\mathcal{P} \setminus P_0$ is homeomorphic to the Euclidean plane we may regard $\mathcal{B}_{P_0}$ as an arrangement in this plane.

It has already been noted by Levi that arrangements of pseudolines are a proper generalization of arrangements of lines. This is due to the existence of incidence laws in plane geometry. E.g., the reason for the non-stretchability of the arrangement shown in Figure 5.9 is the Theorem of Pappus. Arrangements of pseudolines have received attention since they provide a generic model for oriented matroids of rank 3. In this context questions of stretchability are of considerable interest. For more about these connections the reader may consult the 'bible of oriented matroids' [28].

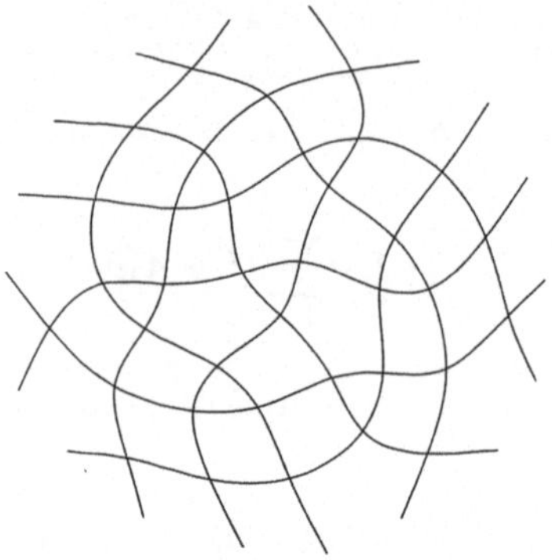

Figure 5.9 A simple arrangement of 9 pseudolines. This is the smallest non-stretchable arrangement.

Closely related is the problem of counting triangles in Euclidean arrangements. Again, the problem is old, in 1889 Roberts asserted that an arrangement of n lines in the plane contains at least $n-2$ triangles if it is simple, i.e., if there is no point of intersection of three or more lines. However, the proof was flawed, the problem remained unsolved for a long time. Ninety years later Shannon [176] proved Roberts theorem using dual configurations. Actually, he proved the analog of Roberts theorem for arbitrary dimensions: Every non-trivial arrangement of n hyperplanes in $\mathbb{R}^d$ contains at least $n-d$ simplicial d-cells, i.e., cells with the structure of a d-simplex.

Here we give a proof for the pseudoline case. As shown by Figure 5.10 in order to obtain Roberts bound it is actually necessary to require that the arrangement is simple in this case. The example belongs to an infinite family of non-simple arrangements with $3n$ lines and $2n$ triangles. In fact, it can be shown that every arrangement of pseudolines has at least $\lceil 3n/2 \rceil$ triangles.

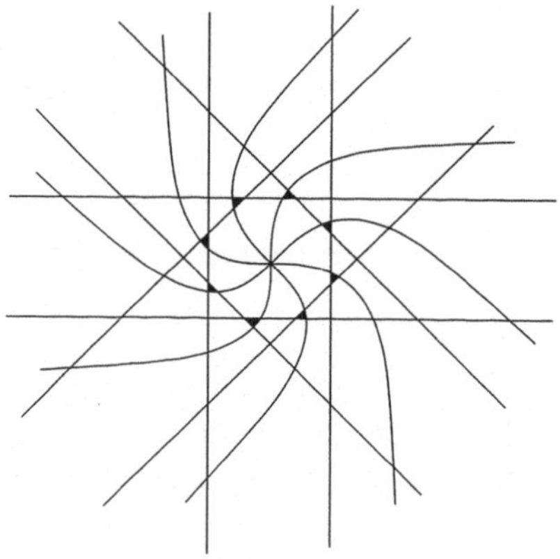

Figure 5.10 A non-simple arrangement of 12 pseudolines with only 8 triangles.

Theorem 5.15 *If $\mathcal{A}$ is a simple arrangement of n lines or pseudolines in the Euclidean plane then $p_3(\mathcal{A}) \geq n-2$.*

Proof. We consider the finite part of $\mathcal{A}$ as a planar graph. Let f_0 be the number of vertices, f_1^b be the number of bounded edges and f_2^b be the number of bounded faces. Since we only consider simple arrangements these statistics can be expressed as functions of the number n of pseudolines.

$$f_0 = \binom{n}{2}, \qquad f_1^b = n\,(n-2), \qquad f_2^b = \binom{n-1}{2}.$$

In this setting Euler's formula is $f_0 - f_1^b + f_2^b = 1$.

We assign labels $\oplus$ or $\ominus$ to the two sides of every edge. Let F be one of the two (possibly unbounded) faces bounded by e and let e' and e'' be the edge-neighbors of e along F. Let l, l' and l'' be the supporting pseudolines of e, e' and e'' respectively. The label of e on the side of F is $\oplus$ if F is contained in the finite triangle T of the arrangement $\{l, l', l''\}$ otherwise the label is $\ominus$. See Figure 5.11 for an illustration of the definition.

Lemma 5.16 *Every edge e of a simple arrangement has a $\oplus$ and a $\ominus$ label.*

Proof. Let F_1 and F_2 be the two faces bounded by e and let e_1', e_1'' and e_2', e_2'' be the edge-neighbors of e in these two faces. Since the arrangement is simple the supporting lines $\{l_1', l_1''\}$ of both pairs of edges are the same. The finite triangular region T of the arrangement $\{l, l', l''\}$ has edge e on its boundary. Therefore, exactly one of the two faces F_1 and F_2 is contained in T. △

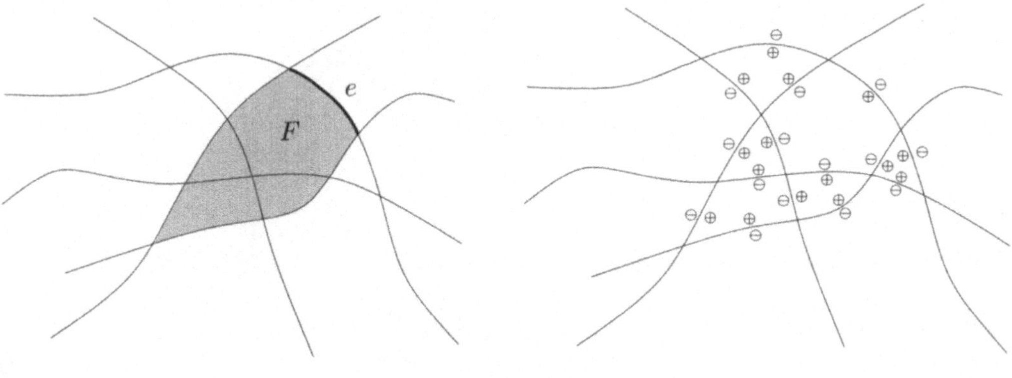

Figure 5.11 The label of e in F is $\oplus$ since F is contained in the shaded triangle. This rule leads to the completed labeling of edges shown on the right.

As seen in the proof of the lemma, the triangular region T used to define the edge label of e on the side of F is independent of F. This allows to adopt the notation $T(e)$ for this region.

Lemma 5.17 *All three edge labels in a triangle are $\oplus$. A quadrangle contains two $\oplus$ and two $\ominus$ labels. For $k \geq 5$ a k sided face contains at most two $\oplus$ labels.*

Proof. If F is a triangle, then for each of its edges e the triangular region $T(e)$ is F itself.

Let F be a quadrangle and e, $\overline{e}$ be a pair of opposite edges of F. Both edges have the same neighboring edges, hence, two of the lines bounding the triangles $T(e)$ and $T(\overline{e})$ are equal. Either $T(e) = F \cup T(\overline{e})$ or $T(\overline{e}) = F \cup T(e)$. In the first case e has label $\oplus$ and $\overline{e}$ has label $\ominus$ in F, in the second case the labels are exchanged. The second pair of opposite edges also has one label $\oplus$ and the other $\ominus$.

It remains to consider the case where F be a face with $k \geq 5$ sides.

Claim. *Any two edges with label $\oplus$ in F are neighbors, i.e., share a common vertex.*

Let $e_1, e_2, \ldots, e_k$ be the edges of F numbered in counterclockwise direction along F and let l_i be the supporting line of e_i. Let e_1 have label $\oplus$ and consider an edge e_i with $4 \leq i \leq k-2$. We show that the label of e_i is $\ominus$:

Face F is contained in $T(e_1)$ and line l_i has to leave $T(e_1) \setminus F$ through l_k and l_2. Figure 5.12 is a generic sketch of the situation.

Consider line l_{i-1}. This line enters the region R_1 bounded by l_2, l_i and the chain of edges $e_3, e_4, \ldots, e_{i-1}$ at the vertex $e_{i-1} \cap e_i$. To leave region R_1 line l_{i-1} has to cross l_2. Therefore, l_{i-1} has to leave the region R_2 bounded by l_i, l_2 and l_k through l_k. Symmetrically, l_{i+1} has a crossing with l_k to leave the region bounded by l_k, l_i and the chain of edges $e_{i+1}, e_{i+2}, \ldots, e_k$. Therefore, to leave region R_2 line l_{i+1} has to cross l_2. This shows that l_{i-1} and l_{i+1} cross inside region R_2. Hence, $T(e_i)$ is contained in R_2 and e_i has label $\ominus$ in F.

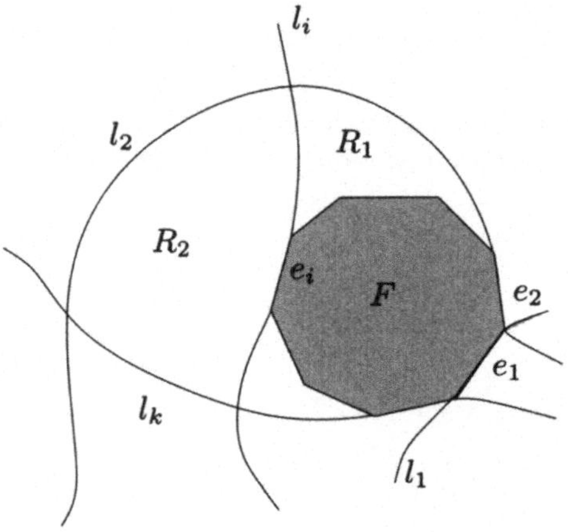

Figure 5.12 Edge e_1 has label $\oplus$ in F so e_i must have $\ominus$.

It remains to show that if e_1 is labeled $\oplus$ then neither e_3 nor e_{k-1} are. Considering the crossing of lines l_4 and l_2 observe that $T(e_3)$ is contained in $T(e_1) \setminus F$. Hence, the label of e_3 in F in $\ominus$. A symmetric argument applies to e_{k-1}. This completes the proof of the claim. △

We use the two lemmas to count the number of $\oplus$ labels in different ways:

$$f_1^b = \sum_F \#\{\oplus \text{ labels in } F\} \leq 2f_2^b + p_3.$$

With $f_1^b = n(n-2)$ and $2f_2^b = (n-1)(n-2)$ this implies

$$p_3 \geq n - 2.$$

□

Back to straight lines we now give a proof for the 2-dimensional case of Shannon's Theorem. For simple arrangements the result is contained in the more general Theorem 5.15. However, the proof extending to the non-simple case is really lovely and it contains all the ingredients required for Shannon's Theorem in all dimensions.

Theorem 5.18 *Every non-trivial Euclidean arrangement $\mathcal{A}$ of n lines contains at least $n-2$ triangles.*

Proof. Let line L_i of the arrangement be given by the equation $a_ix + b_iy = c_i$, for $i = 1, \ldots, n$. The intersection of lines L_i and L_j is at the point $e_{ij} = \left(\frac{c_ib_j - b_ic_j}{a_ib_j - b_ia_j}, \frac{a_ic_j - c_ia_j}{a_ib_j - b_ia_j}\right)$. The area of a triangle T formed by lines L_i, L_j, L_k can be expressed by a determinant:

$$\text{Area}(T) = \frac{g(a,b)}{3} \begin{vmatrix} c_ib_j - b_ic_j & c_jb_k - b_jc_k & c_ib_k - b_ic_k \\ a_ic_j - c_ia_j & a_jc_k - c_ja_k & a_ic_k - c_ia_k \\ a_ib_j - b_ia_j & a_jb_k - b_ja_k & a_ib_k - b_ia_k \end{vmatrix}.$$

Here $g(a,b)$ is the product of the expressions in the last row of the matrix. Via some calculations this can be transformed to the much nicer formula

$$\text{Area}(T) = \frac{g(a,b)}{3} \begin{vmatrix} a_i & a_j & a_k \\ b_i & b_j & b_k \\ c_i & c_j & c_k \end{vmatrix}^2.$$

Assign some real s_i to every line L_i and imagine that starting from its initial position in $\mathcal{A}$ line L_i is shifting parallel with speed s_i. For every real t this gives an arrangement $\mathcal{A}(t)$ whose lines are given by the equations $a_ix + b_iy = c_i + s_it$. By linearity of the determinant the area of the triangle $T(t)$ formed in $\mathcal{A}(t)$ by lines $L_i(t), L_j(t), L_k(t)$ is given by

$$\text{Area}(T(t)) = \frac{g(a,b)}{3} \left(\begin{vmatrix} a_i & a_j & a_k \\ b_i & b_j & b_k \\ c_i & c_j & c_k \end{vmatrix} + t \begin{vmatrix} a_i & a_j & a_k \\ b_i & b_j & b_k \\ s_i & s_j & s_k \end{vmatrix} \right)^2.$$

Thus Area($T(t)$) is a quadratic function of t and it will be constant if and only if the coefficient of t in the above expression vanishes. The condition

$$\begin{vmatrix} a_i & a_j & a_k \\ b_i & b_j & b_k \\ s_i & s_j & s_k \end{vmatrix} = 0$$

is a homogeneous linear equation in the variables s_i, s_j, s_k with coefficients determined by the equations of L_i, L_j, L_k.

Now assume that $\mathcal{A} = \mathcal{A}(0)$ has fewer than $n-2$ triangles. Requiring that the area of these triangular face remains constant in all $\mathcal{A}(t)$ gives at most $n-3$ equations in the n variables $s_1, \ldots, s_n$. Adding the two equations $s_1 = 0$ and $s_2 = 0$ the homogeneous system still has a non-trivial solution $s_1, \ldots, s_n$. In the following we will lead this to a contradiction.

Consider the parameterized family $\mathcal{A}(t)$ of arrangements. By construction, the area of every triangle T of $\mathcal{A} = \mathcal{A}(0)$ remains constant, i.e., Area($T(t)$) = Area($T(0)$) for all t.

The lines L_1 and L_2 remain fixed ($s_1 = 0 = s_2$) but since we have chosen a non-trivial solution some lines are moving. In particular for every moving line L_i there is a t such that in $\mathcal{A}(t)$ the crossing of L_1 and L_2 is on $L_i(t)$. Let $t^* > 0$ be the least with the property that some set of at least three lines which are not concurrent in $\mathcal{A}(0)$ intersect at a vertex of $\mathcal{A}(t)$. By the choice of t^* the combinatorial type of all arrangements $\mathcal{A}(t)$ with t in the open interval $(0, t^*)$ is the same. Most importantly, the set of triangles of these arrangements is the same.

It is tempting to claim that the area of all triangles in $\mathcal{A}(t)$, $t \in (0, t^*)$ is constant. However, there may be sets of $m \geq 3$ concurrent lines in $\mathcal{A}(0)$ which move apart. For a triangle formed by lines which were concurrent in $\mathcal{A}(0)$ the area must be increasing.

Now consider the set $\mathcal{L}$ of newly concurrent lines in $\mathcal{A}(t^*)$ and let $\mathcal{A}_{\mathcal{L}}(t)$ be the sub-arrangement formed by these lines at time t. By induction the sub-arrangement $\mathcal{A}_{\mathcal{L}}(t)$, $t \in (0, t^*)$, contains some triangles. Since the combinatorial type of all arrangements $\mathcal{A}(t)$ with t in the open interval $(0, t^*)$ is the same, these triangles are triangular faces of $\mathcal{A}(t)$. For t close to t^* the area of these triangles has to decrease.

The contradiction is that there is at least one triangle whose area is simultaneously decreasing and non-decreasing when t is approaching t^*. The contradiction was reached by assuming that $\mathcal{A}$ has fewer than $n-2$ triangles. □

5.5 Notes and References

Detailed summaries of the state of knowledge about arrangements in the early 1970s have been given by Grünbaum. In [107] he surveys arrangements in arbitrary dimensions and in the valuable monograph [108] he presents a vast number of results and problems for the two-dimensional case.

Recently, survey articles focusing on different aspects of point configurations and arrangements have appeared in several handbooks. Erdős and Purdy [77] have a rich collection of problems and more than 250 references. A more compact collection is due to Pach [147]. Goodman [101] gives a good overview of the state of the art in arrangements of pseudolines. Agarwal and Sharir [4] emphasize on the complexity of substructures of arrangements, the survey is enriched by a huge bibliography.

Steiner's result on the number of cells in simple arrangements of hyperplanes has found generalizations in many directions. In the past 20 years the interest in the complexity of arrangements and substructures of arrangements was stimulated by the observation that these quantities have immediate consequences in the analysis of algorithms working with geometric structures. These connection between computational and combinatorial geometry are emphasized in the book of Edelsbrunner [65] and in [4]. To hint on the importance of results like Theorem 5.1 we point to another connection. For general set systems there is the notion of *VC-dimension*, in many cases bounds for approximation results or complexity of algorithms can be given in terms of the VC-dimension. Let $\Phi_d(n)$ denote the maximum cardinality of a set system on n elements with VC-dimension at most d, then $\Phi_d(n)$ is exactly the quantity of Theorem 5.1. Gärtner and Welzl [97] show that set systems, realizing the $\Phi_d(n)$ bound, are geometric. More about VC-dimension can be found in the books of Matoušek [137] or Agarwal and Pach [148].

Chazelle, Guibas and Lee [48] exemplify the power of geometric duality by some nice applications in computational geometry. For further reading concerning transformations of planar configurations and duality transforms we refer to Coxeter [52] and Preparata and Shamos [159].

Sylvester was probably motivated to pose his problem by the observation that the statement is false in the complex projective plane. In this plane there are configurations of 9 points arising in connection with cubic curves which have the property that every line containing two of them contains a third point. In finite geometry this structure is now known as the affine plane of order 3. The affirmative answer to Sylvester's problem shows that it is impossible to draw this finite geometry in the plane using only straight lines. Grünbaum [108] has many references for contributions related to Sylvester's problem. A beautiful theorem about colored configurations is due to Chakerian [46]: A non-trivial two-colored configuration always has a monochromatic line. Pach and Pinchasi [149] deal with the colored analog of Sylvester's problem. They show that in a non-trivial two-colored configuration there need not be a monochromatic ordinary line.

Motzkin [144] gave the first non-constant bound for the number of ordinary lines, he proves a lower bound of order $\sqrt{n}$. Kelly and Moser [120] proved the $3n/7$ lower bound. The example shown in Figure 5.5 found by McKee was published in 1968. In 1981 S. Hansen claimed to have a proof for the conjectured $n/2$ result, except for the known special configurations. The work of Hansen, however, was in general incomprehensible and contained errors. Csima and Sawyer [53], see also [54], found a proof for the weaker result: $t_2 \geq 6n/13$.

As already mentioned the statement that n points define at least n lines is valid in many more general situations: Let $\mathcal{L}$ be a set of subsets of a set P. We call the elements of $\mathcal{L}$ lines and the elements of P points. If any two lines from $\mathcal{L}$ have at most one point in common and every pair of points is covered by exactly one line of $\mathcal{L}$, then either $\mathcal{L}$ consists of a single line covering all points or $|\mathcal{L}| \geq n$. A sharpening of this result together with many references is given in [94]. Two classical proofs of the result can be found in The Book [8].

The proof presented for Theorem 5.11 is essentially due to Kelly and Moser. Elliott [70] has an improvement, he showed that the conclusion holds for $n \geq 10$. An example of Kelly and Moser, see Figure 5.13, shows that Elliott's result is best possible.

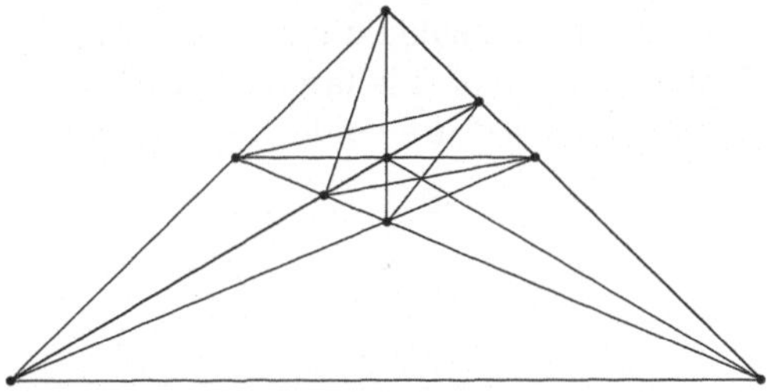

Figure 5.13 A configuration of 9 points spanning only 13 lines.

Along the lines of the given proof it is possible to show that if at most $n-k$ points of $\mathcal{P}$ are collinear then the number of lines is at least $kn-p(k)$, whenever $n \geq q(k)$, for suitable quadratic functions $p(k)$ and $q(k)$. Comprehensive surveys on Sylvester's problem and its relatives are Borwein and Moser [33] and a chapter by Brass and Pach in their announced monograph [35].

To summarize todays knowledge about the number p_3 of triangles in arrangements we first consider the projective case. If $\mathcal{A}$ is a projective arrangement, then:

(1) Every pseudoline is incident with at least three triangles. Therefore, $p_3(\mathcal{A}) \geq n$. Equality is possible for all $n \geq 4$.

(2) $p_3(\mathcal{A}) \leq \frac{1}{3}n(n-1)$ for $n \geq 9$. Equality holds for some arrangements of n lines for infinitely many values of n.

The lower bound is due to Levi [132], see Theorem 5.14. Grünbaum [108] has the following argument for $p_3 \leq \frac{1}{3}n(n-1)$ in simple arrangements: Since $\mathcal{A}$ is simple only one of the cells bounded by an edge can be a triangle. There are $n(n-1)$ edges and every triangle uses three of them. Grünbaum conjectured that the bound remains true for non-simple arrangements of lines with sufficiently large n. A series of papers proving weaker bounds or special cases was published before Roudneff [167] proved Grünbaum's conjecture for $n \geq 9$. Arrangements of pseudolines achieving equality were constructed by Harborth [112]. Recently, Forge and Ramírez-Alfonsín [91] managed to construct families of straight line examples achieving the bound.

If $\mathcal{A}$ is an arrangement of n pseudolines in the Euclidean plane, then:

(1) If $\mathcal{A}$ is simple or stretchable then $p_3(\mathcal{A}) \geq n-2$. Equality is possible for all $n \geq 3$.

(2) If $n \geq 6$ then $p_3(\mathcal{A}) \geq \frac{2}{3}n$. Equality is possible for all $n = 0 \pmod 3$.

(3) $p_3(\mathcal{A}) \leq \frac{1}{3}n(n-2)$. Equality is possible for infinitely many values of n.

The stretchable case of (1) is the 2-dimensional case of Shannon's theorem [176]. With Theorem 5.18 we gave a proof of this fact. The idea for the proof is due to A.Ya Belov, we have learned it through the writeup of Grünbaum [109]. Felsner and Kriegel [85] prove the pseudoline case of (1), Theorem 5.15, as well as (2). The upper bound (3) can be obtained along the lines of Roudneff's proof for the projective case. The examples achieving the bound can also be borrowed from the projective case.

Among the many problems in the area one of the most challenging goes back to Dirac [62]. Let $r^*(n)$ be the smallest integer such that in every nontrivial arrangement of n lines there is a line with at least $r^*(n)$ vertices. Dirac observed that $r^*(n) > \sqrt{n}$ and conjectured $r^*(n) \geq \lceil n/2 \rceil$. Grünbaum [108] (page 25) gives counterexamples for values of n up to 37. There is a family of arrangements showing that $r^*(n) \leq \lceil n/2 \rceil - 2$ for all n of the form $n = 12k+7$, the member of this family with $k = 2$ is shown in Figure 5.14. A reasonable adaption of Dirac's conjecture could be as follows:

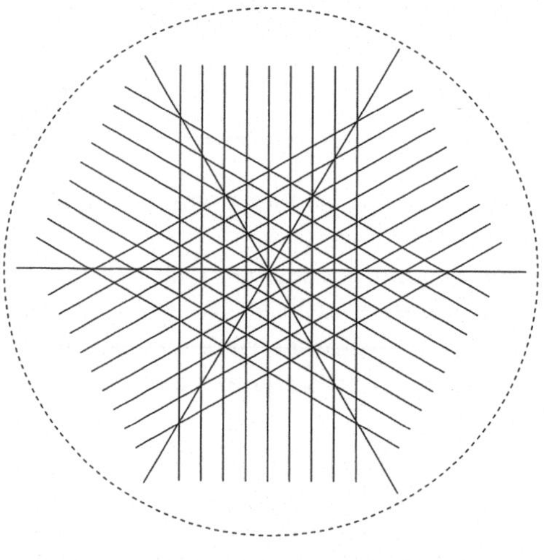

Figure 5.14 An arrangement of 31 lines (including the line at infinity) and at most 14 vertices on each line.

$$\liminf_{n \to \infty} \frac{r^*(n)}{n} = \frac{1}{2}.$$

Erdős had a much weaker question, he asked for the existence of $c > 0$ such that $r^*(n) > cn$. This was answered affirmatively by Szemerédi and Trotter [195] as an application of their incidence theorem (Theorem 3.6), see also Beck [22].

6 Combinatorial Representations of Arrangements of Pseudolines

It can be very useful to have combinatorial representations of geometric objects. The combinatorial structure of such an encoding may be easier to analyze and manipulate than the original object.

In this chapter we focus on combinatorial representations for arrangements of pseudolines. Corresponding results for arrangements of lines and configurations of points follow by specialization and dualization. Working with arrangements of pseudolines instead of arrangements of lines is advisable because it is possible to decide efficiently whether a given combinatorial structure represents an arrangement of pseudolines, but the corresponding question for arrangements of lines is an NP-complete problem.

The workhorse in our approach to combinatorial encodings of arrangements is the Sweeping Lemma (Lemma 6.1). Section 6.2 introduces allowable sequences and wiring diagrams. The Slope Theorem (Theorem 6.4) is a surprising application of this encoding. In Section 6.3 we discuss variants of local sequences and give an application to the enumeration of isomorphism classes of marked arrangements. The study of zonotopal tilings in Section 6.4 yields a standardized way of drawing arrangements of pseudolines. Section 6.5 deals with a representation for arrangements by triangle sign functions. These sign functions induce an order relation on the set of all Euclidean arrangements of n pseudolines. The notion of a signotope is introduced as a generalization of triangle sign functions. In Section 6.6 we study an order $S_r(n)$ on all r-signotopes on n elements.

6.1 Marked Arrangements and Sweeps

An Euclidean arrangement of pseudolines can be defined using the projective definition. To do so take a projective arrangement of $n+1$ pseudolines and declare one of the pseudolines as the line at infinity. Assume that the chosen line intersects the other lines in n different point. Removing this 'line at infinity' from the projective plane leaves an object homeomorphic to the Euclidean plane carrying n pseudolines, an Euclidean arrangement of pseudolines.

Probably, more convenient is a direct definition: A *pseudoline* in the Euclidean plane is a simple curve which approaches a point at infinity in either direction. An *arrangement of pseudolines* is a family of pseudolines with the property that each pair of pseudolines has a unique point of intersection, where the two pseudolines cross. Throughout this chapter arrangements of pseudolines lacking further specification will be assumed to be Euclidean.

An arrangement $\mathcal{A}$ is *simple* if no three pseudolines of $\mathcal{A}$ have a common point of intersection. An arrangement partitions the plane into cells of dimensions 0, 1 or 2, the *vertices*, *edges* and *faces* of the arrangement. Two arrangements are *isomorphic* (*combinatorially equivalent*) if there is an isomorphism (incidence and dimension preserving bijection) between the induced cell decompositions. Edges and faces of the arrangement may either be bounded or unbounded. Let F be an unbounded face of an arrangement

$\mathcal{A}$ and let $\overline{F}$ be the *complementary face of* F, i.e., the face separated from F by all pseudolines. We may orient all pseudolines such that F is in the left half-space and $\overline{F}$ in the right half-space of every line. This orientation of pseudolines induces an orientation of the edges of the arrangement. The pair $(\mathcal{A}, F)$ is a *marked arrangement* or an arrangement with *north-face* F and *south-face* $\overline{F}$. If there is no explicit reference to the north-face of a marked arrangement $\mathcal{A}$ embedded in a coordinatized plane we assume that there is a unique unbounded cell containing a ray to $(0, \infty)$ and this cell is the north-face. Two marked arrangements are *isomorphic* if there is an isomorphism of the induced cell decompositions which maps north-face to north-face and respects the induced orientation of edges. See Figure 6.1 for an illustration.

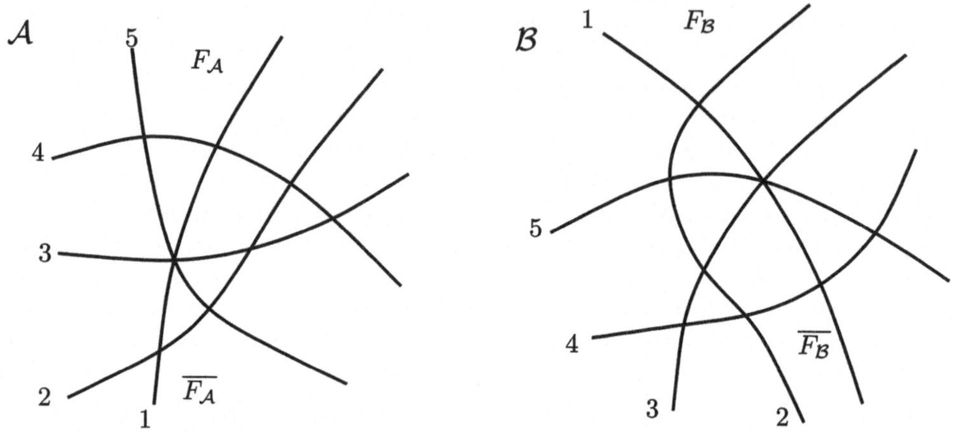

Figure 6.1 Arrangements $\mathcal{A}$ and $\mathcal{B}$ are isomorphic as arrangements but non-isomorphic as marked arrangements.

Sweeping is an important paradigm in algorithmic geometry. The vision is that a line is swept over the plane. In the course of the sweep movement the line will visit all the items of interest, e.g. points, line segments, crossings. In our context the sweeping line is replaced by a pseudoline. This is a variant known as topological sweep. With this idea in mind we begin with a formalization.

Let $(\mathcal{A}, F)$ be a marked arrangement. A *sweep* of $\mathcal{A}$ with north-pole in F is a sequence $c_0, c_1, \dots c_r$ of curves such that each curve c_i has the same endpoints $\overline{x} \in \overline{F}$ and $x \in F$. Further requirements are:

(1) None of the curves c_i contains a vertex of arrangement $\mathcal{A}$.

(2) Each curve c_i has exactly one point of intersection with each line l_j of $\mathcal{A}$.

(3) Any two curves c_i and c_j are interiorly disjoint.

(4) For any two consecutive curves c_i, c_{i+1} of the sequence there is exactly one vertex of arrangement $\mathcal{A}$ between them, i.e., in the interior of the closed curve $c_i \cup c_{i+1}$.

(5) Every vertex of the arrangement is between a unique pair of consecutive curves, hence, the interior of the closed curve $c_0 \cup c_r$ contains all vertices of $\mathcal{A}$.

See Figure 6.2 for an example of a sweep for the arrangement $\mathcal{A}$ of Figure 6.1.

Note that if $c_0, \dots, c_r$ is a sweep for $\mathcal{A}$ then the reversed sequence is also a sweep for $\mathcal{A}$. One of these sweeps is from left to right and the other from right to left. As usual we

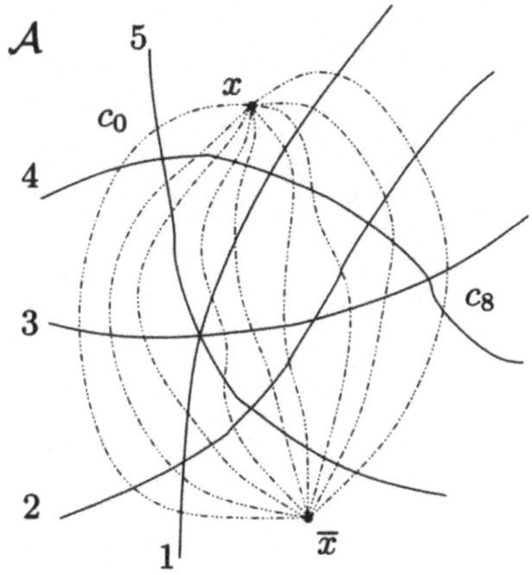

Figure 6.2 A sweep for arrangement $\mathcal{A}$.

will always think of a sweep as a left to right sweep. A discrete sweep as defined here can be transformed into a continuous sweep by appropriate interpolation between each pair c_i, c_{i+1} of curves.

Lemma 6.1 (Sweeping Lemma)
There is a sweep sequence of curves for every marked Euclidean arrangement $(\mathcal{A}, F)$ of pseudolines, i.e., $\mathcal{A}$ can be swept.

Proof. Let $G = (V, E)$ be the graph such that the vertices V of G are the vertices of $\mathcal{A}$ and the edges of G are the bounded edges of the arrangement $\mathcal{A}$. Let $\vec{E}$ be the orientation of the edges of G induced by the orientation of pseudolines (the north-face is in the left halfplane of each pseudoline).

Claim A. The orientation $\vec{E}$ is an acyclic orientation of G.

Walking 'at infinity' and clockwise from $\overline{F}$ to F all pseudolines of $\mathcal{A}$ are met. Let π be the list of lines in the order they are met.

The claim is proved by contradiction: Assume that $\vec{E}$ is not acyclic and choose a cycle C such that the area enclosed by the corresponding curve is minimal. By this choice C corresponds to the boundary of a face of $\mathcal{A}$. With respect to this face the cycle C may be oriented clockwise or counterclockwise. We consider the first case (clockwise), the other is symmetric.

Let $e_1, e_2, \ldots, e_k$ be edges of C in clockwise order and let L_j be the supporting pseudoline of e_j. Since e_j and e_{j+1} are consecutive on C lines L_j and L_{j+1} cross at a vertex of C. From the definition of π and the clockwise orientation of C it follows that $L_j <_\pi L_{j+1}$ (see Figure 6.3). Hence $L_1 <_\pi L_2 <_\pi \ldots <_\pi L_k <_\pi L_1$, a contradiction. $\triangle$

Since $\vec{G} = (V, \vec{E})$ is acyclic there exists a *topological sorting* of $\vec{G}$, i.e., an ordering $v_1, v_2, \ldots, v_r$ of the vertices of the graph such that all edges are directed from left to right. Formally, $i < j$ for every directed edge $v_i \to v_j$ of $\vec{G}$.

Claim B. There are curves $c_0, c_1, \ldots, c_r$ such that vertices $v_1, \ldots, v_i$ are to the left of c_i and vertices $v_{i+1}, \ldots, v_r$ are to the right of c_i for all $i = 1, \ldots, r$.

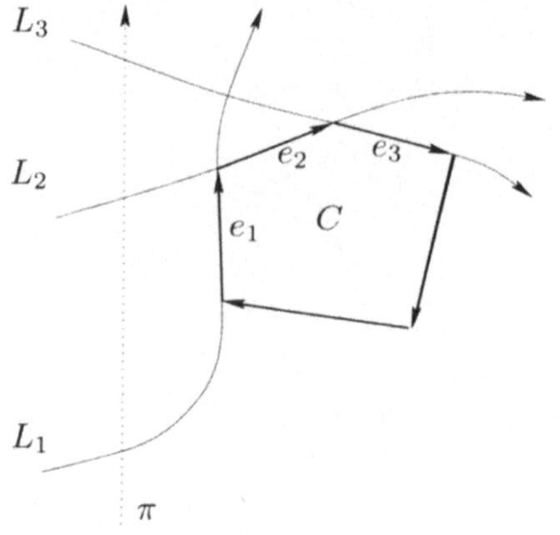

Figure 6.3 Assuming an oriented cycle.

Select arbitrary points $x \in F$ and $\overline{x} \in \overline{F}$. Let R be the union of the closed bounded cells of $\mathcal{A}$. Disregarding some trivial cases the region R is homeomorphic to a disk. Define c_0 as the union of three curves. The first and the second connect x to R within F and $\overline{x}$ to R within $\overline{F}$, the third is the left boundary of an ϵ-tube of the left boundary of R and connected to the two other curves. For an appropriate ϵ this gives a curve as required.

Now suppose that for some $i \leq r$ curve c_{i-1} is already defined. Consider vertex v_i (the numbering of vertices is a topological sort) and let $L_1^i, \ldots, L_t^i$ be the lines of $\mathcal{A}$ containing v_i such that $L_1^i <_\pi \ldots <_\pi L_t^i$. Let T be the triangle defined by curve c_{i-1} and the two lines L_1^i and L_t^i. Since v_i is a source (minimal) in the restriction of $\overrightarrow{G}$ to $v_i, \ldots, v_r$ and $v_1, \ldots, v_{i-1}$ are left of c_{i-1} vertex v_i is the unique vertex of $\mathcal{A}$ in the triangular region T. Define c_i as the right boundary of an ϵ-tube around c_{i-1} and T. For an appropriate ϵ this gives a curve as required, see Figure 6.4. $\triangle$

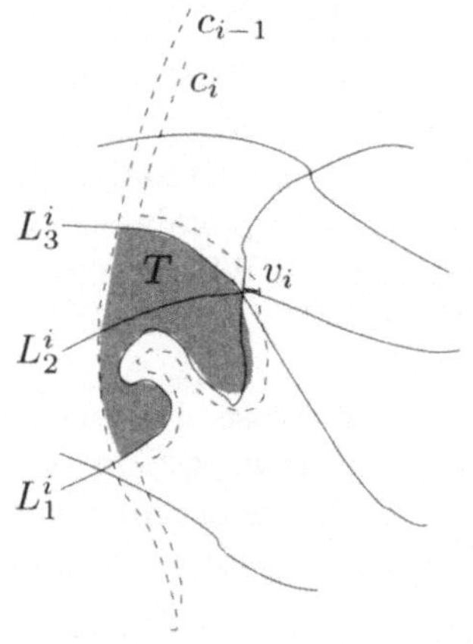

Figure 6.4 Defining c_i based on c_{i-1} and the shaded triangular region T.

The curves $c_0, c_1, \ldots, c_r$ constructed according to Claim B have the five properties of a sweep of $\mathcal{A}$ with north-pole F. This concludes the proof of the lemma. $\square$

Levi's Extension Lemma

As an application of the Sweeping Lemma we derive Levi's extension lemma. This is a fundamental lemma and the proof is a nice game swinging between the projective and the Euclidean plane.

Lemma 6.2 (Levi's extension lemma)
Let $\mathcal{A}$ be an arrangement of n pseudolines and let p, q be two points in the plane which are not both contained in a single line of $\mathcal{A}$. Then there is a pseudoline c containing p and q such that $\mathcal{A} \cup c$ is an arrangement of $n+1$ pseudolines.

Proof. We detail the proof for the case where p and q are not contained in lines of $\mathcal{A}$.

Let p be contained in the face F_p of $\mathcal{A}$. Let $L_1, \ldots, L_n$ be the pseudolines of $\mathcal{A}$ and assume that L_1 contains an edge e of the boundary of F_p. Add the line at infinity L_∞ to the arrangement and map it back to Euclidean space so that L_1 becomes the line at infinity thus obtaining an arrangement $\mathcal{A}'$ with lines $L_\infty, L_2, \ldots, L_n$. Mark $\mathcal{A}'$ such that $p \in F_p$ is the north-pole. Apply the Sweeping Lemma to find a curve c crossing the face F_q containing q. Line c can be bent in F_q to make q a point on c. Extending c from the poles to infinity we obtain an arrangement $\mathcal{A}' \cup c$ of $n+1$ lines. Together with L_1 we have a projective arrangement of $n+2$ lines which can be mapped back to the Euclidean plane using L_∞ as line at infinity. This results in an Euclidean arrangement with lines $L_1, \ldots, L_n, c$ such that the points p and q are both on pseudoline c. □

6.2 Allowable Sequences and Wiring Diagrams

Let $c_0, c_1, \ldots, c_r$ be a sweep for the marked arrangement $(\mathcal{A}, F)$ of n pseudolines. We assume that the lines are labeled $1, \ldots, n$ and identify the lines with their labels. Traversing curve c_i from $\overline{x}$ to x we meet the lines of $\mathcal{A}$ in some order. Since each line is met by c_i exactly once the order of the crossings corresponds to a permutation π_i of $[n]$. Relabeling the lines of $\mathcal{A}$ appropriately we may assume that π_0 is the identity permutation. This labeling of the lines of $\mathcal{A}$ is the *standard labeling*.

Consider the labels of lines crossing at vertex v_i. Since the region T defined in the proof of Claim B is empty of vertices of $\mathcal{A}$ and by property 2 of the sweep curve c_i the lines $L_1^i, \ldots, L_t^i$ containing vertex v_i are a consecutive substring of π_{i-1}. Moreover, in permutation π_{i-1} these lines are in the reversed order and this is the only difference between π_{i-1} and π_i.

Example A. The sequence of permutations obtained from the sweep of Figure 6.2 is $(1,2,3,4,5) \xrightarrow{4,5} (1,2,3,5,4) \xrightarrow{1,2} (2,1,3,5,4) \xrightarrow{1,3,5} (2,5,3,1,4) \xrightarrow{2,5} (5,2,3,1,4) \xrightarrow{1,4}$ $(5,2,3,4,1) \xrightarrow{2,3} (5,3,2,4,1) \xrightarrow{2,4} (5,3,4,2,1) \xrightarrow{3,4} (5,4,3,2,1)$.

Assuming the standard labeling for $\mathcal{A}$ the sequence $\pi_0, \ldots, \pi_r$ has the following properties:

(1) π_0 is the identity permutation and π_r is the reverse permutation on $[n]$.

(2) Each permutation π_i, $1 \leq i \leq r$ is obtained by the reversal of a consecutive substring M_i from the preceding permutation π_{i-1}.

(3) Any two elements $x, y \in [n]$ are joint members of exactly one *move* M_i, i.e., reverse their order exactly once.

A sequence $\Sigma = \pi_0, \ldots, \pi_r$ of permutations with properties (1), (2) and (3) is called a *allowable sequence of permutations.* If each move from π_{i-1} to π_i consists in the reversal of just one pair of elements, i.e., an adjacent transposition, we have $r = \binom{n}{2}$ and the sequence Σ is a *simple allowable sequence.* The existence of sweeps implies that there is an allowable sequence of permutations for every marked arrangement $(\mathcal{A}, F)$. However, more can be said:

Every topological sorting of the graph $\vec{G}$ of $(\mathcal{A}, F)$ induces an allowable sequence. Consider the allowable sequences Σ and Σ' corresponding to topological sortings σ and σ' of $\vec{G}$ with the property that $\sigma = v_1, \ldots, v_i, v_{i+1}, \ldots, v_r$ and $\sigma' = v_1, \ldots, v_{i+1}, v_i, \ldots, v_r$, i.e., σ and σ' differ in an adjacent transposition. It follows that v_i and v_{i+1} are both minimal elements in the restriction of $\vec{G}$ to $\{v_i, v_{i+1}, v_{i+2}, \ldots, v_r\}$. Hence, there is no line in $\mathcal{A}$ that contains vertices v_i and v_{i+1} and the labels of lines involved in the moves $M_i : \pi_{i-1} \rightarrow \pi_i$ and $M_{i+1} : \pi_i \rightarrow \pi_{i+1}$ in Σ are disjoint. In fact for $j \neq i, i+1$ the permutations π_j and π'_j in Σ and Σ' coincide and $M'_i = M_{i+1}$ and $M'_{i+1} = M_i$. Call two allowable sequences Σ and Σ' *elementary equivalent* if Σ can be transformed into Σ' by interchanging two disjoint adjacent moves. Two allowable sequences Σ and Σ' are called *equivalent* if there exists a sequence $\Sigma = \Sigma_1, \Sigma_2, \ldots, \Sigma_m = \Sigma'$ such that Σ_i and Σ_{i+1} are elementary equivalent for $1 \leq i < m$. It is well known that it is possible to transform any topological sorting of a directed acyclic graph $\vec{G}$ into any other by a sequence of adjacent transpositions, i.e., reversals of adjacent pairs of unrelated vertices. Corresponding to the topological sortings of $\vec{G}$ there is an equivalence class of allowable sequences of the arrangement $(\mathcal{A}, F)$. Actually, every allowable sequence obtained from a sweep of $(\mathcal{A}, F)$ belongs to this class.

Theorem 6.3 *There is a bijection between equivalence classes of allowable sequences and marked arrangements of pseudolines. Moreover, this bijection maps simple allowable sequences to simple arrangements.*

Proof. We have already seen how to use $\vec{G}$ to define an equivalence class of allowable sequences corresponding to a marked arrangement $(\mathcal{A}, F)$.

Let Σ be an allowable sequence. The following construction yields an arrangement $\mathcal{A}$ such that Σ corresponds to a sweep of $\mathcal{A}$: Start drawing n horizontal lines called *wires* and let p_j be the vertical line at $x = j$. On the ith wire from below label the crossing with p_j with $\pi_j(i)$, for $j = 0, \ldots, r$. Draw pseudoline i such that it interpolates the crossings labeled i. For an example see Figure 6.5.

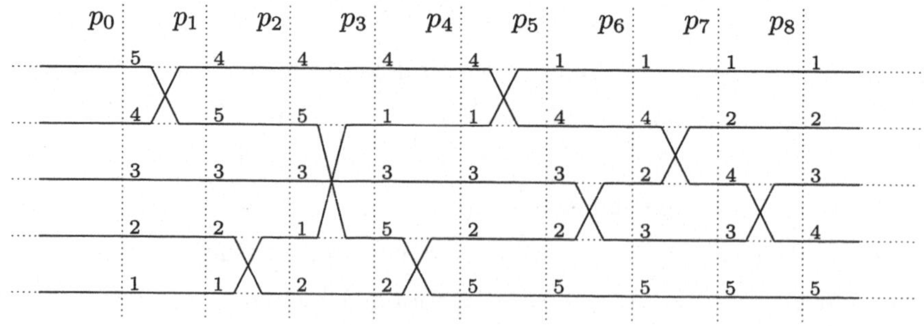

Figure 6.5 A wiring diagram for the arrangement of Figure 6.2

The arrangement thus obtained is the *wiring diagram* for Σ. Since the vertical lines $p_0, \ldots, p_r$ essentially are a sweep sequence of curves for the wiring diagram we see that the mapping from arrangements to allowable sequences is surjective. It remains to show injectivity. Let $(\mathcal{A}, F)$ be any marked arrangement $(\mathcal{A}, F)$ such that Σ corresponds to a sweep of $c_0, \ldots, c_r$ of $\mathcal{A}$. It is obvious that the part of $\mathcal{A}$ between c_{i-1} and c_i is isomorphic to the part of the wiring diagram between p_{i-1} and p_i. These isomorphisms for $i = 1, \ldots, r$ can be glued together to an isomorphism of the arrangements. This proves the first part of the theorem.

For the second part recall that a move $M : \pi_i \to \pi_{i+1}$ reverts as many elements as there are lines crossing at the corresponding vertex in $\mathcal{A}$. □

Application of Allowable Sequences: Ungar's Slope Theorem

From Section 5.3 we know that n points, not all on a line, determine at least n lines. In the Euclidean plane we can ask for the number of parallel classes (slopes) of the connecting lines. The problem was raised by Scott. He conjectured that $2k$ points, not all on a line, determine at least $2k$ slopes. If true, this trivially implies that $2k+1$ points not all on a line also determine at least $2k$ slopes. Configurations of n points (n odd) which determine only $n-1$ slopes exist, they are called *slope critical.* The simplest example, the *bipencil*, consists of $k-1$ points on the positive x-axis, their reflections to the negative x-axis and the 3 additional points $(0,-1), (0,0), (0,1)$. Two other examples taken from infinite families of slope critical configurations are shown in Figure 6.6. The example on the left consists of the vertices of a regular $2k$-gon together with its center. The example on the right is from the family of exponential crosses.

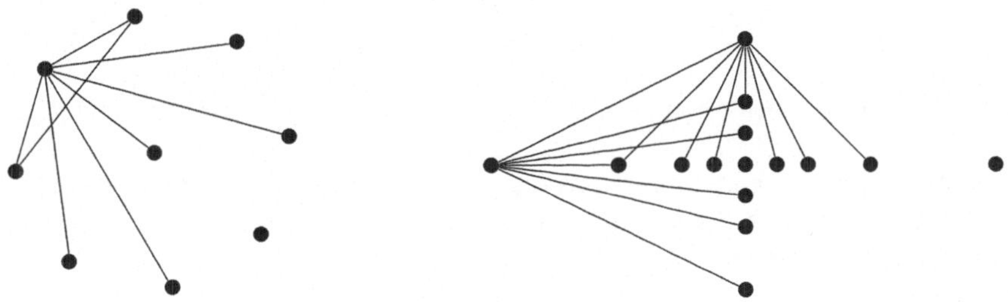

Figure 6.6 Two slope critical configurations.

Theorem 6.4 (Ungar)
A configuration of $2n$ points, not all on a line, determines at least $2n$ slopes.

Proof. A configuration X of points may be dualized yielding an arrangement $\mathcal{A}_X$ of lines. Connecting lines of the points in X dualize to vertices of $\mathcal{A}_X$. Now assume that the duality transform is the polarity at a parabola. In this case connecting lines with the same slope are mapped to vertices with the same x-coordinate. Using a vertical line to sweep the arrangement $\mathcal{A}_X$ we obtain a sequence of permutations $\Sigma_X = \pi_0 \ldots \pi_r$. This sequence is not quite an allowable sequences, it obeys properties (1) and (3) of allowable sequences, but property (2) is not exactly true. The appropriate statement in the given setting is the following:

(2′) Each permutation π_i, $1 \le i \le r$ is obtained by the reversal of one or more disjoint consecutive substrings from the preceding permutation π_{i-1}.

We call a sequence Σ with (1), (2′) and (3) a *generalized allowable sequence*. Actually, this is the original definition of an allowable sequence. The reason is that when coming from point configurations, as Goodman and Pollack did, property (2′) is the natural choice. Property (2) is convenient for pseudoline arrangements and has the advantage that the notion of equivalence of sequences becomes simpler.

Back to the theorem. With point set X we have associated a generalized allowable sequence $\Sigma_X = \pi_0 \ldots \pi_r$ such that the number of slopes determined by X is r, i.e., the number of moves that transform π_0 (the identity) into π_r (the reverse). The following proposition is enough to complete the proof of the theorem.

Proposition 6.5 *Let $\Sigma = \pi_0, \pi_1, \ldots, \pi_r$ be a generalized allowable sequence of permutations of $[2n]$, if $r > 1$ then $r \ge 2n$.*

Proof. Let the *middle-barrier* separate each permutation π_i of $[2n]$ between the first n elements and the last n elements. A move $\pi_i \longrightarrow \pi_{i+1}$ is *crossing* if one of the substrings reversed in the move contains the middle-barrier, i.e., if the move brings some elements from one side of the middle-barrier to the other. A crossing move has *order d* if it brings $2d$ elements across the middle-barrier, see Figure 6.7.

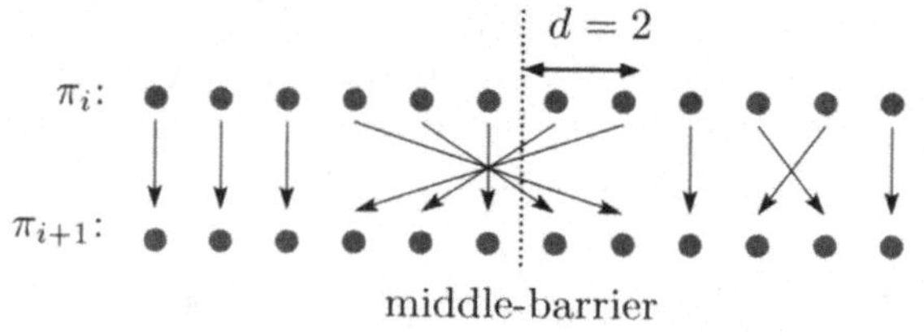

Figure 6.7 A crossing move of order 2.

An allowable sequence with more than one move contains at least two crossing moves. Suppose that there are $t \ge 2$ crossing moves $m_1, \ldots, m_t$ and let d_i be the order of m_i. Each element has to cross the middle-barrier in some crossing move, therefore

$$\sum_{i=1}^{t} 2d_i \ge 2n.$$

Claim. Between consecutive crossing moves m_i and m_{i+1} there are at least $d_i + d_{i+1} - 1$ non-crossing moves.

The basis for the proof is the following simple observation:

($\star$) A move always transforms increasing substrings into decreasing ones.

After move m_i there is a decreasing block of length d_i on either side of the middle-barrier. Alike, before move m_{i+1} there is an increasing block of length d_{i+1} on either side of the middle-barrier. With $\star$ the following statements follow:

- Between m_i and m_{i+1} there is a (unique) move m^* touching the middle barrier, such that middle pair (i.e., the pair of elements left and right of the middle barrier) is changed from decreasing to increasing by m^*.

- Every move between m_i and m^* can strip off at most one element from each of the decreasing blocks on the two sides of the middle-barrier. After m_i the length of the decreasing block is d_i. Move m^* requires that the length of a decreasing block touching the barrier is one.
- Every move between m^* and m_{i+1} can stick on at most one element to each of the increasing blocks on the two sides of the middle-barrier. After m^* the length of such a block is one. Move m_{i+1} requires that the length of this block has grown to d_{i+1}.

Together this shows that there are at least $1 + (d_i - 1) + (d_{i+1} - 1) = d_i + d_{i+1} - 1$ non-crossing moves between m_i and m_{i+1}. △

Claim. Before m_1 and after m_t there are additional $d_1 + d_t - 1$ non-crossing moves.

We reduce this to the previous claim: Let m_1 be the move $\pi_i \longrightarrow \pi_{i+1}$ in the generalized allowable sequence $\Sigma = \pi_0, \pi_1, \ldots, \pi_r$. Consider the mapping taking π to $\pi' = \pi_{i+1}^{-1} \circ \pi$, this is a relabeling such that π'_{i+1} is the identity. Recall that $\overline{\pi}$ is the reverse of π. and consider the sequence $\Sigma' = \pi'_{i+1}, \pi'_{i+2}, \ldots, \pi'_r, \overline{\pi'_1}, \ldots, \overline{\pi'_i}, \overline{\pi'_{i+1}}$. This is a generalized allowable sequence and there is an obvious correspondence between the moves of this sequence and the moves of Σ. In particular, the two last crossing moves m'_{t-1} and m'_t and the moves between them in Σ' are in bijection with m_t and m_1 and the moves before m_1 and after m_t in Σ. The claim now follows from the previous claim applied to Σ'. △

For the total number of moves we have the estimate:

$$t + (d_1 + d_t - 1) + \sum_{i=1}^{t-1} (d_i + d_{i+1} - 1) = \sum_{i=1}^{t} 2d_i \geq 2n.$$

□

6.3 Local Sequences

Representing an arrangement by an allowable sequence can be seen as an encoding by an ordered sequence of vertical cuts through the arrangement. A representation of simple arrangements by a sequence of horizontal cuts can be obtained by associating with line i the permutation α_i of $\{1,..,n\} \setminus i$ reporting the order from left to right in which the other pseudolines cross line i. The family $(\alpha_1, \alpha_2, \ldots, \alpha_n)$ is called the family of *local sequences* of the arrangement. In case of non-simple arrangements local sequences are slightly more general structures than permutations since several lines can cross line i in the same point. For the arrangement of Figure 6.2 the local sequences can be coded as $\alpha_1 = [2, \{3,5\}, 4]$, $\alpha_2 = [1,5,3,4]$, $\alpha_3 = [\{1,5\}, 2, 4]$, $\alpha_4 = [5,1,2,3]$ and $\alpha_5 = [4, \{1,3\}, 2]$.

Later, in Theorem 6.17 we characterize those $(\alpha_i)_{i=1..n}$ corresponding to simple marked arrangements. The next goal (Theorem 6.6), however, is to show that in order to obtain a representation of simple marked arrangements with the standard labeling it is sufficient to know for each i, which entries of α_i are larger than i and which are smaller.

With the local sequence $\alpha_i = (a_1^i, a_2^i, \ldots, a_{n-1}^i)$ corresponding to line i associate a binary vector $\beta_i = (b_1^i, b_2^i, \ldots, b_{n-1}^i)$ such that $b_j^i = 1$ if $a_j^i > i$ and $b_j^i = 0$ if $a_j^i < i$. Since α_i is a permutation of $\{1,..,n\} \setminus i$ exactly $n - i$ entries of β_i are 1, i.e., $\sum_{j=1}^{n-1} b_j^i = n - i$ for all i.

Assume that $\mathcal{A}$ is given with the standard labeling of lines, i.e., $i < j$ implies that far enough to the left the line with label i is below the line with label j. We can use this to reinterpret the meaning of b^i_j: If the j-th crossing on line i in the wiring diagram is a move of line i up into the next wire, then $b^i_j = 1$ and if line i is moving down at its j-th crossing, then $b^i_j = 0$. Hence we can get the value of b^i_j directly from the arrangement, $b^i_j = 1$ iff at its j-th crossing line i is crossing another line from below, i.e., line i is moving up.

Let $\mathcal{B}(n)$ be the set of n-tuples $(\beta_1, \beta_2, \ldots, \beta_n)$ with $\beta_i = (b^i_1, b^i_2, \ldots, b^i_{n-1})$ a binary vector and $\sum_{j=1}^{n-1} b^i_j = n - i$ for all i. Above we have described a mapping Φ from arrangements of n pseudolines to $\mathcal{B}(n)$. Figure 6.8 shows an example.

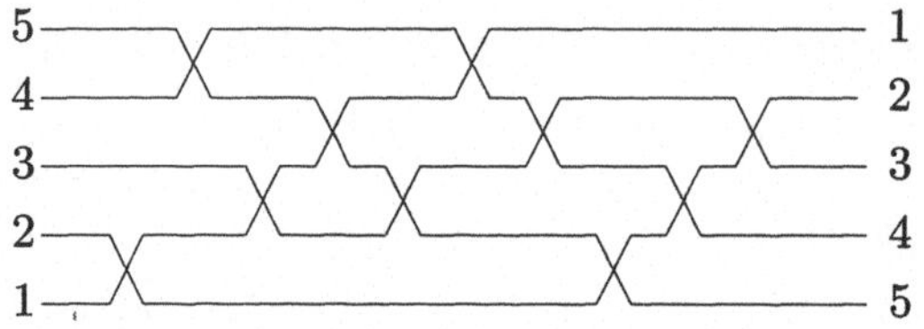

Figure 6.8 A wiring diagram $\mathcal{A}$ of five lines. The corresponding sequence of five binary vectors is $\Phi(\mathcal{A}) = \big((1,1,1,1),(0,1,1,1),(0,1,1,0),(1,0,0,0),(0,0,0,0)\big)$.

Not all elements of $\mathcal{B}(n)$ correspond to arrangements. For $n = 4$ there are 9 elements in $\mathcal{B}(4)$ but only 8 arrangements. The element of $\mathcal{B}(4)$ which is not in the image of Φ is $\big((1,1,1),(1,0,1),(0,1,0),(0,0,0)\big)$.

Theorem 6.6 *The mapping Φ is an injective map from simple arrangements of n pseudolines to $\mathcal{B}(n)$.*

Proof. Below we describe an algorithm *bit-sweep* which takes an element of $\mathcal{B}(n)$ as input and constructs a sequence of permutations of $[n]$. If the element of $\mathcal{B}(n)$ is the $\Phi(\mathcal{A})$ for some simple arrangement $\mathcal{A}$, this sequence of permutations is a simple allowable sequence Σ for $\mathcal{A}$. Since Σ uniquely determines the marked arrangement $\mathcal{A}$, this is also true for $\Phi(\mathcal{A})$. In particular Φ is injective.

Let $(\beta_1, \beta_2, \ldots, \beta_n)$ be the input for the bit-sweep algorithm. The first permutation in the output is the identity, $\pi_0 = (1, 2, \ldots, n)$. For each i there is a *position counter* $p(i)$ which is initialized as $p(i) = 1$. The *state* of the algorithm is a vector $S = (s_1, s_2, \ldots, s_n)$. The state vector is initialized by the first bit of each of the β_i, i.e., $(s_1, s_2, \ldots, s_n) = (b^1_1, b^2_1, \ldots, b^n_1)$. After k steps, i.e., after the output of π_k, the ith entry of the state is

$$s_i = b^{\pi_k(i)}_{p(\pi_k(i))}$$

this is the bit at the actual position $p(\pi_k(i))$ on the line which is the ith element of π_k. In step $k+1$ the algorithm takes the least index i with $s_i = 1$ and $s_{i+1} = 0$. If there is no such index the algorithm stops. If i exists, then π_{k+1} is obtained from the current permutation π_k by an adjacent transposition at positions i and $i+1$. The new permutation is $\pi_{k+1} = (\pi_k(1), \ldots, \pi_k(i-1), \pi_k(i+1), \pi_k(i), \pi_k(i+2), \ldots, \pi_k(n))$. To complete the step the position counters $p(\pi_k(i))$ and $p(\pi_k(i+1))$ are incremented by one and the state vector is updated accordingly, i.e., with new entries s_i and s_{i+1} taken from the appropriate positions of $\beta_{\pi_{k+1}(i)}$ and $\beta_{\pi_{k+1}(i+1)}$. If a position counter reaches the value n the β-vector has no corresponding entry, in that case write a symbol ∞ into the state vector. Figure 6.9 indicates a run of the algorithm.

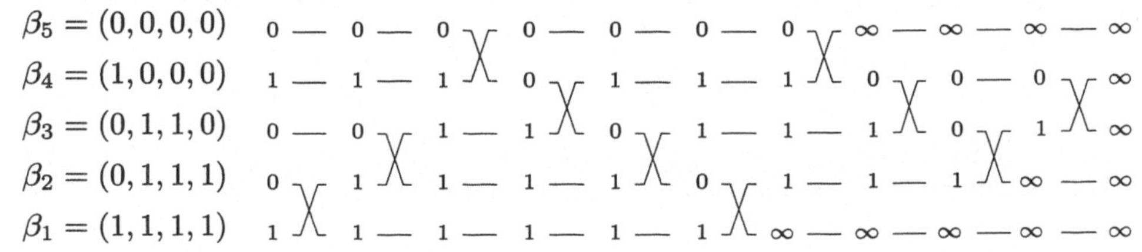

Figure 6.9 The columns show the state vectors in a run of the bit-sweep algorithm.

The *canonical allowable sequence* of an arrangement $\mathcal{A}$ is the allowable sequence which results from a sweep which always picks the lowest admissible vertex for the advance. The claim is that with input $\Phi(\mathcal{A})$ the bit-sweep algorithm, produces the canonical allowable sequence of $\mathcal{A}$.

The proof is by induction. We observe both algorithms, the canonical sweep and the bit-sweep algorithm with input $\Phi(\mathcal{A})$ step by step. Each step for both algorithms consists in the move from one output permutation to the next. The invariant for the proof is that after k steps:

(1) The current permutation in both algorithms is the same π_k.

(2) The sweep has passed $p(i) - 1$ vertices on line i for each $i \in [n]$.

(3) $s_i = 1$ iff at the next crossing the line $\pi_k(i)$ is moving up.

This is obvious for $k = 0$, both algorithms start with the identity permutation. Assume the invariant after k steps. To show that it holds after $k + 1$ steps we need that

($\star$) lines $\pi_k(i)$ and $\pi_k(i+1)$ cross at a vertex which is admissible for the sweep if and only if $s_i = 1$ and $s_{i+1} = 0$.

If $s_i = 1$ and $s_{i+1} = 0$ then (by definition and invariant) line $\pi_k(i)$ is moving up at its next crossing while line $\pi_k(i+1)$ is moving down at its next crossing. Since line $\pi_k(i)$ is below line $\pi_k(i+1)$ and they border a common face in $\mathcal{A}$, the vertex where they cross is admissible for the sweep. Conversely, if lines $\pi_k(i)$ and $\pi_k(i+1)$ cross at an admissible vertex, then line $\pi_k(i)$ is moving up at its next crossing and line $\pi_k(i+1)$ is moving down at its crossing. Therefore, $s_i = 1$ and $s_{i+1} = 0$. □

Application of Local Sequences: Counting Arrangements

Let B_n be the number of simple marked arrangements of n pseudolines. Combinatorial encodings are a tool to get hand on this number. With todays methods exact enumeration is by far out of reach, but there are some asymptotic bounds.

Let A_n be the number of simple allowable sequences. Stanley found a remarkable exact formula for A_n. From that formula the asymptotic growth of A_n is known to be $2^{\Theta(n^2 \log n)}$. Clearly $B_n < A_n$, we will show that B_n is substantially smaller than A_n.

Theorem 6.7 *The number B_n of non-isomorphic simple marked arrangements of n pseudolines is at most $2^{0.72n^2}$.*

Proof. By Theorem 6.6 we have an injective mapping from arrangements of n lines to $\mathcal{B}(n)$, i.e., $|B_n| \leq |\mathcal{B}(n)|$. Counting elements of $\mathcal{B}(n)$ is a simple task:

$$|\mathcal{B}(n)| = \binom{n-1}{0}\binom{n-1}{1}\binom{n-1}{2}\cdots\binom{n-1}{n-1}.$$

Let $f(n) = |\mathcal{B}(n)|$, from $\binom{n}{k} = \frac{n}{n-k}\binom{n-1}{k}$ we obtain the recursion $f(n+1) = \frac{n^n}{n!} f(n)$. Applying Stirling's formula and taking logarithms gives

$$\log_2 f(n+1) = n \log_2 e + \log_2 f(n) + O(\log n) = \sum_{k=1}^{n} k \log_2 e + O(n \log n).$$

With a table lookup we find that $\binom{n}{2} \log_2 e \approx 0.7213(n^2 - n)$. □

To improve the bound on the number of arrangements it would be interesting to have some tools to discriminate between those members from $\mathcal{B}(n)$ which are in the image of Φ and those which are not. At this time we have little more than the bit-sweep algorithm from the proof of Theorem 6.6. We can take an arbitrary element $B \in \mathcal{B}(n)$ as input to this algorithm. The two possible outcomes are.

(1) The algorithm gets stuck before $\binom{n}{2}$ moves have been made, i.e., in the current state $(s_1, \ldots, s_n)$ there is no index i with $s_i = 1$ and $s_{i+1} = 0$.

(2) B indeed corresponds to an arrangement.

Recall the element $B = \big((1,1,1),(1,0,1),(0,1,0),(0,0,0)\big)$ of $\mathcal{B}(4)$ not in the image of Φ. Trying to sweep B we get stuck after three moves. At the second move we may already note that something goes wrong, the lines involved in the crossing of the first move cross back. An *immediate back-cross* is a situation where two lines cross twice in a row. When sweeping $B \in \mathcal{B}(n)$, as in the proof of Theorem 6.6, we recognize an immediate back-cross when the pair $(s_i, s_{i+1}) = (1,0)$ of the move is replaced by $(s'_i, s'_{i+1}) = (1,0)$, i.e., the state vectors S and S' before and after the move are identical.

The sweep like algorithm used for the proof of Theorem 6.6 took an element $B \in \mathcal{B}(n)$ as input and in case $B = \Phi(\mathcal{A})$ for some arrangement $\mathcal{A}$ it produced the canonical allowable sequence for $\mathcal{A}$ as output. Internally, the algorithm also produced a sequence of state vectors. It is quite obvious that the algorithm can be modified such that it takes the sequence $S_0, S_1, \ldots, S_{\binom{n}{2}}$ of state vectors as input and the output produced is the same.

Two successive state vectors S_{i-1} and S_i only differ in two positions, namely, in the least two positions on which the entries of S_{i-1} are $(1,0)$. Therefore, S_i is completely determined by two new bits, call the pair $w_i = (w_i^1, w_i^2)$ of these bits the *replace pair* for the transition from S_{i-1} to S_i.

Hence, the sweep corresponding to $B \in \mathcal{B}(n)$ is completely determined by the initial state $S = S_0$ and a sequence of replace pairs $w_1, w_2, \ldots, w_{(n^2-n)/2}$. A sequence of replace pairs leads to an immediate back-cross exactly iff one of the pairs w_j is $(1,0)$. The number of back-cross free elements of $\mathcal{B}(n)$ and, hence, the number of arrangements can be estimated from above by the number of initial states S and the number of $(1,0)$ free sequences of replace pairs. For S there are $\leq 2^n$ choices and for each pair w_j there remain 3 choices*, therefore:

$$B_n \leq 2^n 3^{\binom{n}{2}} = 2^{\frac{\log_2(3)}{2} n^2 + O(n)} = 2^{0.7924\, n^2 + O(n)}.$$

* Here we disregard the special entries ∞ in the state vectors. Actually, these special entries not necessary, see the proof of Lemma 6.8

This new bound is weaker than the bound from Theorem 6.7. However, the new bound was only using the forbidden immediate back-crossings. The result of Theorem 6.7 only made use of the number of zeros and ones in each β_j. With the replace matrix we next define a representation that takes care of both aspects.

A *replace matrix* is a binary $n \times n$ matrix M with two properties:

$$\text{(1)}\ \sum_{j=1}^{n} m_{ij} = n - i \text{ for } i = 1, .., n,$$

$$\text{(2)}\ m_{ij} \geq m_{ji} \text{ for all } i < j.$$

Lemma 6.8 *There is an injective mapping Ψ from simple arrangements of n pseudolines to $n \times n$ replace matrices.*

Proof. Consider the local bit encoding $\Phi(\mathcal{A})$ (see Theorem 6.6) and let $m_{ii} = b_1^i$, that is, we record the initial state vector S of the sweep along the diagonal of M. If in the kth step of the sweep of $\Phi(\mathcal{A})$ lines i and j cross we define $m_{ij} = 1$ if the next crossing (after the crossing with line j) of line i goes up and $m_{ij} = 0$ if the next crossing of line i goes down. If $i < j$ then at their crossing line i is going up and line j is going down. Since there is no immediate back-cross of lines i and j we conclude $(m_{ij}, m_{ji}) \neq (0, 1)$ or equivalently $m_{ij} \geq m_{ji}$. After the complete sweep of $\Phi(\mathcal{A})$ we remain with a single undefined entry in each row of M. Let this entry be 0.

Property (1) of replace matrices is easily seen to hold for M as defined above. The entries in row i of M are a permutation of the entries of β_i and an additional 0. Property (2) is only in question at a pair i, j with $i < j$ and $m_{ij} = 0$ because it was the last undefined entry of its row. In this case, after crossing line j from below line i was not involved in further crossings. Suppose that line j has a crossing after the crossing with line i. At the first of these later crossings line j has to move down. This is because the position above j is occupied by i, hence, $m_{ji} = 0$. The other possibility is that line j has no further crossings and again $m_{ji} = 0$. In both cases $m_{ij} \geq m_{ji}$.

Hence, $M = \Psi(\mathcal{A})$ is a well defined replace matrix. The mapping Ψ is injective because $M = \Psi(\mathcal{A})$ can guide a sweep to reconstruct (the canonical allowable sequence of) $\mathcal{A}$.

The replace matrix corresponding to the arrangement of Figure 6.8 illustrates this encoding.

$$M = \begin{pmatrix} 1 & 1 & 1 & 0 & 1 \\ 1 & 0 & 0 & 1 & 1 \\ 1 & 0 & 0 & 0 & 1 \\ 0 & 0 & 0 & 1 & 0 \\ 0 & 0 & 0 & 0 & 0 \end{pmatrix}.$$

To obtain an estimate for the number of replace matrices, a probabilistic argument can be used. Consider the probability space Ω of all binary $n \times n$ matrices with $\sum_{j=1}^{n} m_{ij} = n - i$ for $i = 1, \ldots, n$ and let M be a uniformly distributed random matrix from Ω. Let p_i be the probability that a fixed entry in row i of M is 0, i.e., $p_i = \frac{i}{n}$, and $q_i = 1 - p_i$ be the probability that this entry is 1, i.e., $q_i = \frac{n-i}{n}$.

For $i < j$ let E_{ij} be the event $m_{ij} \geq m_{ji}$. Since $m_{ij} \not\geq m_{ji}$ iff $(m_{ij}, m_{ji}) = (0, 1)$ the probability of event E_{ij} is $\text{Prob}[E_{ij}] = (1 - p_i q_j)$. The number R_n of replace matrices can be written as $R_n = |\Omega| \text{Prob}[\bigwedge_{i<j} E_{ij}]$.

For independent events E_{ij} we would have $\text{Prob}[\bigwedge_{i<j} E_{ij}] = \prod_{i<j} \text{Prob}[E_{ij}]$. This would allow to estimate R_n as $\prod_{k=0}^{n-1} \binom{n}{k} \prod_{i<j}(1 - \frac{i(n-j)}{n^2})$. The base 2 logarithm of this function behaves like $0.66n^2$.

Unfortunately, due to the fixed row sums of replace matrices the E_{ij} are not independent. There are positively and negatively correlated pairs. Therefore is not obvious in which direction the error made by ignoring dependencies goes. Nevertheless, this probabilistic approach can be exploited to prove:

Theorem 6.9 *The number R_n of $n \times n$ replace matrices and, hence, the number B_n of non-isomorphic simple marked arrangements of n pseudolines is at most $2^{0.69n^2}$.*

6.4 Zonotopal Tilings

A particularly nice representation of arrangements of pseudolines is the representation by 'zonotopal tilings'. This can be considered as a standardized drawing of the 'dual graph' of the arrangement. Figure 6.10 shows an example.

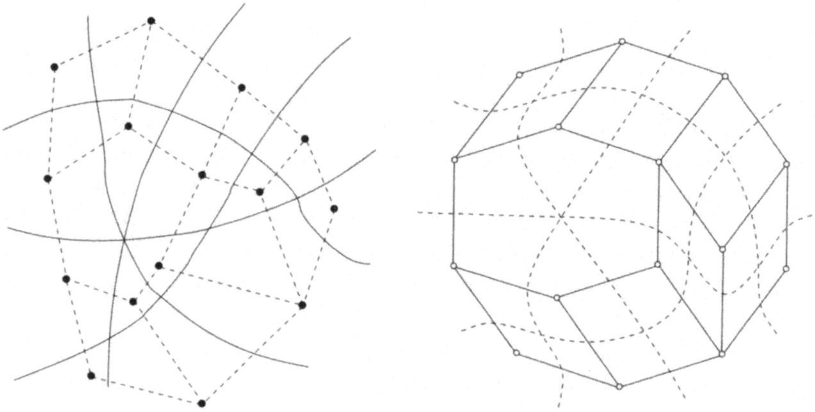

Figure 6.10 An arrangement with its dual graph and the dual graph as zonotopal tiling.

A 2-dimensional *zonotope* is the Minkowski sum of a set of n line segments in $\mathbb{R}^2$, this is a centrally symmetric $2n$-gon. With a vector v_i associate the line segment $[-v_i, +v_i]$, the Minkowski sum of the line segments corresponding to $V = \{v_1, \ldots, v_n\}$ is the set

$$Z(V) = \left\{ \sum_{i=1}^{n} c_i\, v_i : -1 \le c_i \le 1 \text{ for all } 1 \le i \le n \right\}.$$

A *zonotopal tiling T* is a tiling of $Z(V)$ by translates of zonotopes $Z(V_i)$ with $V_i \subset V$. A zonotopal tiling is a *simple zonotopal tiling* if all tiles are rhombi, i.e., $|V_i| = 2$ for all i. A zonotopal tiling together with a distinguished vertex x of the boundary of $Z(V)$ is a *marked zonotopal tiling.*

The next theorem is a precise statement for the correspondence suggested by Figure 6.10. The proof of the theorem is based on a Sweeping Lemma for zonotopal tilings.

Theorem 6.10 *Let V be a set of n pairwise non-collinear vectors in $\mathbb{R}^2$.*

(1) There is a bijection between marked zonotopal tilings of $Z(V)$ and marked arrangements of n pseudolines.

(2) Via this bijection simple tilings correspond to simple arrangements.

We first give the mapping from zonotopal tilings to equivalence classes of allowable sequences. Let $Z(V)$ be a marked zonotope generated by a set V of n pairwise non-collinear vectors. The zonotope $Z = Z(V)$ is a centrally symmetric $2n$-gon. Rotate Z such that the distinguished vertex x is the unique highest vertex of Z, in particular the boundary of Z has no horizontal edge. Assume a labeling of the vectors in V, such that along the left boundary of Z, i.e., on the left path from the lowest vertex $\overline{x}$ to x, the segments correspond to $v_1, v_2, \ldots, v_n$ in this order. This labeling is the *standard labeling* of V.

Given a zonotopal tiling $\mathcal{T}$ of Z, consider the set of y-monotone paths along segments of $\mathcal{T}$ from $\overline{x}$ to x. We define a *sweep of $\mathcal{T}$ with north-pole x* as a sequence $p_0, p_1, \ldots, p_r$ of y-monotone paths from $\overline{x}$ to x in $\mathcal{T}$ with the following properties.

(1) Any two consecutive paths p_i, p_{i+1} of the sequence have exactly one tile T_i of tiling $\mathcal{T}$ between them, i.e., in the interior of the closed curve $p_i \cup p_{i+1}$.

(2) Every tile is between a unique pair of consecutive paths, therefore, $p_0 \cup p_r$ is the boundary of $Z(V)$.

As we did for sweeps of arrangements, we further assume that the sweep of $\mathcal{T}$ is from left to right, i.e., p_0 is the left boundary of $Z(V)$.

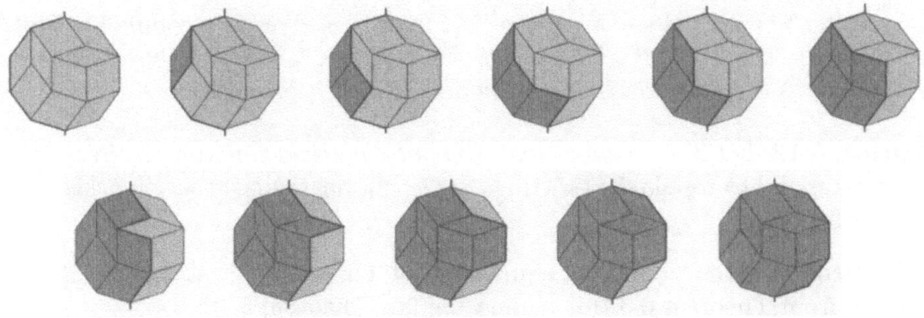

Figure 6.11 A sweep of a zonotopal tiling.

Remark. There is some interest in the maximum number $m(n)$ of y-monotone $\overline{x}$ to x paths a marked zonotopal tiling can have. Knuth ([123] page 39) conjectures that $m(n) \leq n2^{n-2}$. By an inductive argument this would imply that the number of marked arrangements of n pseudolines is bounded by $\Pi_{k=1}^{n} m(k)$. Therefore, the conjectured bound would show that this number is at most $2^{n^2/2+o(n^2)}$, an improvement over the bound of Theorem 6.9.

A sweep of tiling $\mathcal{T}$ induces a total order $T_1, T_2, \ldots, T_r$ on the tiles of $\mathcal{T}$ with the property that any initial segment $T_1, \ldots, T_i$ can be separated from the remaining tiles $T_{i+1}, \ldots, T_r$ by a horizontal translation. Conversely, an order $T_1, T_2, \ldots, T_r$ of the tiles with this separation property corresponds to a sweep: Define path p_i as the right boundary of the union of $T_1, \dots, T_i$. The following lemma implies that a zonotopal tiling $\mathcal{T}$ can be swept.

Lemma 6.11 *Let $\mathcal{C}$ be a family of disjoint convex sets in the plane. There is a $C \in \mathcal{C}$ that can be translated to the left without colliding with another object from $\mathcal{C}$.*

Proof. Define a directed graph $G_{\mathcal{C}}$ whose vertices are the sets in $\mathcal{C}$. Arcs of $G_{\mathcal{C}}$ are the pairs (C, C') such that there is a horizontal line segment in $\mathbb{R}^2 \setminus \mathcal{C}$ with left endpoint at C and right endpoint at C', in other words, C' is visible from C horizontally to the right. By definition, elements that can be translated collision free to the left are exactly the sources of $G_{\mathcal{C}}$.

Imagine the plane being slanted such that a marble placed at a point (x, y) would roll in direction of $(x, -\infty)$ and think of the elements of $\mathcal{C}$ as elevated obstacles. Choose $C \in \mathcal{C}$ arbitrarily, place a marble at the lowest point of C and let it roll, this procedure yields a y-monotone path $p^{\downarrow}$ connecting C to vertical $-\infty$, (note that here we use the convexity of the obstacles). A symmetrical procedure yields a path $p^{\uparrow}$ from the highest point of C to vertical $+\infty$. Let p be the concatenation of $p^{\uparrow}$, a line segment through C and $p^{\downarrow}$, this is a y-monotone path. Let $\mathcal{C}^l$ be the set of elements of $\mathcal{C}$ which are left of p and $\mathcal{C}^r$ the subset right of p. All arcs of $G_{\mathcal{C}}$ go left to right in the ordered partition $[\mathcal{C}^l, C, \mathcal{C}^r]$ of $\mathcal{C}$. Therefore either $\mathcal{C}^l = \emptyset$ and C is a source of $G_{\mathcal{C}}$ or induction yields a source C' of $G_{\mathcal{C}}[\mathcal{C}^l]$. This element C' is also a source of $G_{\mathcal{C}}$. △

A total ordering $C_1, C_2, \dots, C_r$ of the elements of $\mathcal{C}$ has the the property that $C_1 \dots, C_i$ can be separated from $C_{i+1}, \dots, C_r$ by a horizontal translation exactly if $C_1, C_2, \dots, C_r$ is a topological sorting of $G_{\mathcal{C}}$.

The tiles of a tiling $\mathcal{T}$ are a family of disjoint convex sets. The graph $G_{\mathcal{T}}$ has the tiles of $\mathcal{T}$ as vertices. The arcs of $G_{\mathcal{T}}$ are pairs (T, T') of tiles sharing a common segment and oriented from the tile T on the left side of the segment $T \cap T'$ to the tile on the right side. The previous considerations are summarized with a proposition:

Proposition 6.12 *Let $\mathcal{T}$ be a zonotopal tiling of a marked zonotope Z. Sweeps of $\mathcal{T}$ bijectively correspond to topological sortings of $G_{\mathcal{T}}$. In particular, every marked zonotopal tiling $\mathcal{T}$ can be swept.*

The next lemma is the 'zonotopal equivalent' of Theorem 6.3. Actually, together with the bijection from Theorem 6.3 this lemma implies Theorem 6.10.

Lemma 6.13 *There is a bijection between marked zonotopal tilings and equivalence classes of allowable sequences. Moreover, this bijection maps simple zonotopal tilings to classes of simple allowable sequences.*

Proof. First we show how to associate an allowable sequence to every sweep of a zonotopal tiling. Given a sweep sequence $p_0, \dots, p_r$ of paths we associate to each path p_i a sequence π_i recording the labels of the vectors which define the segments along the path in the order of the path from $\overline{x}$ to x. The sequence π_0 is a permutation, the identity. Any two consecutive sequences π_i and π_{i+1} only differ in a substring where path p_i takes the left boundary and path p_{i+1} takes the right boundary of tile T_i. Since T_i is a zonotope, the same labels appear on both boundaries but in reversed order. Hence, all π_i are permutations, moreover, $\pi_i \to \pi_{i+1}$ is a move as in part (2) of the definition of allowable sequences. We also note that π_r is the reverse permutation.

It remains to prove property (3) of allowable sequences: Any two elements $i, j \in [n]$ are reversed in exactly one move or equivalently, that there is a unique tile T in the tiling $\mathcal{T}$ which has boundary segments corresponding to v_i and v_j. The labeling of the vectors is such that on the left border of $Z = Z(v_1, \ldots, v_n)$ we have v_i below v_j iff $i < j$. On the right border of Z the vectors appear in the reverse order, therefore, every pair i, j has to be exchanged by some move. To prove that every pair is exchanged exactly once we use the observation that a sub-zonotope $Z(W)$ with $W \subset V$ also has the vectors in increasing order on its left border and in decreasing order on its right border. The tiles $T \in \mathcal{T}$ are sub-zonotopes $Z(W)$ of Z for suitable $W \subset V$. Hence, the move corresponding to a tile $T \in \mathcal{T}$ is the replacement of an increasing substring by its reverse decreasing substring. If $i < j$ and the order of i and j has been changed by a move, then these two elements are a decreasing pair. Therefore, they cannot participate in an increasing substring and, hence, cannot be exchanged by later moves induced by tiles $T \in \mathcal{T}$.

We know (Proposition 6.12) that the sweeps of $\mathcal{T}$ are in one-to-one correspondence with topological sortings of $G_{\mathcal{T}}$. On the other hand, an equivalence class of allowable sequences corresponds to the topological sortings of the graph $\overrightarrow{G}$ of a marked arrangement (page 92). It is not hard to verify that $G_{\mathcal{T}}$ and $\overrightarrow{G}$ are isomorphic.

For the inverse mapping we have to associate a marked zonotopal tiling to an equivalence class of allowable sequences. Let $\Sigma = \pi_0, \ldots, \pi_r$ be any member of the equivalence class and build the tiling from left to right starting with the left boundary of $Z(V)$. After placing i tiles, three properties remain invariant:

(1) The union of the already placed tiles together with the left boundary of Z is a simply connected region.

(2) The right boundary of this region is a y-monotone path p_i.

(3) The segments along path p_i are in the order given by π_i.

The crucial observation is that when it comes to place tile T_{i+1} path p_i contains a part s with the shape of the left border of T_{i+1}. Only this part of the path is affected by the placement of T_{i+1}. This effects in the replacement of the points on s by points which are further right. From the invariant it follows that the tiles are placed without overlap. Since the last permutation π_r is the reverse of the identity path p_r is the right boundary of $Z(V)$ and zonotope $Z(V)$ is completely covered by tiles. Therefore, the placement of tiles $T_1, \ldots, T_r$ is a tiling $\mathcal{T}$ of $Z(V)$. □

Application of Zonotopal Tilings: k-Sets and $<k$-Sets

Let $\mathcal{A}$ be a marked Euclidean arrangement of pseudolines. Define the kth level of an arrangement $\mathcal{A}$ of n lines as the set of points p in the plane such that exactly k lines of $\mathcal{A}$ pass below p, i.e., p is separated from the south-face $\overline{F}$ by k lines. We are mainly interested in the vertices of $\mathcal{A}$ which belong to the kth level. In the wiring diagram of a simple arrangement these are the vertices of crossings between wires $k+1$ and $k+2$. Recall from Chapter 4, Lemma 4.13, that the k-edges of a set of n points in general position correspond dually to the vertices of the kth level and the $(n-k-2)$th level of the dual arrangement.

Let V_k be the set of vertices of the kth level, the number $|V_k|$ is the complexity of the kth level. We always assume $k \leq n/2$, by symmetry every bound on $|V_k|$ is a bound on $|V_{n-k-2}|$ and vice versa.

A lower bound for the complexity of the kth level is easily obtained. Lines $1, \ldots, k$ start below the kth level and end above it. Assuming that $\mathcal{A}$ is a simple every vertex can only bring one of the lines across, therefore $|V_k| \geq k$.

For an upper bound we consider the kth level of a zonotopal tiling of a zonotope $Z = Z(v_1, \ldots, v_n)$. As generating vectors we choose $v_i = (i, 1)$, for $i = 1, \ldots, n$. All tiles of a simple zonotopal tiling of Z have hight two. Let the belt of a tile be the horizontal segment at height one, i.e., the widest horizontal segment that fits into the tile. Let $\mathcal{T}$ be the zonotopal tiling of Z corresponding to a simple arrangement $\mathcal{A}$ of n pseudolines.

($\star$) Vertices of the kth level of $\mathcal{A}$ correspond bijectively to tiles of $\mathcal{T}$ whose belts lie on the horizontal line $y = k + 1$, see Figure 6.12.

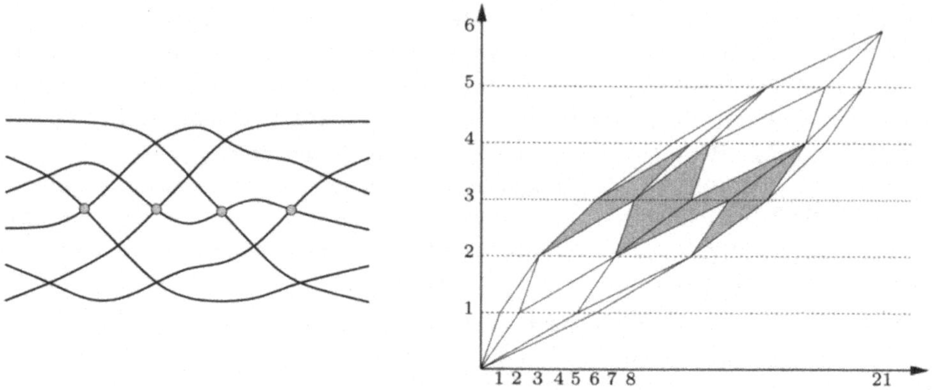

Figure 6.12 An arrangement with a corresponding zonotopal tiling, the 2nd level emphasized.

This observation allows to derive an upper bound for the complexity of the kth level from a one-dimensional packing problem: How many belts of different tiles can be packed disjointly in the interval of intersection of Z with a given horizontal line?

In the following we investigate this later quantity. We obtain a bound for the complexity of the $(k-1)$st level by considering the width of tiles and comparing this to the width of Z at height k.

- The intersection of Z with the line $y = k$ is the interval reaching from $1+2+\ldots+k$ to $n + (n-1) + \ldots + (n-k+1)$. We call this interval the k-interval of Z, it has length $kn - k^2$.

- The width $w(T)$ of a tile T is the length of its belt. A simple zonotopal tiling of Z contains exactly $n - i$ tiles of width i, for $i = 1, \ldots, n-1$, these are the tiles $Z(v_{j_1}, v_{j_2})$ with $|j_1 - j_2| = i$.

Let $T_1, T_2, \ldots T_{\binom{n}{2}}$ be a list of all tiles sorted by increasing width. Take the belts of the tiles in this order and place them disjointly in an interval of length $nk - k^2$. Let T_{t_k} be the last tile whose belt can be used in this process. This index t_k is the bound for $|V_{k-1}|$ we strive for. Formally, t_k is the largest value such that

$$\sum_{i=1}^{t_k} w(T_i) \leq kn - k^2.$$

For $i = 1, \ldots, n-1$ let B_i be the set of tiles of width i. The width of tiles in B_i together is $ni - i^2$. The width of tiles in $\bigcup_{i \le j} B_i$ is $\sum_{i \le j}(ni - i^2) > n\frac{j^2}{2} - \frac{j^3}{3}$. Since $n\frac{j^2}{2} - \frac{j^3}{3} \ge kn - k^2$ for $j = \sqrt{2k}$, there will be no tiles of width $\sqrt{2k}$ or more in the packing. Therefore $t_k \le \sum_{i<\sqrt{2k}} |B_i| = \sum_{i<\sqrt{2k}}(n-i) \le \sqrt{2k}\,n - k$. This proves the theorem:

Theorem 6.14 *The complexity of the kth level of an arrangement of n pseudolines is at most $\sqrt{2(k+1)}\,n - (k+1)$.*

The argument above can be used to bound the complexity of a union of levels. Let $K \subset \{0, 1, \ldots, \lfloor \frac{n}{2} \rfloor\}$. The tiles whose belts are packed into the k-intervals, $k \in K$, are all distinct. Let t_K be maximal such that $\sum_{i=1}^{t_K} w(T_i) \le \sum_{k \in K}(k+1)n - (k+1)^2$. A simple computation yields a bound of $O(n\sqrt{\sum_{k \in K} k})$ for the total complexity $|\bigcup_{k \in K} V_k|$ of the levels.

A particularly nice case is $K = \{0, 1, \ldots, k-1\}$ this means that we are interested in the complexity of the $<k$-levels. Recall that the width of all tiles in B_i is $ni - i^2$ and this equals the length of the i-interval. Therefore, the number of tiles that can be placed on the $<k$-levels is at most $|\bigcup_{i \le k} B_i| = \sum_{i \le k} n - i = kn - \binom{k+1}{2}$. This bound is tight as can be seen from the arrangement generated by the lines $y = -ax + a^2$ for $a = 1, \ldots, n$. This is an example of a *cyclic arrangement*.

Proposition 6.15 *The complexity of the $<k$-levels of an arrangement of n pseudolines is at most $kn - \binom{k+1}{2}$ and this bound is tight.*

6.5 Triangle Signs

So far we have studied arrangements of pseudolines as individual objects. Now we change the focus and consider the set of all arrangements. We consider a graph $\mathcal{G}_n$ whose vertices are all combinatorially different simple marked arrangements of n pseudolines in the Euclidean plane. The edges of $\mathcal{G}_n$ correspond to triangular flips: Let T be a triangular face of an arrangement $\mathcal{A}$. Cutting closely around T the triangle can be replaced by a triangle with the opposite orientation T (see Figure 6.13), this replacement is the *triangular flip* at triangle T. Figure 6.14 shows the graph $\mathcal{G}_n$ for $n = 5$, in the figure the arrangements are represented by their corresponding zonotopal tilings.

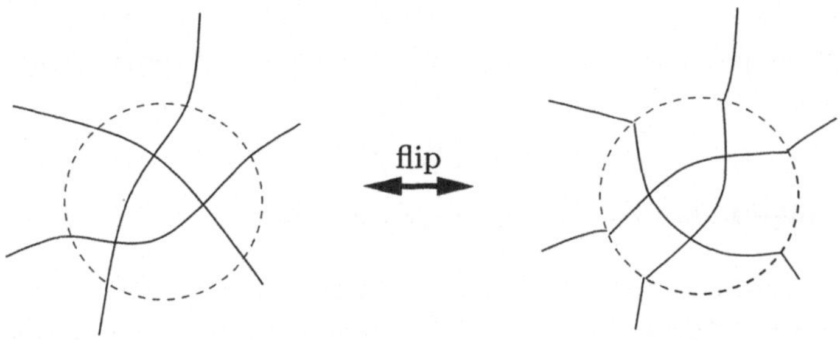

Figure 6.13 A triangular flip.

Below we study an encoding of arrangements by triangle orientations. This encoding imposes a natural orientation on $\mathcal{G}_n$. In Section 6.6 we generalize the patterns and define an order $S_r(n)$, for all $1 \leq r \leq n$, such that $S_1(n)$ is the Boolean lattice, $S_2(n)$ is the weak Bruhat order of the symmetric group, i.e., the elements of $S_2(n)$ are the permutations of $[n]$, and $S_3(n)$ is the abovementioned orientation of $\mathcal{G}_n$. The representation theorem (Theorem 6.16) for arrangements of pseudolines is obtained as a byproduct of a more general correspondence between maximum chains in $S_{r-1}(n)$ and elements of $S_r(n)$ (Theorem 6.21).

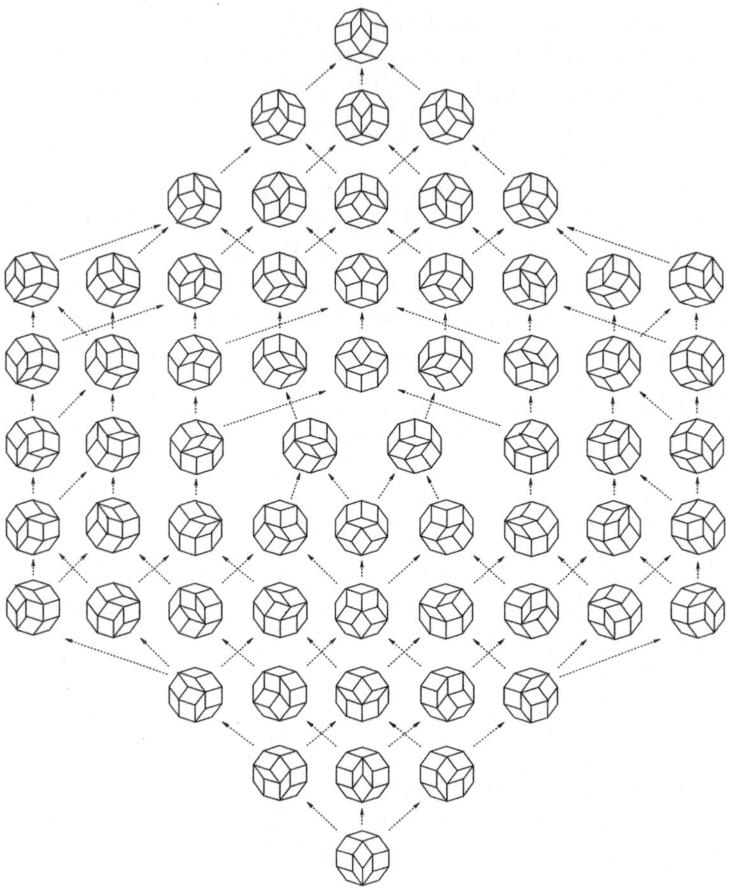

Figure 6.14 The graph $\mathcal{G}_5$ oriented as diagram of the signotope order $S_3(n)$.

Encoding Arrangements by Triangle Orientations

Flips are nicely described in the different encodings of arrangements. In the encoding by zonotopal tilings the projection of a cube is replaced by the view of the cube from the other side. In the encoding by local sequences an adjacent transposition of elements i and j is applied to the local sequence α_k of line k and similarly to local sequences α_i and α_j when the flip-triangle is bounded by lines i, j and k.

Let $(\mathcal{A}, F)$ be a simple marked arrangement of n pseudolines. Consider the arrangement induced by a triple of $\{i,j,k\}$ of lines of $\mathcal{A}$, where we assume $i < j < k$. These three lines can induce two combinatorial different arrangements. Either the crossing of lines i and k is above line j, denote this by the symbol $-$, or the crossing is below line j, denoted by $+$. With this convention a function $f : \binom{[n]}{3} \to \{-,+\}$ is associated with every marked simple arrangement. This function is called the *triangle-sign function* of the arrangement.

For $i < j$ and all $k \neq i,j$ we have $f(\{i,j,k\}) = +$ iff on line k, the crossing with line j precedes the crossing with line i, i.e., on the local sequence α_k the pair (i,j) is an inversion. Permutations are uniquely determined by their sets of inversions (see e.g., [184]). Therefore, the local sequences of an arrangement are uniquely determined by the triangle sign function. From Theorem 6.6 we know that local sequences encode marked simple arrangements, i.e., arrangements with the same local sequences are isomorphic. It follows that triangle-sign functions $f : \binom{[n]}{3} \to \{-,+\}$ also encode marked simple arrangements of pseudolines.

Not every possible sign pattern $f : \binom{[n]}{3} \to \{-,+\}$ can correspond to an arrangement, there are simply too many such functions. Below we derive an obvious necessary condition on the sign patterns of arrangements. Later it will be shown that this condition is also sufficient.

Consider a quadruple of pseudolines h,i,j,k of $\mathcal{A}$. These lines induce a marked arrangement of four pseudolines. There is only one (unmarked) simple arrangement of four lines. This arrangement has eight unbounded faces, each of these can be chosen for the marking. The eight resulting marked simple arrangements of four lines and their triangle-sign functions are shown in Figure 6.15. The sign functions are given as a vector showing the signs of 3-sets in lexicographic order, i.e, as $(\text{sign}(1,2,3), \text{sign}(1,2,4), \text{sign}(1,3,4), \text{sign}(2,3,4))$.

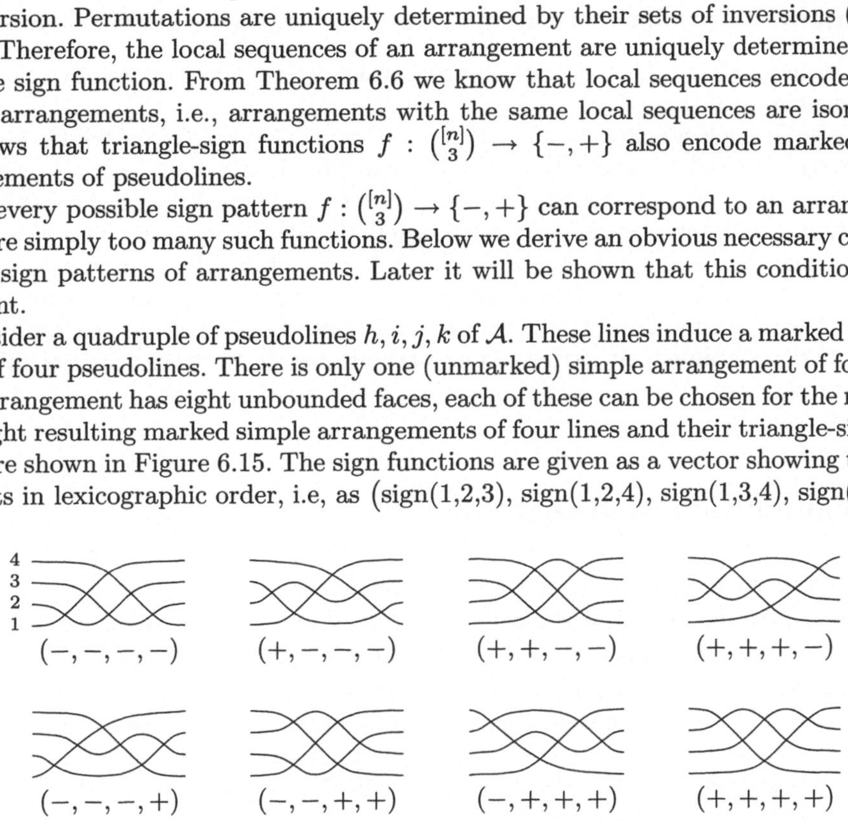

Figure 6.15 The marked simple arrangements of four lines and their triangle-sign functions.

For $A \in \binom{[n]}{4}$ and $1 \leq i \leq 4$ let $A^{\lfloor i \rfloor}$ denote the set A minus the ith element of A in sorted order, e.g., $\{2,4,5,9\}^{\lfloor 3 \rfloor} = \{2,4,9\}$. Restricting an arrangement $\mathcal{A}$ to a subset A of four lines we obtain a restricted sign-pattern $(\text{sign} A^{\lfloor 4 \rfloor}, \text{sign} A^{\lfloor 3 \rfloor}, \text{sign} A^{\lfloor 2 \rfloor}, \text{sign} A^{\lfloor 1 \rfloor})$. This pattern has to be one of the eight triangle-sign functions from Figure 6.15. Order the set $\{-,+\}$ of signs by $- \prec +$. Inspecting the above list we see that the legal sign patterns are characterized by the following property: For every 4 element subset A of $[n]$ either $f(A^{\lfloor 1 \rfloor}) \preceq f(A^{\lfloor 2 \rfloor}) \preceq f(A^{\lfloor 3 \rfloor}) \preceq f(A^{\lfloor 4 \rfloor})$ or $f(A^{\lfloor 1 \rfloor}) \succeq f(A^{\lfloor 2 \rfloor}) \succeq f(A^{\lfloor 3 \rfloor}) \succeq f(A^{\lfloor 4 \rfloor})$. This property is called *monotonicity*.

Theorem 6.16 *A function $f : \binom{[n]}{3} \to \{-,+\}$ is the triangle-sign function of a marked simple arrangement $\mathcal{A}$ of n pseudolines if and only if f is monotone.*

This theorem is a special instance of a more general Theorem 6.21 about signotopes. Translated back, the proof of Theorem 6.21 shows how to construct a simple allowable sequence from a monotone function $f : \binom{[n]}{3} \to \{-,+\}$. Since allowable sequences encode arrangements this gives a mapping $f \to \mathcal{A}$ from monotone functions $f : \binom{[n]}{3} \to \{-,+\}$ to arrangements. This mapping is the inverse of the mapping from $\mathcal{A}$ to the triangle-sign function of $\mathcal{A}$.

Theorem 6.17 *A set $(\alpha_i)_{i=1..n}$ with α_i a permutation of $[n] \setminus \{i\}$ is the set of local sequences of a simple marked arrangement of order n if and only if for all $i < j < k$ the pairs (i,j), (i,k), (j,k) are inversions in α_k, α_j, α_i or they are all three non-inversions.*

Proof. The necessity of the condition on local sequences can be checked by considering sub-arrangements of three lines.

For the sufficiency we proceed in two steps. First, the property on triples is exactly the property required to associate a triangle-sign function f with $(\alpha_i)_{i=1..n}$ such that $f(i,j,k) = +$ iff (i,j) is an inversion of α_k. In the second step it has to be verified that the function f is monotone on 4-element subsets of $[n]$. If not, then there is an h and $i < j < k$ such that either $f(h,i,j) = +$, $f(h,i,k) = -$, $f(h,j,k) = +$ or $f(h,i,j) = -$, $f(h,i,k) = +$, $f(h,j,k) = -$. Hence, either α_h has inversions i,j and j,k but i,k is a non-inversion or α_h has inversion j,k but non-inversions i,j and j,k. In both cases a contradiction is obtained, since there is no permutation α_h with appropriate inversions and non-inversions. □

6.6 Signotopes and their Orders

In this section we generalize the concept of triangle-sign functions. Let $[n] = \{1,\dots,n\}$ be equipped with the natural linear order. The set of all r element subsets of $[n]$ is $\binom{[n]}{r}$. For $A \in \binom{[n]}{r}$ with $r \geq i$ we let $A^{\lfloor i \rfloor}$ denote the set A minus the ith element of A in sorted order. The set $\{-,+\}$ of signs is ordered by $- \prec +$.

For integers $1 \leq r \leq n$ an *r-signotope* on $[n]$ is a function σ from the r element subsets of $[n]$ to $\{-,+\}$ such that for every $r+1$ element subset P of $[n]$ and all $1 \leq i < j < k \leq r+1$ either $\sigma(P^{\lfloor i \rfloor}) \preceq \sigma(P^{\lfloor j \rfloor}) \preceq \sigma(P^{\lfloor k \rfloor})$ or $\sigma(P^{\lfloor i \rfloor}) \succeq \sigma(P^{\lfloor j \rfloor}) \succeq \sigma(P^{\lfloor k \rfloor})$. We refer to this property as *monotonicity*.

Let $S_r(n)$ denote the set of all r-signotopes on $[n]$ equipped with the order relation $\sigma \leq \tau$ if $\sigma(A) \preceq \tau(A)$ for all $A \in \binom{[n]}{r}$. Call $S_r(n)$ the *r-signotope order.*

Note that for $r = 3$ the definitions reflect our observations for the encodings of marked simple arrangements of pseudolines made in the previous section. In particular Theorem 6.16 implies that $S_3(n)$ is a partial order on the set of marked arrangements of n pseudolines. Indeed $S_3(n)$ is an orientation of the graph $\mathcal{G}_n$, see Figure 6.14.
The list below collects some other special cases and easy observations.

(1) For $r = 1$ monotonicity is vacuous and $S_1(n)$ is just the lattice of subsets of $[n]$, i.e., the Boolean lattice.

(2) For all $n \geq r \geq 1$ there is a unique minimal and a unique maximal element in $S_r(n)$, namely the constant $-$ and the constant $+$ function.

(3) The diagram of $S_r(r+1)$ is a $(2r+2)$-gon for all $r \geq 1$.

(4) There is a natural correspondence between 2-signotopes on $[n]$ and permutations of n. Permutation π and 2-signotope σ correspond to each other if a pair (i, j) is an inversion of π iff $\sigma(i, j) = +$. For the proof that this is a bijection, note that monotonicity of σ corresponds to transitivity of the inversion relation and transitivity of the non-inversion relation for π (see e.g., [95] or [78]). In the *weak Bruhat order* of the symmetric group, the permutations are ordered by inclusion of their inversion sets. With the indicated correspondence between 2-signotopes and permutations, $S_2(n)$ equals the weak Bruhat order of S_n.

The following two constructions of a new signotope from a given one are useful.

Let σ be an r-signotope on the set $X \subset \mathbb{N}$ with $|X| \geq r+1$. For $x \in X$ the *deletion* $\sigma\!\uparrow_x$ is the induced function on $\binom{X\setminus x}{r}$, i.e., $\sigma\!\uparrow_x (A) = \sigma(A)$. This is a r-signotope on $X \setminus x$. The *contraction* of $x \in X$ in σ is the function $\sigma\!\downarrow_x$ on $\binom{X\setminus x}{r-1}$ with $\sigma\!\downarrow_x (A) = \sigma(A \cup x)$. This is a $(r-1)$-signotope on $X \setminus x$.

Maximum Chains of Signotopes

With an r-signotope σ on $[n]$ associate a directed graph. The vertices are the $r-1$ element subsets of $[n]$. Given two $(r-1)$ subsets A and A' of $[n]$ let $P = A \cup A'$. If $|P| > r$ then there is no edge between A and A'. Otherwise $|A \cap A'| = r-2$ and there are i, j such that $A = P^{\lfloor i \rfloor}$ and $A' = P^{\lfloor j \rfloor}$, we assume that $i < j$. If $\sigma(P) = +$, orient the edge as $P^{\lfloor i \rfloor} \xrightarrow{\sigma} P^{\lfloor j \rfloor}$ and otherwise, if $\sigma(P) = -$, orient it as $P^{\lfloor j \rfloor} \xrightarrow{\sigma} P^{\lfloor i \rfloor}$.

Lemma 6.18 *Let $r \geq 2$ and σ be an r-signotope on $[n]$. The graph with vertices $\binom{[n]}{r-1}$ and edges $\xrightarrow{\sigma}$ is acyclic.*

Proof. For $r = 2$ and arbitrary n, relation $\xrightarrow{\sigma}$ is the transitive tournament corresponding to the permutation whose inversion set is the set of pairs (i, j) with $\sigma(i, j) = +$.

For $n = r$: If $\sigma([r]) = -$, then relation $\xrightarrow{\sigma}$ is the lexicographic order on the $r-1$ subsets of $[r]$. Otherwise, $\sigma([r]) = +$ and $\xrightarrow{\sigma}$ it is the reverse-lexicographic order.

Let $n > r > 2$ and let $\tau = \sigma\!\uparrow_n$ be the signotope obtained from σ by deletion of n. By induction $\xrightarrow{\tau}$ is acyclic on $\binom{[n-1]}{r-1}$. Let $\gamma = \sigma\!\downarrow_n$ be the signotope obtained from σ by contraction of n. The vertices of the graph $\xrightarrow{\gamma}$ are the elements of $\binom{[n-1]}{r-2}$. By induction $\xrightarrow{\gamma}$ is acyclic. In the following we use a copy of $\xrightarrow{\gamma}$ on the vertex set $Y = \{A \in \binom{[n]}{r-1} : n \in A\}$. For emphasis we repeat: n is an element of every $A \in Y$

Let $X^- = \{A \in \binom{[n-1]}{r-1} : \sigma(A \cup \{n\}) = -\}$ and $X^+ = \{A \in \binom{[n-1]}{r-1} : \sigma(A \cup \{n\}) = +\}$. The three sets X^-, X^+, Y partition the $r-1$ element subsets of $[n]$. The subgraph of $\xrightarrow{\sigma}$ induced by each of the three blocks of the partition is acyclic: It agrees with the subgraph induced by $\xrightarrow{\tau}$ in case of X^- and X^+ and with the subgraph induced by $\xrightarrow{\gamma}$ in the case of Y. Now consider the edges of $\xrightarrow{\sigma}$ between the blocks. By definition of X^- all edges with one end in X^- and the other end in Y are oriented from X^- to Y. Also all edges with one end in X^+ and the other end in Y are oriented from Y to X^+. The following claim implies that there are no edges from X^+ to X^- in $\xrightarrow{\tau}$. Since on $\binom{[n-1]}{r-1} = X^+ \cup X^-$ the graphs $\xrightarrow{\tau}$ and $\xrightarrow{\sigma}$ agree the proof of the claim completes the proof of the lemma.

Claim. $A \in X^-$ and $B \xrightarrow{\tau} A$ implies $B \in X^-$, i.e., X^- is an ideal in the partial order defined by the transitive closure of $\xrightarrow{\tau}$.

From $B \xrightarrow{\tau} A$ it follows that $P = A \cup B$ is an r-subset of $[n]$. Let i, j be such that $B = P^{\lfloor i \rfloor}$ and $A = P^{\lfloor j \rfloor}$. For $Q = P \cup \{n\}$ we then obtain $Q^{\lfloor i \rfloor} = B \cup \{n\}$, $Q^{\lfloor j \rfloor} = A \cup \{n\}$ and $Q^{\lfloor r+1 \rfloor} = A \cup B = P$. We use the monotonicity of σ on Q and distinguish two cases: (1) If $i < j$, then $B \xrightarrow{\tau} A$ implies $\tau(P) = \sigma(Q^{\lfloor r+1 \rfloor}) = +$. From $A \in X^-$ it follows that $\sigma(Q^{\lfloor j \rfloor}) = \sigma(A \cup \{n\}) = -$. Monotonicity forces $\sigma(Q^{\lfloor i \rfloor}) = \sigma(B \cup \{n\}) = -$, i.e., $B \in X^-$. (2) If $j < i$, then $B \xrightarrow{\tau} A$ implies $\tau(P) = \sigma(Q^{\lfloor r+1 \rfloor}) = -$. From $A \in X^-$ it follows that $\sigma(Q^{\lfloor j \rfloor}) = \sigma(A \cup \{n\}) = -$. Monotonicity forces $\sigma(Q^{\lfloor i \rfloor}) = \sigma(B \cup \{n\}) = -$, i.e., $B \in X^-$.

□

Let σ be an r-signotope on $[n]$ and $A_1, A_2, \ldots, A_{\binom{n}{r-1}}$ be a topological sorting of $\xrightarrow{\sigma}$. For $0 \leq t \leq \binom{n}{r-1}$ define $\tau_t : \binom{n}{r-1} \to \{+, -\}$ by $\tau_t(A) = -$ if $A = A_i$ for some $i > t$ and $\tau_t(A) = +$ if $A = A_i$ for some $i \leq t$.

Proposition 6.19 *For an r-signotope σ on $[n]$ the above construction yields a chain $\tau_0 < \tau_1 < \ldots < \tau_{\binom{n}{r-1}}$ of $(r-1)$-signotopes in $S_{r-1}(n)$ such that for $t = 1, \ldots, \binom{n}{r-1}$ the signs of τ_{t-1} and τ_t differ at a single $(r-1)$-set.*

Proof. By definition the signs of τ_{t-1} and τ_t only differ at the $(r-1)$-set A_t where the sign of τ_{t-1} is $-$ and the sign of τ_t is $+$. Given that the τ_t are signotopes they form a chain as claimed.

It remains to show that each τ_t is an $(r-1)$-signotope. For every r element set P and all i, j, k with $1 \leq i < j < k \leq r$ we either have $P^{\lfloor i \rfloor} \xrightarrow{\sigma} P^{\lfloor j \rfloor} \xrightarrow{\sigma} P^{\lfloor k \rfloor}$ or $P^{\lfloor k \rfloor} \xrightarrow{\sigma} P^{\lfloor j \rfloor} \xrightarrow{\sigma} P^{\lfloor i \rfloor}$. In the first case we have $\tau_t(P^{\lfloor i \rfloor}) \succeq \tau_t(P^{\lfloor j \rfloor}) \succeq \tau_t(P^{\lfloor k \rfloor})$ for all t and in the second case $\tau_t(P^{\lfloor i \rfloor}) \preceq \tau_t(P^{\lfloor j \rfloor}) \preceq \tau_t(P^{\lfloor k \rfloor})$ for all t. This proves monotonicity for τ_t. □

With this preparation we are ready for the proof of Theorem 6.16.

Proof. [Theorem 6.16] Let σ be a 3-signotope, i.e., a function $\sigma : \binom{[n]}{3} \to \{-, +\}$ obeying monotonicity on 4-subsets of $[n]$. By Proposition 6.19 there is a chain $\tau_0, \ldots, \tau_{\binom{n}{2}}$ in $S_2(n)$ corresponding to σ. Each τ_t, $t = 0, \ldots, \binom{n}{2}$, encodes a permutation of $[n]$. τ_0 is the identity permutation and $\tau_{\binom{n}{2}}$ is the reverse. Moreover, two permutations τ_t and τ_{t+1} differ in a single sign where τ_t is $-$ and τ_{t+1} is $+$. This means that there is a single pair (i, j) which is a non-inversion of τ_t but an inversion of τ_{t+1}. This pair has to be an adjacent pair of both permutations. We conclude that $\tau_0, \ldots, \tau_{\binom{n}{2}}$ is a simple allowable sequence. By Theorem 6.3 this allowable sequence encodes an arrangement $\mathcal{A}$. From the construction it is easily verified that the triangle induced by lines i, j, k in $\mathcal{A}$ is a $+$ triangle iff $\sigma(ijk) = +$. This shows that the $\mathcal{A}$ only depends on σ and not on the choice of the sequence $\tau_0, \ldots, \tau_{\binom{n}{2}}$ for σ. This proves a bijection between 3-signotopes and arrangements. □

Simple allowable sequences encode simple marked arrangements (Theorem 6.3). This is generalized with the next proposition it shows that saturated chains of $(r-1)$-signotopes encode r-signotopes.

Proposition 6.20 *Let $1 < r \leq n$ and $\tau_0 < \tau_1 < \ldots < \tau_{\binom{n}{r-1}}$ be a maximum chain in $S_{r-1}(n)$. For $t = 1, \ldots, \binom{n}{r-1}$ let A_t be the unique $(r-1)$-set with $\tau_{t-1}(A_t) = -$ and $\tau_t(A_t) = +$. There exists a r-signotope σ on $[n]$ so that $A_1, \ldots, A_{\binom{n}{r-1}}$ is a topological sorting of $\xrightarrow{\sigma}$.*

Proof. For a set $A \in \binom{[n]}{r-1}$ let $\rho(A)$ be the index of A in the list $A_1, \ldots, A_{\binom{n}{r-1}}$. Note that the monotonicity of all τ_t's implies that for all $D \in \binom{[n]}{r}$ either $\rho(D^{\lfloor 1 \rfloor}) < \rho(D^{\lfloor 2 \rfloor}) < \ldots < \rho(D^{\lfloor r \rfloor})$ or $\rho(D^{\lfloor 1 \rfloor}) > \rho(D^{\lfloor 2 \rfloor}) > \ldots > \rho(D^{\lfloor r \rfloor})$. In the first case let $\sigma(D) = +$ in the second case $\sigma(D) = -$. We have to show that σ is a r-signotope, i.e., that σ is monotone at $r+1$ sets. Let $Q \in \binom{[n]}{r+1}$ and consider indices $1 \le i < j < k \le r+1$. Suppose $\sigma(Q^{\lfloor i \rfloor}) = \sigma(Q^{\lfloor k \rfloor}) = +$. Let $Q^{\lfloor i,j \rfloor}$ denote the set Q minus the ith largest and the jth largest element of Q, e.g., $\{1,2,5,7,8\}^{\lfloor 2,3 \rfloor} = \{1,7,8\}$. From $\sigma(Q^{\lfloor i \rfloor}) = +$ we obtain $\rho(Q^{\lfloor i,j \rfloor}) < \rho(Q^{\lfloor i,k \rfloor})$. From $\sigma(Q^{\lfloor k \rfloor}) = +$ we obtain that $\rho(Q^{\lfloor i,k \rfloor}) < \rho(Q^{\lfloor j,k \rfloor})$. Hence $\rho(Q^{\lfloor i,j \rfloor}) < \rho(Q^{\lfloor j,k \rfloor})$ which implies $\sigma(Q^{\lfloor j \rfloor}) = +$ as required. The argument for $\sigma(Q^{\lfloor i \rfloor}) = \sigma(Q^{\lfloor k \rfloor}) = -$ is symmetric. Therefore, σ is an r-signotope. From the definition of σ based on $A_1, \ldots, A_{\binom{[n]}{r-1}}$ it follows that this sequence is a topological sorting of the relation $\xrightarrow{\sigma}$. □

Propositions 6.19 and 6.20 together imply the structure theorem for signotopes:

Theorem 6.21
There is a surjective mapping from maximum chains in $S_{r-1}(n)$ to $S_r(n)$.

It seems appropriate to point to two special cases of this theorem:

(1) The case $r = 2$: There is a surjective mapping from maximum chains in the Boolean lattice of subsets of $[n]$ to permutations of $[n]$. Actually, this mapping is a bijection.

(2) Maximum chains in $S_2(n)$ are the same as simple allowable sequences. The case $r = 3$ of the theorem is a reformulation of the statement: Every simple allowable sequence of permutations of $[n]$ encodes a simple marked arrangement of n pseudolines.

6.7 Notes and References

With this chapter I have tried to illustrate the beauty and usefulness of combinatorial encodings of geometric data. For the sake of a compactness and to avoid confusion the presentation was essentially restricted to arrangements of pseudolines in the Euclidean plane. Encodings of projective arrangements and configurations of points are closely related. The following is a sample of more or less general sources on the topic: The book of Björner et al. [28] covers oriented matroids in depth. A more concise source on oriented matroids is the handbook article of Richter-Gebert and Ziegler [163]. In his monograph [123] Knuth takes an axiomatic and self contained approach to combinatorial point configurations. Grünbaum's booklet [108], though not emphasizing encodings, has to be mentioned. Closest to the content of this chapter is Goodman's handbook article [101] about arrangements of pseudolines. Large parts of this chapter have been adapted from the author's articles [80, 88].

For the sweeping lemma (Lemma 6.1) there are at least two sources. Snoeyink and Hershberger [180] have a theorem implying the Sweeping Lemma for simple arrangements. In the book on oriented matroids [28] a result equivalent to the Sweeping Lemma is derived as a consequence of Levi's extension lemma. Here we have reverted the direction. Conbsequently Levi's extension lemma and the Sweeping Lemma are essentially equivalent. Levi [132] originally stated his lemma for projective arrangements, an English transcription can be found in Grünbaum [108]. In higher dimensional arrangements of pseudoplanes sweeps and extensions are much more involved: It is known that sweeps can get stuck (Richter-Gebert [161]) and extensions need not exist (Goodman and Pollack [102]).

Sequences of permutations as encodings for point configurations were first used by Perrin in 1882. The modern definition of allowable sequences goes back to Goodman and Pollack [103], this paper also contains the connection with local sequences. The wiring diagram is defined in Goodman [100]. An overview on applications of allowable sequences is given by Goodman and Pollack [104] and more recently in Goodman's handbook article [101].

The slope theorem (Theorem 6.4) is due to Ungar [205] 1982. With this application of allowable sequences he confirmed a conjecture of Scott 1970. Scott had shown a lower bound of $\sqrt{2n}$ for the number of slopes determined by n points. A significant improvement was made by Burton and Purdy [42] in 1979, they investigated properties of the zonotopal tiling associated with the arrangement generated by n points and obtain a bound of $\frac{n-1}{2}$ slopes. A survey of the slope problem with emphasis to slope critical configurations was given by Jamison [117].

Simple allowable sequences also turn up in the theory of Coxeter groups. The symmetric group S_n of permutations of $[n]$ is generated by the adjacent transpositions $\sigma_i = (i, i+1)$, $i = 1, \ldots, n-1$. These generators satisfy the Coxeter relations:

$$\begin{aligned} \sigma_i^2 &= id & i &= 1, \ldots, n-1 & &\text{(COX 0)} \\ \sigma_i \sigma_j &= \sigma_j \sigma_i & |i-j| &\geq 2 & &\text{(COX 1)} \\ \sigma_i \sigma_{i+1} \sigma_i &= \sigma_{i+1} \sigma_i \sigma_{i+1} & i &= 1, \ldots, n-2 & &\text{(COX 2)} \end{aligned}$$

For $\pi \in S_n$ the least k such that π can be written as $\pi = \sigma_{i_1} \sigma_{i_2} \cdots \sigma_{i_k}$ is called the length of π and the word $\sigma_{i_1}, \sigma_{i_2}, \ldots, \sigma_{i_k}$ is called a reduced decomposition for π. Two reduced words (decompositions) represent the same permutation iff they are related by a sequence of moves of type COX 1 and COX 2. Note that reduced words of the reverse of the identity and simple allowable sequences are in one-to-one correspondence. Alternatively, these objects can also be seen as maximal chains in the weak Bruhat order of the symmetric group. In this last context their number A_n has been determined by Stanley [183]. His remarkable formula is

$$A_n = \frac{\binom{n}{2}!}{(2n-3) \cdot (2n-5)^2 \cdot (2n-7)^3 \cdot \ldots \cdot 5^{n-3} \cdot 3^{n-2}}.$$

This number A_n is the number of standard Young tableaux of staircase shape. Edelman and Greene [64] prove a combinatorial bijection between reduced decompositions and such tableaux. An alternative proof of that bijection was presented in [82], the argument there involves operations on wiring diagrams. This bijection was used by Edelman [63] to compute the average complexity of the kth level of an allowable sequence.

Exact numbers for different kinds of arrangements and a small number of lines are given by Knuth [123], Björner et al. [28] and Ziegler [218]. The most recent result is the enumeration of order types for $n = 11$ by Aichholzer et al.

Knuth [123] proves lower and upper bounds for the number of arrangements:

$$2^{\frac{n^2}{6} - \frac{5n}{2}} \leq B_n \leq 3^{\binom{n+1}{2}}.$$

This implies $\log_2(B_n) \leq 0.7924\,(n^2+n)$. Knuth also reports on some computations supporting a conjecture of $\log_2(B_n) \leq \binom{n}{2}$ and explains how to derive the slightly weaker bound $\log_2(B_n) \leq 0.7194\,n^2$ from the zone theorem. The bound in Theorem 6.9 is from Felsner [80]. The number of realizable arrangements, i.e., arrangements of lines, is much smaller, Goodman and Pollack [104] show an upper bound in $O(2^{n\log n})$.

Theorem 6.10 is equivalent to the rank 3 version of the Bohne-Dress Theorem which gives a bijection between zonotopal tilings of d-dimensional zonotopes and oriented matroids of rank $d+1$ with a realizable one-element contraction. The correspondence between oriented matroids and arrangements is given by the representation theorem for oriented matroids. This theorem states that oriented matroids of rank $d+1$ are in bijection with arrangements of pseudohyperplanes in d-dimensiononal projective space. An accessible treatment of these connections is given by Ziegler [219]. A more geometric proof of the Bohne-Dress Theorem was given by Richter-Gebert and Ziegler [164]. An elementary proof of the rank 3 version of the Bohne-Dress Theorem is contained in Felsner and Weil [88]. Another proof of that theorem is given by Elnitsky [71] in the context of reduced decompositions.

The bound for the complexity of the k-level of an arrangement given in Theorem 6.14 reproves a bound we have already seen in Chapter 4, e.g. Proposition 4.4. Somehow I feel like defending the decision of including this weak result. Its just, I like the proof. The bound for the complexity of a set K of levels is due to Welzl [211]. In Chapter 4 we have already remarked that this bound also has been improved. An exact bound for $<k$-sets was first obtained by Alon and Györi [11].

Triangle sign encodings of arrangements are related to chirotopes of uniform rank 3 oriented matroids (see [28]). A result similar to Theorem 6.17 was obtained by Streinu [189] in the context of generalized configurations of points.

Manin and Schechtman [136] introduced the *higher Bruhat order* $B(n, r-1)$ which is an order relation on the set of r-signotopes on $[n]$ (the name *signotope*, however, was introduced much later by Felsner and Weil [88]). The higher Bruhat order relation $\leq_{HB}$ is defined as follows: Let σ and τ be two r-signotopes with $\sigma(A) = \tau(A)$ for all r-subsets A with one exception A^* where $\sigma(A^*) = -$ and $\tau(A^*) = +$ in this case we call the pair (σ, τ) a *single-step*. The order relation $\leq_{HB}$ is the transitive closure of the single-step relation, i.e, $\sigma \leq_{HB} \tau$ iff there is a sequence $\sigma = \sigma_0, \sigma_1, \ldots \sigma_t = \tau$ such that for $i = 1, \ldots, t$ the pair (σ_{i-1}, σ_i) is a single-step. Higher Bruhat orders were further studied by Ziegler [218]. In particular, Ziegler showed that the higher Bruhat order $B(n, r-1)$ and the signotope order $S_r(n)$ are not equal in general. His example is $B(8,3) \neq S_4(8)$. For $r \leq 2$, obviously, $B(n, r-1) = S_r(n)$. Ziegler also shows that $B(n, n-k-1) = S_{n-k}(n)$ for $k \leq 3$. Felsner and Weil [87] proved $B(n,2) = S_3(n)$. Theorem 6.21 is the main result on higher Bruhat orders, it can be found in [136, 218, 88].

The combinatorial structure of signotopes and signotope orders has been studied by Ziegler [218] and Felsner and Weil [88]. In these papers it is also shown that r-signotopes encode geometric structures also for r larger than 3. Ziegler [218] gives a geometric interpretation of signotopes as single element extensions of cyclic hyperplane arrangements. In terms of the theory of oriented matroids the pseudohyperplane arrangements associated with signotopes are the adjoints of the duals of Ziegler's single element extensions, see [89]. In the interpretation of Felsner and Weil [88] signotopes correspond to arrangements of pseudohyperplanes in $\mathbb{R}^{r-1}$. However, for $r > 3$ the situation is not as nice as for $r = 3$. In higher dimensions only a restricted class of arrangements of pseudohyperplanes is actually encoded by signotopes. Theorem 6.21 implies these arrangements have the nice property of being sweepable.

7 Triangulations and Flips

Let $\mathcal{P}$ be a set of points in the plane and assume that $\mathcal{P}$ is in general position. A *triangulation* of $\mathcal{P}$ is a maximal non-crossing geometric graph with vertex set $\mathcal{P}$. All bounded faces of a triangulation are triangles (that's why we call them triangulation), the unbounded face is the outside of the convex hull of $\mathcal{P}$. Triangulations play a prominent role in many applicable and applied disciplines like computational geometry, computer graphics and numerical modeling. In this chapter we discuss combinatorial and geometrical properties of triangulations and of the flip-graph on the set of all triangulations of a point set $\mathcal{P}$.

Let pq be an edge of a triangulation T of $\mathcal{P}$ and assume that pq belongs to two triangles pqr and pqs whose union is a convex quadrangle. The *diagonal flip* of edge pq consists in removing edge pq and replacing it by the other diagonal rs of the quadrangle.

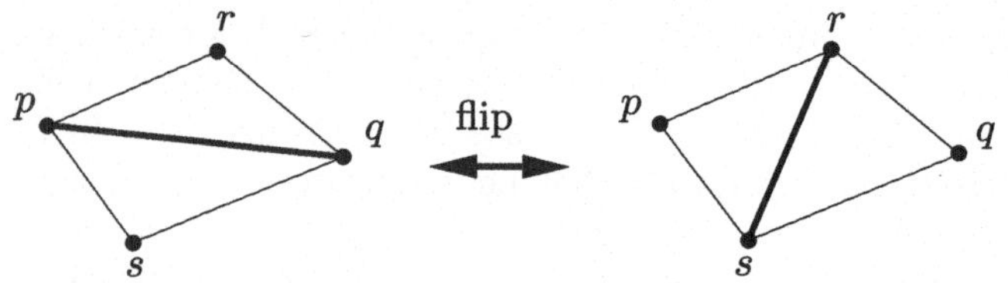

Figure 7.1 A diagonal flip replacing edge pq with rs.

The *flip-graph* $\mathcal{G}(\mathcal{P})$ is the graph whose vertices are the triangulations of $\mathcal{P}$ and two triangulations are adjacent if there is a diagonal flip transforming one into the other. In Section 7.1 we ask about minimal and maximal degree in the flip graph. In Section 7.2 the Delaunay triangulation is introduced. Since every triangulation of $\mathcal{P}$ can be transformed into the Delaunay triangulation of $\mathcal{P}$ by a sequence of Lawson flips (Proposition 7.2) the flip-graph is connected. The proof of Proposition 7.2 makes use of a lifting of $\mathcal{P}$ to a paraboloid in 3-space. In Section 7.3 we study regular triangulations of $\mathcal{P}$, i.e., triangulations that arise as projections of the convex hull of a lifting of $\mathcal{P}$. The secondary polytope of $\mathcal{P}$ is a high dimensional polytope whose vertices are in bijection with regular triangulations of $\mathcal{P}$. In the special case of a point set $\mathcal{P}$ in convex position the number of triangulations is given by a Catalan number and the secondary polytope is the associahedron, this case is subject of Section 7.4. In the last section we deal with the diameter of the flip-graph of a point set in convex position. The lower bound construction of Sleator, Tarjan and Thurston is obtained by an argument involving volumes of ideal polytopes in hyperbolic space.

7.1 Degrees in the Flip-Graph

Among the simplest questions that can be asked about a graph are the questions about maximal and minimal degree. The degree of a triangulation T in $\mathcal{G}(\mathcal{P})$ strongly depends on the set $\mathcal{P}$ of underlying points. Let us ask for extremal degrees just in terms of the number n of points.

Let T be the triangulation of Figure 7.2. Each line of a pentagram consists of a slightly bent chain of three edges. Only the five hull edges of this triangulation are non-flippable. It can be shown that triangulations of $n \geq 5$ points with a convex hull of only 3 or 4 edges have at least 6 non-flippable edges. Hence, examples of nested pentagons as the one in Figure 7.2 are extremal.

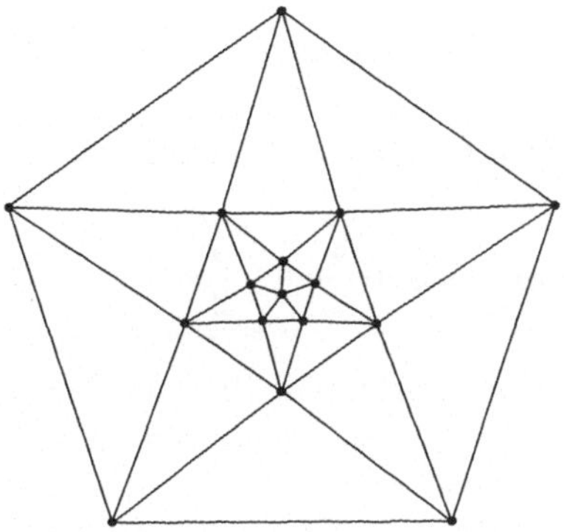

Figure 7.2 A triangulation with only 5 non-flippable edges.

What concerns the minimum degree in flip-graphs the precise answer is given with the following proposition.

Proposition 7.1 *Any triangulation T of a set $\mathcal{P}$ of n points in the plane contains at least $\frac{n}{2} - 2$ flippable edges. The bound is sharp.*

Proof. Let T be a triangulation of a set $\mathcal{P}$ of n points and assume that the convex hull of $\mathcal{P}$ contains γ points. From the counts of edges and faces of maximal planar graphs we deduce that T has $3n-3-\gamma$ edges and $2n-2-\gamma$ triangles, i.e., bounded triangular faces. Clearly, the γ edges of the convex hull are non-flippable. Every non-flippable interior edge e has one endpoint, say u, such that the sum of the two angles adjacent to e at u exceeds π. Define an orientation on the non-flippable interior edges such that e is oriented away from u, see Figure 7.3. We emphasize an important property of this orientation:

- Any two outward oriented edges at a common point u share an angle at u.

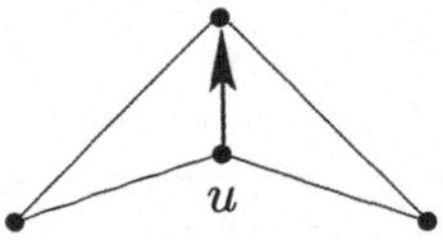

Figure 7.3 The orientation of a non-flippable interior edge.

It follows that a point u can have at most three outward oriented edges and this maximum is only attained by points of degree three in T. Classify the interior points of T by the number of outward oriented edges: Let η_i be the number of interior points with i outward oriented edges, $i = 3, 2, 1, 0$. Clearly

$$\eta_3 + \eta_2 + \eta_1 \leq n - \gamma.$$

A corner u of a triangle Δ in T is the *root* of Δ if the two edges of Δ incident to u are both outward oriented at u. The definition implies that every triangle of T has at most one root.

If u has out-degree two then there is exactly one triangle rooted at u. If u has out-degree three then there are exactly three triangles rooted at u. In all other cases u is not a root. Counting roots of triangles on one side and all triangles of T on the other we obtain

$$3\eta_3 + \eta_2 \leq 2n - 2 - \gamma.$$

Adding the two inequalities with weights, 3/2 for the first and 1/2 for the second, yields:

$$3\eta_3 + 2\eta_2 + \frac{3}{2}\eta_1 \leq \frac{5}{2}n - 2\gamma - 1.$$

Every interior non-flippable edge is oriented. By counting oriented edges at base points their number is seen to be $3\eta_3 + 2\eta_2 + \eta_1$. The above inequality implies that there are at most $\frac{5}{2}n - 2\gamma - 1$ interior non-flippable edges. In total, together with the hull edges, there are at most $\frac{5}{2}n - \gamma - 1$ non-flippable edges. Therefore, the number of flippable edges of T is at least

$$\left(3n - 3 - \gamma\right) - \left(\frac{5}{2}n - \gamma - 1\right) = \frac{1}{2}n - 2.$$

It remains to show that the bound can be attained. Let T be a triangulation of a set of m points in convex position, e.g., of the corners of a regular m-gon. T has $m - 3$ interior edges and $m - 2$ triangles. Subdivide each triangle with a new point connected to the three corners, this gives a triangulation T^* of a set of $n = 2m - 2$ points. The flippable edges are the $m - 3 = \frac{n}{2} - 2$ interior edges of T. Figure 7.4 shows a member from a different family of extremal examples. □

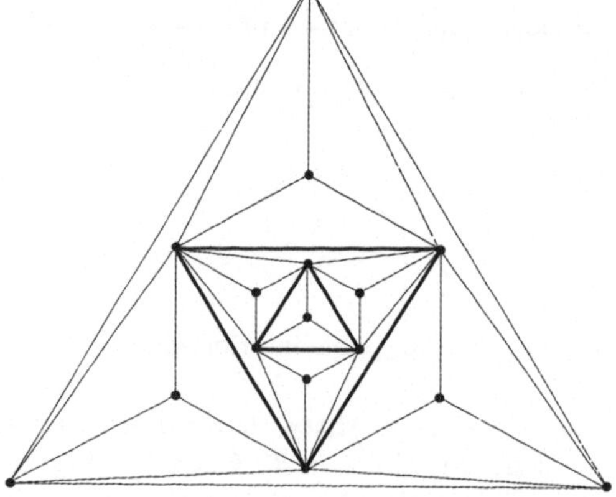

Figure 7.4 A triangulation with $\frac{n}{2} - 2$ flippable (thick) edges.

7.2 Delaunay Triangulations

Let $\mathcal{P}$ be a set of n points in $\mathbb{R}^2$. The *Voronoi region* $V(p)$ of a point $p \in \mathcal{P}$ is the set of all points x that are at least as close to p as to any other point in $\mathcal{P}$; formally

$$V(p) = \{x \in \mathbb{R}^2 : ||x - p|| \leq ||x - q|| \text{ for all } q \in \mathcal{P}\}.$$

Let $V_q(p)$ be the set of points that are at least as close to p as to q. This set $V_q(p)$ is a halfplane defined by the bisecting line of p and q. The Voronoi region of p with respect to $\mathcal{P}$ is the intersection of the halfplanes $V_q(p)$ for $q \in \mathcal{P} \setminus \{p\}$. Therefore, $V(p)$ is a convex polygonal region, possibly unbounded. Every point $x \in \mathbb{R}^2$ has a closest point in $\mathcal{P}$, so it belongs to a Voronoi region. The Voronoi regions of two points lie in opposite halfplanes defined by the bisecting line of the points. It follows that Voronoi regions are internally disjoint. The Voronoi regions together with edges and vertices of their boundaries form the *Voronoi diagram* of $\mathcal{P}$. Figure 7.5 shows an example.

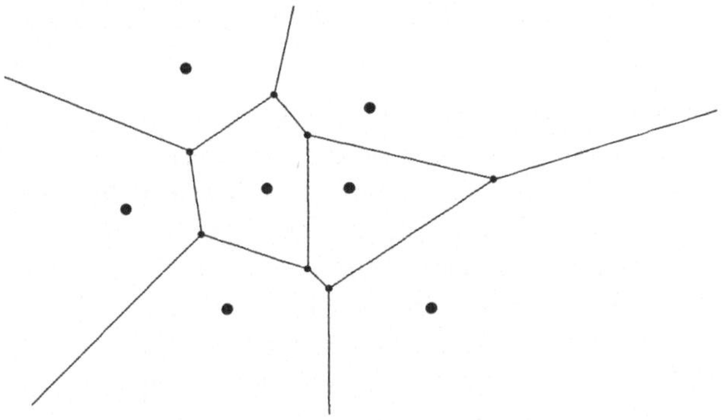

Figure 7.5 The Voronoi diagram of a set of seven points.

If the Voronoi regions of p and q share an edge we call the points $p, q \in \mathcal{P}$ *Delaunay neighbors*. Connecting all pairs of Delaunay neighbors by straight edges we obtain a graph G embedded with vertex set $\mathcal{P}$. This graph is the dual of the planar Voronoi diagram and hence also planar. Duality alone does not yet imply that the embedding of G is plane. However in the embedding of G the edges leaving a vertex p are in the same cyclic order as the dual edges are around $V(p)$. Since the Voronoi edges are non-crossing this indeed implies that G is plane. The faces of G correspond to the vertices of the Voronoi diagram which are of degree 3, usually. The graph G is the *Delaunay triangulation* of $\mathcal{P}$. Figure 7.6 shows an example.

A Voronoi vertex of degree more than 3 corresponds to a point $x \in \mathbb{R}^2$ with more than three nearest neighbors in $\mathcal{P}$. That is there is a circle C with center x such that C contains four or more points from $\mathcal{P}$. This kind of situation can be considered 'unlikely' and we assume that our point sets are free of such degeneracies. A point set is in *general position* if there is no degeneracy, including the degeneracy of 3 or more points on a line.

A triple $p, q, r \in \mathcal{P}$ is a Delaunay triangle if p, q, r are the vertices of a triangular face in the Delaunay triangulation. A circle C is an *empty circle* for $\mathcal{P}$ if there is no point of $\mathcal{P}$ in the interior of C. Dual to a Delaunay triangle p, q, r there is a Voronoi vertex $v = V(p) \cap V(q) \cap V(r)$. Consider the largest empty circle with center v, this circle has p, q and r on the boundary, hence, it is the circumcircle for the triangle p, q, r. Conversely,

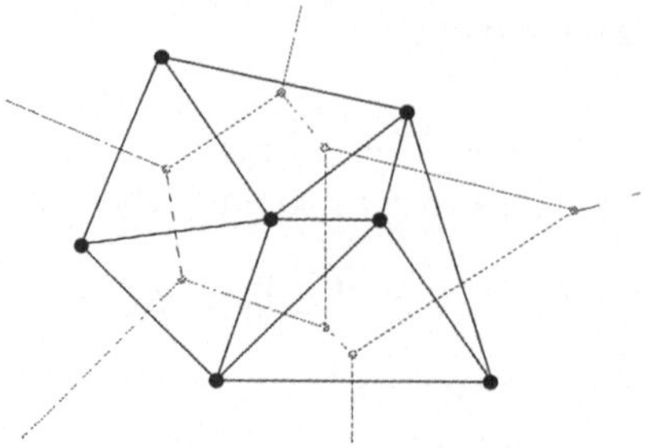

Figure 7.6 The Delaunay triangulation of the point set of Figure 7.5.

let p, q, r be three points of $\mathcal{P}$ with an empty circumcircle. The center v of this circle is in $V(p) \cap V(q) \cap V(r)$, moreover, as a member of three different Voronoi regions v is a Voronoi vertex.

Characterization of Delaunay triangles. Let $\mathcal{P} \subset \mathbb{R}^2$ be a set of n points in general position. A triple $p, q, r \in \mathcal{P}$ is a Delaunay triangle iff the circumcircle of p, q, r is empty.

A Delaunay edge p, q is dual to a Voronoi edge $V(p) \cap V(q)$. For any point x on that Voronoi edge there is an empty circle with center x and p, q on the boundary. Conversely, let p, q be points of $\mathcal{P}$ such that there is an empty circle C touching both. The center of C is on the bisecting line of p and q. Consider the set X of all points x on this bisector such that there is an empty circle with center x and p, q on the boundary. Not all points of $\mathcal{P}$ are on a line, therefore, X is a proper subset of the bisector and there is an extreme point $x_0 \in X$. The empty circle with center x_0 has a third point $r \in \mathcal{P}$ on the boundary. The circumcircle property implies that p, q, r is a Delaunay triangle, hence, p, q is a Delaunay edge.

Characterization of Delaunay edges. Let $\mathcal{P} \subset \mathbb{R}^2$ be a set of n points in general position. A pair $p, q \in \mathcal{P}$ is a Delaunay edge iff there is an empty circle with p and q on the boundary.

Let T be a triangulation of $\mathcal{P}$ containing triangles p, q, r and p, q, s such that s is in the interior of the circumcircle of p, q, r. In that case p, q, r, s form a convex quadrangle and the edge p, q is flippable. Note that p, q is not a Delaunay edge. We call p, q a *weak edge* of the triangulation T. The flip of a weak edge in a triangulation is a *Lawson flip*. Figure 7.7 shows an example and illustrates the fact that the edge replacing the weak edge is not weak.

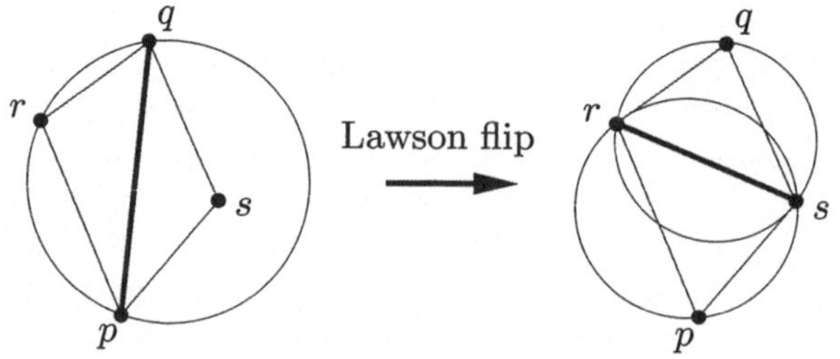

Figure 7.7 A weak edge (p, q) and the Lawson flip replacing (p, q) by (r, s) which is not weak.

Proposition 7.2 *Let T be an arbitrary triangulation of a set $\mathcal{P}$ of n points in general position. Every algorithm that starts with T and repeatedly performs Lawson flips will reach the Delaunay triangulation of $\mathcal{P}$ with at most $\binom{n}{2}$ flips.*

Before proving the proposition we emphasize two implications.

- The Delaunay triangulation is the unique triangulation of $\mathcal{P}$ which has no weak edge.
- The flip-graph $\mathcal{G}(\mathcal{P})$ is connected with diameter at most $n^2 - n$, i.e., given two triangulations T_1, T_2 of $\mathcal{P}$ it takes at most $n^2 - n$ flips to transform T_1 into T_2.

Proof of Proposition 7.2. This works by lifting triangulations into 3-space. A point $p \in \mathcal{P}$ is lifted to $\hat{p} = (p_1, p_2, p_1^2 + p_2^2)$, that is p is lifted to the point $\hat{p}$ on the paraboloid $z = x^2 + y^2$ vertically above p. A triangulation T of $\mathcal{P}$ is lifted to $\hat{T}$ by lifting each triangle p, q, r of T to the spatial triangle with corners $\hat{p}, \hat{q}, \hat{r}$. The crucial property of the lifting is stated in the following lemma whose proof will be given later.

Lemma 7.3 *A point s is in the interior of the circumcircle of p, q, r if and only if $\hat{s}$ is below the plane spanned by $\hat{p}, \hat{q}, \hat{r}$.*

Consider a Lawson flip $T \to T_f$ which replaces p, q by r, s. The lifted triangulations $\hat{T}$ and $\hat{T_f}$ enclose the tetrahedron $\hat{p}, \hat{q}, \hat{r}, \hat{s}$. It follows from the lemma that the lifted triangulation $\hat{T}$ contains the two upper triangles $\hat{p}, \hat{q}, \hat{r}$ and $\hat{p}, \hat{q}, \hat{s}$ of the tetrahedron and $\hat{T_f}$ contains the two lower triangles $\hat{p}, \hat{r}, \hat{s}$ and $\hat{q}, \hat{r}, \hat{s}$. In other words, surface $\hat{T_f}$ is below surface $\hat{T}$ and the edge $\hat{p}, \hat{q}$ of $\hat{T}$ is strictly above $\hat{T_f}$. A sequence of Lawson flips produces a lowering sequence of surfaces and an edge that has once been flipped away can never be reinserted. This implies that there are at most as many Lawson flips as there are edges on a set of n points, namely $\binom{n}{2}$.

It remains to prove that the Delaunay triangulation is the unique Lawson flip free triangulation. We use the characterization of Delaunay triangles. If T is not Delaunay, then T contains a triangle p, q, r which has a circumcircle C with a point $s \in \mathcal{P}$ interior to the cycle. We assume that p, q and r, s are the diagonals of the quadrangle. By Lemma 7.3 the lifted segment $\hat{r}, \hat{s}$ is below $\hat{p}, \hat{q}$. This reveals that $\hat{T}$ is not convex and, hence, contains some non-convex edge $\hat{a}, \hat{b}$. Attached to a, b there are triangles a, b, c and a, b, d. Since $\hat{a}, \hat{b}$ is non-convex the point $\hat{d}$ is below the plane spanned by $\hat{a}, \hat{b}, \hat{c}$. The lemma implies that a, b is a weak edge and allows a Lawson flip. □

Proof of Lemma 7.3. For $a = (a_1, a_2, a_3) \in \mathbb{R}^3$ let $a^+ = (a_1, a_2, a_3, 1)$ be the corresponding homogeneous point in $\mathbb{R}^4$. Four points a, b, c, d in $\mathbb{R}^3$ are coplanar iff the determinant $|a^+, b^+, c^+, d^+|$ vanishes. If $|a^+, b^+, c^+, d^+| < 0$ then looking from d the triangle (a, b, c) appears as a counterclockwise triangle. Let (p, q, r) be a counterclockwise triangle in $\mathbb{R}^2$ with this triangle associate the following mapping $\phi : \mathbb{R}^2 \to \mathbb{R}$

$$\phi(s) = \begin{vmatrix} p_1 & p_2 & p_1^2 + p_2^2 & 1 \\ q_1 & q_2 & q_1^2 + q_2^2 & 1 \\ r_1 & r_2 & r_1^2 + r_2^2 & 1 \\ s_1 & s_2 & s_1^2 + s_2^2 & 1 \end{vmatrix}.$$

This determinant vanishes if the lifted point $\hat{s}$ is in the plane determined by $\hat{p}, \hat{q}, \hat{r}$, otherwise, the sign tells whether $\hat{s}$ is above or below the plane. The lemma is equivalent to the statement $\phi(s) = 0$ iff s is on the circumcircle C of p, q, r and $\phi(s) > 0$ iff s is in the interior of C. Let $m = (m_1, m_2)$ be the center of C and let ϕ_m be the mapping corresponding to the triangle $(p-m, q-m, r-m)$. It follows from basic properties of determinants (linearity and the fact that a determinant with two columns which are multiples of each other vanishes) that $\phi(s) = \phi_m(s-m)$. However, if ρ is the radius of C then

$$\phi_m(s-m) = \begin{vmatrix} p_1 - m_1 & p_2 - m_2 & \rho^2 & 1 \\ q_1 - m_1 & q_2 - m_2 & \rho^2 & 1 \\ r_1 - m_1 & r_2 - m_2 & \rho^2 & 1 \\ s_1 - m_1 & s_2 - m_2 & (s_1 - m_1)^2 + (s_2 - m_2)^2 & 1 \end{vmatrix}.$$

If $s \in C$ then $(s_1 - m_1)^2 + (s_2 - m_2)^2 = \rho^2$ and the last two columns are linearly dependent, i.e., $\phi_m(s-m) = 0$. Since ϕ is a quadric, C is the complete set of zeros of ϕ and the sign of ϕ is the same for all interior points. The value of $\phi(m) = \phi_m(0)$ is ρ^2 times the determinant of the homogenized points p^+, q^+, r^+. Since p, q, r is a counterclockwise triangle, this determinant is positive, hence $\phi(m) = \phi_m(0) > 0$. △

The proof of Proposition 7.2 yields a beautiful characterization of the Delaunay triangulation:

Corollary 7.4 *The Delaunay triangulation of a set $\mathcal{P}$ of n points in general position is the vertical projection of the lower convex hull of the point set lifted to the paraboloid, i.e., of $\hat{\mathcal{P}} = \{\hat{p} = (p_1, p_2, p_1^2 + p_2^2) : p = (p_1, p_2) \in \mathcal{P}\}$.*

7.3 Regular Triangulations and Secondary Polytopes

In the previous section we have investigated the lifting of point sets to the paraboloid $z = x^2 + y^2$. We now consider more general liftings and use them for the construction of an interesting polytope associated to a set of points.

For a set $\mathcal{P} = \{p_1, \ldots, p_n\}$ of n points in the plane and any numbers $w_i \in \mathbb{R}$, $i = 1, \ldots, n$ consider the lifting that takes p_i to the point $\hat{p_i}$ vertically above p_i at height w_i. Let $\hat{\mathcal{P}_w}$ be the set of lifted points. Suppose that $\hat{\mathcal{P}_w}$ is in general position meaning that no four points are coplanar. The convex hull of $\hat{\mathcal{P}_w}$ is a simplicial polytope, i.e., all facets of this polytope are triangles. This polytope is invariant under the addition of a constant to all the weights w_i, therefore, we can assume that all the w_i are positive. The vertical projection of the lower convex hull, i.e., of the faces that are visible from the plane, is a triangulation with vertices in $\mathcal{P}$. The triangulations that can be obtained with this construction are the *regular triangulations* of $\mathcal{P}$.

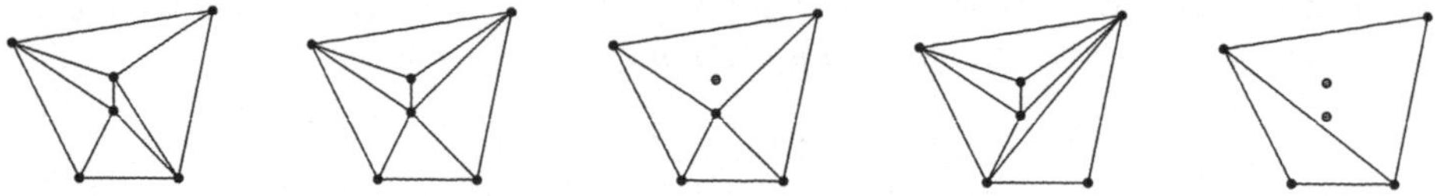

Figure 7.8 Some regular triangulations of a set $\mathcal{P}$ of six points.

- If $p \in \mathcal{P}$ is not a vertex of the convex hull of $\mathcal{P}$, then $\hat{p}$ may be beyond some triangle spanned by points of $\hat{\mathcal{P}_w}$. Consequently, the vertex set $\mathcal{V}$ of a regular triangulation of $\mathcal{P}$ is a subset of $\mathcal{P}$. In fact, every subset $\mathcal{V} \subseteq \mathcal{P}$ that includes all vertices of the convex hull of $\mathcal{P}$ is the vertex set of some regular triangulation.

- The triangulation shown in Figure 7.9 is not regular. Suppose there is a lifting whose lower projection yields the figure. By subtracting a linear function from the weights we can achieve that the inner triangle is in the plane, i.e., $w_4 = w_5 = w_6 = 0$. Clearly $w_1, w_2, w_3 > 0$, the edge needed in the quadrangle p_1, p_2, p_4, p_5 requires $w_2 > w_1$. The two other quadrangles force $w_3 > w_2$ and $w_1 > w_3$. Together these requirements are contradictory.

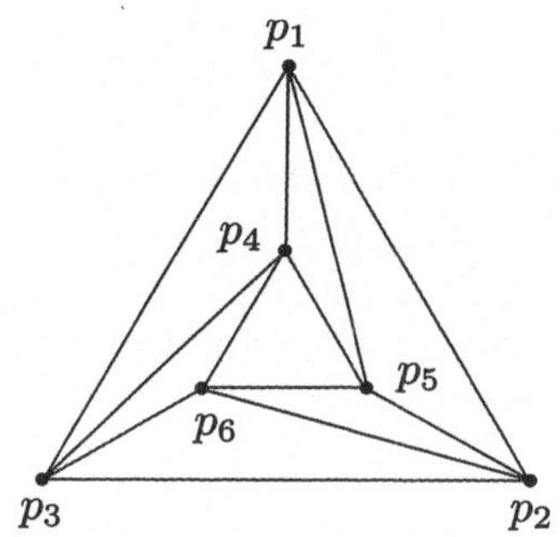

Figure 7.9 A non-regular triangulation.

Let T be a triangulation of $\mathcal{P}$, for a triangle $\Delta \in T$ denote the area with $\mathrm{vol}(\Delta)$. For a point $p \in \mathcal{P}$ let

$$\varphi(p) = \sum_{p \in \Delta \in T} \mathrm{vol}(\Delta).$$

be the sum of the areas of triangles having p as a vertex. The *volume vector* of T is the vector

$$\varphi(T) = \big(\varphi(p_1), \varphi(p_2), \ldots, \varphi(p_n)\big) \in \mathbb{R}^n.$$

The *secondary polytope* $\Sigma(\mathcal{P})$ of $\mathcal{P} = \{p_1, \ldots, p_n\}$ is the convex span of the volume vectors of all triangulations with a vertex set $\mathcal{V} \subseteq \mathcal{P}$ such that all vertices of the convex hull are in $\mathcal{V}$.

Theorem 7.5 *Let $\mathcal{P} = \{p_1, \ldots, p_n\}$ be a set of points in general position and $\Sigma(\mathcal{P})$ be the secondary polytope.*

(1) The dimension of $\Sigma(\mathcal{P})$ is $n - 3$.

(2) The vertices of $\Sigma(\mathcal{P})$ are the volume vectors of regular triangulations of $\mathcal{P}$.

(3) Faces of $\Sigma(\mathcal{P})$ correspond to regular subdivisions of $\mathcal{P}$, in particular the edges of $\Sigma(\mathcal{P})$ correspond to flips of the two types shown in Figure 7.10.

Proof. (1) A lower bound on the dimension of $\Sigma(\mathcal{P})$ can be obtained with induction. If $|\mathcal{P}| = 3$ there is a unique triangulation and $\Sigma(\mathcal{P})$ is a point. If $|\mathcal{P}| = 4$ we may have three or four points on the convex hull. In either case there are two triangulations, Figure 7.10 shows the possible configurations. Hence, if $|\mathcal{P}| = 4$, then the polytope $\Sigma(\mathcal{P})$ is one-dimensional. Consider $\mathcal{P} = \{p_1, \ldots, p_n\}$ with $n \geq 4$. Suppose p_n is not a vertex of the convex hull $\mathrm{CH}(\mathcal{P})$ of $\mathcal{P}$. Every triangulation of $\mathcal{P} \setminus p_n$ is a triangulation of $\mathcal{P}$ and every triangulation with $\varphi(p_n) = 0$ is a triangulation of $\mathcal{P} \setminus p_n$. Therefore

$\Sigma(\mathcal{P} \setminus p_n) = \Sigma(P) \cap H(x_n = 0)$ where $H(x_n = 0)$ is the hyperplane with last coordinate zero. Since $\Sigma(\mathcal{P})$ is not contained in $H(x_n = 0)$ the increase of dimension from $\Sigma(\mathcal{P} \setminus p_n)$ to $\Sigma(\mathcal{P})$ is at least one. A similar argument applies when p_n is a vertex of the convex hull $\text{CH}(\mathcal{P})$. In this case $\Sigma(\mathcal{P} \setminus p_n) = \Sigma(P) \cap H(x_n = \delta)$ where $\delta = \text{vol}(\text{CH}(\mathcal{P})) - \text{vol}(\text{CH}\,\mathcal{P} \setminus p_n)$.

To show that the dimension of $\Sigma(\mathcal{P})$ is at most $n - 3$ we exhibit three independent linear identities satisfied by all volume vectors of triangulations of $\mathcal{P}$. A triangulation T contributing to $\Sigma(\mathcal{P})$ is a triangulation of the interior of the convex hull $\text{CH}(\mathcal{P})$. Therefore, $\text{vol}(\text{CH}(\mathcal{P})) = \sum_{\Delta \in T} \text{vol}(\Delta) = \frac{1}{3} \sum_p \sum_{p \in \Delta \in T} \text{vol}(\Delta)$. Written in terms of the volume vector:

$$3\,\text{vol}(\text{CH}(\mathcal{P})) = \sum_{i=1}^{n} \varphi(p_i).$$

Let b be the barycenter of $\text{CH}(\mathcal{P})$. Think of b as the center of gravity of $\text{CH}(\mathcal{P})$ with the uniform mass distribution. A triangulation T can be used to compute b by concentrating the mass of every triangle Δ of T in the barycenter b_Δ of the triangle. This yields $b = \text{vol}(\text{CH}(\mathcal{P}))^{-1} \sum_{\Delta \in T} b_\Delta \text{vol}(\Delta)$. The barycenter of a triangle $\Delta = \Delta(p_i, p_j, p_k)$ is the point $b_\Delta = \frac{1}{3}(p_i + p_j + p_k)$. Therefore,

$$3\,b\,\text{vol}(\text{CH}(\mathcal{P})) = \sum_{\Delta \in T} (p_i + p_j + p_k)\text{vol}(\Delta) = \sum_{i=1}^{n} p_i \sum_{p_i \in \Delta \in T} \text{vol}(\Delta) = \sum_{i=1}^{n} p_i\, \varphi(p_i).$$

Each of the two coordinates of this vector equation gives an affine subspace of $\mathbb{R}^n$ that contains $\Sigma(\mathcal{P})$. Together we have found three linear identities satisfied by volume vectors of triangulations. If $\mathcal{P}$ is non-degenerate, these identities are independent. Together with the lower bound this shows that the dimension of $\Sigma(\mathcal{P})$ is exactly $n - 3$.

(2) Consider a regular triangulation T, we want to prove that $\varphi(T)$ is a vertex of $\Sigma(\mathcal{P})$. This can be done by showing that there is a linear function which attains its unique minimum value over $\Sigma(\mathcal{P})$ at $\varphi(T)$. Since T is regular, there is a vector $w \in \mathbb{R}^n$ such that T is the projection of the lower hull of the lifted point set $\hat{\mathcal{P}}_w$ in $\mathbb{R}^3$. The claim is that this lifting vector w defines the objective function we look for:

$$\langle w, \varphi(T) \rangle < \langle w, \varphi(T') \rangle \text{ for all triangulations } T' \neq T. \tag{7.1}$$

For the proof of the claim we consider 3-dimensional volumes. Enclosed by a triangle $\Delta = \Delta(p_i, p_j, p_k)$ and the lifted triangle $\Delta(\hat{p_i}, \hat{p_j}, \hat{p_k})$ there is a triangular prism with edge length w_i, w_j and w_k. The volume of this prism can be written as $\frac{w_i + w_j + w_k}{3}\,\text{vol}(\Delta)$. Let T' be a triangulation of $\mathcal{P}$ and $\hat{T'}$ be the lifted triangulated surface. The volume between this surface $\hat{T'}$ and the plane $z = 0$ is the sum of the volumes of triangular prisms and can be written as:

$$\sum_{\substack{i,j,k \\ \Delta(p_i,p_j,p_k) \in T'}} \frac{w_i + w_j + w_k}{3}\,\text{vol}(\Delta) = \sum_{i=1}^{n} \frac{w_i}{3} \sum_{p_i \in \Delta \in T'} \text{vol}(\Delta) = \sum_{i=1}^{n} \frac{w_i}{3}\,\varphi(p_i) = \frac{1}{3} \langle w, \varphi(T') \rangle.$$

Fix a lifting vector w. The volume below a surface based on the lifted point set $\hat{\mathcal{P}}_w$ is always at least as large as the volume below the lower convex hull of $\hat{\mathcal{P}}_w$. As shown, the volume below a lifted triangulation $\hat{T'}$ is $\frac{1}{3}\langle w, \varphi(T') \rangle$. This implies equation 7.1, hence, all regular triangulations of $\mathcal{P}$ are vertices of $\Sigma(\mathcal{P})$.

Given a vertex v of $\Sigma(\mathcal{P})$ there is some $w \in \mathbb{R}^n$ such that v is the unique minimum of $x \to \langle w, x\rangle$ over $\Sigma(\mathcal{P})$. Lift $\mathcal{P}$ with this w to $\hat{\mathcal{P}}_w$ and let T be the projection of the lower hull of $\hat{\mathcal{P}}_w$. By the above we know that $\langle w, \varphi(T)\rangle \leq \langle w, x\rangle$ for all $x \in \Sigma(\mathcal{P})$. Hence $v = \varphi(T)$. This proves that the vertices of $\Sigma(\mathcal{P})$ are in bijection with the regular triangulations of $\mathcal{P}$.

Figure 7.10 The two types of 'tetrahedral' flips corresponding to the edges of the secondary polytope $\Sigma(\mathcal{P})$.

(3) Let F be a face of $\Sigma(\mathcal{P})$ and $T_1, \ldots, T_k$ be the triangulations corresponding to the vertices $\varphi(T_i)$ of F. Let w be the normal of a supporting hyperplane of F. All the lifted triangulations $\hat{T}_1, \ldots, \hat{T}_k$ minimize the volume below the surface. Therefore they all coincide with the lower convex hull of $\hat{\mathcal{P}}_w$. Let S be the vertical projection of this hull, S is a subdivision of $\mathcal{P}$ containing all edges of $\textsc{ch}(\mathcal{P})$. The vertices of S are in $\mathcal{P}$ and the faces of S are convex. Every edge of S is an edge of each of the T_i and each (completed) triangulation of S is one of the T_i. This mapping from faces of $\Sigma(\mathcal{P})$ to regular subdivisions of $\mathcal{P}$ is bijective. □

7.4 The Associahedron and Catalan families

A particularly nice family of secondary polytopes are the associahedra. The *associahedron* $\mathcal{A}_n$ is the secondary polytope of a set $\mathcal{P}$ of n points in convex position.

The coordinates of the secondary polytope depend on the coordinates of points of $\mathcal{P}$, however, the combinatorial structure of the polytope remains unaffected by the choice of the set $\mathcal{P}$ of n points in convex position. So the associahedron $\mathcal{A}_n$ is actually an equivalence class of polytopes. The classical realization of $\mathcal{A}_n$ is the realization as secondary polytope of the vertices of a regular n-gon $\mathcal{C}_n$.

A triangulation T of a convex n-gon has $2n-3$ edges. The convex cycle has n edges and $n-3$ edges are interior, we call them *diagonals.* Let $\mathcal{G}_n$ be the flip-graph of $\mathcal{C}_n$, Figure 7.11 displays the graph $\mathcal{G}_6$. All triangulations of $\mathcal{C}_n$ are regular, therefore, $\mathcal{G}_n$ is the graph of the associahedron. The number of triangulations of $\mathcal{C}_n$ is the *Catalan number* C_{n-2}:

$$C_n = \frac{1}{n+1}\binom{2n}{n}.$$

Catalan numbers form a fascinating sequence arising in many counting problems. Stanley [186] (exercise 6.19), gives a list of 66 Catalan families, i.e., combinatorial interpretations of Catalan numbers. Among the most prominent Catalan families we find:

[CF_1] Triangulations of a labeled $(n+2)$-gon.

[CF_2] Binary trees with $n+1$ leaf vertices.

[CF_3] Rooted plane trees with $n+1$ vertices.

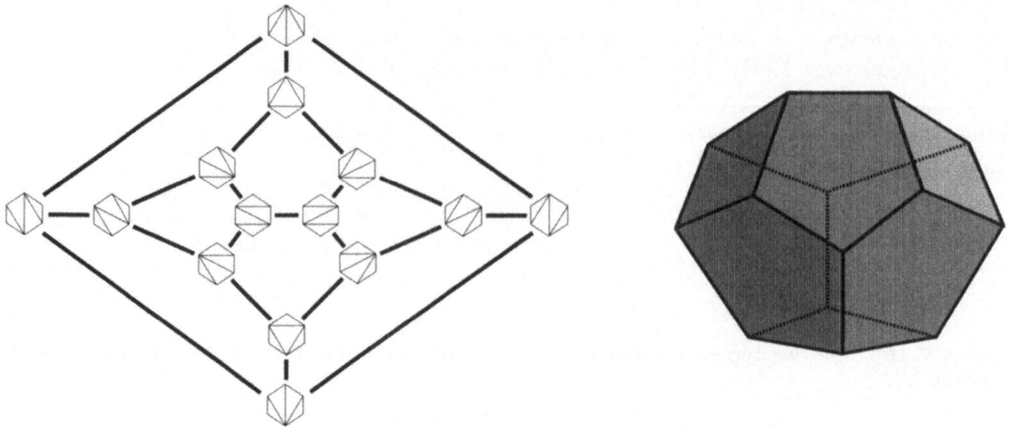

Figure 7.11 The flip-graph $\mathcal{G}_6$ of a hexagon and the associahedron $\mathcal{A}_6$.

[CF_4] Paths in the plane from $(0,0)$ to $(2n,0)$ with steps $(1,1)$ and $(1,-1)$ that never go below the x-axis (*Dyck path*).

[CF_5] Ways to parenthesize a non-associative product $x_0 \cdot x_1 \cdot \ldots \cdot x_n$ with n pairs of parentheses, e.g., $(((x_0 \cdot x_1) \cdot x_2) \cdot ((x_3 \cdot (x_4 \cdot x_5)) \cdot x_6))$.

Figure 7.12 indicates bijections between the first four of these Catalan families. The plane binary tree of the figure is the evaluation tree for the product illustrating family CF_5, this hints a bijection between CF_2 and CF_5.

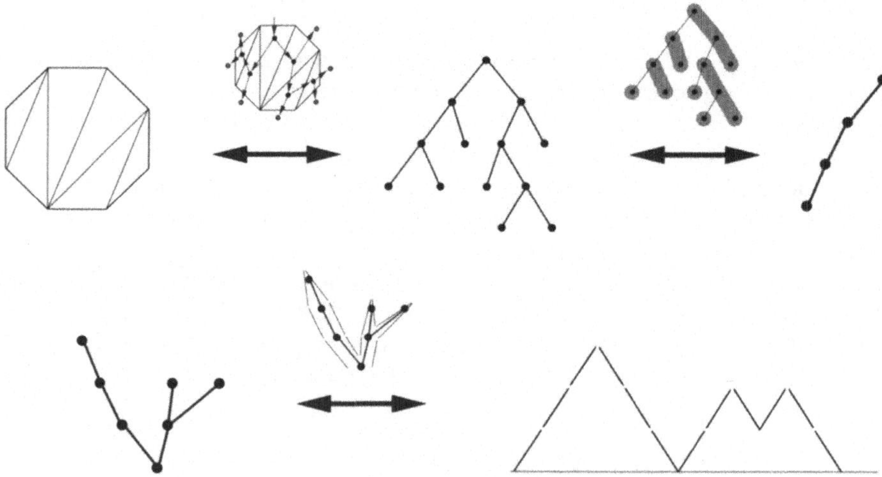

Figure 7.12 Bijections between four Catalan families.

The faces of dimension k of the associahedron $\mathcal{A}_n$ correspond to sets of $n-3-k$ non-crossing diagonals in $\mathcal{C}_n$, this is a consequence of part 3 of Theorem 7.5. The number C_n^k of ways to draw k non-crossing diagonals in a convex n-gon is known to be:

$$C_n^k = \frac{1}{n+k}\binom{n+k}{k+1}\binom{n-3}{k}.$$

7.5 The Diameter of $\mathcal{G}_n$ and Hyperbolic Geometry

We have used Lawson flips and the Delaunay triangulation to show that for every set $\mathcal{P}$ of n points the diameter of $\mathcal{G}(\mathcal{P})$ is at most $n^2 - n$. In the particular case of points in convex position there is a substantially smaller bound:

Proposition 7.6 *The diameter of the flip-graph $\mathcal{G}_n$ of n points in convex position is at most $2n - 10 + \lfloor \frac{12}{n} \rfloor$, hence, at most $2n - 10$ for $n \geq 13$.*

Proof. Let T_1, T_2 be triangulations of $\mathcal{C}_n$. The degree $d_i(x)$ of a point x in T_i is the number of diagonals incident to x. If $d_i(x) < n - 3$ then the degree of x can be increased by an appropriate flip $T_i \to T_i'$. Therefore, T_i can be transformed into the star triangulation S_x which has all its $n - 3$ diagonals incident to x. The number of flips required to get from T_i to S_x is $n - 3 - d_i(x)$. The number of flips required to get from T_1 to T_2 via S_x is at most $2n - 6 - d_1(x) - d_2(x)$. Consequently, T_1 can be transformed into T_2 with no more than $\min_x(2n - 6 - d_1(x) - d_2(x)) = 2n - 6 - \max_x(d_1(x) + d_2(x))$ flips.

A bound on $\max_x(d_1(x)+d_2(x))$ is obtained from the average of $d_1(x)+d_2(x)$ which is $\frac{1}{n}\sum_x(d_1(x) + d_2(x)) = \frac{1}{n}(4n - 12)$. Together this gives the upper bound $2n - 10 + \lfloor \frac{12}{n} \rfloor$, as claimed. □

The bound on $\mathrm{diam}(\mathcal{G}_n)$ given in the proposition is tight for small $n \leq 18$. For $n \leq 8$ this is doable by hand, for the larger values a computer search is reported. Surprisingly, the bound $2n-10$ is also known to be tight for large n. The lower bound was obtained by Sleator, Tarjan and Thurston [179] in 1988. Their exciting analysis is based on volume estimates for hyperbolic polytopes. The full argument is too complex for our context. We constrain the exposition to an outline of the beautiful proof.

(1) If T_1, T_2 is a pair of triangulations which maximizes the flip-distance $\mathrm{dist}(T_1, T_2)$, then the triangulations have no diagonal in common. Henceforth, we assume that T_1, T_2 is such a pair.

(2) The union of T_1 and T_2 is a maximal planar graph $G = G(T_1, T_2)$ and, hence, 3-connected. The Theorem of Steinitz implies that G is the skeleton graph of a convex polytope in $\mathbb{R}^3$. Let P_G be such a polytope with skeleton G.

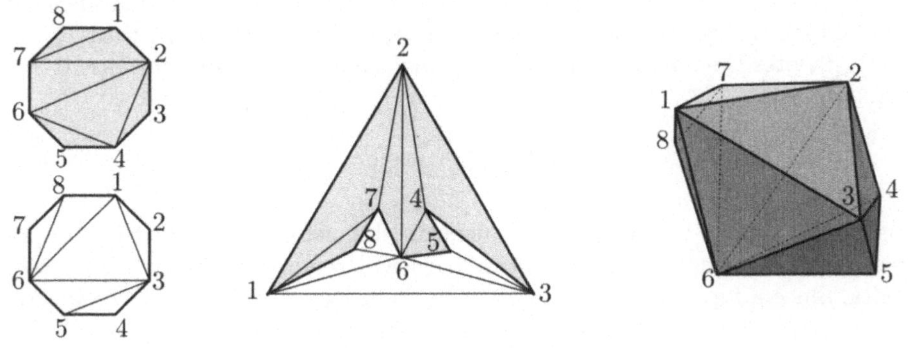

Figure 7.13 From a pair of triangulations to a polytope.

(3) Let $T_1 \to T_1'$ be a flip replacing edge (p, q) by edge (r, s). Let $\hat{p}, \hat{q}, \hat{r}, \hat{s}$ be the corresponding vertices of the polytope P_G. Cutting off the tetrahedron τ spanned by $\hat{p}, \hat{q}, \hat{r}, \hat{s}$

from the polytope P_G leaves a polytope P' such that the skeleton graph of P' is the union of T_1' and T_2.

The vision is to iterate this process: Use each flip of a flip-sequence from T_1 to T_2 to cut off a tetrahedron from the polytope. This should associate a tetrahedral decomposition of P_G with every flip sequence transforming T_1 to T_2.

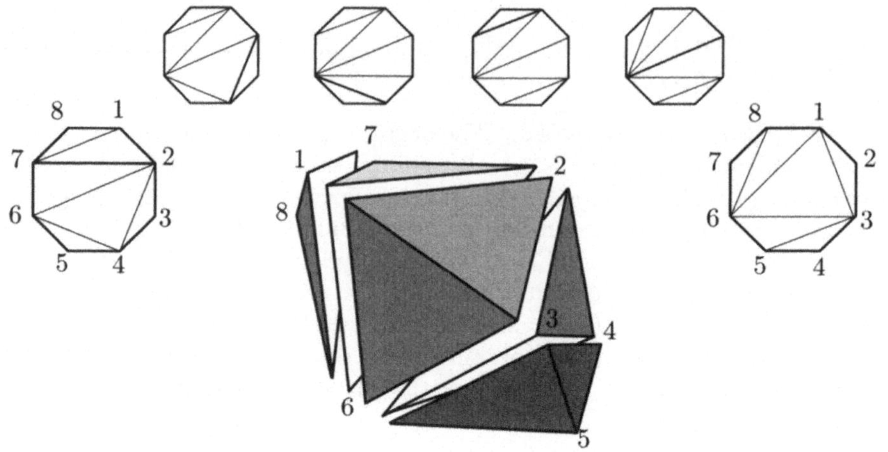

Figure 7.14 A tetrahedral decomposition induced by a flip sequence.

The truth is more delicate: Cutting off a tetrahedron can make the polytope non-convex and later flips may flip away non-convex edges of the polytope. In that case the tetrahedron corresponding to the flip can be glued onto the polytope so that the polytope retains a skeleton graph as required. Actually, this is again a simplification. Gluing a tetrahedron on a non-convex polytope may cause self-intersections. These problems can be bypassed with the use of an appropriate notion of pseudo-polytope.

(4) A flip sequence from T_1 to T_2 corresponds to a sequence of tetrahedral operations (cut off or glue on) transforming the initial polytope P_G with $G = G(T_1, T_2)$ into the polytope with skeleton $G(T_2, T_2)$. The later is a degenerate polytope, actually, it is only a union of triangles, it has no interior points and hence no volume. Let the flip sequence consist of t flips and let $\tau_1 \ldots \tau_t$ be the tetrahedra corresponding to the flips, then $P_G = \bigcup_i \tau_i$. Let vol(P) denotes the volume of a 3 dimensional polytope P. Our considerations imply the inequality

$$\operatorname{vol}(P_G) \leq \sum_{i=1}^{t} \operatorname{vol}(\tau_i).$$

If V_Δ is the maximum volume of a tetrahedron that can be inscribed in the polytope P_G, then every covering of P_G with tetrahedra will require at least $\operatorname{vol}(P_G)/V_\Delta$ many tetrahedra. Since a flip sequence from T_1 to T_2 induces a tetrahedral cover we conclude

$$\operatorname{dist}(T_1, T_2) \geq \operatorname{vol}(P_G)/V_\Delta.$$

(5) At first, this bound for the flip distance of two triangulations seems to be extremely poor. Every polytope P in $\mathbb{R}^3$ has an inscribed tetrahedron of volume $c \cdot \operatorname{vol}(P)$ for a small constant c.

Volumes of polytopes in hyperbolic space behave completely different. In hyperbolic space there is a constant V_0 which is an upper bound for the volume of all tetrahedra. We digress for the introduction of some elements of hyperbolic geometry.

Hyperbolic geometry. In hyperbolic geometry there are many parallels to a line through every given point not on the line. There are several models of hyperbolic space within Euclidean space. There are three important models for the hyperbolic plane.

($\mathbb{K}$) The *Klein model.* The points are the points of an open unit disk D. The lines are chords, i.e., straight in the Euclidean sense. This model has the advantage that distances are easy to compute. Given points a and b consider the line spanned by a and b and let α and β be the limit points on the boundary circle ∂D, see Figure 7.15. If $\overline{pq}$ is the Euclidean distance of p and q, then the hyperbolic distance of a and b can be expressed as:

$$\mathrm{dist}_{\mathbb{K}}(a, b) = \frac{1}{2}\left|\log\left(\frac{\overline{a\alpha}\cdot\overline{b\beta}}{\overline{b\alpha}\cdot\overline{a\beta}}\right)\right|.$$

($\mathbb{P}$) The *Poincaré model.* The points are the points of a disk D. The lines are the diameter of the disc and open arcs of circles orthogonal to ∂D. This model is conformal, i.e., the angle of two intersecting lines is the Euclidean angle of the tangents at the point of intersection.

($\mathbb{H}$) The *half-space model.* The points are all points above the x-axis in the Euclidean plane. Lines are vertical rays emanating upward from a point on the x-axis and halfcircles with center on the x-axis. This model is again conformal.

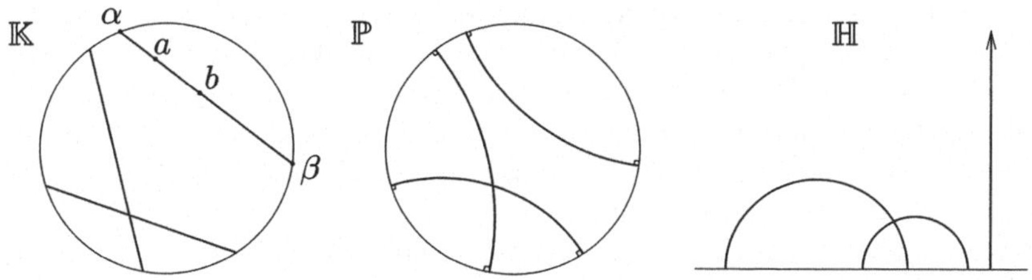

Figure 7.15 Models for the hyperbolic plane.

In Section 5.1 we have used the sphere S^2 as a hub to connect between different geometries. This can be done again. Figure 7.16 should transport the idea of how to construct points $p_{\mathbb{K}}$, $p_{\mathbb{P}}$, $p_{\mathbb{H}}$ in the three models if any of them is given.

Area and volume. Triangles in hyperbolic space have angle sum less than π. In fact, the Gauß-Bonnet theorem states that the area of a triangle with angles α, β and γ is exactly $\pi - \alpha - \beta - \gamma$.

Think of hyperbolic space as enhanced by the sphere at infinity. In the half-space model of the plane this adds the x-axis and a point v_∞ which is an endpoint of all vertical lines. In the three dimensional half-space model the sphere at infinity consists of the plane $z = 0$ and a point v_∞.

An *ideal triangle* is one with three corners in the sphere at infinity. Ideal triangles have area π, they are triangles of maximal area. Another amazing fact about ideal triangles is that they are all congruent, i.e., they can be transformed into each other by an isometry.

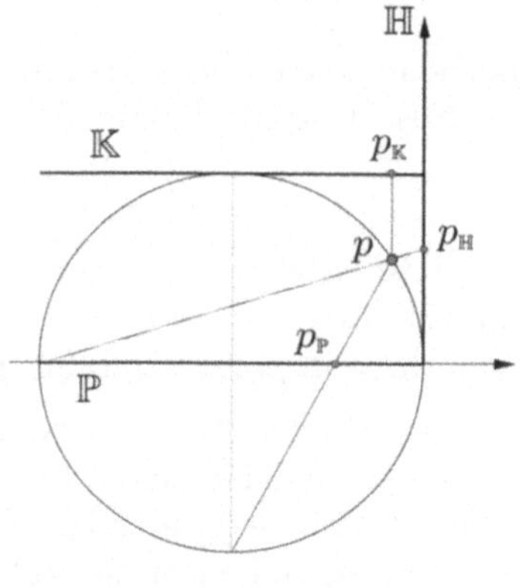

Figure 7.16 A point p on the half-sphere and the corresponding points in $\mathbb{K}$, $\mathbb{P}$ and $\mathbb{H}$.

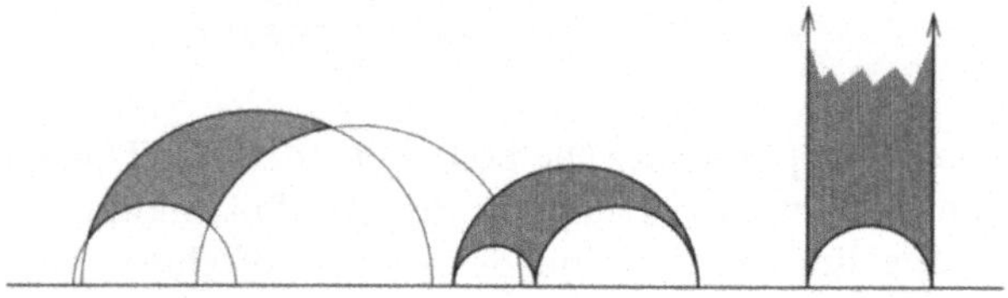

Figure 7.17 A triangle of area 0.78 and two ideal triangles in $\mathbb{H}$.

An *ideal tetrahedron* has all its four vertices on the sphere at infinity. In the half-space model it can be assumed that one vertex of an ideal tetrahedron T is at v_∞ and the other three vertices are in the plane$z = 0$. Let α, β and γ be the angles of the base triangle spanned by the three points in the plane. Let $\mathrm{Lob}(t) = -\int_0^t \log(|2\sin(u)|)du$, Milnor proved that $\mathrm{vol}(T) = \mathrm{Lob}(\alpha) + \mathrm{Lob}(\beta) + \mathrm{Lob}(\gamma)$. This function attains its maximum when the base triangle is equilateral. The maximum is $V_0 = 3 \cdot \mathrm{Lob}(\frac{\pi}{3}) = 1.0149416$.

An *ideal polytope* is a polytope which has all its vertices on the sphere at infinity. Again it can be assumed that v_∞ is one of the vertices of an ideal polytope P in the half-space model. The skeleton of P is just the Delaunay triangulation of the vertices of P in the plane, together with all the edges connecting convex hull vertices with v_∞. The volume of the ideal polytope P is obtained as the sum of the volumina of the ideal tetrahedra spanned by v_∞ and a triangle of the Delaunay triangulation.

Proposition 7.7 *For each k and $n = k^2 + 1$ there exist triangulations T_1, T_2 of C_n with* $\mathrm{dist}(T_1, T_2) \geq 2n - 4\sqrt{n} + O(1)$.

Proof. Consider the section of the triangular grid shown in Figure 7.18. This is the Delaunay triangulation of a set S of k^2 points. All the $2(k-1)^2$ triangles in this triangulation are equilateral. The set S together with v_∞ is the vertex set of an ideal polytope P in the half-space model. This polytope P has $n = k^2 + 1$ vertices and volume $\mathrm{vol}(P) = 2(k-1)^2 V_0$ where V_0 is the volume of an ideal tetrahedron over an equilateral base. Since V_0 is the maximal volume of a hyperbolic tatrahedron every covering of P by tetrahedra will require at least $2(k-1)^2 = 2n - 4\sqrt{n} + O(1)$ tetrahedra.

Given the polytope P from the preceding paragraph, it remains to find triangulations T_1, T_2 of C_n such that $P = P_G$ with $G = G(T_1, T_2)$. The existence of such triangulations is

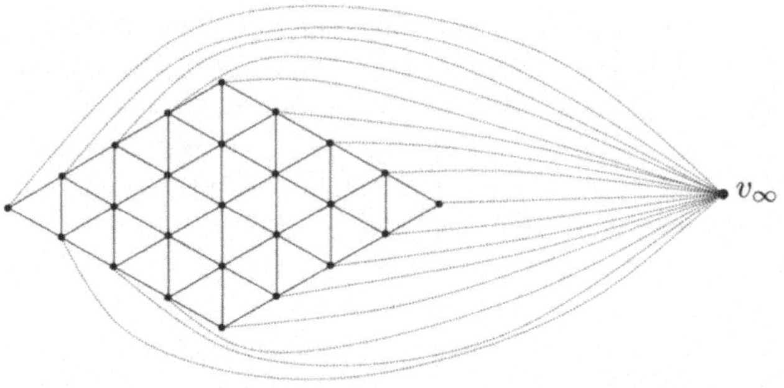

Figure 7.18 Sketch of an ideal polytope composed by $2(k-1)^2$ tetrahedra of volume V_0.

implied by a theorem of Whitney: Every 4-connected planar triangulation has a Hamilton cycle. A Hamilton cycle of the skeleton of P can be identified with the hull of C_n. The cycle induces a partition of the remaining edges: In a planar drawing of G this partition corresponds to the partition into edges in the interior of the cycle and edges in the exterior. Together with the cycle each of these two sets of edges is (equivalent to) a triangulation of C_n. In the case of the graph of Figure 7.18 it is easy to describe a Hamilton cycle explicitly. □

Let P be a simplicial polytope which has a vertex of degree Δ. From the proof of Proposition 7.6 it follows that two triangulations whose union is isomorphic to the skeleton of P have flip-distance at most $2n-4-\Delta$. In the example used for the proof of Proposition 7.7 the degree of v_∞ is $\Delta_\infty \approx 4\sqrt{n}$.

To improve the lower bound on the flip-distance, polytopes with smaller maximum degree are required. The construction of [179] is based on the icosahedron. Each face of this polytope is subdivided into k^2 equilateral triangles. The resulting polytope has $20k^2$ triangles and $n = 10k^2 + 2$ vertices. The ideal polyhedron $P(k)$ is obtained by projecting the edges of the icosahedron out to the sphere at infinity and using a conformal mapping to send the faces of the icosahedron to the corresponding spherical triangles.

By a technically demanding analysis Sleator, Tarjan and Thurston show that for large n the tetrahedralization of $P(k)$ requiring the least number of tetrahedra is of cone type. That is there is a vertex belonging to all the tetrahedra, consequently there are $2n-10$ tetrahedra in a minimal tetrahedralization. Via the Theorem of Whitney this implies that the diameter of $\mathcal{G}_n$ is $2n-10$ for n sufficiently large and of the form $n = 10k^2 + 2$.

7.6 Notes and References

Triangulations span the full range of geometry from pure to applied and from high- to low-dimensional. This span may be illustrated by the following recent contributions: Santos [169] investigates concepts of triangulations for oriented matroids. Edelsbrunner [66] presents algorithmic and structural aspects of triangulations relevant for mesh generation. Remarkably, flips play a prominent role in both. The bound on the number of flippable edges, Proposition 7.1, is from Hurtado, Noy and Urrutia [116]. This paper contains examples of triangulations of a point set which are at flip distance $\Theta(n^2)$. Similar examples are discussed by Santos and Seidel [170] in the context of counting triangulations.

Voronoi diagrams and Delaunay triangulations are named after the Russian mathematicians Georgii Feodosevich Voronoi (1868-1908) and Boris Nikolaevich Delone (1890-1980). The concepts themselves have been studied earlier among others by Dirichlet, Gauß and Descartes. The quadrangular flip (*Lawson flip*) was studied in a paper by Lawson [129]. He proved that the flip graph $\mathcal{G}(\mathcal{P})$ is connected for all point sets $\mathcal{P}$ in the plane. More comprehensive accounts on Delaunay triangulations including the analysis of various algorithms for their construction can be found in the books of Edelsbrunner [65] and [66] and in the handbook article of Aurenhammer and Klein [18]. These sources also contain generalizations, e.g., higher order Voronoi diagrams and power diagrams.

Secondary polytopes where introduced by Gel'fand et al. [98] in the context of generalized hypergeometric functions. A self-contained study of these polytopes is Billera et al. [27]. In Ziegler's book on polytopes [219] secondary polytopes are investigated as a subclass of fiber polytopes. The name *associahedron* was coined by Lee [130]. Some authors remark that associahedra already appear in work of Stasheff around 1960 and call them *Stasheff polytopes.* Lee gave explicit coordinates for the associahedron $\mathcal{A}_n$, he also investigates the number $f_j = C_n^{n-3-j}$ of j-dimensional faces of $\mathcal{A}_n$. A nice derivation of the formula for C_n^k is given by Stanley [185]. Actually, these formulas have been known for more then hundred years, Stanley [185] contains references to work of Kirkman, Prouhet and Cayley.

We make no attempt to extract references from the vast literature on Catalan numbers. The extensive collection for pointers to this topic is the second volume of Richard Stanley's Enumerative Combinatorics [186].

The bounds for the diameter of the flip-graph $\mathcal{G}_n$ of a convex polygon were obtained by Sleator, Tarjan and Thurston [179]. Most of our exposition is close to the lines of their brilliant paper. For additional aspects of hyperbolic geometry we recommend Milnor [141] and Cannon et al. [44].

8 Rigidity and Pseudotriangulations

A *framework* $G[\mathbf{p}]$ is a graph $G = (V, E)$ and an embedding $\mathbf{p} : V \to \mathbb{R}^d$. The straight edges of the framework are thought of as rigid bars connecting vertices (joints) where incident bars are connected flexibly. An important problem for civil engineers, is the question: "is a given framework rigid?"

The first part of this chapter is about rigidity of plane frameworks. The concept of infinitesimal motions allows to deal with the problem in terms of linear algebra. Stress is introduced as the dual notion of motion. Theorem 8.9 collects three characterizations for minimal generically rigid graphs, mgr-graphs. These characterizations nicely generalize well known characterizations of trees which happen to be the mgr-graphs in one dimension.

In Section 8.2 we define pseudotriangulations and show that minimal pseudotriangulations constitute a particular class of planar mgr-graphs. A special class of motions, expansive motions, are used in Section 8.3 to define a polyhedron whose vertices are in bijection to minimal pseudotriangulations. The edges of the polyhedron correspond to edge flips between two minimal pseudotriangulations. As an application of this structure we indicate a solution to the Carpenter's Rule Problem: "Can a linkage in the plane be moved continuously to a position where all its vertices are on a line, so that during the motion the linkage remains non-crossing and edge lengths are preserved?"

8.1 Rigidity, Motion and Stress

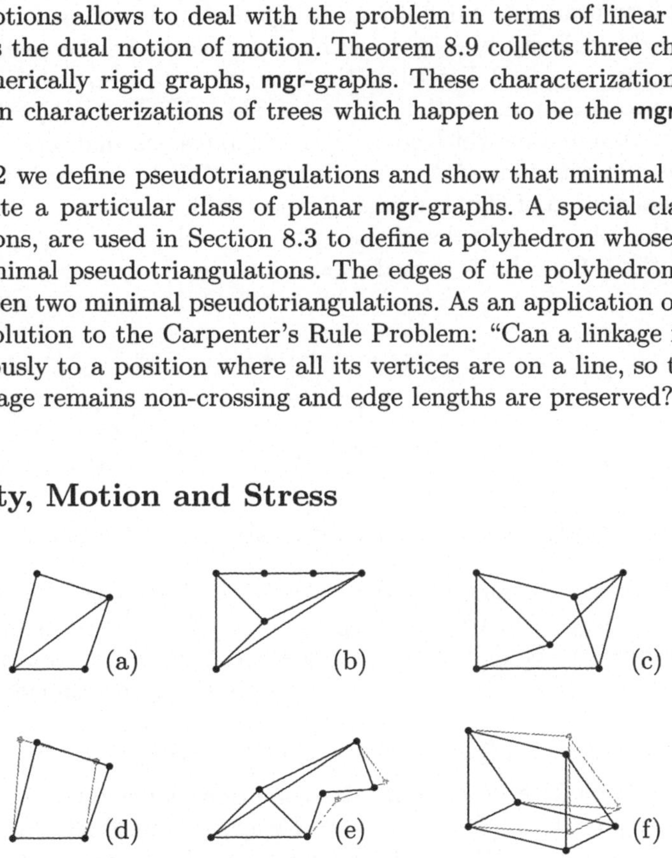

Figure 8.1 Frameworks (a,b,c) are rigid, (d,e,f) are flexible.

Figure 8.1 shows that there are some subtleties to questions of rigidity: At first note that the notion of rigidity depends on the ambient space. All six frameworks are flexible as frameworks in space but as plane frameworks those in the upper row are rigid. The underlying graph for frameworks (b) and (e) is the same, even the length of the edges are equal, still (b) is rigid and (e) flexible, i.e., non-rigid. The graphs for (c) and (f) are isomorphic and again one of the frameworks is rigid the other flexible. The difference between the two pairs is that almost all embeddings of (e) are flexible while almost all embeddings of (c) are rigid.

Let a framework $G[\mathbf{p}]$ be given. If $G[\mathbf{p}]$ is flexible, then there is a motion which moves vertices $\mathbf{p}_i = (x_i, y_i)$ of the framework along paths $\mathbf{p}_i(t) = (x_i(t), y_i(t))$. Throughout the motion the length $||\mathbf{p}_i - \mathbf{p}_j||$ of every edge $\{i,j\}$ has to remain constant and the same holds for the square of the length, $||\mathbf{p}_i - \mathbf{p}_j||^2 = \langle\, \mathbf{p}_i(t) - \mathbf{p}_j(t)\,,\, \mathbf{p}_i(t) - \mathbf{p}_j(t)\,\rangle$. By differentiation, this condition becomes

$$\frac{d}{dt}\langle\, \mathbf{p}_i(t) - \mathbf{p}_j(t)\,,\, \mathbf{p}_i(t) - \mathbf{p}_j(t)\,\rangle = 2\,\langle\, \mathbf{p}_i(t) - \mathbf{p}_j(t)\,,\, \mathbf{p}'_i(t) - \mathbf{p}'_j(t)\,\rangle = 0.$$

In particular the initial velocities $\mathbf{v}_i := \mathbf{p}'_i(0)$ of the motion satisfy

$$\langle\, \mathbf{p}_i - \mathbf{p}_j\,,\, \mathbf{v}_i - \mathbf{v}_j\,\rangle = 0 \quad \text{for all} \quad \{i,j\} \in E.$$

We turn the observation into a definition. An *infinitesimal motion* of a plane framework is an assignment of a velocity $\mathbf{v}_i \in \mathbb{R}^2$ to each vertex i such that for each edge $\{i,j\} \in E$ we have $\langle\, \mathbf{p}_i - \mathbf{p}_j\,,\, \mathbf{v}_i - \mathbf{v}_j\,\rangle = 0$. Rewriting the equation as $\langle\, \mathbf{p}_i - \mathbf{p}_j\,,\, \mathbf{v}_i\,\rangle = \langle\, \mathbf{p}_i - \mathbf{p}_j\,,\, \mathbf{v}_j\,\rangle$ we note that the condition means that $\mathbf{v}_i$ and $\mathbf{v}_j$ have the same projection on $\mathbf{p}_i - \mathbf{p}_j$.

A *trivial motion* is a motion which come from a rigid transformation of the whole plane, i.e., translations and rotations. A framework $G[\mathbf{p}] = (V, E, \mathbf{p})$ is *infinitesimally rigid* if every infinitesimal motion of $G[\mathbf{p}]$ is trivial. Figure 8.2 illustrates non-trivial infinitesimal motions. Note that the frameworks of examples (b) and (c) are rigid, still they admit infinitesimal motions. Our aim is to understand infinitesimal rigidity of frameworks.

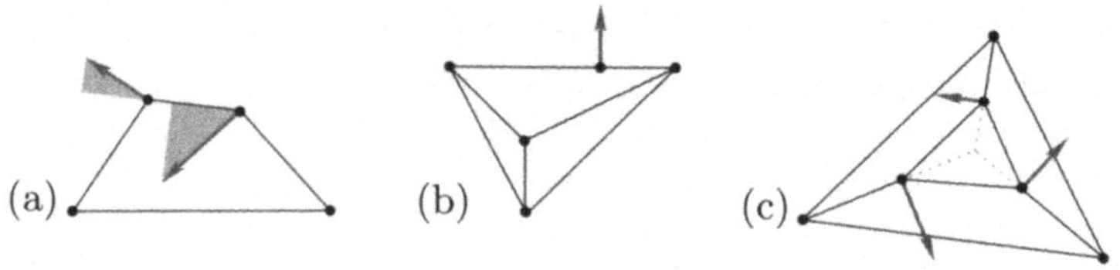

Figure 8.2 Arrows indicate the velocity vectors of non-trivial infinitesimal motions.

The *rigidity matrix* of a plane framework $G[\mathbf{p}] = (V, E, \mathbf{p})$ with $n = |V|$ and $m = |E|$ is a $m \times 2n$ matrix $\mathbf{R}_{G[\mathbf{p}]}$. The rows of $\mathbf{R}_{G[\mathbf{p}]}$ are indexed by the edges of G. Each vertex of G has two columns in $\mathbf{R}_{G[\mathbf{p}]}$ representing the two coordinates. Let $\mathbf{p}_i = (x_i, y_i)$ and $\mathbf{p}_j = (x_j, y_j)$ and suppose $\{i,j\}$ is an edge. The row of $\mathbf{R}_{G[\mathbf{p}]}$ corresponding to edge $\{i,j\}$ is

$$[\,0\,0\ldots0\,0\,(x_i - x_j)\,(y_i - y_j)\,0\,0\ldots0\,0\,(x_j - x_i)\,(y_j - y_i)\,0\,0\ldots0\,0\,]$$

The rigidity matrix enables us to write the conditions for a infinitesimal motion $\mathbf{v} : V \to \mathbb{R}^2$ in a compact form: $\mathbf{R}_{G[\mathbf{p}]} \cdot \mathbf{v} = \mathbf{0}$. Hence, infinitesimal motions are exactly the elements of the kernel of $\mathbf{R}_{G[\mathbf{p}]}$. The following velocity vectors correspond to trivial motions of $G[\mathbf{p}]$:

$$\mathbf{t}_x = [\,1\,0\,1\,0\ldots1\,0\,], \quad \mathbf{t}_y = [\,0\,1\,0\,1\ldots0\,1\,], \quad \mathbf{t}_r = [\,-y_1\;x_1\;-y_2\;x_2\ldots\;-y_n\;x_n\,].$$

$\mathbf{t}_x$ and $\mathbf{t}_y$ are translations in x and y direction, $\mathbf{t}_r$ is a counterclockwise rotation around $\mathbf{0}$. Check that these vectors satisfy $\langle\, \mathbf{r}\,,\, \mathbf{t}\,\rangle = 0$ for every row $\mathbf{r}$ of $\mathbf{R}_{G[\mathbf{p}]}$. It follows that the rank of $\mathbf{R}_{G[\mathbf{p}]}$ is at most $2n - 3$.

The rigid transformations of $\mathbb{R}^2$ form a 3-dimensional vector space. Hence, the vectors $\mathbf{t}_x, \mathbf{t}_y, \mathbf{t}_r$ constitute a basis of the space of trivial motions for the framework $G[\mathbf{p}]$. This implies that $G[\mathbf{p}]$ is infinitesimally rigid iff $\dim(\mathsf{ker}(\mathbf{R}_{G[\mathbf{p}]})) = 3$. Recall the dimension-formula from linear algebra:

$$\dim(\mathsf{ker}(\mathbf{R}_{G[\mathbf{p}]})) + \dim(\mathsf{im}(\mathbf{R}_{G[\mathbf{p}]})) = \dim(\mathsf{domain}(\mathbf{R}_{G[\mathbf{p}]})) = 2n.$$

If $G[\mathbf{p}]$ is infinitesimally rigid the formula implies $\mathsf{rank}(\mathbf{R}_{G[\mathbf{p}]})) = \dim(\mathsf{im}(\mathbf{R}_{G[\mathbf{p}]})) = 2n-3$. If $|E| > 2n-3$ then it is possible to delete a row from $\mathbf{R}_{G[\mathbf{p}]}$ without affecting the rank. Therefore, a minimal infinitesimally rigid graph G has exactly $2n-3$ edges. We summarize the findings:

Proposition 8.1 *The rank of the rigidity matrix $\mathbf{R}_{G[\mathbf{p}]}$ of a framework $G[\mathbf{p}] = (V, E, \mathbf{p})$ is at most $2|V|-3$. Moreover, $G[\mathbf{p}]$ is minimal infinitesimally rigid iff $|E| = 2|V|-3$ and* $\mathsf{rank}(\mathbf{R}_{G[\mathbf{p}]}) = 2|V|-3$.

Row- and column-rank of a matrix are the same. In the context of the row-rank of the rigidity matrix some additional terminology is in use. A *self-stress* on a framework $G[\mathbf{p}] = (V, E, \mathbf{p})$ is an assignment of forces $\omega : E \to \mathbb{R}$ to the edges, subject to the condition that all vertices remain in equilibrium:

$$\sum_{\substack{j \\ \{i,j\}\in E}} \omega_{ij}(\mathbf{p}_i - \mathbf{p}_j) = 0 \quad \text{for all } i \in V.$$

Self-stresses of $G[\mathbf{p}]$ are the solutions of $\omega \cdot \mathbf{R}_{G[\mathbf{p}]} = \mathbf{R}^T_{G[\mathbf{p}]} \cdot \omega = \mathbf{0}$.

A subset E' of edges of $G[\mathbf{p}] = (V, E, \mathbf{p})$ is *dependent* if there is a non-trivial self-stress ω with support in E', i.e., $\omega(e) = 0$ for all $e \in E \setminus E'$. If E' is not dependent it is *independent.* If $E' = E$, we also speak of a dependent or independent graph or framework.

The following operation $\mathbf{H}_2$ allows to build larger independent graphs from smaller ones.

($\mathbf{H}_2$) The graph $G^+ = (V^+, E^+)$ is produced from $G = (V, E)$ by a $\mathbf{H}_2$-addition if $V^+ = V \cup \{v_0\}$ with $v_0 \notin V$ and there are two vertices v_i, v_j in V such that $E^+ = E \cup \{(v_0, v_i), (v_0, v_j)\}$. If $G[\mathbf{p}]$ is a framework with graph G, we require that the point $\mathbf{p}_0$ of $G^+[\mathbf{p}^+]$ is chosen such that $\mathbf{p}_0, \mathbf{p}_i, \mathbf{p}_j$ are not collinear.

Lemma 8.2 *Let $G^+[\mathbf{p}^+]$ be produced by a sequence of $\mathbf{H}_2$-additions from $G[\mathbf{p}]$. The framework $G^+[\mathbf{p}^+]$ is independent iff $G[\mathbf{p}]$ is independent.*

Proof. Let ω be a self-stress for $G^+[\mathbf{p}^+]$. The equilibrium condition for vertex v_0 is the equation $\omega_{0i}(\mathbf{p}_0 - \mathbf{p}_i) + \omega_{0j}(\mathbf{p}_0 - \mathbf{p}_j) = \mathbf{0}$. Since $\mathbf{p}_0, \mathbf{p}_i, \mathbf{p}_j$ are not collinear, $(\mathbf{p}_0 - \mathbf{p}_i)$ and $(\mathbf{p}_0 - \mathbf{p}_j)$ are linearly independent. This enforces $\omega_{0i} = \omega_{0j} = 0$. Therefore, ω is a non-trivial self-stress on $G^+[\mathbf{p}^+]$ only if its restriction to $G[\mathbf{p}]$ is a non-trivial self-stress.

Conversely, augmenting a self-stress ω of $G[\mathbf{p}]$ with $\omega_{0i} = 0$ and $\omega_{0j} = 0$ gives a self-stress of $G^+[\mathbf{p}^+]$. □

Proposition 8.3 *Let $G[\mathbf{p}] = (V, E, \mathbf{p})$ be generated from a single edge by a sequence of $\mathbf{H}_2$-additions. The framework $G[\mathbf{p}]$ is independent and* $\mathsf{rank}(\mathbf{R}_{G[\mathbf{p}]}) = |E| = 2|V| - 3$.

Proof. The rigidity matrix for a single edge is a 1×4 matrix. The rank is 1 and there is no non-trivial self-stress.

Let $G[\mathbf{p}]$ be generated from a single edge by a sequence of $\mathbf{H}_2$-additions. Each $\mathbf{H}_2$-step adds two edges, thus $|E| = 2|V| - 3$. Inductive application of Lemma 8.2 shows that the rows of $\mathbf{R}_{G[\mathbf{p}]}$ are linearly independent. Differently stated: The kernel of $\mathbf{R}^T_{G[\mathbf{p}]} : \mathbb{R}^{|E|} \to \mathbb{R}^{2|V|}$ is trivial, hence, $\mathsf{rank}(\mathbf{R}_{G[\mathbf{p}]}) = \mathsf{rank}(\mathbf{R}^T_{G[\mathbf{p}]}) = |E| = 2|V| - 3$. □

By definition, a framework $G[\mathbf{p}]$ is independent iff $\dim(\mathsf{ker}(\mathbf{R}^T_{G[\mathbf{p}]})) = 0$. From

$$\dim(\mathsf{ker}(\mathbf{R}^T_{G[\mathbf{p}]})) + \dim(\mathsf{im}(\mathbf{R}^T_{G[\mathbf{p}]})) = \dim(\mathsf{domain}(\mathbf{R}^T_{G[\mathbf{p}]})) = |E|.$$

it follows that $\dim(\mathsf{ker}(\mathbf{R}^T_{G[\mathbf{p}]})) = |E| - \mathsf{rank}(\mathbf{R}^T_{G[\mathbf{p}]})$. Since $\mathsf{rank}(\mathbf{R}^T_{G[\mathbf{p}]}) \leq 2|V|-3$ (Proposition 8.1) every graph with $|E| > 2|V| - 3$ is dependent. Assume that E is independent and $|E| < 2|V| - 3$. Let $\mathbf{p}$ be in general position* so that the rank of the rigidity matrix of the complete graph on $\mathbf{p}$ is $2|V| - 3$. In that case it is possible to add edges to E so that with each added edge the rank of $\mathbf{R}^T_{G[\mathbf{p}]}$ increases by one, until $|E| = 2|V| - 3$. This proves a dual to Proposition 8.1:

Proposition 8.4 *Let* $\mathbf{p}$ *be in general position. A framework* $G[\mathbf{p}]$ *is maximal independent iff* $\mathsf{rank}(\mathbf{R}_{G[\mathbf{p}]}) = 2|V| - 3$ *and* $|E| = 2|V| - 3$.

Let G be a graph with n vertices. With a given embedding $\mathbf{p}$ we have $\mathsf{rank}(\mathbf{R}_{G[\mathbf{p}]}) = 2n - 3$ iff the determinant of a $(2n-3) \times (2n-3)$ submatrix is non-zero. Each of these determinants is a polynomial in the coordinates of the embedding. Consequently, each of the determinants is either the zero polynomial or its set of non-zeros is open and dense in $\mathbb{R}^{2n}$. Therefore, the graph G either has $\mathsf{rank}(\mathbf{R}_{G[\mathbf{p}]}) < 2n - 3$ for all embeddings $\mathbf{p}$ or for almost all embeddings $\mathbf{p}$ of G the rigidity matrix has rank $2n - 3$. In the later case we call the graph *generically rigid.* The nice thing about generic rigidity is that it is a property of the graph G alone. If there is need for a embedding $\mathbf{p}$ of G, this can can be chosen *generically*, i.e., such that the ranks of all submatrices of $\mathbf{R}_{G[\mathbf{p}]}$ are constant in some neighborhood of $\mathbf{p}$. Recall the examples from Figure 8.1. From the graphs of (a) and (d) the first is rigid, the second non-rigid, the graph of (c) is generically rigid but the graph of (b) is not generically rigid.

Minimal generically rigid graphs in the plane, mgr-graphs for short, are generically rigid, but they lose this property upon removal of any edge. We summarize our knowledge about mgr-graphs:

(I) G is a mgr-graph and $\mathbf{p}$ a generic embedding $\Longrightarrow$ $G[\mathbf{p}]$ is minimal infinitesimally rigid (Proposition 8.1).

(II) G is a mgr-graph and $\mathbf{p}$ a generic embedding $\Longrightarrow$ $G[\mathbf{p}]$ is maximal independent, i.e., self-stress free (Proposition 8.4).

A necessary condition for mgr-graphs is given with the next proposition.

Proposition 8.5 (Edge count)
If $G = (V, E)$ *is a* mgr-graph, *then* $|E| = 2|V|-3$ *and* $|E'| \leq 2|V[E']|-3$ *for all* $\emptyset \neq E' \subset E$ *and the set* $V[E']$ *of all vertices incident to edges in* E'.

Proof. By definition $G = (V, E)$ is a mgr-graph iff R_G has generic rank $2|V| - 3$ and E is minimal, i.e., $\mathsf{rank}(R_{G'}) < 2|V| - 3$ for all $G' = (V, E')$ with $E' \subset E$. This implies $|E| = 2|V| - 3$.

If there exists an $E' \subset E$ with $|E'| > 2|V[E']| - 3$, then there is a non-trivial self-stress ω' on $(V[E'], E')$. With $\omega_e = 0$ for all $e \in E \setminus E'$ this extends to a non-trivial self-stress ω on G. However, in (II) we have noted that the mgr-graph G is independent, i.e., has no non-trivial self-stress. □

* It suffices that no line contains all points of $\mathbf{p}$.

Figure 8.3 shows mgr-graphs which cannot be constructed from a single edge with $\mathbf{H}_2$-additions. There is a second operation $\mathbf{H}_3$, such that the two operations together suffice to generate every mgr-graph from a single edge.

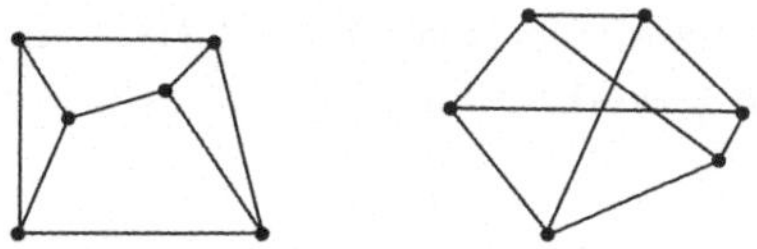

Figure 8.3 Minimal rigid graphs which cannot be constructed from $\mathbf{H}_2$-additions.

($\mathbf{H}_3$) Given a graph $G = (V, E)$ with an edge $e = \{v_i, v_j\}$ and a vertex $v_k \neq v_i, v_j$. A new graph $G^+ = (V^+, E^+)$ is produced from G by a $\mathbf{H}_3$-addition if $V^+ = V \cup \{v_0\}$ with $v_0 \notin V$ and $E^+ = \big(E \setminus e\big) \cup \{(v_0, v_i), (v_0, v_j), (v_0, v_k)\}$.

Let $G^+ = (V, E)$ be a graph and suppose that G^+ has a vertex v_0 of degree three adjacent to v_i, v_j, v_k. Let G_{ij} be the (multi)-graph obtained by deleting v_0 and its edges and adding the edge $\{v_i, v_j\}$, graphs G_{ik} and G_{jk} are defined alike.

Lemma 8.6 *The graph G^+ is independent iff at least one of G_{ij}, G_{ik} and G_{jk} is independent.*

Proof. Suppose G^+ is independent and has m edges. A generic embedding $\mathbf{p}$ of G^+ has $\mathsf{rank}(\mathbf{R}_{G^+[\mathbf{p}]}) = m$. If the rank of the rigidity matrix of one of G_{ij}, G_{ik}, G_{jk} is $m-2$, then this graph is independent.

Assume that all these ranks are smaller, then the frameworks have non-trivial self-stresses α, β and γ, such that:

$$\alpha_{ij} R_{ij} = \sum_{e \in E^*} \alpha_e R_e\ , \qquad \beta_{ik} R_{ik} = \sum_{e \in E^*} \beta_e R_e\ , \qquad \gamma_{jk} R_{jk} = \sum_{e \in E^*} \gamma_e R_e\ .$$

Here $E^* = E \setminus \{(v_0, v_i), (v_0, v_j)(v_0, v_k)\}$ and $\alpha_{ij} \neq 0$, $\beta_{ik} \neq 0$ and $\gamma_{jk} \neq 0$ because G^+ is independent.

Consider the complete graph on $\{v_0, v_i, v_j, v_k\}$ embedded at $\mathbf{p}_0, \mathbf{p}_i, \mathbf{p}_j, \mathbf{p}_k$. This graph has 4 vertices and 6 edges, since $6 > 2 \cdot 4 - 3$ there is a non-trivial self-stress

$$\omega_{0i} R_{0i} + \omega_{0j} R_{0j} + \omega_{0k} R_{0k} + \omega_{ij} R_{ij} + \omega_{ik} R_{ik} + \omega_{jk} R_{jk} = \mathbf{0}.$$

Substituting from above, yields:

$$\omega_{0i} R_{0i} + \omega_{0j} R_{0j} + \omega_{0k} R_{0k} + \sum_{e \in E^*} \Big(\frac{\omega_{ij}}{\alpha_{ij}} \alpha_e + \frac{\omega_{ik}}{\beta_{ik}} \beta_e + \frac{\omega_{jk}}{\gamma_{jk}} \gamma_e\Big) R_e = \mathbf{0}.$$

This is a non-trivial self-stress on $G^+[\mathbf{p}]$, contradicting our assumption. Therefore, at least one of G_{ij}, G_{ik} and G_{jk} has a rigidity matrix of rank $m-2$ and is independent.

For the converse suppose that G_{ij} is independent and let $\mathbf{p}$ be generic embedding of G_{ij} so that $\mathsf{rank}(\mathbf{R}_{G_{ij}[\mathbf{p}]}) = m - 2$. Extend $\mathbf{p}$ by specifying $\mathbf{p}_0 = \frac{\mathbf{p}_i + \mathbf{p}_j}{2}$, i.e., place vertex v_0 on the midpoint of $\{\mathbf{p}_i, \mathbf{p}_j\}$.

Assume that G_{ij} is independent and G^+ dependent, then $\mathsf{rank}(\mathbf{R}_{G^+[\mathbf{p}]}) < m$ and there is a non-trivial self-stress ω. The equation for vertex v_0 reads:

$$\omega_{0i}(\mathbf{p}_0 - \mathbf{p}_i) + \omega_{0j}(\mathbf{p}_0 - \mathbf{p}_j) + \omega_{0k}(\mathbf{p}_0 - \mathbf{p}_k) = 0.$$

Since $(\mathbf{p}_0 - \mathbf{p}_i)$ and $(\mathbf{p}_0 - \mathbf{p}_j)$ are parallel and $(\mathbf{p}_0 - \mathbf{p}_k)$ is a different direction we conclude that $\omega_{0k} = 0$ and $-\omega_{0i} = \omega_{0j}$. Define ω' for G_{ij} by $\omega'_{ij} = \omega_{0j}/2$ and $\omega'_e = \omega_e$ for all other edges e of G_{ij}. The definition is so that $\omega'_{ij}(\mathbf{p}_i - \mathbf{p}_j) = \omega_{0j}(\mathbf{p}_0 - \mathbf{p}_j)$, This and the assumption that ω is non-trivial implies that ω' is a non-trivial self-stress of G_{ij}, a contradiction. □

A graph G has a *Henneberg construction* iff G can be produced from a single edge by a sequence of $\mathbf{H}_2$ and $\mathbf{H}_3$-additions.

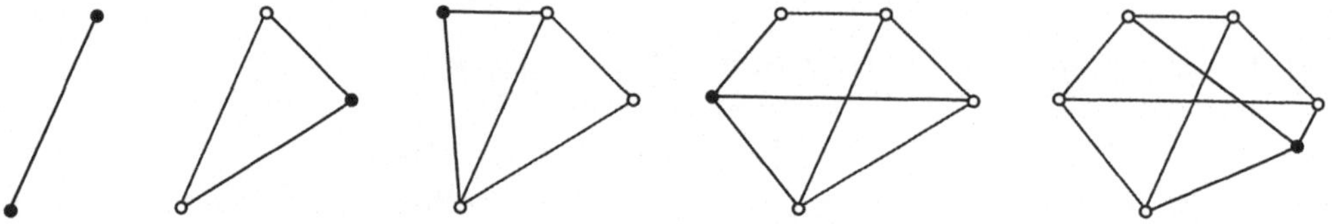

Figure 8.4 A Henneberg construction for $K_{3,3}$.

Theorem 8.7 *A graph $G = (V, E)$ is* mgr *iff G has a Henneberg construction.*

Proof. Suppose there is a Henneberg construction for G. A single edge is independent and $\mathbf{H}_2$ and $\mathbf{H}_3$-additions preserve generic independence (Lemma 8.2 and 8.6). Hence, G is generically independent. Since $|E| = 2|V| - 3$ the graph is mgr.

For the converse, let $G = (V, E)$ be a mgr-graph and $d(v_i)$ be the degree of vertex v_i. From $\sum d(v_i) = 2|E| = 4|V| - 6$ it follows that there is a vertex of degree at most 3 in G. A vertex of degree 0 or 1 would violate Proposition 8.5. Let v_0 be a vertex with $d(v_0) = 2$ or 3.

If $d(v_0) = 2$, then by Lemma 8.2 the graph G' induced by $V' = V \setminus \{v_0\}$ is independent, since it has two edges less than G it is mgr. A Henneberg construction of G' is guaranteed by induction. The $\mathbf{H}_2$-addition of v_0 to G' gives a Henneberg construction of G.

If $d(v_0) = 3$, then by Lemma 8.6 there are neighbors v_i, v_j of v_0 such that the graph G_{ij} is mgr. A Henneberg construction of G_{ij} is extended by the $\mathbf{H}_3$-addition of v_0 to G_{ij} which replaces the edge $\{v_i, v_j\}$. This gives, again with induction, a Henneberg construction of G. □

Actually it is possible to prescribe the two vertices of the initial edge of a Henneberg construction. This additional property will be helpful later.

Lemma 8.8 *Let $G = (V, E)$ be a* mgr*-graph and v_i, v_j be any two vertices of G, then there is a Henneberg construction of G beginning with the edge $\{v_i, v_j\}$.*

Proof. In the proof of Theorem 8.7 the vertex v_0 was selected subject to the condition that $d(v_0) = 2$ or 3. From $\sum d(v_i) = 2|E| = 4|V| - 6$ it follows that there are at least three vertices with this property. Therefore, it is possible to choose a vertex different from v_i and v_j in every step. At the end of the recursion there is a single edge which must be $\{v_i, v_j\}$. □

The following theorem gives two additional characterizations of mgr-graphs.

Theorem 8.9 *Each of the following conditions on a graph $G = (V, E)$ is a characterization of* mgr-*graphs in the plane:*

(1) G can be produced from a single edge by a sequence of $\mathbf{H}_2$ *and* $\mathbf{H}_3$*-additions. (Henneberg construction)*

(2) For any two vertices $v \neq w$ the (multi)-graph with edges $E \cup \{(v, w)\}$ is the union of two disjoint spanning trees. (Recski Theorem)

(3) $|E| = 2|V| - 3$ *and* $|E'| \leq 2|V[E']| - 3$ *for all* $\emptyset \neq E' \subset E$. *(Laman Condition)*

Proof. We know that (1) is a characterization of mgr-graphs (Theorem 8.7), it remains to prove that conditions (1), (2) and (3) are equivalent.

(1⇒2) The idea is to build trees T_1 and T_2 along a Henneberg construction. From Lemma 8.8 we take the existence of a Henneberg construction of G starting with edge $\{v, w\}$. Let G_0 be this initial graph, duplicate edge $\{v, w\}$. At this stage $T_1 = \{(v, w)\}$ and $T_2 = \{(v, w)\}$ is a decomposition of $E(G_0) \cup \{(v, w)\}$ into two spanning trees.

Let $G^+ = (V \cup \{v_0\}, E^+)$ with $E^+ = E \cup \{(v_0, v_i), (v_0, v_j)\}$ be a $\mathbf{H}_2$-addition from $G = (V, E)$. Assume that $E \cup \{(v, w)\}$ is the union of two spanning trees T_1 and T_2. Define $T_1^+ = T_1 \cup \{(v_0, v_i)\}$ and $T_2^+ = T_2 \cup \{(v_0, v_j)\}$, this is a partition of $E^+ \cup \{(v, w)\}$ into two spanning trees.

Let $G^+ = (V \cup \{v_0\}, E^+)$ with $E^+ = (E \setminus \{(v_i, v_j)\}) \cup \{(v_0, v_i), (v_0, v_j), (v_0, v_k)\}$ be a $\mathbf{H}_3$-addition from $G = (V, E)$. Assume that $E \cup \{v, w\}$ is the union of two spanning trees T_1 and T_2 and that the edge $\{v_i, v_j\}$ is in T_1. Let $T_1^+ = (T_1 \setminus \{(v_i, v_j)\}) \cup \{(v_0, v_i), (v_0, v_j)\}$ and $T_2^+ = T_2 \cup \{(v_0, v_k)\}$, this is a partition of $E^+ \cup \{(v, w)\}$ into two spanning trees.

The inductive definition yields a partition of $E \cup \{(v, w)\}$ into two spanning trees T_1 and T_2.

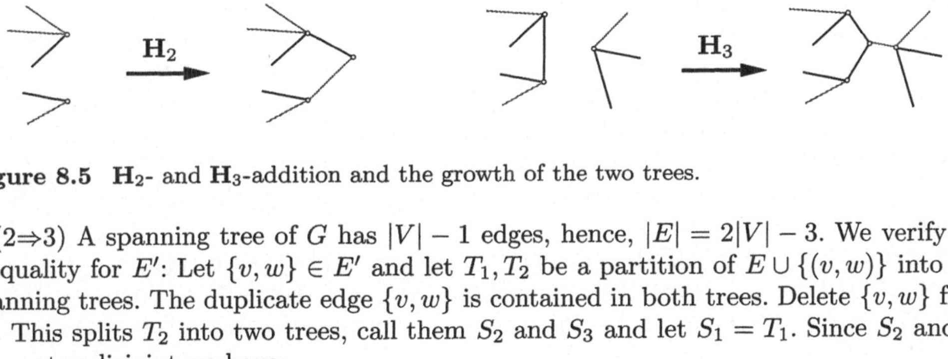

Figure 8.5 $\mathbf{H}_2$- and $\mathbf{H}_3$-addition and the growth of the two trees.

(2⇒3) A spanning tree of G has $|V| - 1$ edges, hence, $|E| = 2|V| - 3$. We verify the inequality for E': Let $\{v, w\} \in E'$ and let T_1, T_2 be a partition of $E \cup \{(v, w)\}$ into two spanning trees. The duplicate edge $\{v, w\}$ is contained in both trees. Delete $\{v, w\}$ from T_2. This splits T_2 into two trees, call them S_2 and S_3 and let $S_1 = T_1$. Since S_2 and S_3 are vertex disjoint we have:

- Each vertex v is incident to exactly two of the three trees S_1, S_2, S_3.

Back to E'. Let $F_i = E' \cap S_i$ and $V_i = V[E'] \cap S_i$. Since $v, w \in V[E']$ our construction guarantees that $|V_i| \geq 1$ for $i = 1, 2, 3$. Each (V_i, F_i) is a forest with non-empty vertex-set, therefore, $|F_i| \leq |V_i| - 1$. Each vertex of $V[E']$ is contained in at most two of the forests, therefore, $\sum |V_i| \leq 2|V[E']|$. This gives the inequality

$$|E'| = \sum |F_i| \leq \sum (|V_i| - 1) \leq 2|V[E']| - 3.$$

(3⇒1) Let G be a graph with the Laman Property. The claim is that there is a graph G' with one vertex less, such that G' has the Laman Property and G can be obtained from G' by a $\mathbf{H}_2$- or a $\mathbf{H}_3$-addition.

The Laman Property implies (see the proof of Theorem 8.7) that G has a vertex v_0 with $d(v_0) = 2$ or 3.

If $d(v_0) = 2$, then removing v_0 and the two incident edges gives a graph G' with the Laman Property. Clearly, G is obtained from G' by $\mathbf{H}_2$-addition of v_0.

If $d(v_0) = 3$, then consider the (multi)-graphs $G_{i,j}$, $G_{i,k}$ and $G_{j,k}$ as before. If one of them, say $G_{i,j}$, has the Laman Property, we choose $G' = G_{i,j}$ and G can be obtained from G' by $\mathbf{H}_3$-addition of v_0.

Suppose non of $G_{i,j}$, $G_{i,k}$ and $G_{j,k}$ has the property. Choose $E_i \subseteq E(G_{j,k})$ as a minimal set violating the Laman Property. That is $|E_i| > 2|V[E_i]| - 3$ and $|E_i'| \leq 2|V[E_i']| - 3$ for all $E_i' \subsetneqq E_i$. It follows that $|E_i| = 2|V[E_i]| - 2$ and $\{v_j, v_k\} \in E_i$. Sets $E_j \subseteq E(G_{i,k})$ and $E_k \subseteq E(G_{i,j})$ are chosen alike.

First assume that $V[E_i]$, $V[E_j]$ and $V[E_k]$ have pairwise exactly one vertex in common. This implies $|V[E_i]| + |V[E_j]| + |V[E_k]| = |V[E_i] \cup V[E_j] \cup V[E_k]| + 3 = |V[E_i \cup E_j \cup E_k]| + 3$. It follows that $|E_i \cup E_j \cup E_k| = |E_i| + |E_j| + |E_k| = 2(|V[E_i]| + |V[E_j]| + |V[E_k]|) - 6 = 2|V[E_i \cup E_j \cup E_k]|$. In G we consider the set $E^* = (E_i \cup E_j \cup E_k) \setminus \{\{v_i, v_j\}, \{v_i, v_k\}, \{v_j, v_k\}\} \cup \{\{v_0, v_i\}, \{v_0, v_j\}, \{v_0, v_k\}\}$. This set E^* contains v_0, hence, $|V[E^*]| = |V[E_i \cup E_j \cup E_k]| + 1$. For $|E^*|$ we compute: $|E^*| = |E_i \cup E_j \cup E_k| = 2|V[E_i \cup E_j \cup E_k]| = 2|V[E^*]| - 2$. Consequently, E^* violates the Laman Property of G, a contradiction.

For the other case we prepare with the following observation:

($\star$) Suppose A and B are sets of edges with $|V[A] \cap V[B]| \geq 2$. Suppose further that $|A| = 2|V[A]| - a$ and $|B| = 2|V[B]| - b$ and $|A \cap B| \leq 2|V[A \cap B]| - c$. Then $|A \cup B| \geq 2|V[A \cup B]| - a - b + c$.

For the proof of ($\star$) note that $|V[A \cap B]| \leq |V[A] \cap V[B]|$ and compute: $|A \cup B| = |A| + |B| - |A \cap B| \geq (2|V[A]| - a) + (2|V[B]| - b) - (2|V[A \cap B]| - c) \geq 2(|V[A]| + |V[B]| - |V[A] \cap V[B]|) - a - b + c = 2|V[A] \cup V[B]| - a - b + c = 2|V[A \cup B]| - a - b + c$.

Assume that the two sets $V[E_i]$ and $V[E_j]$ have at least two vertices in common. Use ($\star$), with $A = E_i$, $a = 2$, $B = E_j$, $b = 2$ and $c = 3$, this choice of c is legitimized by the fact that G is Laman. This yields $|E_i \cup E_j| \geq 2|V[E_i \cup E_j]| - 1$. Since G is Laman and $E_i \cup E_j$ has exactly two edges not from G, the inequality is tight: $|E_i \cup E_j| = 2|V[E_i \cup E_j]| - 1$.

Using ($\star$), with $A = E_i \cup E_j$, $a = 1$, $B = E_k$, $b = 2$ and $c = 3$, gives $|E_i \cup E_j \cup E_k| \geq 2|V[E_i \cup E_j \cup E_k]|$. Since G is Laman and $E_i \cup E_j \cup E_k$ has exactly three edges not from G, the inequality is again tight.

As before $E^* = (E_i \cup E_j \cup E_k) \setminus \{\{v_i, v_j\}, \{v_i, v_k\}, \{v_j, v_k\}\} \cup \{\{v_0, v_i\}, \{v_0, v_j\}, \{v_0, v_k\}\}$, the set E^* contains v_0, hence, $|V[E^*]| = |V[E_i \cup E_j \cup E_k]| + 1$. For $|E^*|$ we compute: $|E^*| = |E_i \cup E_j \cup E_k| = 2|V[E_i \cup E_j \cup E_k]| = 2|V[E^*]| - 2$. This is again contradicting the Laman Property of G. □

The 1-Dimensional Case

It is instructive to compare the result for 2-dimensional rigidity with the much simpler case of 1-dimension rigidity. In the 1-dimensional setting a framework $G[\mathbf{p}]$ is a graph $G = (V, E)$ with an embedding $\mathbf{p} : V \to \mathbb{R}$.

An assignment of velocities $\mathbf{v} : V \to \mathbb{R}$ is an infinitesimal motion of $G[\mathbf{p}]$ iff $\mathbf{v}_i = \mathbf{v}_j$ for every edge $\{i, j\} \in E$. Trivial motions of the line are translations, therefore, $\mathbf{v}$ is a trivial infinitesimal motion iff all velocities $\mathbf{v}_i$ are equal. With these observations it is easy to deduce:

- The rank of the rigidity matrix of a 1-dimensional framework $G[\mathbf{p}] = (V, E, \mathbf{p})$ is at most $|V| - 1$.
- A 1-dimensional framework $G[\mathbf{p}]$ is (infinitesimally) rigid iff the graph G of the framework is connected.
- A 1-dimensional framework $G[\mathbf{p}] = (V, E, \mathbf{p})$ is minimally rigid iff G is a spanning tree of V.

A self-stress of $G[\mathbf{p}]$ is an assignment of forces $\omega : E \to \mathbb{R}$ to the edges so that all vertices remain in equilibrium: $\sum_{\{i,j\}\in E} \omega_{ij}(\mathbf{p}_j - \mathbf{p}_i) = 0 \quad$ for all $i \in V$. If i is a vertex with $d(i) = 1$ and $\{i, j\}$ is the edge incident it i then $\omega_{ij} = 0$ in every self-stress. If C is a cycle of G, then there is a self-stress with $\omega_e \neq 0$ for all $e \in C$. This gives the result corresponding to Proposition 8.4.

- A 1-dimensional framework $G[\mathbf{p}] = (V, E, \mathbf{p})$ is maximal independent if G is cycle free and $|E| = |V| - 1$.

On the line every embedding that puts all vertices to different points is generic. The characterization of mgr-graphs for the line is very easy, they are trees.

With the following definition we can give a complete analog to Theorem 8.9.

($\mathbf{H}_1$) The graph $G^+ = (V^+, E^+)$ is produced from $G = (V, E)$ by a $\mathbf{H}_1$-addition if $V^+ = V \cup \{v_0\}$ with $v_0 \notin V$ and there is a vertex v_i in V such that $E^+ = E \cup \{(v_0, v_i)\}$.

Theorem 8.10 *Each of the following conditions is a characterization of mgr-graphs in 1-dimension:*

(1) G can be produced from a single vertex by a sequence of $\mathbf{H}_1$-additions.

(2) G is a tree.

(3) $|E| = |V| - 1$ and $|E'| \leq |V_{[E']}| - 1$ for all $E' \subset E$.

8.2 Pseudotriangles and Pseudotriangulations

A *pseudotriangle* is a simple polygon with exactly three convex vertices. The convex vertices are called *corners* of the pseudotriangle. Between any two corners there is a polygonal chain with only concave vertices, i.e., a chain bent towards the interior of the pseudotriangle, see Figure 8.6.

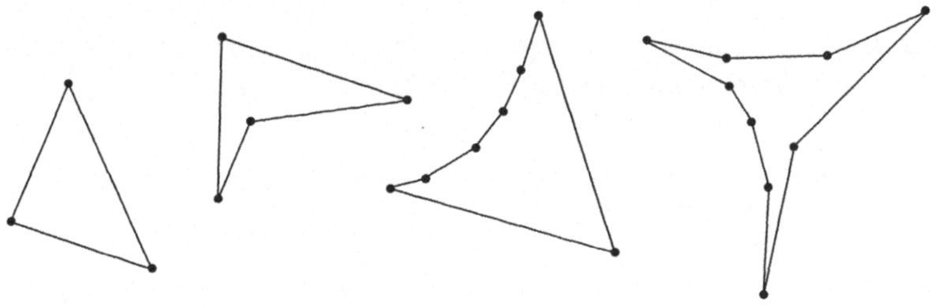

Figure 8.6 A collection of pseudotriangles.

Let $\mathcal{P}$ be a set of points in the plane. A *pseudotriangulation* of $\mathcal{P}$ is a non-crossing geometric graph with vertex set $\mathcal{P}$ which includes all edges of the convex hull and such that all the bounded faces are pseudotriangles (implying that there are no isolated points $p \in \mathcal{P}$ in the pseudotriangles). In this section we are mainly interested in minimum pseudotriangulations, i.e., pseudotriangulations with a minimal number of edges. In Theorem 8.11 it is shown that minimum pseudotriangulations are a particular class of mgr-graphs.

A geometric graph is *pointed at a vertex v* if one of the angles between neighboring edges at v is of size at least π. A geometric graph is *pointed* if it is pointed at each vertex. Figure 8.7 illustrates these definitions.

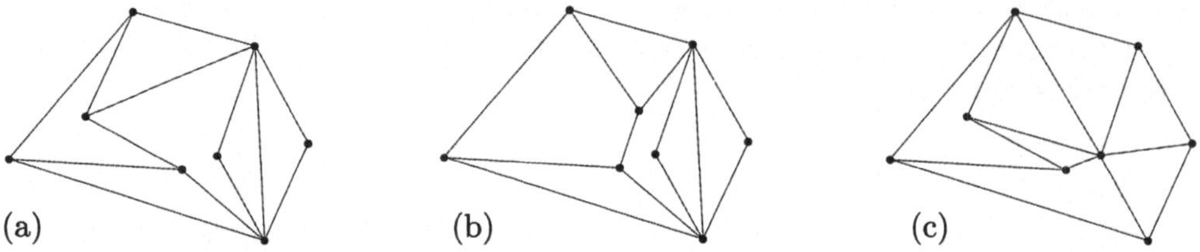

Figure 8.7 A pointed pseudotriangulation (a), the same graph but not a pseudotriangulation (b) and a non-pointed pseudotriangulation (c).

Note that the pseudotriangulation (c) in the figure is minimal (it has no removable edge) but not minimum (with the same number of vertices it has more edges than pseudotriangulation (a)).

Theorem 8.11 *Let G be a geometric graph on a set $\mathcal{P}$ of n points in general position. The following properties are equivalent:*

(1) G is a pointed pseudotriangulation of $\mathcal{P}$.

(2) G is a pseudotriangulation of $\mathcal{P}$ with $2n-3$ edges.

(3) G is a non-crossing and pointed graph on $\mathcal{P}$ with $2n-3$ edges.

(4) G is a minimum pseudotriangulation of $\mathcal{P}$.

(5) G is a non-crossing, pointed mgr-graph *on $\mathcal{P}$.*

(6) G is maximal non-crossing and pointed, i.e., upon addition of an edge the graph will lose one of the properties.

Before going into the proof of the theorem we have a lemma.

Lemma 8.12 *Let G be a non-crossing geometric graph with n vertices and e edges. Any two of the following three properties imply the third,*

(1) G is pointed.

(2) G is a pseudotriangulation.

(3) $e = 2n-3$

Proof. Euler's formula gives: $e = n + f^b - 1$, where f^b is the number of bounded faces. Recall the fact that every simple polygon and hence every bounded face of G, has at least three convex corners.

Let $c_{\mathbf{p}}$ be the number of angles of size less than π between neighboring edges at $\mathbf{p}$, i.e., $c_{\mathbf{p}}$ is the number of faces which have $\mathbf{p}$ as a convex corner. Let $d_{\mathbf{p}}$ be the degree of vertex $\mathbf{p}$. Note that $c_{\mathbf{p}} = d_{\mathbf{p}} - 1$ if $\mathbf{p}$ is pointed and otherwise $c_{\mathbf{p}} = d_{\mathbf{p}}$. The total number of pointed vertices is n^*. Then

$$2e = \sum_{\mathbf{p}} d_{\mathbf{p}} = \sum_{\mathbf{p}} c_{\mathbf{p}} + n^* = \sum_{F} (\#\text{ convex corners of } F) + n^* \geq 3f^b + n^*.$$

Equality holds iff G is a pseudotriangulation. Subtracting the inequality from Euler's formula multiplied by three gives:

$$e \leq 3n - n^* - 3.$$

Again, equality holds iff G is a pseudotriangulation. Based on this the three implications required for the lemma are immediate. △

Proof of Theorem 8.11. The lemma implies that (1), (2) and (3) are equivalent. Moreover, a pseudotriangulation has $3n - n^* - 3 \geq 2n - 3$ edges. Therefore, a pseudotriangulation with $e = 3n - 3$ is a minimum pseudotriangulation, i.e., (2⇒4). For the converse we need that a pseudotriangulation of $\mathcal{P}$ with $2n-3$ edges exists. This can be shown by induction: Let $\mathbf{p}_0$ be a point from the convex hull of $\mathcal{P}$. Consider a pseudotriangulation of $\mathcal{P} \setminus \{\mathbf{p}_0\}$ with $2(n-1) - 3$ edges and add $\mathbf{p}_0$ back, together with the two tangents to the convex hull of $\text{CH}(\mathcal{P} \setminus \{\mathbf{p}_0\})$.

The characterization of mgr-graphs (Theorem 8.9) implies $e = 2n - 3$, hence, (5⇒3). For the converse consider a set E' of edges of a pointed pseudotriangulation G. The graph $G' = (V[E'], E')$ is planar and pointed. The inequality $e \leq 3n - n^* - 3$ from the proof of the lemma implies $|E'| \leq 2|V[E']| - 3$. This is the Laman condition, hence, G is a mgr-graph.

A pointed planar graph has $e \leq 2n - 3$, therefore, (3⇒6). For the converse assume that G is maximal planar and pointed with $e < 2n - 3$. This implies that there is a bounded face F with at least four convex corners. Consider a geodesic, that is a shortest path, in the interior of F connecting two convex corners which are not adjacent in the cyclic order of convex corners induced by the boundary cycle of F, see Figure 8.8. Such a geodesic contains a straight segment s traversing the interior of F. The two endpoints of s are vertices of G. Moreover, either these endpoints are the endpoints of the path or

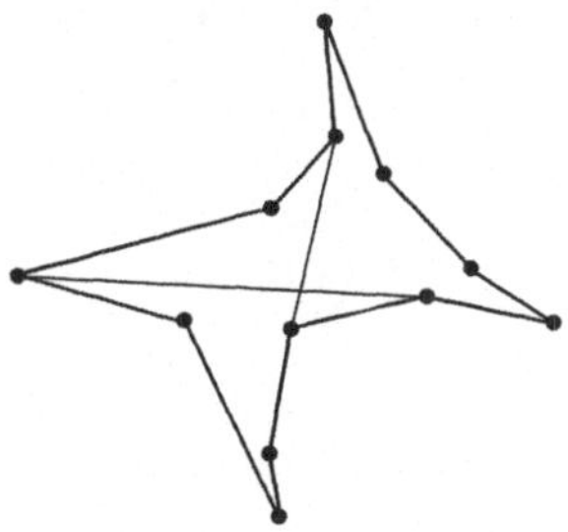

Figure 8.8 A pseudoquadrilateral with two geodesic paths, between pairs of opposite corners.

the path continues with a boundary edge of F such that the angle between this edge and s is more than π, i.e., upon addition of s this vertex of s remains pointed. This shows that s can be added to G as an edge. This addition preserves planarity and pointedness and, therefore, contradicts the maximality of G. □

As in the context of triangulations, flips play a prominent role in studies of pointed pseudotriangulations. Let e be an arbitrary interior edge of a pointed pseudotriangulation P. Removing e from P we have a planar pointed graph with $2n-4$ edges. A combination of Euler's formula with a double counting of corners (as in the proof of Lemma 8.12) implies that the new face created by the removal of e must be a pseudoquadrilateral.

In a pseudoquadrilateral there are the two geodesics each connecting a pair of opposite corners, Figure 8.8. Each of these two geodesics contains a single line segment which can be added such that the pseudoquadrilateral is subdivided into two pseudotriangles. The exchange between these two edges is the *flip*-operation in the context of pseudotriangulations, for examples see Figure 8.9.

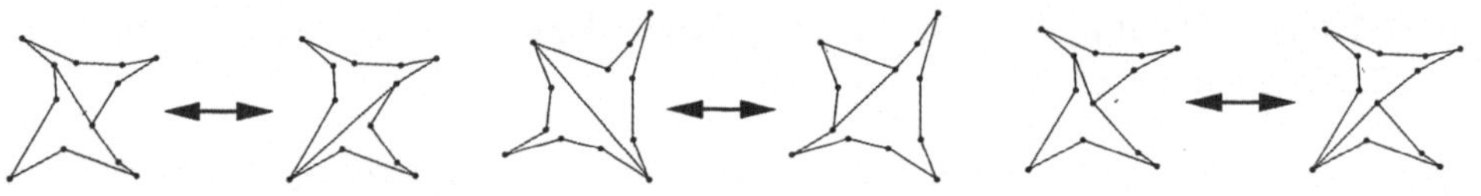

Figure 8.9 Three examples for a flip of edges.

The *flip-graph* $\mathcal{G}^{\curlyvee}(\mathcal{P})$ is the graph whose vertices are the pointed pseudotriangulations of $\mathcal{P}$ the edges of $\mathcal{G}^{\curlyvee}(\mathcal{P})$ are pairs of pseudotriangulations such that there is a flip operation transforming one into the other. An example of a flip-graph is shown in Figure 8.10.

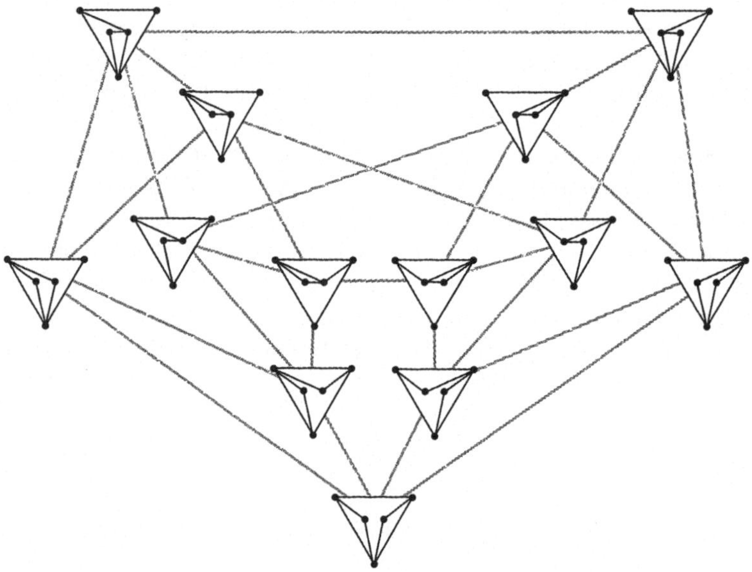

Figure 8.10 The flip-graph $\mathcal{G}^{\curlyvee}(\mathcal{P})$ of pointed pseudotriangulations of a set $\mathcal{P}$ of 5 points.

Proposition 8.13 *Let $\mathcal{G}^{\curlyvee}(\mathcal{P})$ be the flip-graph of pointed pseudotriangulations of a set $\mathcal{P}$ of n points in general position.*

(a) *$\mathcal{G}^{\curlyvee}(\mathcal{P})$ is regular of degree $2n-3-k$, where k is the number of points on $\mathrm{CH}(\mathcal{P})$.*

(b) *$\mathcal{G}^{\curlyvee}(\mathcal{P})$ is connected.*

Proof. We have already observed that every interior edge of a pointed pseudotriangulation P of $\mathcal{P}$ can be flipped. By Theorem 8.11 P has exactly $2n-3$ edges and k of these edges are on the convex hull, hence, not interior and non-flippable. This proves part (a).

For (b) we use an inductive argument. Let $\mathbf{p}_0$ be a point from the convex hull of $\mathcal{P}$. Every pointed pseudotriangulation of $\mathcal{P} \setminus \mathbf{p}_0$ can be extended by adding $\mathbf{p}_0$ and the two tangents connecting $\mathbf{p}_0$ to the convex hull $\text{CH}(\mathcal{P} \setminus \mathbf{p}_0)$. We assume, by induction, that $\mathcal{G}^{\curlyvee}(\mathcal{P} \setminus \mathbf{p}_0)$ is connected. To complete the proof we show that starting from an arbitrary pointed pseudotriangulation we can perform flips to reduce the degree of $\mathbf{p}_0$ until this degree is two: Let e be an interior edge incident to $\mathbf{p}_0$. The pseudoquadrilateral created by removing e has $\mathbf{p}_0$ as one of its four corners. The geodesic connecting the two neighboring corners of $\mathbf{p}_0$ avoids $\mathbf{p}_0$, hence, the edge e' replacing e after the flip also avoids $\mathbf{p}_0$. □

From the results of the following section it will follow that there is a $(2n-3)$-dimensional polytope whose vertices correspond to pointed pseudotriangulations of $\mathcal{P}$ and whose skeleton-graph is the flip-graph $\mathcal{G}^{\curlyvee}(\mathcal{P})$. Here, of course, $\mathcal{P}$ denotes a set of n points in general position in the plane. In particular the graph of Figure 8.10 is the graph of a 4-dimensional polytope.

8.3 Expansive Motions

An expansive motion on a point set is a motion which is increasing (or at least non-decreasing) on the distance of every pair of points. An *infinitesimal expansive motion* of a set $\mathcal{P}$ of n points in the plane is an assignment $\mathbf{v} : \mathcal{P} \to \mathbb{R}^2$ of a velocity to each point such that:

$$\langle \mathbf{p}_i - \mathbf{p}_j \,,\, \mathbf{v}_i - \mathbf{v}_j \rangle \geq 0 \quad \text{for all } 1 \leq i < j \leq n.$$

In terms of the rigidity matrix $\mathbf{R}_{\mathcal{P}}$ of the complete graph on $\mathcal{P}$ this reads $\mathbf{R}_{\mathcal{P}} \cdot \mathbf{v} \geq 0$. Taking the velocities as variables this system of inequalities defines a polyhedral cone. This cone contains the trivial motions, i.e., those corresponding to rigid transformations of the whole plane. These trivial motions can be excluded if we fix three velocity variables:

$$v_1^1 = v_1^2 = v_2^1 = 0 \quad \text{in other words:} \quad \mathbf{v}_1 = (0,0)\,, \;\; \mathbf{v}_2 = (0, v_2^2).$$

The normalized cone thus obtained is the *expansion cone* of $\mathcal{P}$, we denote this cone by $\bar{X}_0(\mathcal{P})$.

Lemma 8.14 *Let $\mathcal{P}$ be a set of n points in general position. The expansion cone $\bar{X}_0(\mathcal{P})$ is a pointed $(2n-3)$-dimensional cone.*

Proof. Translate $\mathcal{P}$ such that $\mathbf{p}_1 = (0,0)$ and $\mathbf{p}_2 = (0, y_2)$. The motion $\mathbf{v}$ with $\mathbf{v}_i = \mathbf{p}_i$ for all i is strictly expanding on all i,j with $1 \leq i < j \leq n$. Therefore, $\mathbf{v}$ is an interior point of $\bar{X}_0(\mathcal{P})$ and the cone is full-dimensional, i.e., $(2n-3)$-dimensional.

A cone is pointed if it contains no line. The pointedness of $\bar{X}_0(\mathcal{P})$ follows from the fact that the intersection of its facets is the origin: A motion $\mathbf{v}$ belongs to the facet defined by points $\mathbf{p}_i, \mathbf{p}_j$ iff the distance between these points remains invariant under the motion. Therefore, a motion $\mathbf{v}$ is in the intersection of all facets iff it keeps the length of all edges of the complete graph $K_n(\mathcal{P})$ invariant. Since this graph is rigid only trivial motions keep all edge length invariant but the only trivial motion with $\mathbf{v}_1 = (0,0)$ $\mathbf{v}_2^1 = 0$ corresponds to the origin. □

The expansion cone is a highly degenerate object. It is a polyhedron in $(2n-3)$ dimensional space, yet an extreme ray of the polyhedron is, in general, contained in a quadratic number of facets. Below we will investigate this in detail, an illustration for this fact is given with Figure 8.11. The left part of the figure indicates an expansive motion. The edges shown correspond to rigid pairs, i.e., pairs with $\langle \mathbf{p}_i - \mathbf{p}_j , \mathbf{v}_i - \mathbf{v}_j \rangle = 0$. Any additional edge would yield a rigid framework, therefore, this expansive motion corresponds to a ray of the cone. The right part of the figure shows a framework with $2n-4$ edges which determines the same rigid components and hence the same expansive motion.

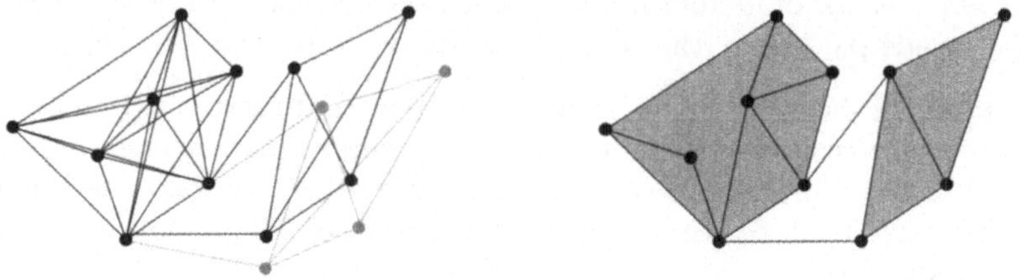

Figure 8.11 An expansive motion corresponding to a ray of the expansion cone and a planar framework with the same flexibility.

8.4 The Polyhedron of of Pointed Pseudotriangulations

We study the expansion cone $\bar{X}_0(\mathcal{P})$ and perturbations of this cone. The perturbations will be given by a scalar f_{ij} for every pair i, j and the translated inequalities:

$$\langle \mathbf{p}_i - \mathbf{p}_j , \mathbf{v}_i - \mathbf{v}_j \rangle \geq f_{ij} \quad \text{for all } 1 \leq i < j \leq n.$$

For any choice of scalars f_{ij} the inequalities define a polyhedron $\bar{X}_f(\mathcal{P})$. The expansion cone $\bar{X}_0$ is the special case $f \equiv 0$. The remarkable fact is that there is a choice of f such that the vertices of the perturbed cone $\bar{X}_f$ are in one to one correspondence to pointed pseudotriangulations of $\mathcal{P}$. Given such a choice of f we call the polyhedron $\bar{X}_f(\mathcal{P})$ the *polyhedron of pointed pseudotriangulations* of $\mathcal{P}$.

Lemma 8.15 *Let $\mathcal{P}$ be a set of n points in general position. For every choice of f the perturbed expansion cone $\bar{X}_f(\mathcal{P})$ is $(2n-3)$-dimensional and has at least one vertex.*

Proof. In the proof of Lemma 8.14 we found a strictly interior point of $\bar{X}_0$. Some positive multiple of this point is strictly interior in $\bar{X}_f$.

The recession cone of $\bar{X}_f$ is

$$\text{rec}(\bar{X}_f) = \{\mathbf{y} \in \mathbb{R}^{2n-3} : \mathbf{x} + t\,\mathbf{y} \in \bar{X}_f \text{ for all } \mathbf{x} \in \bar{X}_f, t > 0\}.$$

If $\bar{X}_f$ had no vertex then this would also be true for the recession cone $\text{rec}(\bar{X}_f)^{\dagger}$. The recession cone $\text{rec}(\bar{X}_f)$ equals $\bar{X}_0$ which has a vertex by Lemma 8.14. □

† Sorry, this is not elementary. Consult [219] for some background.

A point $\mathbf{v} \in \bar{X}_f(\mathcal{P})$ is a solution to the system $\mathbf{R}_{\mathcal{P}} \cdot \mathbf{v} \geq f$ of inequalities. Each inequality corresponds to a pair of points, i.e., an edge. An edge is *rigid* with respect to $\mathbf{v}$ iff $\mathbf{v}$ attains equality in the corresponding inequality. Let $E_{\mathbf{v}}$ denote the set of edges which are rigid with respect to $\mathbf{v}$. A face K of $\bar{X}_f(\mathcal{P})$ determines the set E_K of edges which are rigid for every $\mathbf{v} \in K$.

Lemma 8.16 *Consider the set $E_{\mathbf{v}}$ of rigid edges of a point $\mathbf{v} \in \bar{X}_0(\mathcal{P})$. If $E_{\mathbf{v}}$ contains (a) two crossing edges, (b) a non-pointed vertex (c) a convex subpolygon, then it contains the complete graph on the endpoints of all involved edges. In the case of (c) the complete graph also includes all points of $\mathcal{P}$ which are in the interior of the subpolygon. Figure 8.12 illustrates the three cases.*

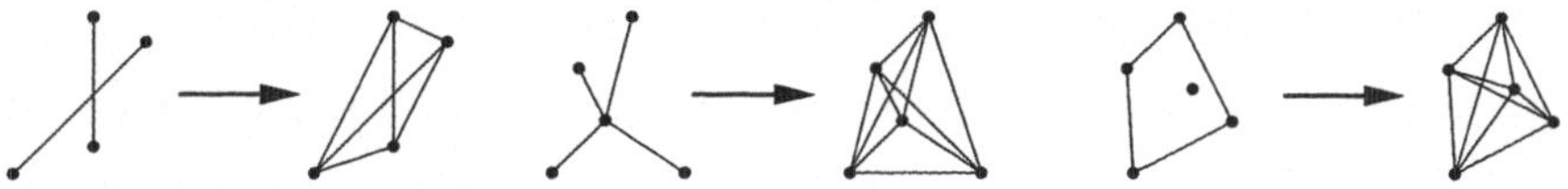

Figure 8.12 The closure for rigid sets of edges $E_{\mathbf{v}}$.

Proof. (a) Let $(\mathbf{p}_i, \mathbf{p}_j)$ and $(\mathbf{p}_k, \mathbf{p}_l)$ be a pair of crossing edges in $E_{\mathbf{v}}$ and let $\mathbf{x}$ be the point of intersection of the two edges. Suppose that the distance between $\mathbf{p}_i$ and $\mathbf{p}_k$ is strictly increasing. Adding a trivial motion to $\mathbf{v}$ we obtain a motion $\mathbf{v}'$ with $\mathbf{v}'_i = \mathbf{v}'_j = (0,0)$. The motion $\mathbf{v}'$ keeps the edge $(\mathbf{p}_i, \mathbf{p}_j)$ invariant, it is expanding on the pair i, k and non-decreasing on j, k. Since $\mathbf{x}$ is between $\mathbf{p}_i$ and $\mathbf{p}_j$ also $\langle \mathbf{p}_k - \mathbf{x}, \mathbf{v}'_k \rangle > 0$, see Figure 8.13. The vector $\mathbf{p}_k - \mathbf{p}_l$ is a positive scale of $\mathbf{p}_k - \mathbf{x}$, therefore, $\langle \mathbf{p}_k - \mathbf{p}_l, \mathbf{v}'_k \rangle > 0$. Similarly,

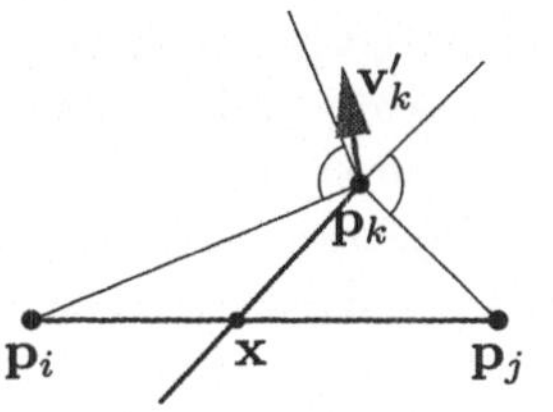

Figure 8.13 Crossing edges in $E_{\mathbf{v}}$.

$\langle \mathbf{p}_l - \mathbf{p}_k, \mathbf{v}'_l \rangle \geq 0$. Together, this implies that the length of $(\mathbf{p}_k, \mathbf{p}_l)$ is not preserved under $\mathbf{v}'$, hence, not under $\mathbf{v}$. This is in contradiction to $(\mathbf{p}_k, \mathbf{p}_l) \in E_{\mathbf{v}}$.

(b) Let $\mathbf{p}_i$ be a non-pointed vertex in $E_{\mathbf{v}}$. Suppose the neighbors of $\mathbf{p}_i$ do not move rigidly with $\mathbf{p}_i$. Since the distances are fixed some angles between neighboring edges at $\mathbf{p}_i$ change. Since the sum of angles around $\mathbf{p}_i$ is constant, a change is only possible if there is a decreasing angle. This, however, implies that the distance between the endpoints decreases, a contradiction.

(c) This time we consider the inner angles at the vertices of a convex polygon in $E_{\mathbf{v}}$. If the polygon does not move rigidly, some of these angles change. Since the sum of the inner angles is constant a change is only possible if one of them decreases. This, however, would imply that the distance between the endpoints decreases. Therefore, the the complete graph on the vertices of the polygon belongs to $E_{\mathbf{v}}$.

Now consider a point interior to the polygon. Adding a trivial motion to $\mathbf{v}$ we obtain a motion $\mathbf{v}'$ which keeps the polygon stationary. If the interior point moves it decreases its distance to one of the polygon edges and, hence, to a polygon vertex. □

Theorem 8.17 *For every set $\mathcal{P}$ of $n \geq 3$ points in general position there exist perturbations f_{ij} such that there is a bijection between the faces of the polyhedron $\bar{X}_f(\mathcal{P})$ and pointed non-crossing graphs on $\mathcal{P}$. The bijection $K \leftrightarrow E_K$ is order reversing, i.e., $K \subset K'$ if and only if $E_K \supset E_{K'}$. In particular:*

(a) *Vertices of $\bar{X}_f(\mathcal{P})$ correspond to pointed pseudotriangulations of $\mathcal{P}$.*

(b) *The graph of bounded edges of $\bar{X}_f(\mathcal{P})$ is the flip-graph $\mathcal{G}^{\curlyvee}$.*

(c) *Extreme rays correspond to pointed pseudotriangulations with one convex hull edge removed.*

A choice of perturbations $f = (f_{ij}) \in \mathrm{IR}^{\binom{n}{2}}$ is called *valid* if the statement of Theorem 8.17 becomes true.

Lemma 8.18 *A set of perturbations f_{ij} is valid if and only if the graph with edge set $E_{\mathbf{v}}$ is non-crossing and pointed for every $\mathbf{v} \in \bar{X}_f(\mathcal{P})$.*

Proof. For every vertex $\mathbf{x}$ of $\bar{X}_f(\mathcal{P})$ the clained properties of $E_{\mathbf{x}}$ follow from the definition of valid. If $\mathbf{v} \in \bar{X}_f(\mathcal{P})$ is a convex combination of vertices $\mathbf{x}_i$, $i \in I$, then $E_{\mathbf{v}} = \bigcap_{i \in I} E_{\mathbf{x}_i}$. This implies that the graph with edge set $E_{\mathbf{v}}$ is non-crossing and pointed. The harder part is the converse.

The polyhedron $\bar{X}_f$ is $(2n-3)$-dimensional, Lemma 8.15. Hence, every vertex $\mathbf{v}$ of $\bar{X}_f$ is incident to at least that many facets and these facets correspond to rigid edges, i.e., edges in $E_{\mathbf{v}}$. The graph with edge set $E_{\mathbf{v}}$ is pointed and non-crossing by assumption. With Theorem 8.11 it follows that it is a pointed pseudotriangulation, in particular $E_{\mathbf{v}} = 2n-3$. It follows that $\mathbf{v}$ is incident to $2n-3$ facets and, by definition, the polyhedron is a simple polyhedron. Every vertex figure of a simple polyhedron is a simplex. Therefore, the faces incident to a vertex $\mathbf{v}$ of $\bar{X}_f$ are in bijection with subsets of facets incident to $\mathbf{v}$. These subsets of facets in turn correspond to subgraphs of the pointed pseudotriangulation $E_{\mathbf{v}}$.

Since the polyhedron $\bar{X}_f$ is simple, every vertex $\mathbf{v}$ is incident to $2n-3$ edges of the polyhedron. These edges of $\bar{X}_f$ correspond to the subgraphs of $E_{\mathbf{v}}$ obtained by deleting a single edge. Let K_{ij} be the polyhedral edge corresponding to the removal of the edge $\{i, j\}$ from $E_{\mathbf{v}}$. If K_{ij} is bounded, then there is a second pointed pseudotriangulation containing $E_{\mathbf{v}} \setminus \{i, j\}$, this must be the pointed pseudotriangulation obtained from $E_{\mathbf{v}}$ by flipping the edge $\{i, j\}$. Hence, the bounded edges of $\bar{X}_f$ at vertex $\mathbf{v}$ correspond to the flips of the pointed pseudotriangulation $E_{\mathbf{v}}$. Since the flip-graph is connected (Proposition 8.13.b) we can reach every vertex of $\bar{X}_f$ along bounded edges from an arbitrary initial vertex. Since $\bar{X}_f$ has a vertex, this implies that all pointed pseudotriangulations appear as vertices. Moreover, the graph of bounded edges is exactly the flip-graph $\mathcal{G}^{\curlyvee}$. The simplicity of $\bar{X}_f$ implies that all pointed and non-crossing graphs appear as the graphs of rigid edges corresponding to faces of $\bar{X}_f$.

As for the extreme rays, they also correspond to subgraphs of $E_{\mathbf{v}}$ obtained by deleting a single edge. In this case there is exactly one completion to a pointed pseudotriangulation. Therefore, the removed edge must be an edge from the convex hull. □

The following lemma provides a simple criterion for valid perturbations.

Lemma 8.19 *A set of perturbations* $(f_{ij}) \in \mathbb{R}^{\binom{n}{2}}$ *is valid if and only if its restriction to every four point subset of* $\mathcal{P}$ *is valid.*

Proof. The implication "$\Longrightarrow$" is obvious. For the converse we use the previous lemma. There it was shown that $f = (f_{ij})$ is valid iff $E_{\mathbf{v}}$ is non-crossing and pointed for every $\mathbf{v} \in \bar{X}_f$. If this condition is violated then there is a subsets $\mathcal{P}'$ of four points in $\mathcal{P}$ such that the subgraph of $E_{\mathbf{v}}$ induced by these four points violates either the non-crossing or the pointedness condition. Let $\mathbf{v}'$ and f' be the restrictions of $\mathbf{v}$ and f to $\mathcal{P}'$. Then $\mathbf{v}' \in \bar{X}_{f'}(\mathcal{P}')$ and the graph with edge set $E_{\mathbf{v}'}$ is crossing or non-pointed, hence f' is not valid for $\mathcal{P}'$. □

The following theorem completes the proof of Theorem 8.17.

Theorem 8.20 *Let* $\mathbf{a}$ *and* $\mathbf{b}$ *be any two points in the plane. For every set of* n *points* $\{\mathbf{p}_1, \ldots, \mathbf{p}_n\}$ *in general position in the plane the following perturbations* $f = (f_{i,j})$ *are valid:*

$$f_{ij} = \det\begin{pmatrix} \mathbf{a} & \mathbf{p}_i & \mathbf{p}_j \\ 1 & 1 & 1 \end{pmatrix} \cdot \det\begin{pmatrix} \mathbf{b} & \mathbf{p}_i & \mathbf{p}_j \\ 1 & 1 & 1 \end{pmatrix}.$$

The full proof can be found in the original source, Rote, Streinu and Santos [166]. There it is verified that the perturbations of the theorem are valid on every subset of four points. By Lemma 8.19 this implies that f is valid for all n points. We indicate the main stations in the argument for 4-element point sets. Let $\mathcal{P} = (\mathbf{p}_1, \mathbf{p}_2, \mathbf{p}_3, \mathbf{p}_4)$ be in general position.

- Let $\gamma_{ij} = \det\begin{pmatrix} \mathbf{p}_i & \mathbf{p}_j & \mathbf{p}_k \\ 1 & 1 & 1 \end{pmatrix} \cdot \det\begin{pmatrix} \mathbf{p}_i & \mathbf{p}_j & \mathbf{p}_l \\ 1 & 1 & 1 \end{pmatrix}$ and $\omega_{ij} = \dfrac{1}{\gamma_{ij}}$, here k and l are the two indices other than i and j. The weights ω_{ij} on the edges of the complete graph K_4 on $\mathcal{P}$ define a self-stress.
- The signs of the determinants in the definition of ω_{ij} correspond to the orientations of the triangles $\{\mathbf{p}_i, \mathbf{p}_j, \mathbf{p}_k\}$ and $\{\mathbf{p}_i, \mathbf{p}_j, \mathbf{p}_l\}$. Therefore, $\omega_{ij} > 0$ if $\{i, j\}$ is a convex hull edge and $\omega_{ij} < 0$ if $\{i, j\}$ is an interior edge. Equivalently: $\omega_{ij} < 0$ iff $K_4 \backslash \{i, j\}$ is a pointed pseudotriangulation.
- Let $R = \sum_{1 \le i < j \le 4} \omega_{ij} f_{ij}$ with ω_{ij} and f_{ij} as above. For every edge $\{k, l\}$ of K_4 the following statements are equivalent:

  (1) The cone $\bar{X}_f(\mathcal{P})$ has a vertex $\mathbf{v}$ such that $E_{\mathbf{v}}$ is $K_4 \setminus \{k, l\}$.

  (2) R and ω_{kl} have opposite signs.
- With the above the claim of the theorem can be reduced to the inequality $R > 0$. For ω_{ij} and f_{ij} as above, the stronger statement $\sum\limits_{1 \le i < j \le 4} \omega_{ij} f_{ij} = 1$ can be verified:

  The idea is to consider $R = \sum_{1 \le i < j \le 4} \omega_{ij} f_{ij}$ as a function $R(\mathbf{a}, \mathbf{b})$ of $\mathbf{a}$ and $\mathbf{b}$. For fixed $\mathbf{a}$ this is an affine function $R(\mathbf{a}, \mathbf{b}) = R_{\mathbf{a}}(\mathbf{b})$ of $\mathbf{b}$. Since $R(\mathbf{p}_i, \mathbf{p}_j) = 1$ for all $i \neq j$, it follows that $R_{\mathbf{p}_i}(\mathbf{b}) \equiv 1$. Exchanging the roles of $\mathbf{a}$ and $\mathbf{b}$ we note that $R(\mathbf{a}, \mathbf{b}) = R_{\mathbf{b}}(\mathbf{a})$ is affine and attains the value 1 whenever $\mathbf{a} = \mathbf{p}_i$, therefore, $R(\mathbf{a}, \mathbf{b}) \equiv 1$.

This completes the sketch for the proof of Theorem 8.20. The conclusion is that the polyhedron of pointed pseudotriangulations $\bar{X}_f$ with properties as listed in Theorem 8.17 exists.

The defining inequalities of $\bar{X}_f$ which correspond to convex hull edges can be set to equality:

$$\langle \mathbf{p}_i - \mathbf{p}_j \, , \, \mathbf{v}_i - \mathbf{v}_j \rangle = f_{ij} \quad \text{for all convex hull edges } ij.$$

These equality constraints make the set of solutions $\mathbf{v}$ bounded. But still, with the above choice of perturbations f the vertices of this polytope $X_f(\mathcal{P})$ are in bijection to pointed pseudotriangulations of $\mathcal{P}$. The polytope $X_f(\mathcal{P})$ is the *polytope of pointed pseudotriangulations.* It should be remarked that in the case of a set $\mathcal{P}$ of points in convex position the polytope $X_f(\mathcal{P})$ is an associahedron. More precisely, the combinatorial structure of $X_f(\mathcal{P})$ is the structure of the associahedron, the representation by inequalities is not equivalent to the representation of the associahedron as secondary polytope.

8.5 Expansive Motions and Straightening Linkages

A *linkage* is a planar (non-crossing) framework whose graph is a path. *Straightening a linkage* means to apply a motion to the framework which ends in an embedding of the path on a line. The problem gets interesting by the requirement that throughout the motion the linkage remains planar, that is free of self-intersections.

The Carpenter's Rule Problem: Is it always possible to straighten a linkage?

Based on the theory presented in this chapter the above problem has an elegant affirmative answer. Actually, expansive motions were invented for the proof that every linkage can be straightened. The essential observation in this context is the fact that the application of an expansive motion to a non-crossing framework will preserve planarity.

Let L be a linkage which is not straight and let $\mathcal{P}$ be the points/vertices of L. In the convex hull $\text{CH}(\mathcal{P})$ there is an edge $\{i,j\}$ which does not belong to L. The expansive cone $\bar{X}_0(\mathcal{P})$ contains a ray $\vec{r}$ such that all edges of L belong to $E_{\vec{r}}$, the set of rigid edges of motions in $\vec{r}$, but $\{i,j\} \notin E_{\vec{r}}$. This means that $\{i,j\}$ is infinitesimally expanding.

The crucial result is that there is a real motion corresponding to the infinitesimal motion $\vec{r}$. The real motion is curved, see Figure 8.14. The infinitesimal motion only gives

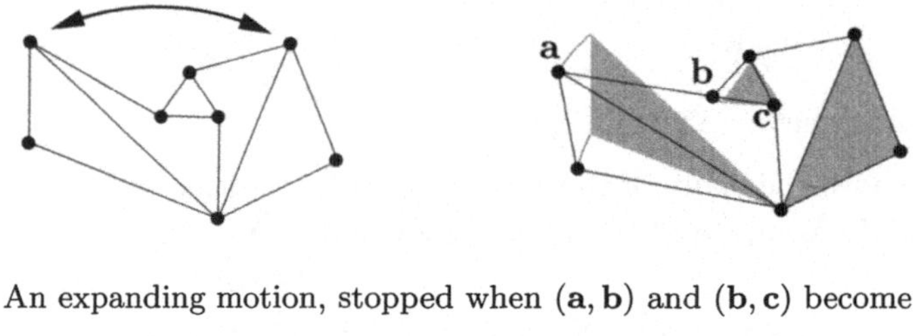

Figure 8.14 An expanding motion, stopped when $(\mathbf{a}, \mathbf{b})$ and $(\mathbf{b}, \mathbf{c})$ become collinear.

the initial velocities. The motion from the framework in the left part of the figure to the framework at the right can be interpreted as a smooth trajectory belonging to a certain initial infinitesimal motion. In the configuration on the right the edges $\mathbf{a}, \mathbf{b}$ and $\mathbf{b}, \mathbf{c}$ are collinear and the distance of $\mathbf{a}$ and $\mathbf{c}$ is at its extreme. Still this framework has an expansive motion but with a different set of rigid components. These cursory remarks may serve as an indication of how to combine small expanding motions whose existence comes from infinitesimal expanding motions to global straightening motion for a linkage.

8.6 Notes and References

The first major result in rigidity theory was a theorem of Cauchy.

> If two combinatorially equivalent 3-polyhedra P and P' are realized in $\mathbb{R}^3$ as convex polytopes such that corresponding faces are congruent, then the two polytopes are congruent.

A nice proof of Cauchy's theorem can be found in The Book [8]. Figure 8.15 illustrates that the conditions in the theorem are necessary. Dropping the convexity condition or the condition on the faces allows non-congruent realizations. In particular it is not enough to require that the edges of P and P' are of the same length to make them congruent.

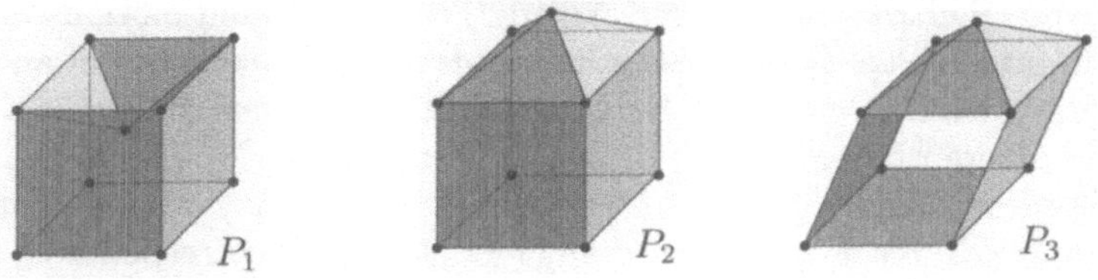

Figure 8.15 Polyhedron P_1 is non-convex, P_2 and P_3 have edges of equal length but non-congruent faces.

Asimow and Roth [17] showed that the 1-skeleton of a strictly convex polytope in 3-space which has at least one non-triangular face is not rigid.

It is not hard to see that the graphs of Asimow and Roth are not generically rigid in $\mathbb{R}^3$. This follows from a generalization of Proposition 8.5: If a graph $G = (V, E)$ is minimally generically rigid in dimension 3, then $|E| = 3|V| - 6$ and $|E'| \leq 3|V_{[E']}| - 6$ for all $E' \subset E$.

Laman [128] proved that in 2 dimensions the counting condition is also sufficient for generic rigidity. In 3 dimensions there are graphs with the 'right number of edges' which are non-rigid. It remains a major open problem to find a combinatorial characterization of generic 3-rigidity of graphs.

As shown in Theorem 8.9 both Laman property and Henneberg construction can be interpreted as characterizations of graphs which are the union of two spanning trees after adding any new edge. Frank and Szegő [93] extend this to a characterization of graphs which are the union of k spanning trees after addition of a new edge.

The articles of Whiteley [215] and [216] and the book by Graver, Servatius and Servatius [105] give comprehensive introductions into rigidity theory. These sources also provide lots of pointers into the huge literature on this topic.

A set of edges of a framework $G_{[\mathbf{p}]} = (V, E, \mathbf{p})$ is *independent* iff the corresponding rows of the rigidity matrix are independent. Therefore, the independent subsets of E are a matroid. If the embedding $\mathbf{p}$ is generic, this is the *generic rigidity matroid* of G. The bases of the generic rigidity matroid of a complete graph are exactly the mgr-graphs. In the 1-dimensional case the generic rigidity matroid of a connected graph G coincides with the cycle matroid of G (see page 138).

Pseudotriangulations have been used and studied in the context of ray-shooting by Chazelle et al. [47] and of visibility by Pocchiola and Vegter [158, 157] and others [146]. A paper of Streinu [190] where pseudotriangulations were applied to the problem of straightening linkages made them very popular. Since then new applications in polygon guarding, Speckmann and Tóth [181] and kinetic data structures, Kirkpatrick et al. [122], have been found.

There has also been a lot of research around questions of generating and enumerating pseudotriangulations, see [6, 7, 23, 41, 55]. The most prominent open problem in the area is the question whether every plane set of points has at least as many pseudotriangulations as triangulations.

Kettner et al. [121] have shown that every point set has a pseudotriangulation P such that the maximum degree of a vertex in P is 5, this result is best possible. Bespamyatnikh [24] shows that the diameter of the flip graph is of order $O(n \log n)$.

Haas et al. [111] have the following result: If G is a planar graph with the Laman property, then there is an embedding of G as a pointed pseudotriangulation.

Connelly, Demaine, and Rote [51] proved that linkages in the plane can be straightened and closed polygonal chains can be convexified. Their unlocking motions are expansive. The proof relies on the duality of stresses and motions. The existence of an infinitesimal expansive is shown via the Maxwell-Cremona theorem. The real motion is then obtained as a solution to differential equations.

Streinu [190] proved that a pointed pseudotriangulation is rigid and that removing a convex hull edge yields one degree of freedom (1-DOF) which corresponds to an expansive motion. Adding bars to a given collection of polygonal chains to produce a pointed pseudotriangulation and using flips to get from one pseudotriangulation to another, leads to a piecewise-algebraic straightening motion.

Rote, Streinu and Santos [166] introduce the expansive cone $\bar{X}_0(\mathcal{P})$ and the perturbed cone $\bar{X}_f(\mathcal{P})$ of a point set $\mathcal{P}$. Section 8.3 follows closely along the lines of that paper. The cited paper contains the full proof of Theorem 8.20. Moreover, the authors discuss connections of the polytope of constrained expansions with the associahedron. A surprising appearance of the associahedron is found as follows: They consider a special perturbation of the cone of expansive motions in 1 dimension. The vertices of this perturbed cone are shown to be in in bijection with non-crossing alternating trees. Non-crossing alternating trees are a Catalan family and indeed by adding a bounding constraint they obtain a polytope isomorphic to the associahedron.

Orden and Santos [145], building on ideas from [166], find a more general polyhedron whose faces correspond to non-crossing marked geometric graphs on a planar point set $\mathcal{P}$. In this context a marking of a graph is a subset of its pointed vertices.

Bibliography

[1] P. K. Agarwal, B. Aronov, T. Chan, and M. Sharir, *On levels in arrangements of lines, segments, planes, and triangles*, Discr. and Comput. Geom., 19 (1998), pp. 315–331. (**67**)

[2] P. K. Agarwal, B. Aronov, J. Pach, R. Pollack, and M. Sharir, *Quasi-planar graphs have a linear number of edges*, Combinatorica, 17 (1997), pp. 1–9. (**15**)

[3] P. K. Agarwal, B. Aronov, and M. Sharir, *On levels in arrangements of lines, segments, planes, and triangles*, in Proc. 13th Symp. Comput. Geom., 1997. (**66**)

[4] P. K. Agarwal and M. Sharir, *Arrangements and their applications*, in Handbook of Computational Geometry, J.-R. Sack et al., ed., North-Holland, 2000, pp. 49–119. (**67, 84**)

[5] O. Aichholzer, F. Aurenhammer, and H. Krasser, *On the crossing number of complete graphs*, in Proc. 18th Symp. Comput. Geom., 2002, pp. 19–24. (**68**)

[6] O. Aichholzer, F. Aurenhammer, H. Krasser, and B. Speckmann, *Convexity minimizes pseudo-triangulations*, in Proc. 14th Canad. Conf. Comput. Geom., 2002, pp. 158–161. (**150**)

[7] O. Aichholzer, G. Rote, B. Speckmann, and I. Streinu, *The zigzag path of a pseudo-triangulation*, in Proc. 8th . Workshop Algo. and Data Struct., vol. 2748 of Lect. Notes Comput. Sci., Springer-Verlag, 2003, pp. 377–389. (**150**)

[8] M. Aigner and G. M. Ziegler, *Proofs from* The Book, Springer-Verlag, 1998. (**13, 85, 149**)

[9] M. Ajtai, V. Chvátal, M. Newborn, and E. Szemerédi, *Crossing-free subgraphs*, Ann. Discrete Math., 12 (1982), pp. 9–12. (**51**)

[10] J. Akiyama and N. Alon, *Disjoint simplices and geometric hypergraphs*, in Combinatorial Mathematics;, G. S. Blum et al., ed., vol. 555, Annals of the New York Academy of Sciences, 1989, pp. 1–3. (**14**)

[11] N. Alon and E. Györy, *The number of small semispaces of a finite set of points in the plane*, J. Combin. Theory Ser. A, 41 (1986), pp. 154–157. (**67, 113**)

[12] H. Alt, S. Felsner, F. Hurtado, M. Noy, and E. welzl, *Point-sets with few k-sets*, Comp. Geom.: Theory and Appl., 16 (2000), pp. 95–101. (**67**)

[13] A. Andrzejak, B. Aronov, S. Har-Peled, R. Seidel, and E. Welzl, *Results on k-sets and j-facets via continuous motions*, in Proc. 14th Symp. Comput. Geom., 1998. (**66, 67**)

[14] A. ANDRZEJAK AND E. WELZL, *Halving point sets*, in Proc. ICM Berlin 1998, vol. III, Doc. Math., J. DMV, 1998, pp. 471–478. (**67**)

[15] B. ARONOV, B. CHAZELLE, H. EDELSBRUNNER, L. J. GUIBAS, M. SHARIR, AND R. WENGER, *Points and triangles in the plane and halving planes in space*, Discr. and Comput. Geom., 6 (1991), pp. 435–442. (**67**)

[16] B. ARONOV AND M. SHARIR, *Cutting circles into pseudo-segments and improved bounds for incidences*, Discr. and Comput. Geom., 28 (2002), pp. 475–490. (**51**)

[17] L. ASIMOV AND B. ROTH, *The rigidity of graphs*, Trans. Amer. Math. Soc., 245 (1978), pp. 279–289. (**149**)

[18] F. AURENHAMMER AND R. KLEIN, *Voronoi diagrams*, in Handbook of Computational Geometry, J. R. Sack and J. Urrutia, eds., Elsevier, 2000, pp. 201–290. (**130**)

[19] I. BÁRÁNY, Z. FÜREDI, AND L. LOVÁSZ, *On the number of halving planes*, Combinatorica, 10 (1990), pp. 175–183. (**67**)

[20] I. BÁRÁNY AND W. STEIGER, *On the expected number of k-sets*, Discr. and Comput. Geom., 11 (1994), pp. 243–263. (**67**)

[21] D. BAYER, I. PEEVA, AND B. STURMFELS, *Monomial resolutions*, Math. Res. Lett., 5 (1998), pp. 31–46. (**41**)

[22] J. BECK, *On the lattice property of the plane and some problems of Dirac, Motzkin and Erdős in combinatorial geometry*, Combinatorica, 3 (1983), pp. 281–297. (**51, 86**)

[23] S. BESPAMYATNIKH, *Enumerating pseudo-triangulations in the plane*, in Proc. 14th Canad. Conf. Comput. Geom., 2002, pp. 162–166. (**150**)

[24] S. BESPAMYATNIKH, *Transforming pseudo-triangulations*, in Int. Conf. on Computational Sci., P.M.A. Sloot et al., ed., vol. 2657 of Lect. Notes Comput. Sci., Springer-Verlag, 2003, pp. 533–539. (**150**)

[25] D. BIENSTOCK AND N. DEAN, *Bounds for rectilinear crossing numbers*, J. Graph Theory, 17 (1993), pp. 333–348. (**50**)

[26] N. L. BIGGS, E. K. LLOYD, AND R. J. WILSON, *Graph Theory 1736–1936*, Claredon Press, 1976. (**13**)

[27] L. J. BILLERA, P. FILLIMAN, AND B. STURMFELS, *Constructions and complexity of secondary polytopes*, Adv. Math., 83 (1990), pp. 155–179. (**130**)

[28] A. BJÖRNER, M. LAS VERGNAS, N. WHITE, B. STURMFELS, AND G. M. ZIEGLER, *Oriented Matroids*, Cambridge University Press, 1993. (**79, 111, 112, 113**)

[29] B. BOLLOBÁS, *Extremal graph theory*, in Handbook of Combinatorics, Vol II, L. e. Graham, Grötschel, ed., North-Holland, 1995, pp. 1231–1292. (**13**)

[30] N. BONICHON, *A bijection between realizers of maximal plane graphs and pairs of non-crossing Dyck paths*, in Proc. Formal Power Series and Alg. Combin., 2002. (**42**)

[31] N. BONICHON, C. GAVOILLE, AND N. HANUSSE, *An information-theoretic upper bound of planar graphs using triangulation*, in 20th Annual Symposium on Theoretical Aspects of Computer Science (STACS), vol. 2607 of Lect. Notes Comput. Sci., Springer-Verlag, 2003, pp. 499 – 510. (**42**)

[32] N. BONICHON, B. LE SAËC, AND M. MOSBAH, *Wagners theorem on realizers*, in Proc. 29th Int. Colloq. Automata, Lang, Progr., Lect. Notes Comput. Sci., Springer-Verlag, 2002, pp. 1043 – 1053. (**42**)

[33] P. BORWEIN AND W. O. J. MOSER, *A survey of Sylvester's problem and its generalizations*, Aequat. Math., 40 (1990), pp. 111–135. (**85**)

[34] P. BRASS, G. KÁROLYI, AND P. VALTR, *On a Turán-type extremal theory for convex geometric graphs*, in Discrete and Computational Geometry, The Goodman and Pollack Festschrift, vol. 25 of Algorithms and Combinatorics, Springer Verlag, 2003, pp. 275–300. (**16**)

[35] P. BRASS, W. O. J. MOSER, AND J. PACH, *Research Problems in Discrete Geometry*, Monograph, 2004. (**85**)

[36] E. BREHM, *3-orientations and Schnyder 3–tree–decompositions*, master's thesis, Freie Universität Berlin, Germany, 2000 . http://www.math.tu-berlin.de/~felsner/Diplomarbeiten/brehm.ps.gz. (**41**)

[37] G. BRIGHTWELL AND W. T. TROTTER, *The order dimension of convex polytopes*, SIAM J. Discr. Math., 6 (1993), pp. 230–245. (**41**)

[38] G. BRIGHTWELL AND W. T. TROTTER, *The order dimension of planar maps*, SIAM J. Discr. Math., 10 (1997), pp. 515–528. (**41**)

[39] G. R. BRIGHTWELL AND E. R. SCHEINERMAN, *Representations of planar graphs*, SIAM J. Discr. Math., 6 (1993), pp. 214–229. (**39**)

[40] T. BRITZ AND S. FOMIN, *Finite posets and Ferrer shapes*, Adv. Math., 158 (2001), pp. 86–127. (**14**)

[41] H. BRÖNNIMANN, L. KETTNER, M. POCCHIOLA, AND J. SNOEYINK, *Counting and enumerating pseudo-triangulations with the greedy flip algor ithm*, in manuscript, 2001. (**150**)

[42] G. R. BURTON AND G. PURDY, *The directions determined by n points in the plane*, J. Lond. Math. Soc., II. Ser, 20 (1979), pp. 109–114. (**112**)

[43] G. CAIRNS AND Y. NIKOLAYEVSKY, *Bounds for generalized thrackles*, Discr. and Comput. Geom., 23 (2000), pp. 191–206. (**15**)

[44] J. W. CANNON, W. J. FLOYD, R. KENYON, AND W. R. PARRY, *Hyperbolic geometry*, in Flavors of Geometry, S. Levy, ed., vol. 31 of Math. Sci. Res. Inst. Publ., Cambridge University Press, 1997, pp. 59–115. (**130**)

[45] V. Capoyleas and J. Pach, *A Turán-type theorem on chords of a convex polygon*, J. Combin. Theory Ser. B, 56 (1992), pp. 9–15. (**14, 15**)

[46] D. Chakerian, *Sylvester's problem on collinear points and a relative*, Amer. Math. Monthly, 77 (1970), pp. 164–167. (**84**)

[47] B. Chazelle, H. Edelsbrunner, M. Grigni, L. Guibas, J. Hershberger, M. Sharir, and J. Snoeyink, *Ray shooting in polygons using geodesic triangulations*, Algorithmica, 12 (1994), pp. 54–68. (**149**)

[48] B. Chazelle, L. J. Guibas, and D. T. Lee, *The power of geometric duality*, BIT, 25 (1985), pp. 76–90. (**84**)

[49] M. Chrobak and G. Kant, *Convex grid drawings of 3-connected planar graphs*, Int. J. Comput. Geom. Appl., 7 (1997), pp. 211–223. (**38**)

[50] K. Clarkson, H. Edelsbrunner, L. Guibas, M. Sharir, and E. Welzl, *Combinatorial complexity bounds for arrangements of curves and spheres*, Discr. and Comput. Geom., 5 (1990), pp. 99–160. (**51**)

[51] R. Connelly, E. Demaine, and G. Rote, *Straightening polygonal arcs and convexifying polygonal cycles*, in Proc. 41st IEEE Symp. Found. Comput. Sci., 2000, pp. 432–442. (**150**)

[52] H. S. M. Coxeter, *Introduction to Geometry*, John Wiley & Sons, 2nd ed., 1969. (**84**)

[53] J. Csima and E. T. Sawyer, *There exist* $6n/13$ *ordinary points*, Discr. and Comput. Geom., 9 (1993), pp. 187–202. (**85**)

[54] J. Csima and E. T. Sawyer, *The* $6n/13$ *theorem revisited*, in Graph theory, combinatorics, algorithms and applications. Vol. 1, Y. Alavi et al., ed., Wiley, 1995, pp. 235–249. (**85**)

[55] F. S. D. Randall, G. Rote and J. Snoeyink, *Counting triangulations and pseudo-triangulations of wheels*, in Proc. 13th Canad. Conf. Comput. Geom., 2001, pp. 149–152. (**150**)

[56] H. de Fraysseix, J. Pach, and R. Pollack, *Small sets supporting Fary embeddings of planar graphs*, in Proc. 20th ACM Symp. Theory Comput., 1988, pp. 426–433. (**37**)

[57] H. de Fraysseix, J. Pach, and R. Pollack, *How to draw a planar graph on a grid*, Combinatorica, 10 (1990), pp. 41–51. (**37**)

[58] P. O. de Mendez, *Orientations bipolaires*, PhD thesis, Paris, 1994. (**41**)

[59] T. K. Dey, *Improved bounds on planar* k*-sets and related problems*, Discr. and Comput. Geom., 19 (1998), pp. 373–382. (**66**)

[60] T. K. Dey and H. Edelsbrunner, *Counting triangle crossings and halving planes*, Discr. and Comput. Geom., 12 (1994), pp. 281–289. (**67**)

[61] G. Di Battista, P. Eades, R. Tamassia, and I. G. Tollis, *Graph Drawing*, Prentice Hall, 1999. (**37**)

[62] G. A. DIRAC, *Collinearity properties of sets of points*, Quart. J. Math. Ser. 2, 2 (1951), pp. 221–227. (**86**)

[63] P. EDELMAN, *On the average number of k-sets*, Discr. and Comput. Geom., 8 (1992), pp. 209–213. (**67, 112**)

[64] P. EDELMAN AND C. GREENE, *Balanced tableaux*, Adv. Math., 63 (1987), pp. 42–99. (**112**)

[65] H. EDELSBRUNNER, *Algorithms in Combinatorial Geometry*, vol. 10 of EATCS Monographs on Theoretical Computer Science, Springer-Verlag, 1987. (**84, 130**)

[66] H. EDELSBRUNNER, *Geometry and Topology for Mesh Generation*, Cambridge University Press, 2001. (**129, 130**)

[67] H. EDELSBRUNNER, P. VALTR, AND E. WELZL, *Cutting dense point sets in half*, Discr. and Comput. Geom., 17 (1997), pp. 243–255. (**67**)

[68] H. EDELSBRUNNER AND E. WELZL, *On the number of line separations of a finite set in the plane*, J. Combin. Theory Ser. A, 40 (1985), pp. 15–29. (**66**)

[69] G. ELEKES, *On the number of sums and products*, Acta Arithm., 81 (1997), pp. 365–367. (**51**)

[70] P. D. T. A. ELLIOTT, *On the number of circles determined by n points*, Acta Math. Acad. Sci. Hung, 18 (1967), pp. 181–188. (**85**)

[71] S. ELNITSKY, *Rhombic tilings of polygons and classes of reduced words in Coxeter groups*, J. Combin. Theory Ser. A, 77 (1997), pp. 193–221. (**113**)

[72] D. EPPSTEIN, *Fifteen proofs of Euler's formula:* $V - E + F = 2$. http://www.ics.uci.edu/˜eppstein/junkyard/euler. (**13**)

[73] D. EPPSTEIN, *Geometric lower bounds for parametric matroid optimization*, Discr. and Comput. Geom., 20 (1998), pp. 463–476. (**67**)

[74] P. ERDŐS, *On a set of distances of n points*, Amer. Math. Monthly, 53 (1946), pp. 248–250. (**14, 51**)

[75] P. ERDŐS AND R. K. GUY, *Crossing number problems*, Amer. Math. Monthly, 80 (1973), pp. 52–58. (**50, 51**)

[76] P. ERDÖS, L. LOVÁSZ, A. SIMMONS, AND E. G. STRAUSS, *Dissection graphs of planar point sets*, in A Survey of Combinatorial Theory, North-Holland, 1973, pp. 139–149. (**66**)

[77] P. ERDÖS AND G. PURDY, *Extremal problems in combinatorial geometry*, in Handbook of Combinatorics, Vol I, Graham, Grötschel, and Lovász, eds., North-Holland, 1995, pp. 809–874. (**51, 84**)

[78] S. EVEN, A. PNUELI, AND A. LEMPEL, *Permutation graphs and transitive graphs*, J. of the ACM, 19 (1972), pp. 400–410. (**109**)

[79] I. FARY, *On straight lines representation of planar graphs*, Acta Sci. Math. Szeged, 11 (1948), pp. 229–233. (**13, 37**)

[80] S. FELSNER, *On the number of arrangements of pseudolines*, Discr. and Comput. Geom., 18 (1997), pp. 257–267. (**111, 113**)

[81] S. FELSNER, *Convex drawings of planar graphs and the order dimension of 3-polytopes*, Order, 18 (2001), pp. 19–37. (**37, 41**)

[82] S. FELSNER, *The skeleton of a reduced word and a correspondence of Edelman and Greene*, Elec.. J. Combin., 8 (2001), p. 21p. (**112**)

[83] S. FELSNER, *Lattice structures from planar graphs*, 2002. submitted. (**41, 42**)

[84] S. FELSNER, *Geodesic embeddings and planar graphs*, 2004. to appear. (**41**)

[85] S. FELSNER AND K. KRIEGEL, *Triangles in Euclidean arrangements*, Discr. and Comput. Geom., 22 (1999), pp. 429–438. (**86**)

[86] S. FELSNER AND W. T. TROTTER, *Posets and planar graphs*, J. Graph Theory, (2004). to appear. (**41**)

[87] S. FELSNER AND H. WEIL, *A theorem on higher Bruhat orders*, Discr. and Comput. Geom., 23 (2000), pp. 121–127. (**113**)

[88] S. FELSNER AND H. WEIL, *Sweeps, arrangements and signotopes*, Discr. Appl. Math., 18 (2001), pp. 257–267. (**111, 113**)

[89] S. FELSNER AND G. M. ZIEGLER, *Zonotopes associated with higher Bruhat orders*, Discr. Math., 241 *Tverberg Festschrift* (2001), pp. 301–312. (**113**)

[90] W. FENCHEL, *Lösung Aufgabe 167 von Hopf und Pannwitz*, Jahresber. der Deutschen Mathem. Vereinig., 45 (1935), pp. 34–35. (**14**)

[91] D. FORGE AND J. L. R. ALFONSÍN, *Straight line arrangements in the real projective plane*, Discr. and Comput. Geom., 20 (1998), pp. 155–161. (**86**)

[92] A. FRANK, *On chain and antichain families of partially ordered sets*, J. Combin. Theory Ser. B, 29 (1980), pp. 176–184. (**14**)

[93] A. FRANK AND L. SZEGŐ, *Constructive characterizations for packing and covering with trees*, Discr. Appl. Math., 131 (2003), pp. 347–371. (**149**)

[94] P. FRANKL AND Z. FÜREDI, *A sharpening of Fisher's inequality*, Discr. Math., 90 (1991), pp. 103–107. (**85**)

[95] T. GALLAI, *Transitiv orientierbare graphen*, Acta Math. Acad. Sci. Hung, 18 (1967), pp. 25–66. (**109**)

[96] M. R. GAREY AND D. S. JOHNSON, *Crossing number is NP-complete*, SIAM J. Alg. Discr. Meth., 4 (1983), pp. 312–316. (**51**)

[97] B. GÄRTNER AND E. WELZL, *Vapnik-Chervonenkis dimension and (pseudo-) hyperplane arrangements*, Discr. and Comput. Geom., 12 (1994), pp. 399–432. (**84**)

[98] I. M. GEL'FAND, A. V. ZELEVINSKIJ, AND M. M. KAPRANOV, *Newton polyhedra of principal a-determinants*, Sov. Math., Dokl., 40 (1990), pp. 278–281. (**130**)

[99] W. GODDARD, M. KATCHALSKI, AND D. J. KLEITMAN, *Forcing disjoint segments in the plane*, Europ. J. Combin., 17 (1997), pp. 391–395. (**14**)

[100] J. E. GOODMAN, *Proof of a conjecture of Burr, Grünbaum and Sloane*, Discr. Math., 32 (1980), pp. 27–35. (**112**)

[101] J. E. GOODMAN, *Pseudoline arrangements*, in Handbook of Discrete and Computational Geometry, Goodman and O'Rourke, eds., CRC Press, 1997, pp. 83–110. (**84, 111, 112**)

[102] J. E. GOODMAN AND R. POLLACK, *Three points do not determine a (pseudo-) plane*, J. Combin. Theory Ser. A, 31 (1981), pp. 215–218. (**112**)

[103] J. E. GOODMAN AND R. POLLACK, *Semispaces of configurations, cell complexes of arrangements*, J. Combin. Theory Ser. A, 37 (1984), pp. 257–293. (**112**)

[104] J. E. GOODMAN AND R. POLLACK, *Allowable sequences and order types in discrete and computational geometry*, in New Trends in Discrete and Computational Geometry, J. Pach, ed., vol. 10 of Algorithms and Combinatorics, Springer-Verlag, 1993, pp. 103–134. (**112, 113**)

[105] J. GRAVER, B. SERVATIUS, AND H. SERVATIUS, *Combinatorial Rigidity*, Graduate Studies in Mathematics, 2., American Math. Soc., 1993. (**149**)

[106] P. GRITZMANN, B. MOHAR, J. PACH, AND R. POLLACK, *Embedding a planar triangulation with vertices at specified points*, Amer. Math. Monthly, 98 (1991), pp. 165–166. (**16**)

[107] B. GRÜNBAUM, *Arrangements of hyperplanes*, Congr. Numer., 3 (1971), pp. 41–106. (**84**)

[108] B. GRÜNBAUM, *Arrangements and Spreads*, Regional Conf. Ser. Math., American Mathematical Society, 1972 (reprinted 1980). (**84, 85, 86, 111, 112**)

[109] B. GRÜNBAUM, *How many triangles?*, Geombinatorics, 8 (1998), pp. 154–159. (**86**)

[110] R. K. GUY, *Crossing numbers of graphs*, in Graph Theory and Applications, vol. 303 of Lect. Notes in Math., Springer-Verlag, 1972, pp. 111–124. (**50, 68**)

[111] R. HAAS, D. ORDEN, G. ROTE, F. SANTOS, B. SERVATIUS, H. SERVATIUS, D. SOUVAINE, I. STREINU, AND W. WHITELEY, *Planar minimally rigid graphs and pseudo-triangulations*, in Proc. 19th Symp. Comput. Geom., 2003. (**150**)

[112] H. HARBORTH, *Some simple arrangements of pseudolines with a maximum number of triangles*, in Discrete geometry and convexity, vol. 440, Ann. N. Y. Acad. Sci., 1985, pp. 31–33. (**86**)

[113] X. HE, *Grid embeddings of 4-connected plane graphs*, Discr. and Comput. Geom., 17 (1997), pp. 339–358. (**37**)

[114] S. HOŞTEN AND W. D. MORRIS, *The order dimension of the complete graph*, Discr. Math., 201 (1999), pp. 133–139. (**41**)

[115] J. E. HOPCROFT AND P. J. KAHN, *A paradigm for robust geometric algorithms*, Algorithmica, 7 (1992), pp. 339–380. (**38**)

[116] F. Hurtado, M. Noy, and J. Urrutia, *Flipping edges in triangulations*, Discr. and Comput. Geom., 22 (1999), pp. 333–346. (**129**)

[117] R. E. Jamison, *A survey of the slope problem*, in Discrete Geometry and Convexity, J. Goodman et al., ed., vol. 440, Ann. NY Acad. Sci., 1985, pp. 34–51. (**112**)

[118] G. Kant, *Drawing planar graphs using the canonical ordering*, Algorithmica, 16 (1996), pp. 4–32. (**38**)

[119] M. Katchalski and H. Last, *On geometric graphs with no two edges in convex position*, Discr. and Comput. Geom., 19 (1998), pp. 399–404. (**15**)

[120] L. M. Kelly and W. Moser, *On the number of ordinary lines determined by n points*, Canad. J. Math., 10 (1958), pp. 210–219. (**85**)

[121] L. Kettner, A. Mantler, J. Snoeyink, B. Speckmann, and F. Takeuchi, *Bounded-degree pseudo-triangulations of points*, Comp. Geom.: Theory and Appl., 25 (2003). (**150**)

[122] D. Kirkpatrick, J. Snoeyink, and B. Speckmann, *Kinetic collision detection for simple polygons*, Int. J. Comput. Geom. Appl., 12 (2002), pp. 3–27. (**150**)

[123] D. E. Knuth, *Axioms and Hulls*, vol. 606 of Lect. Notes Comput. Sci., Springer-Verlag, 1992. (**101, 111, 112**)

[124] A. Kotlov, L. Lovász, and S. Vampala, *The Colin de Verdière number and sphere representations of graphs*, Combinatorica, 17 (1997), pp. 483–521. (**39**)

[125] Y. S. Kupitz, *Extremal problems in combinatorial geometry*, Aarhus University Lecture Notes Series 53, Aarhus University, 1979. (**14**)

[126] Y. S. Kupitz, *On pairs of disjoint segments in convex position in the plane*, Annals of Discr. Math. 20, Rosenfeld, Zaks (eds) (1984), pp. 203–208. (**15**)

[127] Y. S. Kupitz and M. Perles, *Extremal theory for convex matchings in convex geometric graphs*, Discr. and Comput. Geom., 15 (1996), pp. 195–220. (**14, 15**)

[128] G. Laman, *On graphs and rigidity of plane skeletal structures*, J. Engineering Math., 4 (1970), pp. 331–340. (**149**)

[129] C. L. Lawson, *Transforming triangulations*, Discr. Math., 3 (1972), pp. 365–372. (**130**)

[130] C. W. Lee, *The associahedron and triangulations of the n-gon*, Europ. J. Combin., 10 (1989), pp. 551–560. (**130**)

[131] F. T. Leighton, *New lower bound techniques for VLSI*, Math. Syst. Theory, 17 (1984), pp. 47–70. (**51**)

[132] F. Levi, *Die Teilung der projektiven Ebene durch Gerade oder Pseudogerade*, Ber. Math.-Phys. Kl. sächs. Akad. Wiss. Leipzig, 78 (1926), pp. 256–267. (**69, 79, 85, 112**)

[133] L. Lovász, *On the number of halving lines*, Ann. Univ. Sci. Budapest, Eötvös, Sect. Math., 14 (1971), pp. 107–108. (**66**)

[134] L. LOVÁSZ, J. PACH, AND M. SZEGEDY, *On Conway's thrackle conjecture*, Discr. and Comput. Geom., 18 (1997), pp. 369–376. (**15**)

[135] L. LOVÁSZ, K. VESZTERGOMBI, U. WAGNER, AND E. WELZL, *Convex quadrilaterals and k-sets*, in Towards a Theory of Geometric Graphs, J. Pach, ed., Contemp. Math., Amer. Math. Soc, 2004. (**67, 68**)

[136] Y. I. MANIN AND V. V. SCHECHTMAN, *Arrangements of hyperplanes, higher braid groups and higher Bruhat orders*, in Algebraic Number Theory – in honour of K. Iwasawa, J. Coates et al., ed., vol. 17 of Advanced Studies in Pure Mathematics, Kinokuniya Company/Academic Press, 1989, pp. 289–308. (**113**)

[137] J. MATOUŠEK, *Geometric Discrepancy*, vol. 18 of Algorithms and Combinatorics, Springer-Verlag, 1999. (**84**)

[138] J. MATOUŠEK, *Lectures on Discrete Geometry*, vol. 212 of Graduate Texts in Mathematics, Springer-Verlag, 2002. (**51, 66, 67**)

[139] E. MILLER, *Planar graphs as minimal resolutions of trivariate monomial ideals*, Documenta Math., 7 (2002), pp. 43–90. (**41**)

[140] E. MILLER AND B. STURMFELS, *Monomial ideals and planar graphs*, in Proc. Appl. Alg., Alg. Algo., Error-Corr. Codes, M. Fossorier et al., ed., vol. 1719 of Lect. Notes Comput. Sci., Springer-Verlag, 1999, pp. 19–28. (**41**)

[141] J. W. MILNOR, *Hyperbolic geometry: the first 150 years*, Bull. Amer. Math. Soc., 6 (1982), pp. 9–24. (**130**)

[142] B. MOHAR AND C. THOMASSEN, *Graphs on Surfaces*, John Hopkins University Press, 2001. (**13**)

[143] J. W. MOON, *On the distribution of crossings in random complete graphs*, J. Soc. Ind. Appl. Math., 13 (1965), pp. 506–510. (**68**)

[144] T. MOTZKIN, *The lines and planes connecting the points of a finite set*, Trans. Amer. Math. Soc., 70 (1951), pp. 451–464. (**85**)

[145] D. ORDEN AND F. SANTOS, *The polytope of non-crossing graphs on a planar point set*, in http://arxiv.org/abs/math.CO/0302126, 2003. (**150**)

[146] J. O'ROURKE AND I. STREINU, *Vertex-edge pseudo-visibility graphs: Characterization and recognition*, in Proc. 13th Symp. Comput. Geom., 1997, pp. 119–128. (**149**)

[147] J. PACH, *Finite point configurations*, in Handbook of Discrete and Computational Geometry, Goodman and O'Rourke, eds., CRC Press, 1997, pp. 3–18. (**84**)

[148] J. PACH AND P. K. AGARWAL, *Combinatorial Geometry*, John Wiley & Sons, 1995. (**14, 39, 40, 51, 84**)

[149] J. PACH AND R. PINCHASI, *Bichromatic lines with few points*, J. Combin. Theory Ser. A, 90 (2000), pp. 326–335. (**84**)

[150] J. PACH, R. RADOICIC, G. TARDOS, AND G. TÓTH, *Graphs drawn with at most 3 crossings per edge*, 2004. in preparation. (**51**)

[151] J. Pach, J. Spencer, and G. Tóth, *New bounds on crossing numbers*, in ACM Symp. Computational Geometry, 1999, pp. 124–133. (**51**)

[152] J. Pach, W. Steiger, and E. Szemerédi, *An upper bound on the number of planar k-sets*, Discr. and Comput. Geom., 7 (1992), pp. 109–123. (**66**)

[153] J. Pach, T. Thiele, and G. Tóth, *Three-dimensional grid drawings of graphs*, in Proc. Graph Drawing '97, G. Di Battista, ed., vol. 1353 of Lect. Notes Comput. Sci., Springer-Verlag, 1997, pp. 47–51. (**38**)

[154] J. Pach and J. Törőcsik, *Some geometric applications of Dilworth's theorem*, Discr. and Comput. Geom., 12 (1993), pp. 1–7. (**14**)

[155] J. Pach and G. Tóth, *Graphs drawn with few crossings per edge*, Combinatorica, 17 (1997), pp. 427–439. (**51**)

[156] J. Pach and G. Tóth, *Which crossing number is it, anyway?*, in Proc. 39th IEEE Symp. Found. Comput. Sci., 1998, pp. 617–627. (**51**)

[157] M. Pocchiola and G. Vegter, *Pseudo-triangulations: Theory and applications*, in Proc. 12th Symp. Comput. Geom., 1996, pp. 291–300. (**149**)

[158] M. Pocchiola and G. Vegter, *Topologically sweeping visibility complexes via pseudo-triangulations*, Discr. and Comput. Geom., 16 (1996), pp. 419–453. (**149**)

[159] F. P. Preparata and M. I. Shamos, *Computational Geometry: An Introduction*, Springer-Verlag, 1985. (**84**)

[160] K. Reuter, *On the order dimension of convex polytopes*, Europ. J. Combin., 11 (1990), pp. 57–63. (**41**)

[161] J. Richter-Gebert, *Oriented matroids with few mutations*, Discr. and Comput. Geom., 10 (1993), pp. 251–269. (**112**)

[162] J. Richter-Gebert, *Realization spaces of polytopes*, vol. 1643 of Lect. Notes in Math., Springer-Verlag, 1997. (**13, 38**)

[163] J. Richter-Gebert and G. Ziegler, *Oriented matroids*, in Handbook of Discrete and Computational Geometry, Goodman and O'Rourke, eds., CRC Press, 1997, pp. 111–132. (**111**)

[164] J. Richter-Gebert and G. M. Ziegler, *Zonotopal tilings and the Bohne-Dress theorem*, in Jerusalem combinatorics '93, H. Barcelo, ed., vol. 178 of Contemp. Math., Amer. Math. Soc., 1994, pp. 211–232. (**113**)

[165] P. Rosenstiehl and R. E. Tarjan, *Rectilinear planar layouts and bipolar orientations of planar graphs*, Discr. and Comput. Geom., 1 (1986), pp. 343–353. (**37**)

[166] G. Rote, I. Streinu, and F. Santos, *Expansive motions and the polytope of pointed pseudo-triangulations*, in Discrete and Computational Geometry, The Goodman and Pollack Festschrift, vol. 25 of Algorithms and Combinatorics, Springer Verlag, 2003, pp. 699–736. (**147, 150**)

[167] J.-P. Roudneff, *The maximum number of triangles in arrangements of pseudo-lines*, J. Combin. Theory Ser. B, 66 (1996), pp. 44–74. (**86**)

[168] H. SACHS, *Coin graphs, polyhedra and conformal mappings*, Discr. Math., 134 (1994), pp. 133–138. (**39**)

[169] F. SANTOS, *Triangulations of oriented matroids*, Mem. Am. Math. Soc., 741 (2002), p. 80 p. (**129**)

[170] F. SANTOS AND R. SEIDEL, *A better upper bound on the number of triangulations of a planar point set*, J. Combin. Theory Ser. A, 102 (2003), pp. 186–193. (**129**)

[171] E. R. SCHEINERMAN AND H. S. WILF, *The rectilinear crossing number of a complete graph and Sylvester's "four point problem" of geometric probability*, Amer. Math. Monthly, 101 (1994), pp. 939–943. (**68**)

[172] W. SCHNYDER, *Planar graphs and poset dimension*, Order, 5 (1989), pp. 323–343. (**37, 40**)

[173] W. SCHNYDER, *Embedding planar graphs on the grid*, in Proc. 1st ACM-SIAM Symp. Discr. Algo., 1990, pp. 138–148. (**37, 40**)

[174] W. SCHNYDER AND W. T. TROTTER, *Convex embeddings of 3-connected plane graphs*, Abstr. of the AMS, 13 (1992), p. 502. (**37**)

[175] O. SCHRAMM, *How to cage an egg*, Invent. Math., 107 (1992), pp. 534–560. (**39**)

[176] R. W. SHANNON, *Simplicial cells in arrangements of hyperplanes*, Geom. Dedicata, 8 (1979), pp. 179–187. (**80, 86**)

[177] M. SHARIR, S. SMORODINSKY, AND G. TARDOS, *An improved bound for k-sets in three dimensions*, Discr. and Comput. Geom., 26 (2001), pp. 195–204. (**67**)

[178] M. SHARIR AND E. WELZL, *Balanced lines, halving triangles and the generalized lower bound theorem*, in Proc. 17th Symp. Comput. Geom., 2001, pp. 315–318. (**67**)

[179] D. D. SLEATOR, R. E. TARJAN, AND W. P. THURSTON, *Rotation distance, triangulations, and hyperbolic geometry*, J. Amer. Math. Soc., 1 (1988), pp. 647–682. (**125, 129, 130**)

[180] J. SNOEYINK AND J. HERSHBERGER, *Sweeping arrangements of curves*, in Discrete and Computational Geometry: Papers from the DIMACS Special Year, J. Goodman, R. Pollack, and W. Steiger, eds., American Mathematical Society, 1991, pp. 309–349. (**111**)

[181] B. SPECKMANN AND C. TÓTH, *Allocating vertex π-guards in simple polygons via pseudo-triangulations*, in Proc. 14th ACM-SIAM Symp. Discr. Algo., 2003. (**150**)

[182] J. SPENCER, E. SZEMERÉDI, AND W. T. TROTTER, *Unit distances in the Euclidean plane*, in Graph Theory and Combinatorics, B. Bollobás, ed., Academic Press, 1984. (**51**)

[183] R. STANLEY, *On the number of reduced decompositions of elements of Coxeter groups*, Europ. J. Combin., 5 (1984), pp. 359–372. (**112**)

[184] R. P. STANLEY, *Enumerative Combinatorics*, vol. 1, Wadsworth & Brooks/Cole, 1986. (**107**)

[185] R. P. STANLEY, *Polygon dissections and standard Young tableaux*, J. Combin. Theory Ser. A, 76 (1996), pp. 175–177. (**130**)

[186] R. P. STANLEY, *Enumerative Combinatorics*, vol. 2, Cambridge Univ. Press, 1999. (**123, 130**)

[187] S. K. STEIN, *Convex maps*, Proc. Amer. Math. Soc., 2 (1951), pp. 464–466. (**13, 37**)

[188] J. STEINER, *Einige Gesetze über die Theilung der Ebene und des Raumes*, Crelle's J. für Reine und Angew. Math., 1 (1826), pp. 349–364. (**69**)

[189] I. STREINU, *Clusters of stars*, in Proc. 13th Symp. Comput. Geom., 1997, pp. 439–441. (**113**)

[190] I. STREINU, *A combinatorial approach to planar non-colliding robot arm motion planning*, in Proc. 41st IEEE Symp. Found. Comput. Sci., 2000, pp. 443–453. (**149, 150**)

[191] J. W. SUTHERLAND, *Lösung Aufgabe 167 von Hopf und Pannwitz*, Jahresber. der Deutschen Mathem. Vereinig., 45 (1935), pp. 33–34. (**14**)

[192] J. J. SYLVESTER, *Mathematical question 11851*, Educational Times, 59 (1893), p. 98. (**69**)

[193] L. SZÉKELY, *Crossing numbers and hard Erdős problems in discrete geometry*, Comb., Probab. and Comput., 6 (1997), pp. 353–358. (**51**)

[194] L. SZÉKELY, *Erdős on unit distances and the Szemerédi-Trotter theorems*, in Paul Erdős and his mathematics II, G. Halász et al., ed., Springer-Verlag, 2002, pp. 649–666. (**52**)

[195] E. SZEMERÉDI AND W. T. TROTTER, *Extremal problems in discrete geometry*, Combinatorica, 3 (1983), pp. 381–392. (**51, 86**)

[196] C. THOMASSEN, *Planarity and duality of finite and infinite graphs*, J. Combin. Theory Ser. B, 29 (1980), pp. 244–271. (**35**)

[197] G. TÓTH, *Note on geometric graphs*, J. Combin. Theory Ser. A, 89 (2000), pp. 126–132. (**14**)

[198] G. TÓTH, *Point sets with many k-sets*, Discr. and Comput. Geom., 26 (2001), pp. 187–194. (**66**)

[199] G. TÓTH AND P. VALTR, *Geometric graphs with few disjoint edges*, in Proc. 14th Symp. Comput. Geom., 1998, pp. 184–191. (**14**)

[200] W. T. TROTTER, *Combinatorics and Partially Ordered Sets: Dimension Theory*, Johns Hopkins Series in the Mathematical Sciences, The Johns Hopkins University Press, 1992. (**41**)

[201] P. TURÁN, *A note of welcome*, J. Graph Theory, 1 (1977), pp. 7–9. (**50**)

[202] W. T. TUTTE, *Convex representations of graphs*, Proc. London Math. Soc., 10 (1960), pp. 304–320. (**37**)

[203] W. T. TUTTE, *How to draw a graph*, Proc. London Math. Soc., 13 (1963), pp. 743–768. (**37**)

[204] W. T. TUTTE, *Toward a theory of crossing numbers*, J. Combin. Theory, 8 (1970), pp. 45–53. (**51**)

[205] P. UNGAR, *2n noncollinear points determine at least 2n directions*, J. Combin. Theory Ser. A, 33 (1982), pp. 343–347. (**112**)

[206] P. VALTR, *On geometric graphs with no k pairwise parallel edges*, Discr. and Comput. Geom., 19 (1998), pp. 461–470. (**15**)

[207] P. VALTR, *Generalizations of Davenport-Schinzel sequences*, in Contemporary trends in discrete mathematics, R. Graham et al., ed., vol. 49 of DIMACS, Ser. Discrete Math. Theor. Comput. Sci., DIMACS, 1999, pp. 349–389. (**15**)

[208] K. WAGNER, *Bemerkungen zum Vierfarbenproblem*, Jahresber. der Deutschen Mathem. Vereinig., 46 (1936), pp. 26–32. (**13, 37**)

[209] U. WAGNER, *On the rectilinear crossing number of complete graphs*, in Proc. 14th ACM-SIAM Symp. Discr. Algo., 2003, pp. 583–588. (**68**)

[210] E. WEISSTEIN, *Sylvester's four-point problem* http://mathworld.wolfram.com/SylvestersFour-PointProblem.html. (**68**)

[211] E. WELZL, *More on k-sets in the plane*, Discr. and Comput. Geom., 1 (1986), pp. 95–100. (**67, 113**)

[212] E. WELZL, *Entering and leaving j-facets*, Discr. and Comput. Geom., 25 (2001), pp. 351–364. (**67**)

[213] D. B. WEST, *Parameters of partial orders and graphs: Packing, covering and representation*, in Graphs and Orders, I. Rival, ed., D. Reidel, 1985, pp. 267–350. (**14**)

[214] D. B. WEST, *An Introduction to Graph Theory*, Prentice Hall, 1995. (**13, 14**)

[215] W. WHITELEY, *Matroids and rigid structures*, in Matroid applications, N. White, ed., Encycl. Math. Appl. 40, Cambridge Univ. Press, 1992, pp. 1–53. (**149**)

[216] W. WHITELEY, *Some matroids from discrete applied geometry*, in Matroid theory, J. E. Bonin et al., ed., Contemp. Math. 197, American Math. Soc., 1996, pp. 171–311. (**149**)

[217] H. WHITNEY, *2-isomorphic graphs*, Amer. J. Math., 55 (1933), pp. 245–254. (**13**)

[218] G. M. ZIEGLER, *Higher Bruhat orders and cyclic hyperplane arrangements*, Topology, 32 (1993), pp. 259–279. (**112, 113**)

[219] G. M. ZIEGLER, *Lectures on Polytopes*, vol. 152 of Graduate Texts in Mathematics, Springer-Verlag, 1994. (**13, 39, 67, 113, 130, 144**)

Index

and $$\mu'^{\bar{\alpha}} = \operatorname{Im} \lambda'^{\alpha} = B_{\beta}^{\alpha}\mu^{\beta} + A_{\bar{\beta}}^{\bar{\alpha}}\mu^{\bar{\beta}} .$$

This transformation represents a change of basis in V^*. The corresponding new dual basis $\{E'_{\beta}, E'_{\bar{\beta}}\}$ of V will be given by:

$$E'_{\beta} = C_{\beta}^{\alpha} E_{\alpha} + D_{\bar{\beta}}^{\bar{\alpha}} E_{\bar{\alpha}}$$

$$E'_{\bar{\beta}} = -D_{\beta}^{\alpha} E_{\alpha} + C_{\bar{\beta}}^{\bar{\alpha}} E_{\bar{\alpha}}$$

for some $C_{\beta}^{\alpha} = C_{\bar{\beta}}^{\bar{\alpha}}$ and $D_{\beta}^{\alpha} = D_{\bar{\beta}}^{\bar{\alpha}}$ satisfying

$$C_{\alpha}^{\beta} A_{\beta}^{\gamma} - D_{\alpha}^{\beta} B_{\beta}^{\gamma} = \delta_{\alpha}^{\gamma}$$

$$C_{\alpha}^{\beta} B_{\beta}^{\gamma} + D_{\alpha}^{\beta} A_{\beta}^{\gamma} = 0 ,$$

because then we have:

$$\begin{aligned} \mu'^{\alpha}(E'_{\beta}) &= (A_{\gamma}^{\alpha} \mu^{\gamma} - B_{\bar{\gamma}}^{\bar{\alpha}} \mu^{\bar{\gamma}})(C_{\beta}^{\delta} E_{\delta} + D_{\bar{\beta}}^{\bar{\delta}} E_{\bar{\delta}}) \\ &= C_{\beta}^{\gamma} A_{\gamma}^{\alpha} - 0 + 0 - D_{\bar{\beta}}^{\bar{\delta}} B_{\bar{\delta}}^{\bar{\alpha}} \\ &= \delta_{\beta}^{\alpha} \end{aligned}$$

and similarly:

$$\mu'^{\alpha}(E'_{\bar{\beta}}) = \mu'^{\bar{\alpha}}(E'_{\beta}) = 0$$

$$\mu'^{\bar{\alpha}}(E'_{\bar{\beta}}) = \delta_{\bar{\beta}}^{\bar{\alpha}} .$$

Hence, making use of Lemma II.15, we have for the change of basis in V:

$$\det \begin{pmatrix} C_{\beta}^{\alpha} & D_{\bar{\beta}}^{\bar{\alpha}} \\ -D_{\beta}^{\alpha} & C_{\bar{\beta}}^{\bar{\alpha}} \end{pmatrix} = |\det(C_{\beta}^{\alpha} + iD_{\beta}^{\alpha})|^2 > 0 .$$

This inequality shows that the orientation of the basis $[E'_\alpha, E'_{\bar{\alpha}}]$ of V "induced" by a different choice of functions λ^α is the same as the orientation defined by the original basis $E_\alpha, E_{\bar{\alpha}}$. But the possible choices for the λ^α are limited only by the requirement that they span the subspaces $W(J^*)$ of V^{*c} determined by J (and be $\mathbb{C}$-linearly independent). This, then, is the orientation for V determined by J.

<u>II.17: REMARK:</u> A complex structure J and its conjugate -J define opposite orientations for V if $n = 2m = 4k + 2$ and the same orientation if $n = 2m = 4k$.

<u>II.18: REMARK:</u> It can be shown directly that the relations $J(E_\alpha) = E_\alpha$ and $J(E_{\bar{\alpha}}) = -E_\alpha$ are preserved by a change in basis for V "induced" as above by a different choice of functions λ^α.

CHAPTER III: COMPLEX MANIFOLDS

We are now ready to introduce the concept of a complex manifold. A complex manifold generalizes the idea of a Riemann surface and at the same time specializes the concepts of differentiable and analytic structures on a manifold to an even more restricted type of structure, a complex structure. Later, we shall argue that the use of complex variables in the differential geometry of relativity is justified only after the manifold under consideration has been equipped with a (modified) complex structure. (The intermediate step of introducing an almost complex structure on a manifold is discussed in Chapter V.)

Manifolds

It will be helpful to begin with a brief review of what is meant by a manifold (see Spivak [50] for details):

III.1: DEFINITION: A manifold is a Hausdorff topological space M with the property that each point $p \in M$ has a neighborhood U homeomorphic to $\mathbb{R}^n$ for some value of n (see Figure III.1).

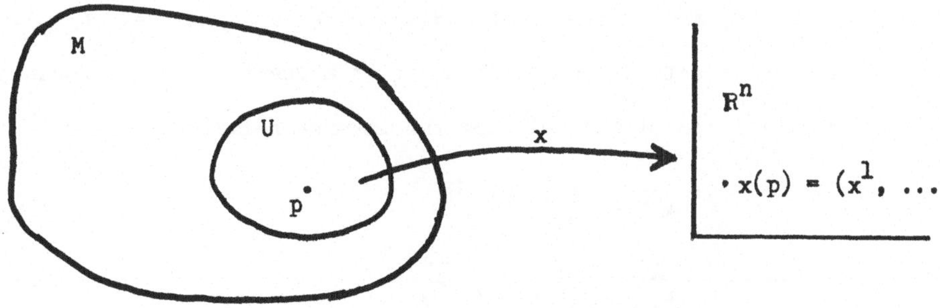

Figure III.1

III.2: REMARK: The neighborhoods U in the above definition must be open sets. (Invariance of Domain is needed for the proof; see Spivak [50].)

The homeomorphisms ("coordinate charts") $x: U \subseteq M \to \mathbb{R}^n$ of the definition provide a system of local coordinates for points of the manifold. The value of n in $\mathbb{R}^n$ is unique for each $p \in M$, and we shall consider only manifolds for which n is the same for all $p \in M$; then n is said to be the dimension of M. In fact, we shall further stipulate that M be even-dimensional, $n = 2m$.

When this is the case, there are two ways to "coordinatize" points of M: referring back to Figure III.1, we have

$$x(p) = (x^1, \ldots, x^n)$$

for any $p \in U \subseteq M$. Since M is assumed to be even-dimensional, we can take $x^1, \ldots, x^m, x^{m+1}, \ldots, x^{2m} = x^n$ to be the real coordinates of p associated with the chart x and $z^1, \ldots, z^m$ to be the complex coordinates of p associated with x, where

$$z^\alpha \equiv x^\alpha + ix^{m+\alpha} \equiv x^\alpha + ix^{\bar{\alpha}}$$

(recall that Latin indices range and sum over $1, \ldots, n$; Greek indices range and sum over $1, \ldots, m$; and barred Greek indices range and sum over $m + 1, \ldots, 2m = n$). These two sets of coordinates are in a one-to-one relationship by virtue of the definition of the z^α and the inverse relations

$$x^\alpha = \frac{1}{2}(z^\alpha + \overline{z^\alpha})$$

$$x^{\bar{\alpha}} = \frac{1}{2i}(z^\alpha - \overline{z^\alpha}) .$$

We have thus arrived at a way of assigning complex coordinates to the points of any even-dimensional manifold. But for this assignment to be useful, we must consider what happens in the intersection of two coordinate patches. Suppose $p \in U \cap U'$, where $x: U \to \mathbb{R}^n$ and $x': U' \to \mathbb{R}^n$ (see Figure III.2):

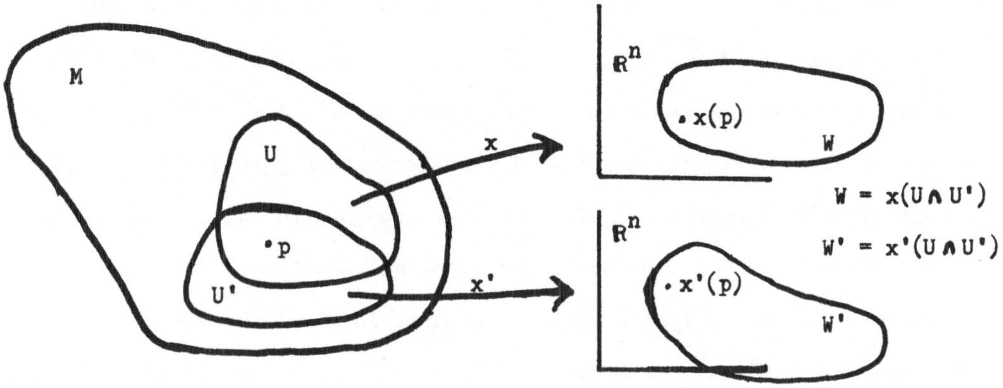

Figure III.2

We can restrict x and x' to functions $x_r: U \cap U' \to W$ and $x_r': U \cap U' \to W'$ with W and W' as in Figure III.2. Then

$$x_r' \circ x_r^{-1}: W \to W' \quad \text{is a homeomorphism.}$$

Accordingly, $p \in U \cap U'$ has two sets of real coordinates, x^a and $x^{a'}$, which are related by one-to-one transformations

$$x^{a'} = x^{a'}(x^b)$$

$$x^a = x^a(x^{b'}) \quad .$$

Similarly, p has two sets of complex coordinates, z^α and $z^{\alpha'}$, which will be related by one-to-one functions

$$z^{\alpha'} = z^{\alpha'}(z^{\beta}, \overline{z^{\beta}})$$

$$z^{\alpha} = z^{\alpha}(z^{\beta'}, \overline{z^{\beta'}}) \quad .$$

Complex Structures

A structure on a manifold is obtained by restricting the allowed sets $U, U', \ldots$ to those for which the associated transformations $x^{a'} = x^{a'}(x^b), \ldots$ or equivalently the transformations $z^{\alpha'} = z^{\alpha'}(z^{\beta}, \overline{z^{\beta}}), \ldots$ belong to some specific pseudogroup of transformations (see Frölicher [17] and Appendix A). (There must, of course, be enough sets $U, U', \ldots$ left to cover the manifold for the structure to be well-defined.) The most familiar example is a differentiable structure, in which the allowed transformations $x^{a'} = x^{a'}(x^b)$ are required to be C^{∞} functions (i.e., infinitely often continuously differentiable). Similarly, for a real analytic structure, the transformations must be real analytic functions of the arguments. In special relativity, the primary interest is in transformations belonging to the Lorentz group, so that we might speak of a manifold with a "Lorentz structure."

In a similar spirit, we say that a manifold admits a complex structure (complex analytic structure, holomorphic structure) if it can be covered with sets $U, U', \ldots$ such that in any intersection $U \cap U'$ the associated transformations $z^{\alpha'} = z^{\alpha'}(z^{\beta}, \overline{z^{\beta}}), \ldots$ of the complex coordinates are complex analytic (holomorphic) functions $z^{\alpha'} = z^{\alpha'}(z^{\beta})$.

Distilling the above comments, we can make the following formal definition.

III.3: DEFINITION: A complex manifold (manifold which admits a complex structure) is a manifold which can be covered with open sets U, U', ... homeomorphic to $\mathbb{R}^n$, n = 2m an even integer, such that if z^α, $z^{\beta'}$ are the complex coordinates associated with U, U' where $U \cap U'$ is non-empty, then

$$z^{\beta'} = z^{\beta'}(z^\alpha) \quad , \quad \det(\partial z^{\beta'}/\partial z^\alpha) \neq 0 \quad ,$$

where the $z^{\beta'}(z^\alpha)$ are complex analytic functions of the z^α.

By "complex structure" we mean the (maximal) atlas of open sets U, U', ... together with their complex coordinates z^α, $z^{\beta'}$, ... We shall denote such a complex structure by $\{(U, z^\alpha), (U', z^{\beta'}), \ldots\}$.

III.4: REMARK: The requirement $\det(\partial z^{\beta'}/\partial z^\alpha) \neq 0$ in the definition of a complex manifold insures that the coordinate transformations $z^{\beta'} = z^{\beta'}(z^\alpha)$ are one-to-one functions and that local inverses exist: $z^\alpha = z^\alpha(z^{\beta'})$.

III.5: REMARK: Making use of the Cauchy-Riemann conditions for complex analytic functions, we can redefine a complex manifold in terms of real coordinate transformations

$$x^{\beta'} = x^{\beta'}(x^\alpha, x^{\bar{\alpha}}) \quad ; \quad x^{\bar{\beta}'} = x^{\bar{\beta}'}(x^\alpha, x^{\bar{\alpha}})$$

which satisfy

$$\frac{\partial x^{\beta'}}{\partial x^\alpha} = \frac{\partial x^{\bar{\beta}'}}{\partial x^{\bar{\alpha}}} \quad \text{and} \quad \frac{\partial x^{\beta'}}{\partial x^{\bar{\alpha}}} = -\frac{\partial x^{\bar{\beta}'}}{\partial x^\alpha} \quad .$$

(See Frölicher [17].)

An alternative definition of a complex manifold can be given by requiring M to be covered with open sets homeomorphic to $\mathbb{C}^m$, the complex number space of m dimensions, and then requiring the transformations between (complex) coordinates associated with intersecting sets to be complex analytic. Since $\mathbb{C}^m$ is homeomorphic to $\mathbb{R}^{2m}$, the two definitions are easily seen to be equivalent. The definition we have chosen emphasizes the fact that M is a "real manifold" in the usual sense.

Necessary Conditions for a Complex Structure to Exist

Let us look at some necessary conditions for a manifold to admit a complex structure. First of all, a complex manifold is by definition even-dimensional. (When we refer to the dimensionality of a complex manifold, we shall always mean the real dimensionaltiy, unless the complex dimension -- half the real dimension -- is explicitly designated.)

On the other hand, curiously enough, we need not start with a (real) analytic manifold in order to demonstrate the existence of a complex structure. An $n = 2m$-dimensional manifold of differentiability class C^k, where $k \geq 2m + 1$, may be shown to admit a complex structure if certain conditions are met. These conditions will be discussed further in Chapter V.

A property which a manifold must have, in addition to even-dimensionality, in order to admit a complex structure is orientability. Recall that a manifold is orientable if it can be covered by open sets such that if $x^{a'} = x^{a'}(x^b)$ are the (real) coordinate transformations for intersecting sets, then $\det(\partial x^{a'}/\partial x^b) > 0$.

III.6: THEOREM: A complex manifold M is orientable.

Proof: If M has a complex structure, then there are open sets covering M such that the corresponding complex coordinate transformations are given by analytic functions $z^{\alpha'} = z^{\alpha'}(z^{\beta})$ with $\det(\partial z^{\alpha'}/\partial z^{\beta}) \neq 0$. For the real coordinates we then have

$$x^{\alpha'} = x^{\alpha'}(x^{\beta}, x^{\bar{\beta}})$$

$$x^{\bar{\alpha}'} = x^{\bar{\alpha}'}(x^{\beta}, x^{\bar{\beta}}) \quad .$$

with the Cauchy-Riemann conditions giving

$$\frac{\partial x^{\alpha'}}{\partial x^{\beta}} = \frac{\partial x^{\bar{\alpha}'}}{\partial x^{\bar{\beta}}} \quad \text{and} \quad \frac{\partial x^{\alpha'}}{\partial x^{\bar{\beta}}} = -\frac{\partial x^{\bar{\alpha}'}}{\partial x^{\beta}}$$

(compare Remark III.5 above). Then we have

$$\det\left(\frac{\partial x^{a'}}{\partial x^{b}}\right) = \det\begin{pmatrix} \frac{\partial x^{\alpha'}}{\partial x^{\beta}} & \frac{\partial x^{\alpha'}}{\partial x^{\bar{\beta}}} \\ \frac{\partial x^{\bar{\alpha}'}}{\partial x^{\beta}} & \frac{\partial x^{\bar{\alpha}'}}{\partial x^{\bar{\beta}}} \end{pmatrix}$$

$$= \det\begin{pmatrix} \frac{\partial x^{\alpha'}}{\partial x^{\beta}} & \frac{\partial x^{\alpha'}}{\partial x^{\bar{\beta}}} \\ -\frac{\partial x^{\alpha'}}{\partial x^{\bar{\beta}}} & \frac{\partial x^{\alpha'}}{\partial x^{\beta}} \end{pmatrix}$$

$$> 0 \quad \text{by Lemma II.15.}$$

Thus the manifold is orientable.

We caution that even-dimensionality and orientability together are not sufficient conditions for a manifold to admit a complex structure. (A counter-example is the four-sphere S^4, which does not admit a complex structure [7].)

Examples of Complex Manifolds

Sufficient conditions for a manifold to admit a complex structure are quite a bit harder to come by. Therefore it behooves us to find examples of complex manifolds.

The simplest such example is just $\mathbb{R}^{2m}$. Since $\mathbb{R}^{2m}$ can be covered by one open set homeomorphic to $\mathbb{R}^{2m}$, it suffices to take the standard coordinates $x^1, \ldots, x^m, x^{m+1}, \ldots, x^{2m}$ everywhere. The complex coordinates are then $z^1 = x^1 + ix^{m+1}, \ldots, z^m = x^m + ix^{2m}$. This coordinate system defines the canonical complex structure for $\mathbb{R}^{2m}$, and $\mathbb{R}^{2m}$ with this complex structure is denoted $\mathbb{C}^m$.

It is important to note that this structure is not the only complex structure which can be put on $\mathbb{R}^{2m}$. An inequivalent complex structure is obtained by choosing real coordinates $x^1, \ldots, x^m, -x^{m+1}, \ldots, -x^{2m}$ with the corresponding complex coordinates $z^{1'} = x^1 - ix^{m+1}, \ldots, x^{m'} = x^m - ix^{2m}$, for example. Then the coordinate transformations are given by

$$z^{\alpha'} = \overline{z^{\alpha}} ,$$

which are not analytic functions. Hence this complex structure is not equivalent to the canonical structure. $\mathbb{R}^{2m}$ with this complex structure (the structure conjugate to the canonical structure; see below) is denoted $\overline{\mathbb{C}^m}$.

The next simplest examples of complex manifolds are the complex projective spaces, $\mathbb{CP}^m$. These are defined by starting with $\mathbb{C}^{m+1}$, removing the point (0, ..., 0), and identifying points which are proportional: $(z^1, \ldots, z^{m+1}) \equiv (\varkappa z^1, \ldots, \varkappa z^{m+1})$, $\varkappa \neq 0$. The resulting space can be covered with local coordinate charts as follows. For those points for which $z^\mu \neq 0$, we form the local coordinates

$$w_{(\mu)}{}^{\alpha} = z^{\alpha}/z^{\mu}, \qquad \alpha = 1, \ldots, \mu-1, \mu+1, \ldots, m+1 \ .$$

The points for which such coordinates are well defined form an open set (homeomorphic to $\mathbb{R}^{2m}$) and the collection of all $m+1$ such open sets covers $\mathbb{CP}^m$. In the intersection of two such sets we have

$$w_{(\nu)}{}^{\alpha} = \frac{z^{\alpha}}{z^{\nu}} = \frac{z^{\alpha}/z^{\mu}}{z^{\nu}/z^{\mu}} = \frac{w^{\alpha}{}_{(\mu)}}{w^{\nu}{}_{(\mu)}} \ ,$$

which are analytic functions (note $w^{\nu}{}_{(\mu)} \neq 0$ in the intersection). Thus we have exhibited a complex structure for $\mathbb{CP}^m$. It should be noted that the spaces $\mathbb{CP}^m$ are not homeomorphic to the even-dimensional real projective spaces $\mathbb{P}^{2m}$; the latter are non-orientable and hence do not admit a complex structure. (For a discussion of the importance of $\mathbb{CP}^m$ in complex manifold theory, see [8] or [30].)

The question of complex structures on the even-dimensional spheres S^{2m} is an interesting one. It has been proven (see Chern [7]) that the only spheres which may admit complex structures are S^2 and S^6. For S^6, the non-existence of a complex structure is a recently proven result [1] (although an almost complex structure is known to exist for S^6; see Chapter V). For S^2, a complex structure may be exhibited quite straightforwardly as follows. We imbed S^2 in $\mathbb{R}^3$ as $\xi^2 + \eta^2 + (\zeta - 1/2)^2 = 1/4$ and consider stereographic projection from the north and south poles as shown in Figure III.3.

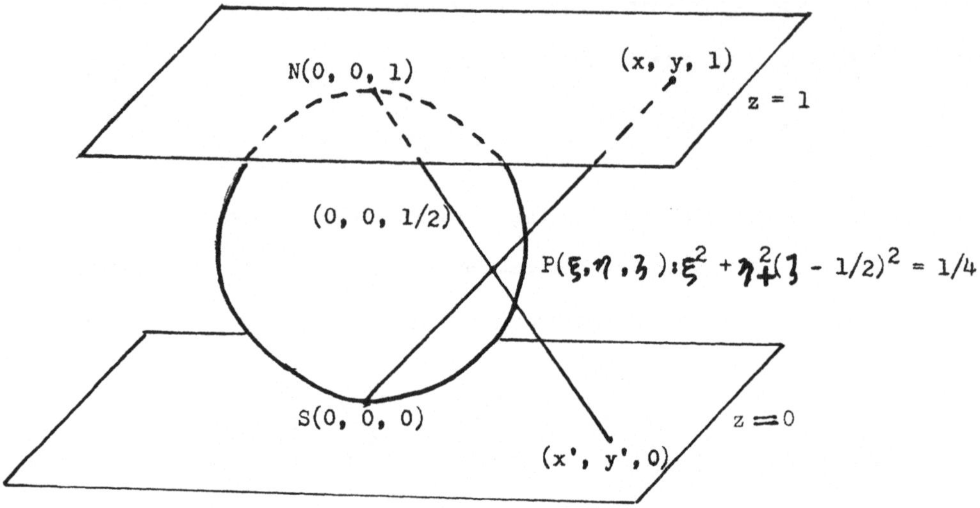

Figure III.3

As complex coordinates we take $z^o = x - iy = \frac{\xi - i\eta}{\zeta}$, valid for $P \neq S$, and $z'^o = x' + iy' = \frac{\xi + i\eta}{1 - \zeta}$, valid for $P \neq N$. In the intersection $S^2 - \{N, S\}$, we then have

$$z'^o = 1/z^o ,$$

which is an analytic function since $z^o \neq 0$ in $S^2 - \{N, S\}$.

III.7: REMARK: S^2 with this complex structure is precisely the complex manifold $\mathbb{CP}^1$ discussed above.

The complex manifolds $\mathbb{C}^1$ and $\mathbb{CP}^1 = S^2$ are examples of orientable surfaces. In fact, any C^∞ orientable 2-surface admits a complex structure

(Riemann surface). Some other examples of complex manifolds are the even-dimensional real tori $\mathbf{T}^{2m} = S^1 \times \ldots \times S^1$; quotient manifolds of $\mathbf{C}^m$ including the complex tori $\mathbf{CT}^m$; the products $S^{2p+1} \times S^{2q+1}$, $p, q \geq 0$, which include the so-called Hopf manifolds homeomorphic to $S^1 \times S^{2m-1}$; and the even-dimensional compact Lie groups such as SU(3). For elaboration, the reader is referred to Chern [8].

The Conjugate Manifold

In the previous section it was noted that $\mathbf{R}^{2m}$ can be equipped with two inequivalent complex structures. The complex coordinates z^α of one structure correspond to the complex conjugates of the coordinates w^α of the other structure. We now generalize this idea to any complex manifold.

III.8: DEFINITION: Given a manifold M with complex structure $\{(U, z^\alpha), (U', z^{\beta'}), \ldots\}$, the conjugate complex structure is the complex structure $\{(U, \overline{z^\alpha}), (U', \overline{z^{\beta'}}), \ldots\}$.

III.9: THEOREM: The conjugate complex structure is indeed a complex structure for M.

Proof: It suffices to show that in an intersection $U \cap U'$ the complex coordinate transformations

$$\overline{z^{\beta'}} = \overline{z^{\beta'}}\,(\overline{z^\alpha})$$

are analytic functions of their arguments. We know that

$$z^{\beta'} = z^{\beta'}\,(z^\alpha)$$

are analytic functions, and hence may be written as power series in the arguments:

$$z^{\beta'} = \Sigma \varkappa_{n_1 \ldots n_m} (z^1 - z^1_{(0)})^{n_1} \ldots (z^m - z^m_{(0)})^{n_m} .$$

Taking the complex conjugate, we have:

$$\overline{z^{\beta'}} = \Sigma \overline{\varkappa}_{n_1 \ldots n_m} (\overline{z^1} - \overline{z^1_{(0)}})^{n_1} \ldots (\overline{z^m} - \overline{z^m_{(0)}})^{n_m} ,$$

which identifies the $\overline{z^{\beta'}}$ as analytic functions of the arguments $\overline{z^{\alpha}}$.

Given a manifold M with a complex structure, we may think of M with the conjugate complex structure as a distinct manifold, the <u>conjugate manifold</u> $\overline{M}$. This idea is embodied in the following definition (see Yano [54]):

<u>III.10: DEFINITION:</u> Given a complex manifold M, the conjugate manifold $\overline{M}$ is that complex manifold for which there exists a homeomorphism $* : M \to \overline{M}$ such that if (U, z^{α}) is a chart of M and $*(U) = V$ then (V, w^{α}) is a chart of $\overline{M}$ with $w^{\alpha}(*(p)) = \overline{z^{\alpha}(p)}$ for all $p \in U$ (see Figure III.4 below).

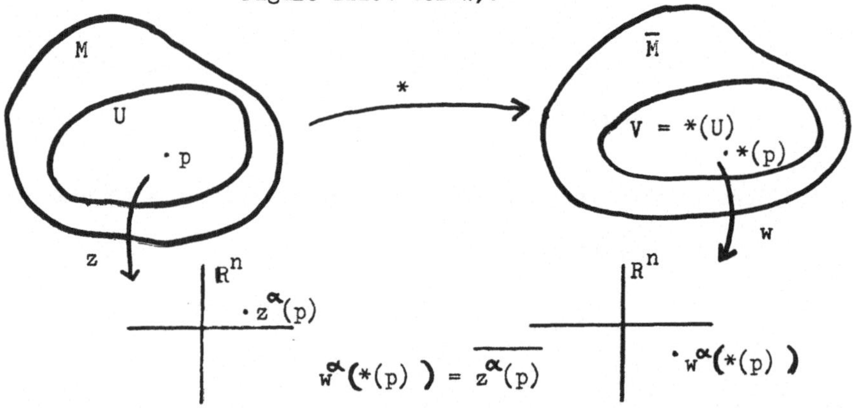

<u>Figure III.4</u>

We shall make use of the following notation. If z^{α} are the coordinates of a point $p \in U \subseteq M$, then the corresponding coordinates w^{α} of the point $*(p) \in *(U) \subseteq \overline{M}$ will be denoted $z^{\overline{\alpha}}$, so that

$$z^{\bar{\alpha}}(*(p)) = \overline{z^{\alpha}(p)} \quad .$$

To be consistent with our previous conventions, we can think of the coordinates in $\bar{M}$ as running from z^{m+1} to z^{2m}.

III.11: REMARK: Given S^2 with the complex structure discussed above, we have, for the conjugate complex structure,

$$z^{\bar{o}} = \frac{\xi + i\eta}{\zeta} \quad \text{and} \quad z^{'\bar{o}} = \frac{\xi - i\eta}{1 - \zeta} \, .$$

The Product Manifold

Given a manifold M with complex structure $\{(U, z^{\alpha}), (U', z^{\alpha'}), \ldots\}$ and its conjugate $\bar{M}$ with complex structure $\{(V, z^{\bar{\alpha}}), (V', z^{\bar{\alpha}'}), \ldots\}$, the product manifold $M \times \bar{M}$ can be given a complex structure in a natural way. The open sets are $U \times V$, $U' \times V$, ... and a point $p \in U \times V$ is coordinatized by the $n = 2m$ complex coordinates z^{α} and $z^{\bar{\alpha}}$, denoted z^a. In an intersection $(U \times V) \cap (U' \times V')$, we have the following analytic coordinate transformations for the z^a:

$$z^{\alpha'} = f^{\alpha'}(z^{\beta})$$

$$z^{\bar{\alpha}'} = \overline{f^{\alpha'}}(\overline{z^{\beta}}) \quad .$$

The function $\overline{f^{\alpha'}}$ is defined by $\overline{f^{\alpha'}}(w^{\beta}) = \overline{f^{\alpha'}(\overline{w^{\beta}})}$, that is to say, if

$$z^{\alpha'} = \Sigma\, \varkappa_{n_1 \ldots n_m} (z^1 - z^1_{(0)})^{n_1} \ldots (z^m - z^m_{(0)})^{n_m} \quad ,$$

then

$$z^{\bar{\alpha}'} = \Sigma\, \overline{\varkappa}_{n_1 \ldots n_m} (z^{\bar{1}} - \overline{z^1_{(0)}})^{n_1} \ldots (z^{\bar{m}} - \overline{z^m_{(0)}})^{n_m} \quad .$$

Then clearly $\overline{f^{\alpha'}}$ is an analytic function of its arguments $z^{\bar{\beta}}$, so that we do indeed have a well-defined complex structure for $M \times \bar{M}$.

<u>III.12: DEFINITION:</u> Given a manifold M with complex structure $\{(U, z^{\alpha}), (U', z^{\beta'}), \ldots\}$ and its conjugate $\bar{M}$ with complex structure $\{(V, z^{\bar{\alpha}}), (V', z^{\bar{\beta}'}), \ldots\}$, the product manifold denotes the manifold $M \times \bar{M}$ with complex structure consisting of the maximal atlas including $\{(U \times V, z^{\alpha}, z^{\bar{\alpha}}), (U' \times V', z^{\beta'}, z^{\bar{\beta}'}), \ldots\}$, where

$$z^{\beta'} = f^{\beta'}(z^{\alpha}) \Leftrightarrow z^{\bar{\beta}'} = \overline{f^{\beta'}}(z^{\bar{\alpha}}) .$$

Note that the maximal atlas for the complex structure on $M \times \bar{M}$ will contain coordinate charts for which the condition

$$z^{\beta'} = f^{\beta'}(z^{\alpha}) \Leftrightarrow z^{\bar{\beta}'} = \overline{f^{\beta'}}(z^{\bar{\alpha}})$$

is not fulfilled: Analytic coordinate transformations exist which mix the barred and unbarred coordinates. However, the product manifold can be covered solely by charts which do satisfy the condition. In fact, these charts give $M \times \bar{M}$ an additional "structure," that of a <u>locally product</u> structure (see Appendix B). It is this locally product complex structure on $M \times \bar{M}$ which will be a major concern in what follows on applications to relativity, since this structure defines the so-called "real subspace" (see below). For the time being, unless otherwise noted, by a chart in $M \times \bar{M}$ we shall mean specifically a chart in the locally product complex structure.

A word of caution is in order concerning dimensionality. If the real dimension of M is $n = 2m$, then the real dimension of $M \times \bar{M}$ is $2n = 4m$. On the other hand, the complex dimension of M is m and the complex dimension of $M \times \bar{M}$ is $n = 2m$. That is, m independent complex coordinates are needed to specify a point of M, whereas $2m$ are needed to designate a point of $M \times \bar{M}$. In particular, a point of a (modified) complex spacetime manifold is determined by two complex coordinates, while a point of the associated product manifold is determined by four complex numbers.

The Real Subspace

When we consider generating new spacetimes from old by "complex coordinate transformations," we shall be very interested in viewing a complex spacetime manifold M as a submanifold of the product manifold $M \times \bar{M}$. We shall now show that M is that submanifold of $M \times \bar{M}$ whose points are given by

$$z^{\bar{\alpha}} = \overline{z^{\alpha}} .$$

III.13: REMARK: This specification of M as a submanifold of $M \times \bar{M}$ is independent of analytic coordinate transformations for charts in the locally product complex structure of $M \times \bar{M}$.

Consider the mapping $f: M \to M \times \bar{M}$ defined by $f(p) = (p, *(p))$, where $*: M \to \bar{M}$ is the homeomorphism given in Definition III.13. For any complex coordinate patch, we will have the corresponding coordinate functions $f(z^{\alpha}) = (z^{\alpha}, \overline{z^{\alpha}})$, or in terms of real coordinates, $f(x^{\alpha}, x^{\bar{\alpha}}) = (x^{\alpha}, x^{\bar{\alpha}}, x^{\alpha}, -x^{\bar{\alpha}})$. This map is easily seen to be regular, since the Jacobean,

$$\frac{\partial(x^{\alpha}, x^{\bar{\alpha}})}{\partial(x^{\beta}, x^{\bar{\beta}}, y^{\beta}, y^{\bar{\beta}})} = \begin{pmatrix} \delta^{\alpha}{}_{\beta} & 0 & \delta^{\alpha}{}_{\beta} & 0 \\ 0 & \delta^{\bar{\alpha}}{}_{\bar{\beta}} & 0 & -\delta^{\bar{\alpha}}{}_{\bar{\beta}} \end{pmatrix} ,$$

is of maximal rank everywhere. The map f identifies M as the submanifold of $M \times \overline{M}$ described by

$$M \equiv f(M) = \{(p, q) \in M \times \overline{M} \mid z^{\bar{\alpha}}(q) = \overline{z^{\alpha}(p)}\} \subseteq M \times \overline{M} .$$

Accordingly, we make the following definition:

III.14: DEFINITION: The real subspace of a product manifold $M \times \overline{M}$ is the submanifold of $M \times \overline{M}$ diffeomorphic to M, given by $f(M)$ as described above.

It is this state of affairs that we loosely describe by saying that M is contained in $M \times \overline{M}$ as those points for which $z^{\bar{\alpha}} = \overline{z^{\alpha}}$ in any complex coordinate system (in the locally product structure).

As an example, consider the product manifold $\mathbb{C}^1 \times \overline{\mathbb{C}^1}$. Any point of $\mathbb{C}^1 \times \overline{\mathbb{C}^1}$ has complex coordinates (z, w) and real coordinates (x, y, u, v), where

$$z = x + iy$$

$$w = u + iv \ .$$

The real subspace $\mathbb{C}^1 = \mathbb{R}^2$ is obtained as those points of $\mathbb{C}^1 \times \overline{\mathbb{C}^1}$ for which $w = \bar{z}$, that is, the points with complex coordinates $(z, \bar{z})$ and real coordinates $(x, y, x, -y)$.

Later we shall be concerned with the question of different real subspaces of $M \times \overline{M}$ being defined by inequivalent locally product structures on $M \times \overline{M}$. This change of real subspaces will form the basis of our interpretation of the Schwarzschild-to-"B" complex coordinate transformation. We postpone the details of this question until a later chapter (Chapter X).

Functions and Analytic Continuation

We now want to consider the matter of functions on a complex manifold M, with real-valued functions $f: M \to R$ looked upon as essentially a special case of complex-valued functions $f: M \to C$. In fact, it will be more helpful to look at the associated coordinate functions $(f \circ x^{-1})(z^{\alpha}, \overline{z^{\alpha}})$ and $(f \circ x^{-1})(x^{\alpha}, x^{\bar{\alpha}})$, where (U, x) is a coordinate chart. For simplicity, we shall denote these functions by $f(z^{\alpha}, \overline{z^{\alpha}})$ and $f(x^{\alpha}, x^{\bar{\alpha}})$.

Note that given a real-valued analytic function $f(x^{\alpha}, x^{\bar{\alpha}})$ of the real coordinates, the corresponding function $f(z^{\alpha}, \overline{z^{\alpha}})$ of the complex coordinates will not in general be a complex analytic function of the z^{α}. However, there is a natural way to associate such a function with a complex analytic function on the product manifold $M \times \bar{M}$ (see Yano [54]):

III.15: THEOREM: Every real analytic function $f(x^{\alpha}, x^{\bar{\alpha}})$ on M gives rise to a complex analytic function $F(z^{\alpha}, z^{\bar{\alpha}})$ on $M \times \bar{M}$.

Proof: Since $f(x^{\alpha}, x^{\bar{\alpha}})$ is an analytic function, it can be written as a power series in the x^{α} and $x^{\bar{\alpha}}$. Then using the relations

$$x^{\alpha} = \frac{1}{2}(z^{\alpha} + \overline{z^{\alpha}}), \qquad x^{\bar{\alpha}} = \frac{1}{2i}(z^{\alpha} - \overline{z^{\alpha}})$$

f can be expressed as a power series in the z^{α} and $\overline{z^{\alpha}}$, $f(z^{\alpha}, \overline{z^{\alpha}})$. Replacing $\overline{z^{\alpha}}$ by $z^{\bar{\alpha}}$ yields a function $F(z^{\alpha}, z^{\bar{\alpha}})$ which is expressed as a power series in the variables $z^{\alpha}, z^{\bar{\alpha}}$. Hence $F(z^{\alpha}, z^{\bar{\alpha}})$ is a complex analytic function on $M \times \bar{M}$.

Clearly, given a complex analytic function $G(z^{\alpha}, z^{\bar{\alpha}})$ on $M \times \bar{M}$, reversing the above procedure produces a complex function whose real and imaginary parts are real analytic functions on M.

Given a real analytic function $f(x^\alpha, x^{\bar{\alpha}})$ on M, we shall refer to the corresponding complex analytic function $F(z^\alpha, \bar{z}^\alpha)$ on $M \times \bar{M}$ as the <u>analytic continuation</u> of $f(x^\alpha, x^{\bar{\alpha}})$.

One of the important results of the theorem above is that we can now define partial derivatives of functions $f(z^\alpha, \overline{z^\alpha})$ on M:

<u>III.16: DEFINITION:</u> Given a real analytic function $f(x^\alpha, x^{\bar{\alpha}})$ of the real coordinates on M and the corresponding function $f(z^\alpha, \overline{z^\alpha})$ of the complex coordinates, we define

$$\frac{\partial f(z^\alpha, \overline{z^\alpha})}{\partial z^\beta} \equiv \left. \frac{\partial F(z^\alpha, z^{\bar{\alpha}})}{\partial z^\beta} \right|_{z^{\bar{\alpha}} = \overline{z^\alpha}}$$

and

$$\frac{\partial f(z^\alpha, \overline{z^\alpha})}{\partial \overline{z^\beta}} \equiv \left. \frac{F(z^\alpha, z^{\bar{\alpha}})}{\partial z^{\bar{\beta}}} \right|_{z^{\bar{\alpha}} = \overline{z^\alpha}} ,$$

where F is the analytic continuation of f.

Essentially, this definition simply states that to take partial derivatives with respect to the z^α and $\overline{z^\alpha}$, we consider the $\overline{z^\alpha}$ as independent of the z^α (i.e., go to the product manifold), differentiate, and re-identify $\overline{z^\alpha}$ as the complex conjugate of z^α (i.e., return down to the real subspace).

CHAPTER IV: VECTORS AND TENSORS ON A COMPLEX MANIFOLD

Having examined complex manifolds and functions on these manifolds, we are in a position to define vector and tensor fields associated with complex manifolds. Our procedure in this chapter is a straightforward application of the material developed in Chapter II. To develop tensor analysis we consider, at a point of a general manifold (independent of any complex structure), the tangent and cotangent vector spaces and their complexifications. We then note the special properties these complexified vector spaces possess when the underlying manifold admits a complex structure. We begin with a review of the ideas of tangent and cotangent space.

The Tangent Space and the Cotangent Space

We start with a manifold M and its differentiable structure. At a point $p \in M$ we consider the ring $\mathcal{F}_p$ of (germs of) differentiable real-valued functions $f: U \to \mathbb{R}$ on neighborhoods U of p. Equivalently, we can look at the corresponding functions $f(x^a)$ of local real coordinates.

A derivation of $\mathcal{F}_p$ is a map $X: \mathcal{F}_p \to \mathcal{F}_p$ which is $\mathbb{R}$-linear and obeys the Leibnitz rule, i.e.,

$$X(\alpha f + \beta g) = \alpha X(f) + \beta X(g)$$

and

$$X(fg) = fX(g) + gX(f)$$

for all $\alpha, \beta \in \mathbb{R}$ and for all functions $f, g \in \mathcal{F}_p$. Thus for a function $f: U \to \mathbb{R}$ with $p \in U$, $X(f)$ is a new function $X(f): U \to \mathbb{R}$.

IV.1: DEFINITION: A tangent vector X at a point $p \in M$ is a derivation of $\mathcal{F}_p$.

In particular, the derivation $\partial/\partial x^a$, defined by its action on coordinate functions $f(x^a)$:

$$(\partial/\partial x^a)\ (f) = \partial f(x^b)/\partial x^a \quad ,$$

is a tangent vector at p. The following theorem, whose proof is omitted, is a familiar result from tensor analysis:

<u>IV.2: THEOREM:</u> The set of all tangent vectors X at a point $p \in M$ forms a real vector space of dimension n, the <u>tangent space</u> M_p at p. The n tangent vectors $\partial/\partial x^a$ form a basis for M_p.

In view of this theorem, and in the notation used in Chapter III (but without reference to any complex structure just yet), any tangent vector $X \in M_p$ of a manifold of even dimension $n = 2m$ may be expanded in terms of the basis $\{\partial/\partial x^\alpha, \partial/\partial x^{\bar{\alpha}}\}$ as

$$X = \xi^\alpha\ \partial/\partial x^\alpha + \xi^{\bar{\alpha}}\ \partial/\partial x^{\bar{\alpha}} \quad .$$

The (real) numbers $\xi^\alpha, \xi^{\bar{\alpha}}$ are referred to as the <u>components</u> of X with respect to the real coordinate basis $\{\partial/\partial x^\alpha, \partial/\partial x^{\bar{\alpha}}\}$ or as the <u>real components</u> of X. Thus we have

$$X(f) = \xi^\beta\ \frac{\partial f(x^\alpha, x^{\bar{\alpha}})}{\partial x^\beta} + \xi^{\bar{\beta}}\ \frac{\partial f(x^\alpha, x^{\bar{\alpha}})}{\partial x^{\bar{\beta}}}$$

for any $X \in M_p$ and any $f \in \mathcal{F}_p$.

Tangent vectors $X \in M_p$ are also referred to as contravariant vectors at p, or simply as vectors at p. For reasons which become clear below, we shall often add the adjective "real" to the phrase "vector at p."

Given the vector space M_p, it is natural to consider the dual space M_p^*, called the <u>cotangent space</u> at p, the space of $\mathbb{R}$-linear operators $\omega: M_p \to \mathbb{R}$. Elements of M_p^* are referred to as cotangent vectors, covariant vectors, or real forms (one-forms) at p.

The basis of M_p^* dual to the basis $\{\partial/\partial x^\alpha, \partial/\partial x^{\bar{\alpha}}\}$ of M_p is denoted by $\{dx^\alpha, dx^{\bar{\alpha}}\}$, so that we have

$$dx^\alpha (\partial/\partial x^\beta) = \delta_\beta{}^\alpha$$

$$dx^\alpha (\partial/\partial x^{\bar{\beta}}) = 0$$

$$dx^{\bar{\alpha}} (\partial/\partial x^\beta) = 0$$

$$dx^{\bar{\alpha}} (\partial/\partial x^{\bar{\beta}}) = \delta_{\bar{\beta}}{}^{\bar{\alpha}} \quad .$$

Thus any element $\omega \in M_p^*$ may be uniquely written

$$\omega = \omega_\alpha dx^\alpha + \omega_{\bar{\alpha}} dx^{\bar{\alpha}} \quad ,$$

where the (real) numbers $\omega_\alpha, \omega_{\bar{\alpha}}$ are called the <u>real components</u> of ω.

<u>The Complexified Tangent Space</u>

We define the <u>complexified tangent space</u> $M_p{}^c$ at a point $p \in M$ as all expressions of the form $X + iY$ where $X, Y \in M_p$ (compare Definition II.1). (This follows because any real vector space has a well-defined complexification.)

Then in a basis $\{\partial/\partial x^\alpha, \partial/\partial x^{\bar{\alpha}}\}$ we have

$$X + iY = (\xi^\alpha + i\eta^\alpha)\, \partial/\partial x^\alpha + (\xi^{\bar{\alpha}} + i\eta^{\bar{\alpha}})\, \partial/\partial x^{\bar{\alpha}} \quad ,$$

so that a complexified tangent vector has complex components with respect to a real coordinate basis. (The nomenclature is somewhat confusing: A complexified tangent vector has complex numbers $\zeta^{\alpha} = \xi^{\alpha} + i\eta^{\alpha}$, $\zeta^{\bar{\alpha}} = \xi^{\bar{\alpha}} + i\eta^{\bar{\alpha}}$ as real components, i.e., components with respect to the real coordinate basis.)

We notice that a complexified tangent vector $Z = Z + iY$ no longer maps $\mathcal{F}_p$ into $\mathcal{F}_p$, because given a real-valued function $f \in \mathcal{F}_p$, its image $Z(f)$ under the action of Z will in general be a complex-valued function. (We could therefore consider instead the ring of (germs of) complex-valued differentiable functions on neighborhoods of $p \in M$.)

In the sequel we shall often refer to a complexified tangent vector simply as a "vector," reserving the phrase "real vector" to designate an element of the (real) tangent space.

Complex Structure on the Tangent Space

It is now natural to consider the matter of complex structures on the real tangent space M_p. Of course, a complex structure is possible only on a manifold M of even dimension. In the case of a complex manifold, we define

IV.3: DEFINITION: The canonical complex structure for the tangent space M_p at a point p of a complex manifold M is the complex structure J specified by the action

$$J(\partial/\partial x^{\alpha}) = \partial/\partial x^{\bar{\alpha}}$$

$$J(\partial/\partial x^{\bar{\alpha}}) = -\partial/\partial x^{\alpha} ,$$

where x^{α}, $x^{\bar{\alpha}}$ are real coordinates of a chart

$(U, x^{\alpha}, x^{\bar{\alpha}})$ belonging to the complex structure of M.

IV.4: REMARK: It is easy to show that the J in the definition above defines a complex structure for M_p.

IV.5: THEOREM: If $(U', x^{\alpha'}, x^{\bar{\alpha}'})$ is another chart of the complex structure of M with $p \in U'$, then

$$J(\partial/\partial x^{\alpha'}) = \partial/\partial x^{\bar{\alpha}'}$$

$$J(\partial/\partial x^{\bar{\alpha}'}) = -\partial/\partial x^{\alpha'} .$$

Proof: Given $J(\partial/\partial x^{\beta}) = \partial/\partial x^{\bar{\beta}}$ and $J(\partial/\partial x^{\bar{\beta}}) = -\partial/\partial x^{\beta}$, we have:

$$J(\partial/\partial x^{\alpha'}) = J\left(\frac{\partial x^{\beta}}{\partial x^{\alpha'}} \frac{\partial}{\partial x^{\beta}} + \frac{\partial x^{\bar{\beta}}}{\partial x^{\alpha'}} \frac{\partial}{\partial x^{\bar{\beta}}}\right)$$

$$= \frac{\partial x^{\beta}}{\partial x^{\alpha'}} \frac{\partial}{\partial x^{\bar{\beta}}} - \frac{\partial x^{\bar{\beta}}}{\partial x^{\alpha'}} \frac{\partial}{\partial x^{\beta}}$$

$$= \frac{\partial x^{\bar{\beta}}}{\partial x^{\bar{\alpha}'}} \frac{\partial}{\partial x^{\bar{\beta}}} + \frac{\partial x^{\beta}}{\partial x^{\bar{\alpha}'}} \frac{\partial}{\partial x^{\beta}}$$ by Remark III.5 (Cauchy-Riemann conditions)

$$= \partial/\partial x^{\bar{\alpha}'} .$$

Similarly,

$$J(\partial/\partial x^{\bar{\alpha}'}) = -\partial/\partial x^{\alpha'} .$$

This theorem states that the canonical complex structure for M_p takes the especially simple form

$$J(\partial/\partial x^{\alpha}) = \partial/\partial x^{\bar{\alpha}}$$

$$J(\partial/\partial x^{\bar{\alpha}}) = -\partial/\partial x^{\alpha}$$

for the real coordinates $x^{\alpha}, x^{\bar{\alpha}}$ of <u>any</u> chart (U, x) containing p and belonging to the complex structure of M. This is an important property of complex manifolds.

<u>IV.6: REMARK:</u> The complexified tangent vectors of type (1,0) with respect to the canonical complex structure of M_p have the form

$$V = \xi^{\alpha} \partial/\partial x^{\alpha} - i \xi^{\alpha} \partial/\partial x^{\bar{\alpha}} ,$$

where the ξ^{α} are any complex numbers. Similarly, the complexified tangent vectors of type (0,1) are given by

$$V = \xi^{\alpha} \partial/\partial x^{\alpha} + i \xi^{\alpha} \partial/\partial x^{\bar{\alpha}}$$

(compare Remark II.8).

<u>Complex Structure on the Cotangent Space</u>

We continue to consider the case of a complex manifold M. We know from Chapter II that the canonical complex structure J on the tangent space M_p induces a complex structure J^* on the cotangent space M_p^*. To determine the action of J^* on the basis vectors $dx^{\alpha}, dx^{\bar{\alpha}}$ of M_p^*, we notice the following:

$$J^* (dx^{\alpha}) (\partial/\partial x^{\beta}) \equiv dx^{\alpha} \big(J (\partial/\partial x^{\beta})\big) = 0$$

$$J^* (dx^{\alpha}) (\partial/\partial x^{\bar{\beta}}) \equiv dx^{\alpha} \big(J (\partial/\partial x^{\bar{\beta}})\big) = - \delta_{\beta}^{\alpha}$$

$$J^* (dx^{\bar{\alpha}}) (\partial/\partial x^{\beta}) \equiv dx^{\bar{\alpha}} \big(J (\partial/\partial x^{\beta})\big) = \delta_{\bar{\beta}}^{\bar{\alpha}}$$

$$J^* (dx^{\bar{\alpha}}) (\partial/\partial x^{\beta}) \equiv dx^{\bar{\alpha}} \left(J (\partial/\partial x^{\beta}) \right) = 0 .$$

From these facts, we conclude:

$$J^* (dx^{\alpha}) = -dx^{\bar{\alpha}}$$

$$J^* (dx^{\bar{\alpha}}) = dx^{\alpha} .$$

IV.7: REMARK: The above characterization of J^* is independent of the choice of local coordinates belonging to a chart of the complex structure of M. That is, if (U', x') is another chart of the complex structure, then

$$J^* (dx^{\alpha'}) = -dx^{\bar{\alpha}'}$$

$$J^* (dx^{\bar{\alpha}'}) = dx^{\alpha'} .$$

The complex structure J^* on the cotangent soace M_p^* provides a direct sum decomposition of the complexified cotangent space M_p^{*c} into type (1,0) and type (0,1) complexified one-forms. From the work of Chapter II, we know that a basis $\{\lambda^{\alpha}\}$ for the complexified one-forms of type (1,0) is given by

$$\begin{aligned} \lambda^{\alpha} &= J^* (dx^{\alpha}) + i dx^{\alpha} \\ &= -dx^{\bar{\alpha}} + i dx^{\alpha} \\ &= i(dx^{\alpha} + i dx^{\bar{\alpha}}) . \end{aligned}$$

Similarly, a basis for the complexified one-forms of type (0,1) is given by

$$\bar{\lambda}^{\alpha} = -i(dx^{\alpha} - i dx^{\bar{\alpha}}) .$$

Relation Between the Canonical Complex Structure and the Manifold Complex Structure

We now make use of the assumption that the manifold M under consideration admits a complex structure in the sense of Chapter III, namely, a covering by analytically related complex coordinate charts. Now the x^{α}, $x^{\bar{\alpha}}$ are real coordinates of a chart of the complex structure of M, so we also have complex coordinates $z^{\alpha} = x^{\alpha} + ix^{\bar{\alpha}}$ with complex conjugate $\overline{z^{\alpha}} = x^{\alpha} - ix^{\bar{\alpha}}$. Hence we can write:

$$\lambda^{\alpha} = id(x^{\alpha} + ix^{\bar{\alpha}}) \equiv idz^{\alpha}$$

$$\bar{\lambda}^{\alpha} = -id(x^{\alpha} - ix^{\bar{\alpha}}) \quad -id\,\overline{z^{\alpha}} = \overline{idz^{\alpha}} \ .$$

Together the complexified one-forms λ^{α} and $\bar{\lambda}^{\alpha}$ comprise a basis for the complexified cotangent space $M_p{}^{*c}$. It is natural to take instead the set $\{dz\ , \overline{dz^{\alpha}}\}$ as an equally valid basis for $M_p{}^{*c}$. Thus a one-form ω may be expressed as

$$\omega = \omega_{\alpha}\, dx^{\alpha} + \omega_{\bar{\alpha}}\, dx^{\bar{\alpha}} = \Omega_{\alpha}\, dz^{\alpha} + \Omega_{\bar{\alpha}}\, \overline{dz^{\alpha}} \ ,$$

where the summation convention in the last term is

$$\Omega_{\bar{\alpha}}\, \overline{dz^{\alpha}} \equiv \sum_{\alpha=1}^{m} \Omega_{\alpha+m}\, \overline{dz^{\alpha}} \ .$$

By virtue of Remark II.10, we may formally identify the space $(M_p{}^{*c})^*$ with the complexified tangent space $M_p{}^c$ $(M_p{}^{*c*} \leftrightarrow M_p{}^{c**} = M_p{}^c)$. Then the basis dual to $\{dz^{\alpha}, \overline{dz^{\alpha}}\}$ is given by $\{\partial/\partial z^{\alpha}, \partial/\partial\overline{z^{\alpha}}\}$, where the operator $\partial/\partial\overline{z^{\alpha}}$ acts in accordance with Definition III.16. We have the following result:

IV.8: THEOREM:

$$\partial/\partial z^{\alpha} = \frac{1}{2}(\partial/\partial x^{\alpha} - i\,\partial/\partial x^{\bar\alpha}),$$

$$\partial/\partial \overline{z^{\alpha}} = \frac{1}{2}(\partial/\partial x^{\alpha} + i\,\partial/\partial x^{\bar\alpha}) = \overline{\partial/\partial z^{\alpha}}\ .$$

Proof: Since $\{\partial/\partial z^{\alpha}, \partial/\partial\overline{z^{\alpha}}\}$ is by definition the basis dual to $\{dz^{\alpha}, d\overline{z^{\alpha}}\}$, we have:

$$\frac{1}{2}(dx^{\alpha} + i dx^{\bar\alpha})(\partial/\partial x^{\beta} - i\,\partial/\partial x^{\bar\beta}) = \delta_{\beta}^{\alpha} \equiv dz^{\alpha}(\partial/\partial z^{\beta})$$

$$\frac{1}{2}(dx^{\alpha} + i dx^{\bar\alpha})(\partial/\partial x^{\beta} + i\,\partial/\partial x^{\bar\beta}) = 0 \equiv dz^{\alpha}(\partial/\partial\overline{z^{\beta}})$$

$$\frac{1}{2}(dx^{\alpha} - i dx^{\bar\alpha})(\partial/\partial x^{\beta} - i\,\partial/\partial x^{\bar\beta}) = 0 \equiv d\overline{z^{\alpha}}(\partial/\partial z^{\beta})$$

$$\frac{1}{2}(dx^{\alpha} - i dx^{\bar\alpha})(\partial/\partial x^{\beta} + i\,\partial/\partial x^{\bar\beta}) = \delta_{\beta}^{\alpha} \equiv d\overline{z^{\alpha}}(\partial/\partial\overline{z^{\beta}})\ .$$

IV.9: REMARK: The content of the above theorem can be obtained by a formal application of the tensor transformation law for the coordinate transformation

$$z^{\alpha} = x^{\alpha} + i x^{\bar\alpha}, \qquad \overline{z^{\alpha}} = x^{\alpha} - i x^{\bar\alpha}\ .$$

Vectors and Tensors

At this point a recapitulation is in order. At a point p of a (complex) manifold M, we have considered four vector spaces. The tangent space M_p consists of the real tangent vectors at p and has the set $\{\partial/\partial x^{\alpha}, \partial/\partial x^{\bar\alpha}\}$ as a basis. The complexified tangent space $M_p{}^c$ contains the (complexified) tangent vectors at p and has the real basis $\{\partial/\partial x^{\alpha}, \partial/\partial x^{\bar\alpha}\}$, and the complex basis $\{\partial/\partial z^{\alpha}, \partial/\partial\overline{z^{\alpha}}\}$. These two bases are connected by the relations $\partial/\partial z^{\alpha} = \frac{1}{2}(\partial/\partial x^{\alpha} - i\,\partial/\partial x^{\bar\alpha})$ and $\partial/\partial\overline{z^{\alpha}} = \frac{1}{2}(\partial/\partial x^{\alpha} + i\,\partial/\partial x^{\bar\alpha})$. The cotangent space $M_p{}^*$ consists of the real one-forms at p with basis $\{dx^\alpha, dx^{\bar\alpha}\}$. The complexified cotangent space $M_p{}^{*c}$ contains the (complexified) one-forms at p, with real basis $\{dx^\alpha, dx^{\bar\alpha}\}$ and complex basis $\{dz^\alpha, d\overline{z^\alpha}\}$ related by $dz^\alpha = dx^\alpha + i dx^{\bar\alpha}$ and $d\overline{z^\alpha} = dx^\alpha - i dx^{\bar\alpha}$.

Now we drop the adjective "complexified" to describe elements of the spaces $M_p{}^c$ and $M_p{}^{*c}$, referring simply to "vectors" and "one-forms." The special case of elements of M_p and $M_p{}^*$ when these are considered as subspaces of $M_p{}^c$ and $M_p{}^{*c}$, respectively, is discussed below.

Tensors of higher rank are defined in the usual way (see for example Nelson [31]) in terms of ($\mathbb{C}$-) multilinear maps of products of the spaces $M_p{}^c$ and $M_p{}^{*c}$ into the complex numbers. Thus, for example, a (complex) tensor T of covariant rank 1 and contravariant rank 1 is a multilinear map $T: M_p{}^c \times M_p{}^{*c} \to \mathbb{C}$, and can be expressed in terms of real and complex bases as

$$\begin{aligned} T &= t_\alpha{}^\beta\, dx^\alpha \otimes \partial/\partial x^\beta + t_{\bar\alpha}{}^\beta\, dx^{\bar\alpha} \otimes \partial/\partial x^\beta \\ &\quad + t_\alpha{}^{\bar\beta}\, dx^\alpha \otimes \partial/\partial x^{\bar\beta} + t_{\bar\alpha}{}^{\bar\beta}\, dx^{\bar\alpha} \otimes \partial/\partial x^{\bar\beta} \\ &= T_\alpha{}^\beta\, dz^\alpha \otimes \partial/\partial z^\beta + T_{\bar\alpha}{}^\beta\, d\overline{z^\alpha} \otimes \partial/\partial z^\beta \\ &\quad + T_\alpha{}^{\bar\beta}\, dz^\alpha \otimes \partial/\partial \overline{z^\beta} + T_{\bar\alpha}{}^{\bar\beta}\, d\overline{z^\alpha} \otimes \partial/\partial \overline{z^\beta} \ . \end{aligned}$$

In this chapter and the next, we adopt the convention that capital letters shall be used for the complex components of tensors and lower-case letters for the real components, as above. We shall also make use of Latin indices to denote either a barred or an unbarred Greek index (so that Latin indices range and sum over $1, \ldots, n = 2m$ as before). A barred Latin index shall have the following meaning:

$$\bar{a} \equiv a + m \ (\text{mod } 2m) \quad .$$

This notation will greatly simplify some matters to follow.

Given any tensor T, the real components $t^{ab\ldots}_{\ \ cd\ldots}$ are related to the complex components $T^{ab\ldots}_{\ \ cd\ldots}$ by the tensor transformation law

$$T^{ab\ldots}_{\ \ cd\ldots} = \frac{\partial z^a}{\partial x^e}\frac{\partial z^b}{\partial x^f} \cdots \frac{\partial x^g}{\partial z^c}\frac{\partial x^h}{\partial z^d} \cdots t^{ef\ldots}_{\ \ gh\ldots}$$

for the coordinate transformation $z^\alpha = x^\alpha + i x^{\bar\alpha}$, $\overline{z^\alpha} = x^\alpha - i x^{\bar\alpha}$, where $z^a \equiv z^\alpha$ for $a = 1, \ldots, m$ and $z^a \equiv \overline{z^\alpha}$ for $a = m+1, \ldots, n = 2m$.

<u>IV.10: REMARK:</u> (i) Given a vector

$$V = V^\alpha \, \partial/\partial z^\alpha + V^{\bar\alpha} \,/\, \overline{z}^{\bar\alpha} = \xi^\alpha \, \partial/\partial x^\alpha + \xi^{\bar\alpha} \, \partial/\partial x^{\bar\alpha} \; ,$$

the real and complex components are related by

$$V^\alpha = \xi^\alpha + i\xi^{\bar\alpha} \; ,$$

$$V^{\bar\alpha} = \xi^\alpha - i\xi^{\bar\alpha} \; .$$

(ii) Given a one-form

$$\omega = \Omega_\alpha \, dz^\alpha + \Omega_{\bar\alpha} \, d\overline{z}^{\alpha} = \omega_\alpha \, dx^\alpha + \omega_{\bar\alpha} \, dx^{\bar\alpha} \; ,$$

the real and complex components are related by

$$\Omega_\alpha = \frac{1}{2}\left(\omega_\alpha - i\omega_{\bar\alpha}\right) \; ,$$

$$\Omega_{\bar\alpha} = \frac{1}{2}\left(\omega_\alpha + i\omega_{\bar\alpha}\right) \; .$$

The following theorem will play an important role in some later considerations (see Theorem V.11 on the integrability of an almost complex structure).

<u>IV.11: THEOREM:</u> In a complex manifold, the canonical complex structure (extended to act on all of M_p^c, see Definition IV.3) can be described by

$$J(\partial/\partial z^\alpha) = i\, \partial/\partial z^\alpha$$

$$J(\partial/\partial \overline{z^\alpha}) = -i\, \partial/\partial \overline{z^\alpha}$$

in any chart of the complex structure. Conversely, if J has this form with respect to both of two intersecting charts, then the complex coordinate transformation for the charts is complex analytic (provided, of course, that it is real analytic).

<u>Proof:</u> The first part of the theorem is a straightforward result of combining Definition IV.3 with Theorems IV.5 and IV.8. As for the converse, assume that for two intersecting charts z^α and $z^{\alpha'}$ we have

$$J(\partial/\partial z^\alpha) = i\, \partial/\partial z^\alpha$$

$$J(\partial/\partial \overline{z^\alpha}) = -i\, \partial/\partial \overline{z^\alpha}$$

and

$$J(\partial/\partial z^{\alpha'}) = i\, \partial/\partial z^{\alpha'}$$

$$J(\partial/\partial \overline{z^{\alpha'}}) = -i\, \partial/\partial \overline{z^{\alpha'}} \quad .$$

Then we have

$$J(\partial/\partial z^\alpha) = J\left(\frac{\partial z^{\alpha'}}{\partial z^\alpha}\, \partial/\partial z^{\alpha'} + \frac{\partial \overline{z^{\alpha'}}}{\partial z^\alpha}\, \partial/\partial \overline{z^{\alpha'}}\right)$$

$$= \frac{\partial z^{\alpha'}}{\partial z^\alpha}\, J(\partial/\partial z^{\alpha'}) + \frac{\partial \overline{z^{\alpha'}}}{\partial z^\alpha}\, J(\partial/\partial \overline{z^{\alpha'}})$$

$$= i \frac{\partial z^{\alpha'}}{\partial z^{\alpha}} \partial/\partial z^{\alpha'} - i \frac{\partial \overline{z^{\alpha'}}}{\partial z^{\alpha}} \partial/\partial \overline{z^{\alpha'}} .$$

On the other hand,

$$i\, \partial/\partial z^{\alpha} = i \frac{\partial z^{\alpha'}}{\partial z^{\alpha}} \partial/\partial z^{\alpha'} + i \frac{\partial \overline{z^{\alpha'}}}{\partial z^{\alpha}} \partial/\partial \overline{z^{\alpha'}} .$$

But $J(\partial/\partial z^{\alpha}) = i\, \partial/\partial z^{\alpha}$ by assumption, so that we obtain

$$\frac{\partial \overline{z^{\alpha'}}}{\partial z^{\alpha}} = 0$$

And similarly,

$$\frac{\partial z^{\alpha'}}{\partial \overline{z^{\alpha}}} = \frac{\partial z^{\alpha}}{\partial \overline{z^{\alpha'}}} = \frac{\partial \overline{z^{\alpha}}}{\partial z^{\alpha'}} = 0 .$$

Hence the coordinate transformations are all complex analytic, as desired.

Real Tensors

We recall that the complexified tangent space M_p^c may be thought of as containing the (real) tangent space M_p as that subpsace whose elements V satisfy $\overline{V} = V$ (compare the proof of Theorem II.7). Then in a real coordinate system, if $V = \xi^{\alpha} \partial/\partial x^{\alpha} + \xi^{\bar{\alpha}} \partial/\partial x^{\bar{\alpha}}$, we have:

$$V = \overline{V} \Longleftrightarrow \xi^{\alpha} = \overline{\xi^{\alpha}}, \qquad \xi^{\bar{\alpha}} = \overline{\xi^{\bar{\alpha}}} .$$

That is to say, an element of the subspace $M_p \subseteq M_p^c$ has real components which are real numbers (i.e., real-valued components with respect to a real basis). We call such a vector a <u>real vector</u>.

For the complex components of a real vector, we have:

$$V = \overline{V} \Longleftrightarrow V^{\alpha}\,\partial/\partial z^{\alpha} + V^{\bar{\alpha}}\,\partial/\partial z^{\bar{\alpha}} = \overline{V^{\alpha}}\;\overline{\partial/\partial z^{\alpha}} + \overline{V^{\bar{\alpha}}}\;\overline{\partial/\partial z^{\bar{\alpha}}}$$

$$\Longleftrightarrow V^{\alpha} = \overline{V^{\bar{\alpha}}}, \qquad V^{\bar{\alpha}} = \overline{V^{\alpha}} .$$

Hence the complex components of a real vector satisfy

$$V^{\bar{a}} = \overline{V^{a}} .$$

We now generalize this idea to any tensor:

IV.12: DEFINITION: A real tensor is a tensor whose real components are real numbers, or equivalently, a tensor whose complex components satisfy

$$\overline{T^{ab\ldots}{}_{cd\ldots}} = T^{\bar{a}\bar{b}\ldots}{}_{\bar{c}\bar{d}\ldots} .$$

IV.13: REMARK: It is easy to verify that the two definitions of a real tensor given above are equivalent by writing out $\overline{T^{ab\ldots}{}_{cd\ldots}} = T^{\bar{a}\bar{b}\ldots}{}_{\bar{c}\bar{d}\ldots}$ in real components and using the tensor transformation law.

Vectors and One-Forms of Type (1,0) and Type (0,1)

In Remark IV.6 it was shown that the real components of a type (1,0) vector satisfy the relation $\xi^{\bar{\alpha}} = -i\xi^{\alpha}$. Hence (using Remark IV.10) the complex components satisfy

$$\frac{V^{\alpha} - V^{\bar{\alpha}}}{2i} = -i\,\frac{V^{\alpha} + V^{\bar{\alpha}}}{2} \Rightarrow V^{\bar{\alpha}} = 0 .$$

Therefore the vectors of type (1,0) are all of the form

$$V = V^{\alpha}\,\partial/\partial z^{\alpha} .$$

Similarly, the vectors of type (0,1) are of the form

$$v = v^{\bar{\alpha}} \partial/\partial z^{\bar{\alpha}} .$$

For one-forms we have an analogous result. We already know that $\{dz^{\alpha}\}$ is a basis for the one-forms of type (1,0), so that these are given by

$$\omega = \Omega_{\alpha} dz^{\alpha} .$$

Likewise the forms of type (0,1) are all of the form

$$\omega = \Omega_{\bar{\alpha}} dz^{\bar{\alpha}} .$$

We can easily extend the notion of "type" from vectors and one-forms to tensors of arbitrary rank. Consider, for example, a tensor T of covariant rank 2. If T can be written as

$$T = T_{\alpha\beta} dz^{\alpha} \otimes dz^{\beta} ,$$

We say T is of covariant type (2,0). If T has the form

$$T = T_{\alpha\bar{\beta}} dz^{\alpha} \otimes dz^{\bar{\beta}} + T_{\bar{\alpha}\beta} dz^{\bar{\alpha}} \otimes dz^{\beta} ,$$

then T is of covariant type (1,1). And if T can be written as

$$T = T_{\bar{\alpha}\bar{\beta}} dz^{\bar{\alpha}} \otimes dz^{\bar{\beta}} ,$$

then T is of covariant type (0,2).

The generalization to arbitrary tensors is straightforward, so that we speak of tensors T of contravariant type (p,q) and covariant type (r,s) (where T is of contravariant rank p+q and covariant rank r+s). Strictly speaking, this concept is not yet well-defined, because we have not shown that type is preserved under a transformation from one complex basis to another. This matter is clarified below.

Given a tensor T of rank 2 (covariant, contravariant, or mixed), T is said to be pure if it can be written as a sum of terms of type (2,0) and type (0,2); T is said to be hybrid if it can be written as a sum of terms of type (1,1). For example, T_{ab} is pure if $T_{\alpha\bar{\beta}} = T_{\bar{\alpha}\beta} = 0$, and T_{ab} is hybrid if $T_{\alpha\beta} = T_{\overline{\alpha\beta}} = 0$.

Complex Tensors and Complex Manifolds

As we have already remarked, the concepts of complexified tangent and cotangent spaces are not dependent upon the existence of a complex structure on the underlying manifold. On any manifold it makes sense to deal with "complexified fields" $X + iY$, where X and Y are tensor fields (of the same rank) in the ordinary sense. (Recall, for example, the "complexified null vectors" of the null-tetrad formalism in relativity.) This consideration will be important when we study almost complex manifolds in the next chapter.

Nevertheless, several nice properties are present only when the manifold is even-dimensional (so that complex structures exist on the tangent and cotangent spaces) and in fact only when the manifold is complex. These properties arise for the most part from the fact that only in a complex manifold can complex coordinates be introduced which transform analytically in an intersection of charts.

Consider, for example, a vector

$$V = V^{\alpha}\,\partial/\partial z^{\alpha} + V^{\bar{\alpha}}\,\partial/\partial \overline{z^{\alpha}} = \xi^{\alpha}\,\partial/\partial x^{\alpha} + \xi^{\bar{\alpha}}\,\partial/\partial x^{\bar{\alpha}} ,$$

where $V^{\alpha} = \xi^{\alpha} + i\xi^{\bar{\alpha}}$ and $V^{\bar{\alpha}} = \xi^{\alpha} - i\xi^{\bar{\alpha}}$ (Remark IV.10). In another chart (U', x'), Z will take the form

$$V = V^{\alpha'} \partial/\partial z^{\alpha'} + V^{\bar{\alpha}'} \partial/\partial \overline{z^{\alpha'}} = \xi^{\alpha'} \partial/\partial x^{\alpha'} + \xi^{\bar{\alpha}'} \partial/\partial x^{\bar{\alpha}'} ,$$

where $V^{\alpha'} = \xi^{\alpha'} + i\xi^{\bar{\alpha}'}$ and $V^{\bar{\alpha}'} = \xi^{\alpha'} - i\xi^{\bar{\alpha}'}$. Then in the intersection $U \cap U'$ of the two charts we will have

$$\begin{aligned} V^{\alpha'} &= \xi^{\alpha'} + i\xi^{\bar{\alpha}'} \\ &= \frac{\partial x^{\alpha'}}{\partial x^{\beta}} \xi^{\beta} + \frac{\partial x^{\alpha'}}{\partial x^{\bar{\beta}}} \xi^{\bar{\beta}} + i \frac{\partial x^{\bar{\alpha}'}}{\partial x^{\beta}} \xi^{\beta} + i \frac{\partial x^{\bar{\alpha}'}}{\partial x^{\bar{\beta}}} \xi^{\bar{\beta}} . \end{aligned}$$

Then if and only if the Cauchy-Riemann conditions $\partial x^{\alpha'}/\partial x^{\beta} = \partial x^{\bar{\alpha}'}/\partial x^{\bar{\beta}}$ and $\partial x^{\bar{\alpha}'}/\partial x^{\beta} = -\partial x^{\alpha'}/\partial x^{\bar{\beta}}$ hold, we can write

$$\begin{aligned} V^{\alpha'} &= \frac{\partial x^{\alpha'}}{\partial x^{\beta}} (\xi^{\beta} + i\xi^{\bar{\beta}}) + i \frac{\partial x^{\bar{\alpha}'}}{\partial x^{\beta}} (\xi^{\beta} + i\xi^{\bar{\beta}}) \\ &= \frac{\partial z^{\alpha'}}{\partial z^{\beta}} V^{\beta} . \end{aligned}$$

Similarly, if and only if the Cauchy-Riemann condition are satisfied, we will find

$$V^{\bar{\alpha}'} = \overline{\left(\frac{\partial z^{\alpha'}}{\partial z^{\beta}}\right)} V^{\bar{\beta}} .$$

In other words, the familiar transformation law for complex coordinate transformations in the intersection of charts will make sense only if the transformations $z^{\alpha'} = z^{\alpha'}(z^{\beta})$ are analytic, so that the derivatives $\partial z^{\alpha'}/\partial z^{\beta}$ are well-defined. But this requirement, of course, is met only by a manifold which admits a complex structure.

In fact, it is easy to see that in a complex manifold, the general tensor transformation law for the complex components of tensors will take the form

$$T^{\alpha'\ldots\bar{\beta}'\ldots}_{\gamma'\ldots\bar{\delta}'\ldots} = \frac{\partial z^{\alpha'}}{\partial z^{\kappa}}\cdots\overline{\left(\frac{\partial z^{\beta'}}{\partial z^{\lambda}}\right)}\cdots\frac{\partial z^{\mu}}{\partial z^{\gamma'}}\cdots\overline{\left(\frac{\partial z^{\nu}}{\partial z^{\delta'}}\right)}\cdots T^{\kappa\ldots\bar{\lambda}\ldots}_{\mu\ldots\bar{\nu}\ldots} \; .$$

That is to say, an unbarred index transforms according to $\partial z^{\alpha'}/\partial z^{\beta}$ or $\partial z^{\beta}/\partial z^{\alpha'}$, and a barred index according to $\overline{\partial z^{\alpha'}/\partial z^{\beta}}$ or $\overline{\partial z^{\beta}/\partial z^{\alpha'}}$.

IV.14: REMARK: From the observation above, it is easy to verify that in a complex manifold, the type of a tensor is independent of the choice of complex coordinates belonging to charts of the complex structure.

Tensor Fields

At this point, having thoroughly examined tensors and tensor spaces at one point of a manifold, it would be natural to make an excursion into the matter of tensor fields, defined over all the points of the manifold. To treat the subject adequately would require a rather lengthy digression on complex vector bundles. We shall content ourselves here by merely remarking that tensor fields can be looked upon as sets of tensors defined over each point of a manifold whose components vary sufficiently smoothly from point to point (i.e., C^{∞} smoothness, analytic smoothness, etc.). A rigorous discussion of the construction of tensor fields in terms of complex vector bundles is given in Appendix C. For a fuller development, see Chern [8] or Morrow and Kodaira [30] .

Analytic Continuation of Tensors

In our discussion of complex manifolds, the associated product manifold has never been far from the surface, since it enters in the definition of the operators $\partial/\partial z^{\overline{\alpha}}$. In fact, given a tensor field on M, it will often be useful to associate with the field a tensor field on $M \times \overline{M}$.

The procedure is as follows. If we have a vector field

$$V_{(M)} = V^{\alpha}(z^{\beta}, \overline{z^{\beta}})\, \partial/\partial z^{\alpha} + V^{\overline{\alpha}}(z^{\beta}, \overline{z^{\beta}})\, \partial/\partial \overline{z^{\alpha}}$$

defined on M, we define the analytic continuation of $V_{(M)}$ to be the vector field $V_{(M \times \overline{M})}$ defined on $M \times \overline{M}$ by

$$V_{(M \times \overline{M})} = \tilde{V}^{\alpha}(z^{\beta}, z^{\overline{\beta}})\, \partial/\partial z^{\alpha} + \tilde{V}^{\overline{\alpha}}(z^{\beta}, z^{\overline{\beta}})\, \partial/\partial z^{\overline{\alpha}} ,$$

where $\tilde{V}^{\alpha}$ and $\tilde{V}^{\overline{\alpha}}$ are the analytic continuations of the functions V^{α} and $V^{\overline{\alpha}}$, respectively (see Chapter III).

The generalization to tensors of arbitrary rank is straightforward: In each case the (complex) component functions are replaced by their analytic continuations and the basis elements $\partial/\partial \overline{z^{\alpha}}$ and $d\overline{z^{\alpha}}$ are replaced by $\partial/\partial z^{\overline{\alpha}}$ and $dz^{\overline{\alpha}}$, respectively.

IV.15: REMARK: It turns out to be very important to note that the analytic continuation of 0 is 0 (for example,

$V_M = 0 \Rightarrow V_{(M \times \bar{M})} = 0$). This, together with the material above, will be important later when we consider the analytic continuation of Einstein's vacuum equations $R_{ab} = 0$ on spacetimes with a (modified) complex structure.

The Exterior Derivative Operator d

We recall [4] that a differential k-form is a completely skew-symmetric tensor of covariant rank k, which is usually written as $\Omega = \Omega_{ab\ldots d}\, dz^a \wedge dz^b \wedge \cdots \wedge dz^d$, where "$\wedge$" ("hook" or "wedge") is the exterior product. The exterior derivative is then defined as follows:

IV.16: DEFINITION: If $\Omega = \Omega_{ab\ldots d}\, dz^a \wedge dz^b \wedge \cdots \wedge dz^d$ is a k-form, then its exterior derivative $d\Omega$ is given by

$$d\Omega = \Omega_{ab\ldots d,e}\, dz^e \wedge dz^a \wedge dz^b \wedge \cdots \wedge dz^d \; .$$

IV.17: REMARK: d is a well-defined operation, i.e., $d\Omega$ is independent of the coordinate system used to apply Definition IV.16.

We note that if Ω is a k-form, then $d\Omega$ is a (k+1)-form. We shall need two more definitions.

IV.18: DEFINITION: A closed k-form Ω is a form for which $d\Omega = 0$.

IV.19: DEFINITION: An exact k-form Ω is a form for which a (k-1)-form ζ exists such that $\Omega = d\zeta$.

We then have this fundamental result:

IV.20: THEOREM: $d^2 = 0$, i.e., every exact form is closed.

Proof: $d^2\Omega = d\,(\Omega_{ab\ldots d,e}\, dz^e \wedge dz^a \wedge dz^b \wedge \ldots \wedge dz^d)$

$$= \Omega_{ab\ldots d,ef}\, dz^f \wedge dz^e \wedge dz^a \wedge dz^b \wedge \ldots \wedge dz^d)$$

$= 0$, because $\Omega_{ab\ldots d,ef}$ is symmetric in e and f while $dz^f \wedge dz^e$ is antisymmetric.

The converse of this theorem (for fields of forms) is in fact false: Not every closed form is exact (a counter example is the polar coordinate form "$d\phi$"; see [14]). However, it is true locally ("closed forms are locally exact"). This result follows from the next Theorem, often referred to as "Poincare's Lemma" [4]:

IV.21: THEOREM (POINCARE'S LEMMA): Let α be a k-form on $\mathbb{R}^n$ such that $d\alpha = 0$ in a neighborhood $U \subseteq \mathbb{R}^n$ of a point. Then there exists a neighborhood $V \subseteq U \subseteq \mathbb{R}^n$ such that on V, $\alpha = d\beta$ for some (k-1)-form β.

Proof: See [4], [50], or any standard text on differential geometry.

IV.22: COROLLARY: On a differentiable manifold, closed forms are locally exact.

Proof: Since a differentiable manifold is locally diffeomorphic to $\mathbb{R}^n$, the result follows at once from Poincare's Lemma.

It can be shown [4] that the exterior derivative operator d is the unique operator possessing the following properties:

i) $df(X) = X(f)$ for any vector field X and scalar function f

ii) $d(\alpha \wedge \beta) = d\alpha \wedge \beta + (-1)^k \alpha \wedge d\beta$ where α is a k-form

iii) $d^2 = 0$

iv) $d(a\alpha + b\beta) = a\, d\alpha + b\, d\beta$ for $a, b \in \mathbb{C}$.

The Operators ∂ and $\bar{\partial}$

On a complex manifold there exist, in addition to d, two other interesting "exterior derivative operators", which we shall now examine.

IV.23: DEFINITION: If, on a complex manifold,

$$\Omega = \Omega_{ab\ldots d}\, dz^a \wedge dz^b \wedge \cdots \wedge dz^d \quad ,$$

then we define

$$\partial\Omega = \Omega_{ab\ldots d,\sigma}\, dz^\sigma \wedge dz^a \wedge \cdots \wedge dz^d$$

and

$$\bar{\partial}\Omega = \Omega_{ab\ldots d,\bar{\sigma}}\, d\overline{z^\sigma} \wedge dz^a \wedge \cdots \wedge dz^d \quad .$$

IV.24: REMARK: It is straightforward to show that ∂ and $\bar{\partial}$ are in fact well-defined, i.e., $\partial\Omega$ and $\bar{\partial}\Omega$ are independent of the complex coordinates used to apply Definition IV.23.

It is clear that if α is a form of type (p,q) then $\partial\alpha$ is a form of type $(p+1,q)$ and $\bar{\partial}\alpha$ is a form of type $(p,q+1)$.

We note that the existence of the operators ∂ and $\bar{\partial}$ does not contradict the uniqueness of the operator d, because in general we will have:

$$\partial f(X) = f_{,\alpha}\, dz^{\alpha}(X) = f_{,\alpha} X^{\alpha}$$

$$\neq f_{,\alpha} X^{\alpha} + f_{,\bar{\alpha}} X^{\bar{\alpha}} ,$$

i.e., $\partial f(X) \neq X(f)$ and similarly $\bar{\partial} f(X) \neq X(f)$. Thus the property (i) of d given above is not satisfied by ∂ or $\bar{\partial}$. (However, properties (ii), (iii), and (iv) do hold for ∂ and $\bar{\partial}$.) In fact, let us develop some of the properties of ∂ and $\bar{\partial}$:

<u>IV.25: DEFINITION:</u> A form Ω is said to be ∂-closed if $\partial\Omega = 0$, and is said to be $\bar{\partial}$-closed if $\bar{\partial}\Omega = 0$.

<u>IV.26: DEFINITION:</u> A k-form α is said to be ∂-exact if there exists a (k-1)-form β such that $\alpha = \partial\beta$; similarly, α is said to be $\bar{\partial}$-exact if there exists a (k-1)-form γ such that $\alpha = \bar{\partial}\gamma$.

(When referring to the operator d, we shall sometimes say d-closed and d-exact to avoid any confusion.)

<u>IV.27: THEOREM:</u> $d = \partial + \bar{\partial}$.

<u>Proof:</u> If $\Omega = \Omega_{ab\ldots d}\, dz^a \wedge dz^b \wedge \cdots \wedge dz^d$,

then $d\Omega = \Omega_{ab\ldots d,e}\, dz^e \wedge dz^a \wedge dz^b \wedge \cdots \wedge dz^d$

$$= \Omega_{ab\ldots d,\epsilon}\, dz^{\epsilon} \wedge dz^a \wedge dz^b \wedge \cdots \wedge dz^d$$

$$+ \Omega_{ab\ldots d,\bar{\epsilon}}\, \overline{dz^{\epsilon}} \wedge dz^a \wedge dz^b \wedge \cdots \wedge dz^d$$

$$= \partial\Omega + \bar{\partial}\Omega .$$

IV.28: THEOREM: $\partial^2 = \bar{\partial}^2 = 0$, i.e., ∂-exact forms are ∂-closed and $\bar{\partial}$-exact forms are $\bar{\partial}$-closed.

Proof: If $\Omega = \Omega_{ab\ldots d}\, dz^a \wedge dz^b \wedge \ldots \wedge dz^d$, then

$$\partial\partial\Omega = \partial(\Omega_{ab\ldots d,\epsilon}\, dz^\epsilon \wedge dz^a \wedge dz^b \wedge \ldots \wedge dz^d)$$

$$= \Omega_{ab\ldots d,\epsilon\bar{\epsilon}}\, d\overline{z^\epsilon} \wedge dz^\epsilon \wedge dz^a \wedge dz^b \wedge \ldots \wedge dz^d$$

$$= (\text{symmetric in } \epsilon, \bar{\epsilon}) \times (\text{antisymmetric in } \epsilon, \bar{\epsilon})$$

$$= 0 ,$$

and $\bar{\partial}\bar{\partial}\Omega = 0$ similarly.

IV.29: THEOREM: $\partial\bar{\partial} + \bar{\partial}\partial = 0.$

Proof: $d^2 = 0 = (\partial + \bar{\partial})^2$

$$= \partial^2 + \partial\bar{\partial} + \bar{\partial}\partial + \bar{\partial}^2$$

$$= 0 + \partial\bar{\partial} + \bar{\partial}\partial + 0 .$$

IV.30: REMARK: Another straightforward proof of the last theorem above can be obtained by writing

$$\Omega = \Omega_{ab\ldots d}\, dz^a \wedge dz^b \wedge \ldots \wedge dz^d$$

and calculating $\partial\bar{\partial}\Omega + \bar{\partial}\partial\Omega$ directly.

The Dolbeault-Grothendieck Lemma

We have seen that d-closed forms are not in general d-exact, but are in fact locally d-exact. In a similar vein it is true that $\bar{\partial}$-closed forms, while not in general $\bar{\partial}$-exact, are nevertheless locally $\bar{\partial}$-exact (with similar remarks holding for ∂-closed forms). On regions of $\mathbf{C}^n$, this result is known as the Dolbeault-Grothendieck Lemma. The proof is an instructive example of complex analysis (to be compared later to our modified complex analysis), and hence will be gone into in some detail. Our proof will be modeled after that of Chern[8]; a similar proof may be found in Morrow and Kodaira [30].

First we shall require two lemmas:

<u>IV.31: LEMMA</u>: Let D be the disc $\{z \mid |z| < r\}$ in the complex line $\mathbf{C}$, and $D' \subsetneq D$ the disc $\{z \mid |z| < r'\}$ where $r' < r$. Let $f(z)$ be a function defined on D. Then for $z \in D'$,

$$\int_{D'} \frac{f(\zeta)}{(\zeta - z)}\, d\zeta \wedge d\bar{\zeta}$$

$$= \int_{\partial D'} f(\zeta)\, \ln|\zeta - z|^2\, d\bar{\zeta}$$

$$- \int_{D'} f_{,\zeta}\, \ln|\zeta - z|^2\, d\zeta \wedge d\bar{\zeta}\,,$$

where $f_{,\zeta} \equiv \partial f/\partial\zeta$.

<u>Proof</u>: Let Δ_ϵ be a disc of radius ϵ centered on $z \in D'$. We have

$$d[f(\zeta)\, \ln|\zeta - z|^2\, d\bar{\zeta}] = [f(\zeta)\, \ln|\zeta - z|^2]_{,\zeta}\, d\zeta \wedge d\bar{\zeta}$$

$$= \{f(\zeta)\,[\ln(\zeta - z) + \ln(\bar{\zeta} - \bar{z})]\}_{,\zeta}\, d\zeta \wedge d\bar{\zeta}$$

$$= [f_{,\zeta}\, \ln|\zeta - z|^2 + \frac{f}{(\zeta - z)}]\; d\zeta \wedge d\bar{\zeta}\,.$$

Therefore, applying Stoke's Theorem to the differential form $F = f(\zeta)\, \ln|\zeta - z|^2\, d\bar{\zeta}$ in the domain $D' - \Delta_\epsilon$ gives:

$$\int_{D'-\Delta_\epsilon} dF = \int_{\partial D' - \partial\Delta_\epsilon} F$$

$$\Rightarrow \int_{D'-\Delta_\epsilon} \left[f_{,\zeta} \ln|\zeta - z|^2 + \frac{f}{(\zeta - z)} \right] d\zeta \wedge d\bar{\zeta}$$

$$= \int_{\partial D' - \partial\Delta_\epsilon} f(\zeta) \ln|\zeta - z|^2 \, d\bar{\zeta} \, .$$

In taking the limit $\epsilon \to 0$, the integral over $\partial\Delta_\epsilon$ will go to zero, because, assuming $|f(\zeta)| \leq B$ on $\partial\Delta_\epsilon$, we will have:

$$\left| \int_{\partial\Delta_\epsilon} f(\zeta) \ln|\zeta - z|^2 \, d\bar{\zeta} \right| \leq 2\pi\epsilon B \ln \epsilon^2$$

$$\to 0 \quad \text{as} \quad \epsilon \to 0 \, .$$

Therefore we are left with the desired result:

$$\int_{\partial D'} f(\zeta) \ln|\zeta - z|^2 \, d\bar{\zeta} - \int_{D'} f_{,\zeta} \ln|\zeta - z|^2 \, d\zeta \wedge d\bar{\zeta}$$

$$= \int_{D'} \frac{f(\zeta)}{(\zeta - z)} \, d\zeta \wedge d\bar{\zeta} \, .$$

IV.32: LEMMA (GENERALIZED CAUCHY INTEGRAL FORMULA): Let D and D' be as in the previous lemma. Then for $z \in D'$ and for a function β defined in D we have:

$$2\pi i \, \beta(z) = \int_{\partial D'} \frac{\beta \, d\zeta}{(\zeta - z)} + \int_{D'} \frac{\beta_{,\bar{\zeta}}}{(\zeta - z)} \, d\zeta \wedge d\bar{\zeta}$$

and

$$-2\pi i \, \beta(z) = \int_{\partial D'} \frac{\beta \, d\bar{\zeta}}{(\bar{\zeta} - \bar{z})} + \int_{D'} \frac{\beta_{,\zeta}}{(\bar{\zeta} - \bar{z})} \, d\bar{\zeta} \wedge d\zeta \, ,$$

where $\beta_{,\zeta} = \partial\beta/\partial\zeta$ and $\beta_{,\bar{\zeta}} = \partial\beta/\partial\bar{\zeta}$.

Proof: Let Δ_ϵ be as in the previous lemma and apply Stoke's Theorem to the form $\frac{\beta d\zeta}{(\zeta - z)}$ in the domain $D' - \Delta_\epsilon$:

$$\int_{\partial D'} \frac{\beta(\zeta)\, d\zeta}{(\zeta - z)} - \int_{\partial \Delta_\epsilon} \frac{\beta(\zeta)\, d\zeta}{(\zeta - z)} = \int_{D' - \Delta_\epsilon} d\left(\frac{\beta(\zeta)\, d\zeta}{(\zeta - z)}\right)$$

$$= \int_{D' - \Delta_\epsilon} \frac{\beta_{,\bar\zeta}\, d\bar\zeta \wedge d\zeta}{(\zeta - z)} .$$

Now we note that the integral

$$\int_{\partial \Delta_\epsilon} \frac{\beta(\zeta)\, d\zeta}{(\zeta - z)}$$

which appears above will approach $2\pi i \beta(z)$ as $\epsilon \to 0$. Therefore we have the first desired result. The second equation is obtained by taking the complex conjugate of the first and then performing the replacement $\bar\beta \to \beta$.

We are now prepared to prove our major result:

IV.33: THEOREM (DOLBEAULT-GROTHENDIECK LEMMA): If α is a form of type (p,q) with $q \geqslant 1$ such that $\bar\partial\alpha = 0$ on some disc $D = \{|z^\alpha| < r^\alpha\}$ in $\mathbb{C}^m$, then there exists on D a form β of type $(p,q-1)$ such that, in a smaller disc $D' \leq D$, we have $\bar\partial\beta = \alpha$.

Proof: We start with the simplest possible case, namely $m = 1$, $p = 0$, $q = 1$, and then proceed by an induction argument. Assume, then, that on the complex line $\mathbb{C}^1$ we have given a form $\alpha = f(z)\, d\bar z$ of type $(0,1)$ (note that $f(z)$ is not necessarily complex analytic) such that

$\bar{\partial}\alpha = 0$ in some open disc $D = \{z \mid |z| < r\}$. That is, we have

$$\bar{\partial}\alpha = \frac{\partial f}{\partial \bar{z}} \, d\bar{z} \wedge d\bar{z} = 0$$

and we want to show that there exists a scalar β (form of type (0,0)) such that

$$\bar{\partial}\beta = \frac{\partial \beta}{\partial \bar{z}} \, d\bar{z} = f(z) \, d\bar{z} ,$$

in a smaller open disc $D' = \{z \mid |z| < r'\} \subsetneq D$. So we are looking for a solution to the equation

$$\frac{\partial \beta}{\partial \bar{z}} = f(z) .$$

We claim that such a solution is given by

$$\beta(z) = \frac{1}{2\pi i} \int_{D'} \frac{f(\zeta) \, d\zeta \wedge d\bar{\zeta}}{(\zeta - z)} + g(z) ,$$

where g is any analytic function of z and D' is any open disc contained in D. To verify this, we compute:

$$\frac{\partial \beta}{\partial \bar{z}} = \frac{1}{2\pi i} \frac{\partial}{\partial \bar{z}} \left[\int_{D'} \frac{f(\zeta) \, d\zeta \wedge d\bar{\zeta}}{(\zeta - z)} \right] + \frac{\partial g}{\partial \bar{z}} .$$

But $\partial g/\partial \bar{z} = 0$ because g is analytic in z, and, using Lemma IV.31, we can rewrite the integral appearing on the right-hand side to obtain:

$$\frac{\partial \beta}{\partial \bar{z}} = \frac{1}{2\pi i} \frac{\partial}{\partial \bar{z}} \left[\int_{\partial D'} f(\zeta) \ln |\zeta - z|^2 \, d\bar{\zeta} - \int_{D} f_{,\zeta} \ln |\zeta - z|^2 \, d\zeta \wedge d\bar{\zeta} \right] .$$

Now we "take $\partial/\partial\bar{z}$ inside the integrals" (which is justified essentially because the resulting integrals exist, see Chern [8]) and we write $\ln|\zeta - z|^2 = \ln(\zeta - z) + \ln(\bar{\zeta} - \bar{z})$ to obtain:

$$\frac{\partial\beta}{\partial\bar{z}} = \frac{1}{2\pi i}\left[\int_{\partial D'} - \frac{f(\zeta)\,d\bar{\zeta}}{(\bar{\zeta} - \bar{z})} + \int_D \frac{f_{,\bar{\zeta}}}{(\bar{\zeta} - \bar{z})}\,d\zeta \wedge d\bar{\zeta}\right] .$$

But, by applying Lemma IV.32 to the right-hand side of this equation, we obtain

$$\frac{\partial\beta}{\partial\bar{z}} = f(z) .$$

This establishes the desired result for the case $m = 1$, $p = 0$, $q = 1$. It is important to notice that if f is holomorphic in some parameters, then β is also holomorphic in the parameters; this fact will be used below.

To prove the general case, we proceed by means of the following induction argument. Let S_κ ($\kappa = 1, \ldots, m$) denote the statement "the theorem is true for a form α which does not contain any terms involving $d\bar{z}^{\kappa+1}, \ldots, d\bar{z}^m$." We shall show that S_0 is true and that $S_{\kappa-1}$ implies S_κ. Since the statement S_m is equivalent to the theorem itself, this induction argument will suffice for the proof.

So first we notice that if α does not contain any terms involving $d\bar{z}^1, \ldots, d\bar{z}^m$, then α must be zero, because by assumption $q \geq 1$. Therefore S_0 is true.

Next suppose that $S_{\kappa-1}$ is true, i.e., that the theorem is true for forms which do not contain any terms involving $d\bar{z}^{\kappa}, \ldots, d\bar{z}^{m}$. Then suppose we have a form α of type (p,q) which does not contain any terms with $d\bar{z}^{\kappa+1}, \ldots, d\bar{z}^{m}$; any such α can be written as

$$\alpha = (d\bar{z}^{\kappa} \wedge \lambda) + \mu ,$$

where λ is a form of type $(p,q-1)$ and μ is a form of type (p,q), and neither λ nor μ contains any terms with $d\bar{z}^{\kappa}, \ldots, d\bar{z}^{m}$. Since $\bar{\partial}\alpha = 0$ by assumption, it follows that the coefficients of λ and μ are all holomorphic in the variables $\bar{z}^{\kappa+1}, \ldots, \bar{z}^{m}$. Let $\lambda_{a\ldots b}$ be one of the coefficients of λ and, for the moment, consider all the coordinates except z^{κ} and $\bar{z}^{\kappa}$ to be fixed. This allows us to apply our result above for the special case $m = 1$, $p = 0$, $q = 1$, by considering $\lambda_{a\ldots b}\, d\bar{z}^{\kappa}$ as a ($\bar{\partial}$-closed) one-form in the variables z^{κ} and $\bar{z}^{\kappa}$. In this way we can obtain $\tilde{\lambda}_{a\ldots b}$ such that

$$\frac{\partial}{\partial \bar{z}^{\kappa}} \tilde{\lambda}_{a\ldots b} = \lambda_{a\ldots b} .$$

Then $\tilde{\lambda} = \tilde{\lambda}_{a\ldots b}\, dz^{a} \wedge \cdots \wedge dz^{b}$ will be a form of type $(p,q-1)$ and, since λ is holomorphic in $\bar{z}^{\kappa+1}, \ldots, \bar{z}^{m}$, so is $\tilde{\lambda}$ (see comment above).

Then $\bar{\partial}\tilde{\lambda}$ contains no terms with $d\bar{z}^{\kappa+1}, \ldots, d\bar{z}^{m}$, and furthermore the only term with $d\bar{z}^{\kappa}$ is the term $d\bar{z}^{\kappa} \wedge \lambda$ (by virtue of the fact that $\partial/\partial\bar{z}^{\kappa}(\tilde{\lambda}_{a\ldots b}) = \lambda_{a\ldots b}$). From this we conclude that

$$\nu \equiv \bar{\partial}\tilde{\lambda} - (d\bar{z}^{\kappa} \wedge \lambda)$$

does not have any terms involving $d\bar{z}^{\kappa}, d\bar{z}^{\kappa+1}, \ldots, d\bar{z}^{m}$. Thus

$$\alpha = (d\overline{z^{\varkappa}} \wedge \lambda) + \mu$$
$$= \bar{\partial}\tilde{\lambda} - \nu + \mu .$$

Then $\bar{\partial}\alpha = 0$

$$\Rightarrow \quad \bar{\partial}^2\tilde{\lambda} - \bar{\partial}(\mu - \nu) = 0$$

$$\Rightarrow \quad \bar{\partial}(\mu - \nu) = 0 .$$

But now neither μ nor ν contains $d\overline{z^{\varkappa}}, \ldots, d\overline{z^{m}}$; and hence by the induction hypothesis $S_{\varkappa-1}$ we can find a form φ in D such that

$$\bar{\partial}\varphi = \mu - \nu$$

and hence

$$\alpha = \bar{\partial}\tilde{\lambda} + \bar{\partial}\varphi$$
$$= \bar{\partial}(\tilde{\lambda} + \varphi)$$

in D'. This establishes the induction step and hence the theorem is proven.

<u>IV.34: REMARK:</u> It is clearly true that an analogous result could be obtained for ∂-closed forms.

<u>IV.35: THEOREM:</u> $\bar{\partial}$-closed forms are locally $\bar{\partial}$-exact (and ∂-closed forms are locally ∂-exact).

<u>Proof:</u> Since the complex structure of a complex manifold is locally equivalent to that of $\mathbb{C}^m$, it suffices to apply the Dolbeault-Grothendieck

Lemma locally (and its analogue for the operator ∂).

As an example of the usefulness of the Dolbeault-Grothendieck Lemma which will be used later (see Chapter VI), we mention the following theorem:

<u>IV.36: THEOREM:</u> A two form Ω_{ab} of type (1,1) (i.e., of the form $\Omega_{\alpha\bar{\beta}}\, dz^{\alpha} \wedge d\overline{z^{\beta}}$) is d-closed if and only if Ω is locally expressible as $\Omega = \partial\bar{\partial} f$, where f is a scalar.

<u>Proof:</u> Suppose $\Omega = \partial\bar{\partial} f$; then

$$\begin{aligned} d\Omega &= (\partial + \bar{\partial})\, \partial\bar{\partial} f \\ &= \partial^2 \bar{\partial} f + \bar{\partial}\partial\, \bar{\partial} f \\ &= 0 - \partial\bar{\partial}^{2} f \\ &= 0\ . \end{aligned}$$

Conversely, suppose $\Omega = \Omega_{\alpha\bar{\beta}}\, dz^{\alpha} \wedge d\overline{z^{\beta}}$ is closed. Then we have

$$d\Omega = \partial\Omega + \bar{\partial}\Omega = 0\ .$$

Since $\partial\Omega$ is of type (2,1) and $\bar{\partial}\Omega$ of type (1,2), both $\partial\Omega$ and $\bar{\partial}\Omega$ must separately be equal to zero. But if $\bar{\partial}\Omega = 0$, then by the Dolbeault-Grothendieck Lemma there exists locally a form α of type (1,0) such that $\bar{\partial}\alpha = \Omega$. So then we have:

$$\begin{aligned} 0 &= \partial\Omega = \partial\bar{\partial}\alpha \\ &= -\bar{\partial}\partial\alpha \\ &\Rightarrow \bar{\partial}(-\partial\alpha) = 0 \end{aligned}$$

then

$$d(-\partial\alpha) = (\partial + \bar{\partial})(-\partial\alpha)$$
$$= -\partial^2\alpha + \bar{\partial}(-\partial\alpha)$$
$$= 0 \, ,$$

so by Poincare's Lemma, there exists a one-form β such that

$$-\partial\alpha = d\beta \, .$$

Since $d(\partial\alpha) = 0 \Rightarrow \bar{\partial}(\partial\alpha) = 0$, it follows that the components of $-\partial\alpha$ are holomorphic in the z^{α}. Therefore, (compare the comment in the proof of the Dolbeault-Grothendieck Lemma), the components of β are also holomorphic in the z^{α}, so that

$$\bar{\partial}\beta = 0$$

and hence

$$-\partial\alpha = \partial\beta \, .$$

Therefore $-\alpha$ and β can differ locally at most by a one-form of the form ∂f, i.e.,

$$-\alpha = \beta + \partial f$$

for some scalar f. Now we have

$$\Omega = \bar{\partial}\alpha = \bar{\partial}(-\beta - \partial f)$$
$$= 0 - \bar{\partial}\partial f = \partial\bar{\partial} f ,$$

so the theorem is proven.

CHAPTER V: ALMOST COMPLEX MANIFOLDS

In applications to relativity, we shall be concerned with the problem of determining whether a given spacetime admits a (modified) complex structure. In this context, the concept of an almost complex manifold arises naturally as an intermediate consideration: Given a manifold with an almost complex structure, a straightforward procedure exists for determining if the almost complex structure is "induced by" a complex structure on the manifold (integrability condition). Later we shall show that quite general spacetimes admit (modified) almost complex structures.

One of the original reasons for studying almost complex manifolds was summarized in Eckmann [15]. Much can be deduced from the existence of an almost complex structure as opposed to a complex structure, since the former is amenable to the application of techniques from fiber bundle theory. However, we shall be concerned with almost complex manifolds mainly in the context of establishing that a given manifold has a complex structure.

(In fact, in modern approaches to complex manifold theory, the underlying real-manifold nature of a complex manifold is ignored completely and the theory is developed *ab initio* from a definition of a complex manifold in terms of local homeomorphism to $\mathbf{C}^m$. Since, in our applications to relativity, the starting point is always a *real* manifold (spacetime), this modern approach is inappropriate for our purposes.)

Almost Complex Structure

For the moment we consider a general manifold M with a differentiable structure, and we make the following definition:

<u>V.1: DEFINITION:</u> An <u>almost</u> <u>complex</u> <u>structure</u> on M is a real differentiable tensor field J of rank $(1,1)$ with the property

$$J(J(\xi)) = -\xi$$

for any differentiable vector field ξ ; that is, a real tensor J whose (real) components $j_a{}^b$ satisfy

$$j_a{}^s j_s{}^b = -\delta_a{}^b .$$

A manifold which admits an almost complex structure is called an almost complex manifold.

Note that a tensor such as $i\delta_a{}^b$ does <u>not</u> define an almost complex structure, because it is not a real tensor.

Equivalently, an almost complex structure J on M is a differentiable field of endomorphisms $J_p : M_p \to M_p$ on each tangent space M_p such that $J_p(J_p(\xi_p)) = -\xi_p$ for all $\xi_p \in M_p$. Then it is immediately clear that an almost complex structure introduces a (vector space) complex structure on the tangent spaces M_p for each $p \in M$. From this observation, we obtain the following result:

<u>V.2: THEOREM:</u> A manifold M with an almost complex structure J is even-dimensional and orientable.

<u>Proof:</u> As noted above, J induces a vector space complex structure J_p on the tangent space M_p at each $p \in M$. Then by Theorem II.4 M_p is an even-dimensional, whence M is even-dimensional. Furthermore, by Theorem II.14, J_p determines an orientation of M_p for all

$p \in M$. Then since the field J varies continuously, it follows (see Spivak [50]) that the tangent bundle TM -- and hence M itself -- is orientable.

Since an almost complex structure J on M induces a (vector space) complex structure on each M_p, it follows from the discussions of Chapter IV that the existence of such a J determines a choice of type (1,0) and type (0,1) complexified one-forms on the space $M_p{}^{*c}$. In fact, Chern [8] gives such a choice of type (1,0) forms everywhere as an alternate definition of an almost complex structure.

To see the connection between the two definitions, suppose we have an almost complex structure J defined in terms of a tensor field whose components with respect to some (real) coordinate patch $j_a{}^b$ satisfy $j_a{}^s\, j_s{}^b = -\delta_a{}^b$. Then at any point $p \in M$, the complex structure for M_p is given, in terms of this basis, by the endomorphism

$$J_p = j_a{}^b\,(p)\, dx^a \otimes \partial/\partial x^b \ .$$

Then a (complex) tangent vector $V = \zeta^a\, \partial/\partial x^a \in M_p{}^c$ is of type (1,0) with respect to J_p if and only if (compare Remark II.8)

$$J_p\,(V) = iV \Leftrightarrow j_a{}^b\,(p)\,\zeta^a = i\zeta^b \ .$$

Similarly, the endomorphism $J_p{}^*$ which acts on $M_p{}^*$ can be described by the same expression:

$$J_p{}^* = j_a{}^b\,(p)\, dx^a \otimes \partial/\partial x^b \, .$$

A (complex) one-form $\omega = \omega_a\, dx^a \in M_p{}^{*c}$ is of type (1,0) with respect to $J_p{}^*$ if and only if

$$J_p^*(\omega) = i\omega \Longleftrightarrow j_a{}^b(p)\,\omega_b = i\omega_a \quad .$$

Now consider the $n = 2m$ one-forms given by

$$\omega^{(b)} = (j_a{}^b + i\delta_a{}^b)\, dx^a = \omega_a{}^{(b)}\, dx^a \quad .$$

Then we have

$$j_c{}^a\, \omega_a{}^{(b)} = -\delta_c{}^b + i\, j_c{}^b = i\,\omega_c{}^{(b)} \quad .$$

Thus the forms $\omega^{(b)}$ are all of type (1,0). From their construction it is clear that exactly m of them are $\mathbb{C}$-linearly independent, and hence m such $\omega^{(b)}$ form a basis for the type (1,0) forms. In this way the almost complex structure J is seen to be equivalent to a choice of type (1,0) forms on the spaces $M_p{}^{*c}$.

<u>V.3: REMARK:</u> Given the manifold $\mathbb{R}^{2m}$ and the standard coordinate chart $(x^1, \ldots, x^{2m})$,

$$j_a{}^b = \begin{pmatrix} 0 & 1 & & & & & \\ -1 & 0 & & & & 0 & \\ & & 0 & 1 & & & \\ & & -1 & 0 & & & \\ & & & & \ddots & & \\ & 0 & & & & 0 & 1 \\ & & & & & -1 & 0 \end{pmatrix}$$

defines an almost complex structure for $\mathbb{R}^{2m}$.

<u>Conditions for Existence of an Almost Complex Structure</u>

We have seen in Theorem V.2 that an almost complex manifold is necessarily even-dimensional and orientable. However, these conditions are not sufficient for a manifold to admit an almost complex structure: Some counterexamples are the spheres S^4 and S^n, $n \geq 8$ (see Chern [7]), which do not admit an almost complex structure. On the other hand, we shall see below that every complex manifold admits an almost complex structure, but that the existence of a complex structure is not a necessary condition for an almost complex structure (a counterexample is the sphere S^6; see below).

Two necessary and sufficient conditions for a manifold to admit an almost complex structure are given by Yano. The proofs, which are omitted, are straightforward and may be found in reference [54].

V.4: THEOREM (YANO): An $n = 2m$-dimensional manifold M admits an almost complex structure if and only if M contains a distribution π_m of complex dimension m and a distribution $\bar{\pi}_m$ which is complex conjugate to π_m, has no common direction with π_m, and spans with π_m a linear space of complex dimension $n = 2m$.

The distribution π_m is just the space spanned by the vectors of type (1,0) with respect to the almost complex structure, i.e., the vectors V satisfying $J(V) = iV$. Similarly, the distribution $\bar{\pi}_m$ is defined in terms of the type (0,1) vectors.

V.5: THEOREM (YANO): An $n = 2m$-dimensional manifold M admits an almost complex structure if and only if the group of the tangent bundle TM of M can be reduced to the unitary group $U(m)$.

Almost Complex Structure on a Complex Manifold

We now begin an examination of the connection between almost complex structure and (manifold) complex structure.

V.6: THEOREM: A complex manifold admits an almost complex structure.

Proof: Let M be a complex manifold. Recall (Definition IV.3) that the canonical complex structure J is well-defined on M_p for all $p \in M$. With respect to any real chart $(U, x^{\alpha}, x^{\bar{\alpha}})$ of the complex structure of M, the components of J are given by

$$j_a{}^b = \begin{pmatrix} j_\alpha{}^\beta & j_\alpha{}^{\bar{\beta}} \\ j_{\bar{\alpha}}{}^\beta & j_{\bar{\alpha}}{}^{\bar{\beta}} \end{pmatrix} = \begin{pmatrix} 0 & \delta_\alpha{}^{\bar{\beta}} \\ -\delta_{\bar{\alpha}}{}^\beta & 0 \end{pmatrix} .$$

Then as a tensor field defined over all of M, J is real, differentiable, and satisfies

$$j_a{}^s \, j_s{}^b = -\delta_a{}^b .$$

Clearly these three properties remain valid in any chart of the underlying differentiable structure of M. Therefore, J is an almost complex structure for M.

The almost complex structure constructed in this way for a complex manifold M is said to be induced by the complex structure of M. An almost complex structure is also said to be integrable if it arises in this way. Clearly, a given almost complex structure may be induced by at most one complex structure.

An equivalent description of an integrable almost complex structure may be given as follows. Given a complex manifold M, we can construct the tensor

$$J = J_\alpha{}^\beta \, dz^\alpha \otimes \partial/\partial z^\beta + J_{\bar\alpha}{}^{\bar\beta} \, d\overline{z^\alpha} \otimes \partial/\partial \overline{z^\beta}$$

with complex components

$$J_a{}^b = \begin{pmatrix} J_\alpha{}^\beta & J_\alpha{}^{\bar\beta} \\ J_{\bar\alpha}{}^\beta & J_{\bar\alpha}{}^{\bar\beta} \end{pmatrix} = \begin{pmatrix} i\,\delta_\alpha{}^\beta & 0 \\ 0 & -i\delta_{\bar\alpha}{}^{\bar\beta} \end{pmatrix} .$$

By Theorem IV.11, we know that J has these same complex components in any chart of the complex structure.

Then clearly J defines the above-mentioned almost complex structure for M. Note that the forms of type (1,0) determined by the almost complex structure are those of the form $\omega = \Omega_\alpha dz^\alpha$, i.e., precisely the forms of type (1,0) as defined by the complex structure.

Conversely, an almost complex structure J is induced by a complex structure on M if M can be covered by complex coordinate charts which transform by analytic functions in intersections, in such a way that the complex components of J take the above form in every chart.

The question arises as to whether every almost complex manifold can be obtained in this way, that is, whether every almost complex structure is induced by some underlying complex structure (is integrable). The answer to this question is "no;" an almost complex structure must satisfy certain "integrability conditions" if it is to be induced by a complex structure. We now turn our attention to these integrability conditions.

The Nijenhuis Tensor

In examining the question of the integrability of an almost complex structure, the following tensor plays a central role:

V.7: DEFINITION: The Nijenhuis tensor N of a manifold with an almost complex structure J is the tensor whose real components are given by

$$n_{ab}{}^{c} \equiv j_a{}^{s}(j_b{}^{c}{}_{,s} - j_s{}^{c}{}_{,b}) - j_b{}^{s}(j_a{}^{c}{}_{,s} - j_s{}^{c}{}_{,a}) ,$$

where $j_a{}^{b}$ are the real components of J.

The Nijenhuis tensor was first presented in the context of a more general problem (see Nijenhuis [38]). Various authors use various multiples on N, and the term torsion tensor of the almost complex structure is often used. Chern [8] uses the tensor $-n_{ab}{}^{c}$ and Frölicher [17] uses $\frac{1}{4} n_{ab}{}^{c}$.

V.8: REMARK: (a) N is a tensor. To see this, introduce an arbitrary symmetric affine connection $\Lambda_{ab}{}^{c}$ on M, calculate

$$\hat{n}_{ab}{}^{c} \equiv j_a{}^{s}(j_b{}^{c}{}_{;s} - j_s{}^{c}{}_{;b}) - j_b{}^{s}(j_a{}^{c}{}_{;s} - j_s{}^{c}{}_{;a}) ,$$

where ";" denotes the covariant derivative with respect to $\Lambda_{ab}{}^{c}$, and obtain $\hat{n}_{ab}{}^{c} = n_{ab}{}^{c}$.

(b) $n_{ab}{}^{c} = -n_{ba}{}^{c}$.

Vanishing of the Nijenhuis Tensor as Necessary and Sufficient Condition for Integrability

The importance of the Nijenhuis tensor is that its vanishing is a necessary and sufficient condition for an almost complex structure to be integrable. We shall prove necessity for any almost complex structure and sufficiency for the special case that the almost complex structure tensor J is real analytic (although the result can be obtained under weaker assumptions; see below).

First we demonstrate necessity:

<u>V.9: THEOREM:</u> In order for an almost complex structure J to be integrable (to be induced by a complex structure) it is necessary that

$$n_{ab}{}^{c} = 0 \ .$$

<u>Proof:</u> Suppose J is induced by a complex structure. Then (compare Definition IV.3) the real components $j_a{}^b$ with respect to a chart $(U, x^\alpha, x^{\bar\alpha})$ of the complex structure are given by

$$j_a{}^b = \begin{pmatrix} 0 & \delta_\alpha{}^{\bar\beta} \\ -\delta_{\bar\alpha}{}^{\beta} & 0 \end{pmatrix} .$$

Since the components of J are constant in this coordinate system, we have at once

$$n_{ab}{}^{c} = 0 \ .$$

To establish the converse of this theorem, we shall need the following lemma:

<u>V.10: LEMMA:</u> If J is an almost complex structure whose Nijenhuis tensor N vanishes, then

$$a_{ab}{}^{c}\,(j_p{}^{a} - i\,\delta_p{}^{a})\,(j_q{}^{b} - i\,\delta_q{}^{b}) = 0 ,$$

where

$$a_{ab}{}^{c} \equiv (j_b{}^{c}{}_{,a} - j_a{}^{c}{}_{,b}) = - a_{ba}{}^{c} .$$

Proof: We first note that the Nijenhuis tensor can be expressed as

$$n_{ab}{}^{c} = j_a{}^{s}\,a_{sb}{}^{c} - j_b{}^{s}\,a_{sa}{}^{c} .$$

Then a straightforward computation gives:

$$a_{ab}{}^{c}\,(j_p{}^{a} - i\,\delta_p{}^{a})\,(j_q{}^{b} - i\,\delta_q{}^{b})$$

$$= a_{ab}{}^{c}\,j_p{}^{a}\,j_q{}^{b} - a_{pq}{}^{c} - i\,(a_{aq}{}^{c}\,j_p{}^{a} + a_{pb}{}^{c}\,j_q{}^{b})$$

$$= j_p{}^{a}\,(a_{ab}{}^{c}\,j_q{}^{b} + a_{bq}{}^{c}\,j_a{}^{b}) - i\,(a_{sq}{}^{c}\,j_p{}^{s} + a_{ps}{}^{c}\,j_q{}^{s})$$

$$= j_p{}^{a}\,n_{aq}{}^{c} - i\,\delta_p{}^{a}\,n_{aq}{}^{c}$$

$$= 0 ,$$

since $n_{aq}{}^{c} \equiv 0$ by assumption.

We now present the major result of this section; the proof given is that of Frölicher [17].

V.11: THEOREM: If (and only if) J is a real analytic almost complex structure whose Nijenhuis tensor vanishes, then J is integrable.

Proof: We must show that local complex coordinates z, which transform analytically, can be introduced everywhere such that the complex components of J take the form

$$J_a^{\ b} = \begin{pmatrix} i\,\delta_\alpha^{\ \beta} & 0 \\ 0 & -i\delta_{\bar{\alpha}}^{\ \bar{\beta}} \end{pmatrix}$$

in each chart. Equivalently, we are looking for complex coordinates z^α such that the forms of type $(1,0)$ determined by J can be written as

$$\omega = \omega_b\,(j_a^{\ b} + i\,\delta_a^{\ b})\,dx^a = \Omega_\alpha\,dz^\alpha \quad .$$

Accordingly, we consider the system of equations

$$f_a^{\ b}\,dx^a \equiv (j_a^{\ b} + i\,\delta_a^{\ b})\,dx^a = 0 \quad . \qquad \text{(i)}$$

We would like to show that this system is equivalent to some system of m independent equations

$$dz^\alpha = 0 \quad , \qquad \text{(ii)}$$

where z^α are some complex functions of the coordinates x^a, whose integrals $z^\alpha(x) = u^\alpha(x) \pm iv^{\bar{\alpha}}(x) = \text{constant}$ provide us with the desired complex coordinates. To show this, we must demonstrate that the Frobenius integrability conditions are satisfied under our assumption of the vanishing of the Nijenhuis tensor. In order to establish these conditions we treat the system (i) formally as a system of equations for the "unknowns" dx^a. It is easily checked that a complete system of solutions to these equations is provided by the n expressions

$$dx^a_{\ (p)} = j_p^{\ a} - i\,\delta_p^{\ a} \quad .$$

Exactly m of these solutions are linearly independent, and these span the complete solution space. Now we apply the assumption that the Nijenhuis tensor of J vanishes, $n_{ab}{}^{c} = 0$: Making use of Lemma V.10, this implies

$$a_{ab}{}^{c}\, dx^{a}{}_{(p)}\, dx^{b}{}_{(q)} = 0 ,$$

and since the $dx^{a}{}_{(p)}$ span the complete solution space, we have more generally

$$a_{ab}{}^{c}\, dx^{a}\, dx^{b} = 0$$

for any solutions dx^{a} of the equations $f_{a}{}^{b}\, dx^{a} = 0$ (with dx^{a} as the unknowns). We now note that

$$a_{ab}{}^{c} \equiv j_{b}{}^{c}{}_{,a} - j_{a}{}^{c}{}_{,b} = f_{b}{}^{c}{}_{,a} - f_{a}{}^{c}{}_{,b} ,$$

so we can write:

$$(f_{a}{}^{c}{}_{,b} - f_{b}{}^{c}{}_{,a})\, dx^{a} \wedge dx^{b} = 0 . \qquad \text{(iii)}$$

This is the point at which essential use is made of the assumption that J is real analytic. If this is the case, then the coordinate functions $j_{a}{}^{b}$ can be extended to complex values of their arguments (by ordinary analytic continuation). Then equation (iii) is precisely the Frobenius integrability condition that m linearly independent complex functions

$$z^{\alpha}(x) = u^{\alpha}(x) + i\, v^{\bar{\alpha}}(x)$$

exist such that the system

$$dz^{\alpha} = \frac{\partial z^{\alpha}}{\partial x^{b}}\, dx^{b} = \left(\frac{\partial u^{\alpha}}{\partial x^{b}} + i\,\frac{\partial v^{\bar{\alpha}}}{\partial x^{b}}\right) dx^{b} = 0$$

is equivalent to our original system $f_a{}^b\, dx^a = 0$. Since we know that $dx^a{}_{(p)} = j_p{}^a - i\,\delta_p{}^a$ is a solution to these equations, we have

$$\frac{\partial z^{\alpha}}{\partial x^{b}} (j_p{}^b - i\,\delta_p{}^b) = 0$$

$$\frac{\partial z^{\alpha}}{\partial x^{b}}\, j_p{}^b = i\,\frac{\partial z^{\alpha}}{\partial x^{p}} . \qquad \text{(iv)}$$

We now choose as new real coordinates the n functions u^{α}, $v^{\bar{\alpha}}$ so that the corresponding complex coordinates are just the z^{α}. (This is valid because the Jacobean determinant $\partial(u^{\alpha}, v^{\bar{\alpha}})/\partial(x^b)$ is non-zero -- essentially due to the lack of real solutions to the original system of equations; see [17].) We compute the components of J with respect to this complex coordinate system:

$$J_{\alpha}{}^{\beta} = \frac{\partial x^{a}}{\partial z^{\alpha}}\, \frac{\partial z^{\beta}}{\partial x^{b}}\, j_a{}^b$$

$$= \frac{\partial x^{a}}{\partial z^{\alpha}}\, i\, \frac{\partial z^{\beta}}{\partial x^{a}} \qquad \text{by (iv) above}$$

$$= i\,\delta_{\alpha}{}^{\beta} .$$

Similarly we find

$$J_{\bar{\alpha}}{}^{\beta} = J_{\alpha}{}^{\bar{\beta}} = 0$$

$$J_{\bar{\alpha}}{}^{\bar{\beta}} = -i\,\delta_{\bar{\alpha}}{}^{\bar{\beta}} .$$

So the components of J with respect to the complex coordinate system (z^{α}) are given by

$$J_a{}^b = \begin{pmatrix} J_{\alpha}{}^{\beta} & J_{\alpha}{}^{\bar{\beta}} \\ J_{\bar{\alpha}}{}^{\beta} & J_{\bar{\alpha}}{}^{\bar{\beta}} \end{pmatrix} = \begin{pmatrix} i\,\delta_{\alpha}{}^{\beta} & 0 \\ 0 & -i\,\delta_{\bar{\alpha}}{}^{\bar{\beta}} \end{pmatrix} .$$

Thus every point of M has a neighborhood in which coordinates z^{α} are valid and the almost complex structure tensor J has the complex components given above. Since J has these components in <u>any</u> such complex coordinate system, we know that the associated complex coordinate transformations must be complex analytic (compare Theorem IV.11). Hence we have demonstrated the existence of a complex structure on M with respect to whose complex coordinates the almost complex structure J takes the required form. Therefore J is integrable, and the theorem is proven.

In virtue of Theorems V.9 and V.11, we refer to the condition $n_{ab}{}^{c} = 0$ as the integrability condition for an almost complex structure. Several equivalent statements of the integrability condition are given by Yano [54]. Among the most useful of these in applications is the following:

<u>V.12: THEOREM (YANO)</u>: If J is a real analytic almost complex structure, then J is integrable if and only if the systems of equations

$$(\delta_a{}^b - i\, j_a{}^b)\, dx^a = 0$$

$$(\delta_a{}^b + i\, j_a{}^b)\, dx^a = 0$$

are both completely integrable.

If these equations are integrable with solutions z^α = constant and $\overline{z^\alpha}$ = constant, respectively, then it suffices to take the z^α as local complex coordinates.

In practice, it is almost always easier to apply the Frobenius Theorem directly to the system of equations in Theorem V.12, rather than to compute the Nijenhuis tensor. It is interesting to note that, according to Theorem V.12, the integrability condition is equivalent to the condition that the distribution spanned by the forms $\omega^{(b)} = (\delta_a{}^b - i\, j_a{}^b)dx^a$ is "complex-surface-forming" (and the same for the distribution spanned by the complex conjugate forms $\overline{\omega}^{(b)} = (\delta_a{}^b + i\, j_a{}^b)dx^a$).

V.13: REMARK: It is important to note that, as is clear from Theorem V.12, the condition that J be a real tensor insures that the coordinates $\overline{z^\alpha}$ are actually (or at least can be chosen to be) the complex conjugates of the z^α.

Relaxation of Differentiability Requirements on J and Integrable Almost Complex Manifolds

The integrability theorems given above depend for their proofs on the assumption that the almost complex structure tensor J is real analytic. However, it has been shown that an almost complex structure which satisfies much weaker differentiability conditions is integrable if and only if its Nijenhuis tensor vanishes. This result was first obtained by Newlander and Nirenberg [32]. The proof is rather long and involved; we merely state the result here and refer the reader to the original paper [32] for the details.

<u>V.14: THEOREM (NEWLANDER-NIRENBERG)</u>: If M is a 2m-dimensional manifold of differentiability class C^{2m+1} which admits an almost complex structure J of class C^{2m}, then J is induced by a complex structure on M if and only if its Nijenhuis tensor vanishes.

In view of the Newlander-Nirenberg Theorem, an integrable almost complex manifold is completely equivalent to a complex manifold. The integrability theorems thus give us a foothold for determining when -- given a real manifold with an almost complex structure -- it is valid to introduce complex coordinates and treat the manifold as a complex manifold.

A simple calculation of the Nijenhuis tensor establishes that every two-dimensional almost complex structure is integrable (assuming it is of class C^2). In higher dimensions, the integrability condition is definitely non-trivial. In fact, Chern [8] claims that every almost complex manifold of dimension $n \geq 4$ possesses a non-integrable almost complex structure, because even if the given almost complex structure J is integrable, a perturbation of this J will certainly spoil the condition $n_{ab}{}^{c} = 0$.

The sphere S^6 is an interesting manifold in that it admits an almost complex structure (see Frölicher [17]), but it does not admit a complex structure (recently proven in Adler [1]). However, a word of caution is in order concerning this point: The non-integrability of a given almost complex structure does not imply the non-existence of a complex structure; a different almost complex structure may exist which is in fact integrable.

CHAPTER VI: HERMITIAN AND KÄHLERIAN MANIFOLDS

If, on a manifold, a complex structure and a Riemannian (or pseudo-Riemannian) structure are both given, it is desirable to relate these two structures by means of a "compatibility" condition. In the positive definite case, the Riemannian structure of a flat manifold is expressed by $ds^2 = dx^2 + dy^2 + \dots$. The possibility (for an even-dimensional manifold) of writing this line element as $ds^2 = dz\,d\bar{z} + \dots$, where $z = x + iy$, provides the motivation for choosing the "Hermitian" compatibility requirement. If the Hermitian condition is met, the line element of a curved space is a natural generalization of the form of this flat metric in complex coordinates. In relativity, the metric has an indefinite character and in a flat spacetime can be written as $ds^2 = du\,dv - dz\,d\bar{z}$. Later, we shall modify the usual Hermitian requirement to handle this case.

"Kählerian" manifolds are a subset of the class of Hermitian manifolds. The condition that a manifold admit a Kählerian structure is much more restrictive than the Hermitian compatibility requirement, but it is desirable to work with a Kählerian structure for several reasons. For one thing, a great deal is known about (positive definite) Kählerian manifolds. Furthermore, the formulas for the metric connection and the resulting curvature tensor are greatly simplified when the metric is Kählerian (due basically to the fact that all but a few of the Christoffel symbols must automatically vanish). In fact, the Riemann tensor and the Ricci tensor are expressible in terms of one scalar function on a Kählerian manifold. Finally, and perhaps most importantly from the standpoint of the mathematician, the most "natural" connection that can be defined on an Hermitian manifold will not be the metric connection unless the metric is Kählerian (see below).

Insofar as applications to relativity are concerned, Kählerian structures are of great interest because twistor space [44] possesses a Kählerian metric which reflects some of the projective properties of twistors. Also, we shall see later that many interesting spacetimes are conformally related to Kählerian manifolds (Kählerian in the sense of our generalization of the usual definition to meet the requirements of an indefinite metric).

Hermitian Structures on Vector Spaces

As with complex structures, we begin our study of Hermitian structures by first considering the situation on a vector space.

VI.1: DEFINITION: An Hermitian structure on a real vector space V with a complex structure J is a map $H: V \times V \to \mathbb{C}$ with the properties

(i) $H(\alpha X_1 + \beta X_2, Y) = \alpha H(X_1, Y) + \beta H(X_2, Y)$,

(ii) $\overline{H(X, Y)} = H(Y, X)$,

(iii) $H(J(X), Y) = i\, H(X, Y)$,

for all $\alpha, \beta \in R$ and $X_1, X_2, X, Y \in V$.

VI.2: REMARK: Given property (ii) of this definition, property (iii) is equivalent to

(iv) $H(X, J(Y)) = -i\, H(X, Y)$.

Given an Hermitian structure $H(X, Y)$, we can split it into its real and imaginary parts:

$$H(X, Y) = F(X, Y) + i\, G(X, Y) ,$$

so that $F, G: V \times V \to R$. Then property (ii) of the definitions becomes:

$$F(X, Y) - i\, G(X, Y) = F(Y, X) + i\, G(Y, X) \quad .$$

Equating real and imaginary parts, we get the symmetry and antisymmetry properties:

$$F(X, Y) = F(Y, X)$$

$$\text{and} \qquad G(X, Y) = -G(Y, X) \quad .$$

VI.3: DEFINITION: We say that $H(X, Y)$ is positive definite if the corresponding function $F(X, Y)$ is positive definite in the usual sense.

Since $G(X, Y)$ is antisymmetric, there is a unique two-form $\hat{H}$ corresponding to $-\frac{1}{2} G(X, Y)$. We call $\hat{H}$ the Kähler form of the Hermitian structure H.

Let $\{E_\alpha, E_{\bar{\alpha}}\}$ be one of the bases of V discussed in Chapter II, i.e., a basis such that $J(E_\alpha) = E_{\bar{\alpha}}$ and $J(E_{\bar{\alpha}}) = -E_\alpha$. Then we can write:

$$\begin{aligned}
H(X, Y) &= H(x^\alpha E_\alpha + x^{\bar{\alpha}} E_{\bar{\alpha}}, Y) \quad \text{(recall summation conventions)} \\
&= H(x^\alpha E_\alpha + x^{\alpha+m} J(E_\alpha), Y) \\
&= (x^\alpha + i\, x^{\alpha+m})\, H(E_\alpha, Y) \quad \text{(by properties (i) and (iii) above)} \\
&= (x^\alpha + i\, x^{\alpha+m})\, (y^\beta - i\, y^{\beta+m})\, H(E_\alpha, E_\beta) \\
&\qquad \text{(by similar manipulations)} \\
&= \lambda^\alpha(X)\, \bar{\lambda}^\beta(Y)\, H(E_\alpha, E_\beta) \quad \text{(with } \lambda^\alpha(X), \bar{\lambda}^\beta(Y) \text{ as in Chapter II; see Remark II.14)} \\
&= h_{\alpha\beta}\, \lambda^\alpha(X) \otimes \bar{\lambda}^\beta(Y) \quad ,
\end{aligned}$$

where $h_{\alpha\beta} \cong H(E_\alpha, E_\beta) = \bar{h}_{\beta\alpha}$. Therefore the Kähler form is given by:

$$\hat{H}(X, Y) = -\frac{1}{2} G(X, Y) = -\frac{1}{2}\frac{1}{2i}\left[H(X, Y) - \overline{H(X, Y)}\right]$$

$$= \frac{i}{4}\left[h_{\alpha\beta}\,\lambda^\alpha \otimes \bar{\lambda}^\beta - \bar{h}_{\alpha\beta}\,\overline{\lambda^\alpha \otimes \bar{\lambda}^\beta}\right](X, Y)$$

$$= \frac{i}{4}\,h_{\alpha\beta}\left[\lambda^\alpha \otimes \bar{\lambda}^\beta - \bar{\lambda}^\beta \otimes \lambda^\alpha\right](X, Y) \quad ,$$

so that $$\hat{H} = \frac{i}{2}\,h_{\alpha\beta}\,\lambda^\alpha \wedge \bar{\lambda}^\beta \;.$$

VI.4: REMARK: $\hat{H}(X, Y)$ is real. That is $\hat{H}$ is in the exterior algebra of the real vector space V^* itself, specifically in $\Lambda^2(V^*)$.

Hermitian Manifolds

We are now prepared to proceed from the study of Hermitian structures on vector spaces to an examination of Hermitian structures on manifolds. As before in the case of an almost complex structure, we shall be able to transfer the facts developed for vector spaces to the tangent spaces of the manifold in question, provided the manifold admits an almost Hermitian structure, which we now define:

VI.5: DEFINITION: Let M be a manifold with a Riemannian metric g and an almost complex structure J. Then M is said to be an almost Hermitian manifold (manifold which admits an almost Hermitian structure) if and only if

$$g(JX, JY) = g(X, Y)$$

for any vectors X and Y, or in terms of components,

$$J_a^{\ m} J_b^{\ n} g_{mn} = g_{ab} .$$

The tensor

$$H_{ab} = g_{ab} - i J_{ab} ,$$

where $J_{ab} = J_a^{\ m} g_{mb}$, is called the almost Hermitian structure tensor, and g_{ab} is said to be an Hermitian metric.

(Note: In this chapter and the sequel, we are dropping the convention of Chapters IV and V of using capital letters for the complex components of tensors and lower case letters for the real components. From now on, one kernel letter, whether capital or lower case, will be used to refer to either complex or real components of a given tensor. We continue to use Latin indices which range and sum over $1, \ldots, n = 2m$, Greek indices which range and sum over $1, \ldots, m$, and barred Greek indices which range and sum over $\overline{1} = 1+m, \ldots, \overline{m} = m+m = 2m$.)

The adjective "Hermitian" is used here because in a complex coordinate system on a complex manifold the almost Hermitian structure tensor has the "Hermitian" property

$$\overline{H_{\alpha\overline{\beta}}} = H_{\beta\overline{\alpha}} .$$

(In fact, if one starts with a complex, rather than merely almost complex manifold, then the above property can be taken as the definition of an Hermitian structure; see [30].)

It should also be pointed out that the almost complex structure tensor of an almost Hermitian manifold defines an isometry, by virtue of the defining property $g(JX, JY) = g(X, Y)$ (see Helgason [22]).

It remains to be seen how the symmetric character of g, $g_{ab} = g_{ba}$, is connected with the idea of an Hermitian structure on a vector space:

VI.6: THEOREM: The tangent space over a point of an almost Hermitian manifold admits a (vector space) Hermitian structure.

Proof: Let X,Y be type (1,0) vectors. Then $H(X,Y) = g(X,\bar{Y}) - i\, J(X,\bar{Y})$

is the desired Hermitian structure, or in terms of components,

$$H_{ab} = g_{ab} - i\, J_{ab} \ .$$

We show that this H satisfies conditions (i), (ii), and (iii) of Definition VI.1:

(i) H_{ab} is clearly linear.

(ii) First we show that J_{ab} is antisymmetric:

$$J_a^{\ m} J_b^{\ n} g_{mn} = g_{ab}$$

$$\Longrightarrow J_c^{\ a} J_a^{\ m} J_b^{\ n} g_{mn} = J_c^{\ a} g_{ab}$$

$$\Longrightarrow -\delta_c^{\ m} J_b^{\ n} g_{mn} = J_{cb} \qquad \text{(since } J \text{ is an almost complex structure)}$$

$$\Longrightarrow -J_{bc} = J_{cb} \ , \qquad \text{as desired.}$$

Then
$$\overline{H_{ab}} = g_{ab} + i\, J_{ab}$$
$$= g_{ba} - i\, J_{ba}$$
$$= H_{ba} \ , \qquad \text{thus verifying (ii).}$$

(iii) Again in terms of components, we have:

$$\begin{aligned} H(J(X), Y) &= H_{ab}\, J_m{}^a\, x^m\, \bar{y}^b \\ &= (g_{ab} - i\, J_{ab})\, J_m{}^a\, x^m\, \bar{y}^b \\ &= (J_{mb} + i\, g_{mb})\, x^m\, \bar{y}^b \\ &= + i(g_{mb} - i\, J_{mb})\, x^m\, \bar{y}^b \\ &= i\, H(X, Y) \quad , \qquad \text{as desired.} \end{aligned}$$

In view of this theorem, it is natural to refer to the field $g_{ab} - i\, J_{ab}$ as the almost Hermitian structure tensor. We note that

$$H_{(ab)} = g_{ab}$$

$$H_{[ab]} = - i\, J_{ab} \quad ,$$

and the Kähler form J is given by

$$J = \frac{i}{2}\, J_{ab}\, dx^a \wedge dx^b \quad .$$

It is now natural to define an Hermitian manifold as follows:

<u>VI.7: DEFINITION:</u> An Hermitian manifold (manifold which admits an Hermitian structure) is an almost Hermitian manifold for which the almost complex structure tensor is integrable. In this case the tensor H_{ab} is called the Hermitian structure tensor.

Then the following result is obvious:

<u>VI.8: THEOREM:</u> A complex manifold with complex structure J is Hermitian if and only if it admits an almost

Hermitian structure (that is, if the manifold admits a metric tensor g satisfying Definition VI.5).

It can be shown (see [30]) that every complex manifold admits an Hermitian structure.

Some of the interesting properties of Hermitian manifolds are contained in the following theorem:

VI.9: THEOREM: If z^{α} and $\overline{z^{\beta}}$ are complex coordinates on an Hermitian manifold, then

(i) $g_{\alpha\beta} = g_{\overline{\alpha\beta}} = g^{\alpha\beta} = g^{\overline{\alpha\beta}} = 0$,

(ii) $J_{\alpha\beta} = J_{\overline{\alpha\beta}} = J^{\alpha\beta} = J^{\overline{\alpha\beta}} = 0$,

(iii) $J_{\alpha\bar{\beta}} = i\, g_{\alpha\bar{\beta}}$,

$J_{\bar{\alpha}\beta} = -i\, g_{\bar{\alpha}\beta}$,

$J^{\alpha\bar{\beta}} = i\, g^{\alpha\bar{\beta}}$,

$J^{\bar{\alpha}\beta} = -i\, g^{\bar{\alpha}\beta}$.

Proof: (i): Recall (see Theorem V.6 and the discussion which follows it) that in any complex coordinate system $J_a{}^b$ has the components

$$J_{\bar{\alpha}}{}^{\beta} = J_{\alpha}{}^{\bar{\beta}} = 0; \quad J_{\alpha}{}^{\beta} = i\delta_{\alpha}{}^{\beta}; \quad J_{\bar{\alpha}}{}^{\bar{\beta}} = -i\delta_{\bar{\alpha}}{}^{\bar{\beta}} .$$

Then an application of Definition VI.5 yields:

$$g_{\alpha\beta} = J_{\alpha}{}^{m} J_{\beta}{}^{n} g_{mn} = J_{\alpha}{}^{\mu} J_{\beta}{}^{\nu} g_{\mu\nu} = (i\,\delta_{\alpha}{}^{\mu})(i\,\delta_{\beta}{}^{\nu})\, g_{\mu\nu} = -g_{\alpha\beta}$$

$$\Longrightarrow g_{\alpha\beta} = 0$$

and $g_{\overline{\alpha\beta}} = g^{\alpha\beta} = g^{\overline{\alpha\beta}} = 0$ similarly.

(ii): $J_{\alpha\beta} = J_{\alpha}{}^{m} g_{m\beta} = J_{\alpha}{}^{\mu} g_{\mu\beta} = 0$ because $g_{\mu\beta} = 0$

and $J_{\overline{\alpha\beta}} = J^{\alpha\beta} = J^{\overline{\alpha\beta}} = 0$ similarly.

(iii): $J_{\alpha\bar{\beta}} = J_{\alpha}{}^{m} g_{m\bar{\beta}} = J_{\alpha}{}^{\mu} g_{\mu\bar{\beta}} = i\,\delta_{\alpha}^{\mu}\, g_{\mu\bar{\beta}} = i\, g_{\alpha\bar{\beta}}$

and $J_{\bar{\alpha}\beta} = -\,i\, g_{\bar{\alpha}\beta}$, $J^{\alpha\bar{\beta}} = i\, g^{\alpha\bar{\beta}}$, $J^{\bar{\alpha}\beta} = -\,i\, g^{\bar{\alpha}\beta}$ similarly.

As mentioned at the beginning of this Chapter, the "hybrid" character of an Hermitian metric (property (i) in the theorem above) is a generalization of the metric form $ds^2 = dz\,d\bar{z} + \dots$ in flat space: The metric will have the form

$$ds^2 = g_{ab}\, dz^a\, dz^b = g_{\alpha\bar{\beta}}\, dz^{\alpha} dz^{\bar{\beta}} + g_{\bar{\alpha}\beta}\, dz^{\bar{\alpha}}\, dz^{\beta}$$

$$= 2\, g_{\alpha\bar{\beta}}\, dz^{\alpha}\, dz^{\bar{\beta}} \ .$$

Kählerian Manifolds

We proceed to the study of an even richer structure on a complex manifold, namely, a Kählerian structure. As usual, we first make the following definition:

<u>VI.10: DEFINITION</u>: An almost Kählerian manifold (manifold which admits an almost Kählerian structure) is an almost Hermitian manifold for which the almost complex structure tensor $J_a{}^b$ satisfies

$$d(J_{ab}\, dx^a \wedge dx^b) = 0 \ ,$$

where "d" denotes the exterior derivation.

We recall from above that the form field $J_{ab}\, dx^a \wedge dx^b$ is proportional to the Kähler form $\frac{i}{2} J_{ab}\, dx^a \wedge dx^b$, so that the almost Kählerian property can be expressed as the condition that the Kähler form be closed. We proceed at once to the following definition:

<u>VI.11: DEFINITION:</u> A Kählerian manifold (manifold which admits a Kählerian structure) is an almost Kählerian manifold whose almost complex structure tensor J is integrable.

<u>VI.12: THEOREM:</u> A Kählerian manifold is an Hermitian manifold.

<u>Proof:</u> This follows directly from the definitions above.

<u>VI.13: REMARK:</u> An Hermitian manifold is a Kählerian manifold:

(i) if and only if

$$\partial_a J_{bc} + \partial_b J_{ca} + \partial_c J_{ab} = 0$$

(ii) if and only if

$$\nabla_a J_{bc} + \nabla_b J_{ca} + \nabla_c J_{ab} = 0 ,$$

where ∂_a denotes partial differentiation and ∇_a denotes differentiation with respect to the Riemannian connection defined by g_{ab}. (These results follow easily from the condition $d(J_{ab}\, dx^a \wedge dx^b) = 0$.)

It should be remarked that, whereas every complex manifold admits an Hermitian structure, it is only a very restricted class of complex manifolds which admit, in addition a Kählerian structure. The best known examples of Kählerian manifolds are the complex projective spaces $\mathbb{CP}^n$ (see Chapter III). For a proof that these manifolds are Kählerian, see sections 7 and 8 of Chern [8].

Conformally Kählerian Manifolds

In applications to relativity, we shall find it necessary to consider (an analogue of) the following question: If an Hermitian manifold with metric g_{ab} is not Kahlerian, does there nevertheless exist another Hermitian metric $\hat{g}_{ab}$ with respect to which the manifold is, in fact, Kählerian? In examining this question, the first step is to note that if a metric g_{ab} is Hermitian with respect to a given structure tensor $J_a{}^b$, then any metric conformally related to g_{ab}, that is, any metric $\hat{g}_{ab}$ such that

$$\hat{g}_{ab} = \Omega^2 g_{ab} ,$$

is also Hermitian with respect to $J_a{}^b$. We therefore formulate the question, under what circumstances is a given Hermitian metric conformally related to a Kählerian metric? The answer is given by Yano [54] in the form of the following theorem:

<u>VI.14: THEOREM (YANO)</u>: Given an Hermitian manifold with metric g_{ab} and (integrable) almost complex structure $J_a{}^b$, then g_{ab} is conformally related to a Kählerian metric $\hat{g}_{ab} = \Omega^2 g_{ab}$ for some Ω if and only if

$$F_{[a,b]} = 0 ,$$

where

$$F_a = (\partial_a J_{bc} + \partial_b J_{ca} + \partial_c J_{ab}) J^{bc} ,$$

and where ∂_a denotes partial differentiation.

<u>Proof:</u> Suppose $\hat{g}_{ab} = \Omega^2 g_{ab}$ is Kählerian for some Ω; then the Kähler form $\hat{J}_{ab}$ is given by

$$\hat{J}_{ab} = J_a{}^m \hat{g}_{mb} = \Omega^2 J_a{}^m g_{mb} = \Omega^2 J_{ab}$$

(we note that $J_a{}^b$ is invariant under the conformal rescaling, whereas J_{ab} is not).

Then by Remark VI.13 we have:

$$\partial_a \hat{J}_{bc} + \partial_b \hat{J}_{ca} + \partial_c \hat{J}_{ab} = 0$$

$$\Longleftrightarrow \partial_a(\Omega^2 J_{bc}) + \partial_b(\Omega^2 J_{ca}) + \partial_c(\Omega^2 J_{ab}) = 0$$

$$\Longrightarrow F_{abc} + 2(\rho_a J_{bc} + \rho_b J_{ca} + \rho_c J_{ab}) = 0 ,$$

where

$$F_{abc} \equiv \partial_a J_{bc} + \partial_b J_{ac} + \partial_c J_{ab}$$

and

$$\rho_a \equiv \Omega^{-1} \Omega_{,a} = (\ln \Omega)_{,a} .$$

Then transvecting with J^{bc} yields:

$$F_a + 2(\rho_a J_{bc} J^{bc} + \rho_b J_{ca} J^{bc} + \rho_c J_{ab} J^{bc}) = 0$$

where $F_a \equiv F_{abc} J^{bc}$;

$$F_a + 2(\rho_a \delta_b{}^b - \rho_b \delta_a{}^b - \rho_c \delta_a{}^c) = 0 ,$$

using $J_a{}^b J_b{}^c = -\delta_a{}^c$ and $J_{cb} = -J_{bc}$;

$$F_a + 2(n - 2)\rho_a = 0 ,$$

where n is the real dimension of the manifold.

Then, since ρ_a is a gradient, we have as a necessary condition

$$F_{[a,b]} = 0 ,$$

and conversely, making use of the Frobenius theorem (assuming analyticity), this condition is also sufficient for the existence of an Ω such that $\rho_a = (\ln\Omega)_{,a}$. Thus the theorem is established.

(It should be noted that the proof of this theorem is good only for manifolds of real dimension $n > 2$. However, in the case $n = 2$, every two-form is closed and hence every two-dimensional Hermitian manifold is Kählerian.)

Differential Geometry on Hermitian and Kählerian Manifolds

In treating the matters of covariant differentiation and curvature on Hermitian manifolds, two distinct points of view can be taken. On the one hand, an Hermitian manifold may simply be thought of as a special case of a real differentiable Riemannian manifold; in this case one may proceed to treat the connection and curvature exactly as in the real case, with the proviso that "i" is to be treated purely formally as a square root of minus one. The drawback to this point of view is that there will be no particularly meaningful relationships between the Hermitian character of the manifold and the Riemannian structure treated in this way. One might as well do "ordinary" differential geometry and then transform to complex coordinates: Nothing is gained or lost.

On the other hand, we may choose as the starting point not the underlying real differentiable structure, but rather the Hermitian structure itself. When this is done, it is found that an affine connection which is not Riemannian with respect to the Hermitian metric $g_{\alpha\bar{\beta}}$ presents itself as the most natural choice. It is found, moreover, that only in the case of a Kählerian structure does this connection coincide with the (torsion-free) Riemannian connection.

Accordingly, we proceed to examine this "natural" connection for Hermitian manifolds; when the Kählerian case is considered, our remarks will apply also to the "ordinary" Riemannian connection and curvature, since the two ideas coincide in this case.

Our discussion will, to a large extent, follow that of Morrow and Kodaira [30]. For a discussion of the ordinary Riemannian differential geometry on any Hermitian manifold, see Yano [54]; and for a modern treatment of connections and curvature in terms of complex vector bundle theory, see Chern [8] and Appendix C.

We consider, then, an Hermitian manifold with metric g_{ab} satisfying (see Theorem VI.9)

$$g_{\alpha\beta} = g_{\overline{\alpha\beta}} = 0$$

and complex structure tensor $J_a{}^b$ satisfying

$$J_{\alpha}{}^{\overline{\beta}} = J_{\overline{\alpha}}{}^{\beta} = 0$$

$$J_{\alpha}{}^{\beta} = i\,\delta_{\alpha}{}^{\beta}$$

$$J_{\overline{\alpha}}{}^{\overline{\beta}} = -i\,\delta_{\overline{\alpha}}{}^{\overline{\beta}} \quad ;$$

and

$$J_{\alpha\beta} = J_{\overline{\alpha\beta}} = 0$$

$$J_{\alpha\overline{\beta}} = i\,g_{\alpha\overline{\beta}}$$

$$J_{\overline{\alpha}\beta} = -i\,g_{\overline{\alpha}\beta} \quad .$$

As usual, our motivation for looking for a connection arises from the fact that partial derivatives of vectors and tensors are not in general tensorial (i.e., they do not follow the tensor transformation law for a change of coordinates). For example, in going from complex coordinates z^{α}, $\overline{z^{\beta}}$ to coordinates $z^{\alpha'}$, $\overline{z^{\beta'}}$, we have:

$$\begin{aligned}
\partial_{\alpha'} V^{\beta'} &= \frac{\partial}{\partial z^{\alpha'}}\left(\frac{\partial z^{\beta'}}{\partial z^{m}}\, V^{m}\right) \\
&= \frac{\partial}{\partial z^{\alpha'}}\left(\frac{\partial z^{\beta'}}{\partial z^{\mu}}\, V^{\mu}\right) \qquad \left(\text{because } \frac{\partial z^{\beta'}}{\partial \overline{z^{\mu}}} = 0 \text{ by analyticity}\right) \\
&= \frac{\partial z^{\nu}}{\partial z^{\alpha'}}\,\frac{\partial}{\partial z^{\nu}}\left(\frac{\partial z^{\beta'}}{\partial z^{\mu}}\, V^{\mu}\right) \\
&= \frac{\partial z^{\nu}}{\partial z^{\alpha'}}\,\frac{\partial z^{\beta'}}{\partial z^{\mu}}\,(\partial_{\nu} V^{\mu}) + \frac{\partial z^{\nu}}{\partial z^{\alpha'}}\,\frac{\partial^{2} z^{\beta'}}{\partial z^{\nu}\,\partial z^{\mu}}\, V^{\mu} \quad ,
\end{aligned}$$

and similarly,

$$\partial_{\overline{\alpha}'} V^{\overline{\beta}'} = \frac{\partial \overline{z^{\nu}}}{\partial \overline{z^{\alpha'}}}\,\frac{\partial \overline{z^{\beta'}}}{\partial \overline{z^{\mu}}}\,(\partial_{\overline{\nu}} V^{\overline{\mu}}) + \frac{\partial \overline{z^{\nu}}}{\partial \overline{z^{\alpha'}}}\,\frac{\partial^{2} \overline{z^{\beta'}}}{\partial \overline{z^{\nu}}\,\partial \overline{z^{\mu}}}\, V^{\overline{\mu}} \quad .$$

However, for the "hybrid" components $\partial_{\bar{\alpha}} V^{\beta}$ and $\partial_{\alpha} V^{\bar{\beta}}$, we have:

$$\partial_{\bar{\alpha}'} V^{\beta} = \frac{\partial}{\partial z^{\bar{\alpha}'}} V^{\beta'}$$

$$= \left(\frac{\partial z^{\bar{\mu}}}{\partial z^{\bar{\alpha}'}} \frac{\partial}{\partial z^{\bar{\mu}}} \right) \left(\frac{\partial z^{\beta'}}{\partial z^{\nu}} V^{\nu} \right)$$

$$= \frac{\partial z^{\bar{\mu}}}{\partial z^{\bar{\alpha}'}} \frac{\partial z^{\beta'}}{\partial z^{\nu}} (\partial_{\bar{\mu}} V^{\nu})$$

and similarly,

$$\partial_{\alpha'} V^{\bar{\beta}'} = \frac{\partial z^{\mu}}{\partial z^{\alpha'}} \frac{\partial z^{\bar{\beta}'}}{\partial z^{\bar{\nu}}} (\partial_{\mu} V^{\bar{\nu}}) .$$

So the hybrid ("mixed") components of the partial derivatives of vectors <u>do</u> transform tensorially. It is not hard to see that a similar state of affairs holds for covariant vectors as well, namely, $\partial_{\alpha} V_{\bar{\beta}}$ and $\partial_{\bar{\alpha}} V_{\beta}$ will transform tensorially while $\partial_{\alpha} V_{\beta}$ and $\partial_{\bar{\alpha}} V_{\bar{\beta}}$ will generally not. In fact, it is clear that for an arbitrary tensor $T^{a\ldots b}{}_{c\ldots d}$, the quantities

$$\partial_{\bar{\mu}} T^{\alpha\ldots\beta}{}_{\gamma\ldots\delta}$$

and

$$\partial_{\mu} T^{\bar{\alpha}\ldots\bar{\beta}}{}_{\bar{\gamma}\ldots\bar{\delta}}$$

will transform tensorially, while the other components, such as $\partial_{\bar{\mu}} T^{\bar{\alpha}\ldots b}{}_{c\ldots d}$, will not. From the example worked out above, it is evident that this state of affairs is brought about by the fact that

$$\frac{\partial z^{\alpha'}}{\partial z^{\bar{\beta}}} = \frac{\partial z^{\bar{\alpha}'}}{\partial z^{\beta}} = 0$$

for analytic coordinate transformations.

With this in mind, we note that raising or lowering an index with $g^{\alpha\bar\beta}$ or $g_{\alpha\bar\beta}$ changes the index from barred to unbarred, or vice versa (compare Theorem VI.9). Therefore, given a tensor $T^{\alpha\ldots\bar\beta\ldots}_{\gamma\ldots\bar\delta\ldots}$, the quantities

$$\partial_\mu (T^{\alpha\ldots\bar\beta\ldots}_{\gamma\ldots\bar\delta\ldots}\; g_{\alpha\bar\alpha_0}\ldots g^{\gamma\bar\gamma_0}\ldots)$$

and
$$\partial_{\bar\mu}(T^{\alpha\ldots\bar\beta\ldots}_{\gamma\ldots\bar\delta\ldots}\; g_{\bar\beta\beta_0}\ldots g^{\bar\delta\delta_0}\ldots)$$

will transform covariantly. These considerations motivate the following definition of covariant differentiation:

<u>VI.15: DEFINITION:</u> On an Hermitian manifold with metric $g_{\alpha\bar\beta}$, the Hermitian covariant derivative $\mathcal{D}_e$ is defined by

$$\mathcal{D}_\lambda T^{\alpha\ldots\bar\beta\ldots}_{\gamma\ldots\bar\delta\ldots} = (g^{\bar\mu\alpha}\ldots g_{\bar\nu\gamma}\ldots)\partial_\lambda (T^{\rho\ldots\bar\beta\ldots}_{\sigma\ldots\bar\delta\ldots}\; g_{\rho\bar\mu}\ldots g^{\alpha\bar\nu}\ldots)$$

and
$$\mathcal{D}_{\bar\lambda} T^{\alpha\ldots\bar\beta\ldots}_{\gamma\ldots\bar\delta\ldots} = (g^{\bar\rho\mu}\ldots g_{\nu\bar\delta}\ldots)\partial_{\bar\lambda}(T^{\alpha\ldots\bar\rho\ldots}_{\gamma\ldots\bar\sigma\ldots}\; g_{\bar\rho\mu}\ldots g^{\bar\sigma\nu}\ldots) .$$

It is easy to see that this is a well-defined operation, i.e., we have:

<u>VI.16: REMARK:</u> If $T^{\alpha\ldots\bar\beta\ldots}_{\gamma\ldots\bar\delta\ldots}$ is a tensor, then $\mathcal{D}_\lambda T^{\alpha\ldots\bar\beta\ldots}_{\gamma\ldots\bar\delta\ldots}$ and $\mathcal{D}_{\bar\lambda} T^{\alpha\ldots\bar\beta\ldots}_{\gamma\ldots\bar\delta\ldots}$ are tensors.

<u>VI.17: THEOREM:</u>

$$\mathcal{D}_\lambda T^{\alpha\ldots\bar\beta\ldots}_{\gamma\ldots\bar\delta\ldots} = \partial_\lambda T^{\alpha\ldots\bar\beta\ldots}_{\gamma\ldots\bar\delta\ldots} + \gamma^\alpha_{\lambda\rho} T^{\rho\ldots\bar\beta\ldots}_{\gamma\ldots\bar\delta\ldots} + \ldots - \gamma^\sigma_{\lambda\gamma} T^{\alpha\ldots\bar\beta\ldots}_{\sigma\ldots\bar\delta\ldots} - \ldots$$

and

$$\mathcal{D}_{\bar\lambda} T^{\alpha\ldots\bar\beta\ldots}_{\gamma\ldots\bar\delta\ldots} = \partial_{\bar\lambda} T^{\alpha\ldots\bar\beta\ldots}_{\gamma\ldots\bar\delta\ldots} + \gamma^{\bar\beta}_{\bar\lambda\bar\rho} T^{\alpha\ldots\bar\rho\ldots}_{\gamma\ldots\bar\delta\ldots} + \ldots - \gamma^{\bar\sigma}_{\bar\lambda\bar\delta} T^{\alpha\ldots\bar\beta\ldots}_{\gamma\ldots\bar\sigma\ldots} - \ldots ,$$

where $\gamma^\alpha_{\lambda\rho} \equiv g^{\alpha\bar\mu}\partial_\lambda g_{\rho\bar\mu}$

and $\gamma^{\bar\beta}_{\bar\lambda\bar\rho} = g^{\bar\beta\mu}\partial_{\bar\lambda} g_{\bar\rho\mu} = \overline{\gamma^\beta_{\lambda\rho}}$.

<u>Proof</u>: A direct computation, using the facts that

$$g^{\alpha\bar\beta}\,\partial_\lambda g_{\gamma\bar\beta} = -\,g_{\gamma\bar\beta}\partial_\lambda g^{\alpha\bar\beta}$$

and

$$g^{\bar\alpha\beta}\partial_{\bar\lambda}\, g_{\bar\gamma\beta} = -\,g_{\bar\gamma\beta}\partial_{\bar\lambda}\, g^{\bar\alpha\beta} ,$$

produces the formulas given in Definition VI.15. And finally, it is clear that $\gamma^{\bar\alpha}_{\bar\beta\bar\gamma} = \overline{\gamma^\alpha_{\beta\gamma}}$

It should be noticed that the formulas for $\mathcal{D}_\ell T^{\cdots}_{\cdots}$ in terms of the $\gamma^{\alpha}_{\beta\gamma}$ and the $\gamma^{\bar\alpha}_{\bar\beta\bar\gamma}$ are formally identical to the formulas for the Riemannian covariant derivative (except that the $\gamma^{\alpha}_{\beta\gamma}, \gamma^{\bar\alpha}_{\bar\beta\bar\gamma}$ are <u>not</u> in general symmetric in β, γ: see below) if we define γ^{a}_{bc} by setting all components other than $\gamma^{\alpha}_{\beta\gamma}$ and $\gamma^{\bar\alpha}_{\bar\beta\bar\gamma}$ equal to zero. The symbols γ^{a}_{bc} for the "Hermitian Chrisoffel symbols" and $\mathcal{D}_\ell$ for the Hermitian covariant derivative are used to distinguish these from the "ordinary Christoffel symbols" Γ^{a}_{bc} and the Riemannian covariant derivative operator ∇_ℓ.

The fact should be stressed that the Hermitian Christoffel symbols γ^{a}_{bc} are <u>not</u> in general symmetric in b and c, and hence the Hermitian connection $\mathcal{D}_\ell$ is not in general torsion-free. However, another motivation for choosing the Hermitian connection is given by:

<u>VI.18: THEOREM:</u> $\quad \mathcal{D}_\alpha g_{\beta\bar\gamma} = \mathcal{D}_{\bar\alpha} g_{\beta\bar\gamma} = 0.$

<u>Proof:</u>

$$\begin{aligned} \mathcal{D}_\alpha g_{\beta\bar\gamma} &= \partial_\alpha g_{\beta\bar\gamma} - \Gamma^{\sigma}_{\alpha\beta}\, g_{\sigma\bar\gamma} \\ &= \partial_\alpha g_{\beta\bar\gamma} - g^{\sigma\bar\mu}(\partial_\alpha g_{\bar\mu\beta})\, g_{\sigma\bar\gamma} \\ &= \partial_\alpha g_{\beta\bar\gamma} - \delta^{\bar\mu}_{\bar\gamma}\, \partial_\alpha g_{\bar\mu\beta} \\ &= \partial_\alpha g_{\beta\bar\gamma} - \partial_\alpha g_{\bar\gamma\beta} \\ &= 0, \end{aligned}$$

since $g_{\beta\bar\gamma}$ is symmetric, i.e., $g_{\beta\bar\gamma} = g_{\bar\gamma\beta}$.

$\mathcal{D}_{\bar\alpha} g_{\beta\bar\gamma} = 0$ similarly.

We are now ready to prove the following important result:

VI.19: THEOREM: An Hermitian manifold is Kählerian if and only if

$$\gamma^{\alpha}_{\beta\gamma} = \gamma^{\alpha}_{\gamma\beta} \quad \text{and} \quad \gamma^{\bar{\alpha}}_{\bar{\beta}\bar{\gamma}} = \gamma^{\bar{\alpha}}_{\bar{\gamma}\bar{\beta}} ,$$

i.e., if and only if

$$\gamma^{a}_{bc} = \gamma^{a}_{cb} .$$

Proof: By definition, the Hermitian structure $H_{ab} = g_{ab} - i\, J_{ab}$ is Kählerian if and only if

$$d(J_{ab}\, dx^{a} \wedge dx^{b}) .$$

In a complex coordinate system, and using Theorem VI.9, this can be written:

$$\begin{aligned} & g_{\alpha\bar{\beta},\gamma}\, dz^{\gamma} \wedge dz^{\alpha} \wedge d\overline{z^{\beta}} \\ + \; & g_{\alpha\bar{\beta},\bar{\gamma}}\, d\overline{z^{\gamma}} \wedge dz^{\alpha} \wedge d\overline{z^{\beta}} = 0 . \end{aligned} \tag{i}$$

Then, since $\gamma^{\alpha}_{\beta\gamma} = g^{\alpha\bar{\mu}}\, \partial_{\beta}\, g_{\bar{\mu}\gamma}$, it follows that

$$g_{\alpha\bar{\nu}}\, \gamma^{\alpha}_{\beta\gamma} = \delta^{\bar{\mu}}_{\bar{\nu}}\, \partial_{\beta}\, g_{\bar{\mu}\gamma}$$

so that

$$\partial_{\beta}\, g_{\bar{\nu}\gamma} = g_{\alpha\bar{\nu}}\, \gamma^{\alpha}_{\beta\gamma}$$

and similarly, $\partial_{\bar{\beta}}\, g_{\nu\bar{\gamma}} = g_{\bar{\alpha}\nu}\, \gamma^{\bar{\alpha}}_{\bar{\beta}\bar{\gamma}}$. Inserting these formulas into (i) above yields that H_{ab} is Kählerian if and only if

$$\begin{aligned} & g_{\mu\bar{\beta}}\, \gamma^{\mu}_{\alpha\gamma}\, dz^{\gamma} \wedge dz^{\alpha} \wedge d\overline{z^{\beta}} \\ + \; & g_{\alpha\bar{\mu}}\, \gamma^{\bar{\mu}}_{\bar{\beta}\bar{\gamma}}\, d\overline{z^{\gamma}} \wedge dz^{\alpha} \wedge d\overline{z^{\beta}} = 0 \end{aligned}$$

$$\Longleftrightarrow \quad g_{\mu\bar\beta}\,(\gamma^{\mu}_{\alpha\gamma} - \gamma^{\mu}_{\gamma\alpha})\, dz^{\gamma} \wedge dz^{\alpha} \wedge d\bar{z}^{\beta}$$

$$+ \; g_{\alpha\bar\mu}\,(\gamma^{\bar\mu}_{\bar\beta\bar\gamma} - \gamma^{\bar\mu}_{\bar\gamma\bar\beta})\, d\bar{z}^{\gamma} \wedge dz^{\alpha} \wedge d\bar{z}^{\beta} = 0 \; .$$

Thus H_{ab} is Kählerian if and only if

$$\gamma^{\mu}_{\alpha\beta} = \gamma^{\mu}_{\beta\alpha}$$

$$\text{and} \quad \gamma^{\bar\mu}_{\bar\alpha\bar\beta} = \gamma^{\bar\mu}_{\bar\beta\bar\alpha} \; ;$$

that is, $\gamma^{m}_{ab} = \gamma^{m}_{ba}$, as desired.

Some of the importance of this result, especially for application to relativity, is contained in the following corollary:

<u>VI.20: COROLLARY:</u> The Hermitian connection γ^{a}_{bc} coincides with the Riemannian connection Γ^{a}_{bc} if and only if the metric is Kählerian.

<u>Proof:</u> Since γ^{a}_{bc} is a connection with respect to which the metric is covariantly constant (Theorem VI.18), and γ^{a}_{bc} is symmetric if and only if the metric is Kählerian, it follows at once from the Fundamental Theorem of Riemannian Geometry that γ^{a}_{bc} is the Riemannian connection if and only if the metric is Kählerian. (Alternatively, the theorem can be demonstrated by direct computation, using the definition

$$\Gamma^{a}_{bc} = \frac{1}{2}\, g^{am}\,(\partial_c\, g_{bm} + \partial_b\, g_{mc} - \partial_m\, g_{bc})$$

of the Riemannian Christoffel symbols and the Kählerian condition expressed in terms of $g_{\alpha\bar\beta}$ by means of Theorem VI.9.)

Later on, we shall have occasion to use the following result:

VI.21: THEOREM: In a Kählerian manifold, we have

$$\gamma^{\alpha}_{\alpha\beta} = g^{-1}\partial_{\beta}\, g$$

$$\text{and } \gamma^{\bar\alpha}_{\bar\alpha\bar\beta} = g^{-1}\partial_{\bar\beta}\, g \quad , \text{ where } g = \det(g_{\alpha\bar\beta}).$$

Proof: (Following Morrow and Kodaira [30]): Let $A_{\alpha\bar\beta}$ be the cofactor of $g_{\alpha\bar\beta}$, treating $g_{\alpha\bar\beta}$ as a matrix with inverse $g^{\bar\beta\alpha}$. Then we can write:

$$g^{\bar\beta\alpha} = A_{\alpha\bar\beta}/g \;.$$

On the other hand, we have:

$$\frac{\partial g}{\partial g_{\alpha\bar\beta}} = A_{\alpha\bar\beta}$$

so that $\dfrac{\partial g}{\partial g_{\alpha\bar\beta}} = g\, g^{\bar\beta\alpha}.$

Using this and the Kählerian property, we compute:

$$\partial_{\gamma} g = \frac{\partial g}{\partial g_{ab}} \cdot \frac{\partial g_{ab}}{\partial z^{\gamma}}$$

$$= \frac{\partial g}{\partial g_{\alpha\bar\beta}} \cdot \frac{\partial g_{\alpha\bar\beta}}{\partial z^{\gamma}}$$

$$= g\, g^{\bar\beta\alpha}\frac{\partial g_{\alpha\bar\beta}}{\partial z^{\gamma}}$$

$$= g\,\gamma^{\alpha}_{\gamma\alpha} \qquad \text{(by definition of } \gamma^{\alpha}_{\beta\gamma}\text{)}$$

$$= g\,\gamma^{\alpha}_{\alpha\gamma} \qquad \text{(since } \gamma^{a}_{bc} \text{ is symmetric in the Kählerian case)} \;.$$

Thus we have:

$$\gamma^{\alpha}_{\alpha\gamma} = \frac{1}{g} \partial_\gamma g ,$$

and similarly it follows that

$$\gamma^{\bar\alpha}_{\bar\alpha\bar\gamma} = \frac{1}{g} \partial_{\bar\gamma} g .$$

The next theorem will be quite useful later on in deciding when an Hermitian manifold is Kählerian:

VI.22: THEOREM: For an Hermitian manifold, $g_{ab} - i\, J_{ab}$ is Kählerian iff

$$\nabla_c J_{ab} = 0$$

where ∇_c is the Riemannian covariant derivative.

Proof: If J_{ab} is Kählerian, then $\nabla_c = \mathcal{D}_c$ and we have, using Theorems VI.9 and VI.18,

$$\nabla_\gamma J_{\alpha\bar\beta} = \mathcal{D}_\gamma J_{\alpha\bar\beta} = i \mathcal{D}_\gamma g_{\alpha\bar\beta} = 0 ,$$

with the other components vanishing similarly. Conversely, if $\nabla_c J_{ab} = 0$, then using Theorem VI.9 again, we have:

$$\nabla_\gamma J_{\alpha\bar\beta} = i g_{\alpha\bar\beta,\gamma} - \Gamma^{\sigma}_{\gamma\alpha}\, i g_{\sigma\bar\beta} - \Gamma^{\bar\sigma}_{\gamma\bar\beta}\, i g_{\alpha\bar\sigma} = 0 .$$

Writing out the other components of $\nabla_c J_{ab} = 0$ in this way, and taking the appropriate linear combinations of these, yields eventually

$$\Gamma^{\gamma}_{\alpha\beta} = g^{\gamma\bar\mu} \partial_\alpha g_{\beta\bar\mu} ,$$

$$\Gamma^{\bar\gamma}_{\bar\alpha\bar\beta} = g^{\mu\bar\gamma} \partial_{\bar\alpha} g_{\mu\bar\beta} ,$$

other components zero.

Thus $\Gamma_{ab}{}^{c} = \gamma_{ab}{}^{c}$, so by Corollary VI.20 above, the theorem follows.

Curvature on Kählerian Manifolds

In the case of a Kählerian manifold, the Riemann and Ricci curvature tensors can be expressed locally with remarkable conciseness in terms of scalar functions on the manifold. The following theorems are based to a large extent on the results of Morrow and Kodaira [30]. We begin with a lemma:

<u>VI.23: THEOREM:</u> On a Kählerian manifold, $g_{\alpha\bar{\beta}}$ is locally expressible as

$$g_{\alpha\bar{\beta}} = \partial_{\alpha}\, \partial_{\bar{\beta}} K \ ,$$

where K is a real scalar function.

<u>Proof:</u> Since J_{ab} is closed, and of the form $J_{\alpha\beta} = J_{\overline{\alpha\beta}} = 0$, Theorem IV.36 applies to give $J_{\alpha\bar{\beta}} = \partial_{\alpha}\partial_{\bar{\beta}} K'$; but by Theorem VI.9 we have $J_{\alpha\bar{\beta}} = i\, g_{\alpha\bar{\beta}}$, and hence the desired result follows with $K = -\, i\, K'$. Since $\overline{g_{\alpha\bar{\beta}}} = g_{\beta\bar{\alpha}}$, we have $\overline{\partial_{\alpha}\partial_{\bar{\beta}} K} = \partial_{\beta}\, \partial_{\bar{\alpha}} K$, whence $K = \bar{K}$, i.e., the function K is real.

We now proceed to enumerate the definitions and properties of the various curvature tensors which are defined on a Riemannian manifold. Throughout, we adopt the sign conventions of Schild [48].

<u>VI.24: DEFINITION:</u> The Riemann curvature tensor $R^{a}{}_{bcd}$ is given by

$$R^{a}{}_{bcd} = \partial_c \Gamma^{a}_{bd} - \partial_d \Gamma^{a}_{bc}$$
$$+ \Gamma^{s}_{bd} \Gamma^{a}_{sc} - \Gamma^{s}_{bc} \Gamma^{a}_{sd}$$

where the Riemannian Christoffel symbols are given by

$$\Gamma^{a}_{bc} = \frac{1}{2} g^{am} (\partial_c g_{bm} + \partial_b g_{mc} - \partial_m g_{bc}) \ .$$

<u>VI.25:</u> <u>LEMMA:</u> The Riemann tensor has the algebraic symmetries

$$R_{abcd} = R_{[ab][cd]} \ ,$$

$$R_{abcd} = R_{cdab} \ ,$$

and $$R_{a[bcd]} = 0 \ .$$

<u>Proof:</u> Follows from Definition VI.24 after setting $R_{abcd} = g_{am} R^{m}{}_{bcd}$.

<u>VI.26:</u> <u>THEOREM:</u> On a Kählerian manifold,

$$R^{\alpha}{}_{\beta\bar{\gamma}\delta} = \partial_{\bar{\gamma}} \gamma^{\alpha}_{\delta\beta} \ ,$$

$$R^{\bar{\alpha}}{}_{\overline{\beta\gamma\delta}} = -\partial_{\delta} \gamma^{\bar{\alpha}}_{\overline{\gamma\beta}} \ ,$$

$$R^{\alpha}{}_{\beta\gamma\bar{\delta}} = -\partial_{\bar{\delta}} \gamma^{\alpha}_{\gamma\beta} \ ,$$

$$R^{\bar{\alpha}}{}_{\bar{\beta}\gamma\bar{\delta}} = \partial_{\gamma} \gamma^{\bar{\alpha}}_{\overline{\delta\beta}} \ ,$$

other components zero (except those obtained from these by symmetry considerations).

<u>Proof:</u> On a Kählerian manifold, we have $\Gamma^{a}_{bc} = \gamma^{a}_{bc}$, so we get, using Definition VI.24:

$$R^{\alpha}{}_{\beta\bar{\gamma}\delta} = \partial_{\bar{\gamma}} \gamma^{\alpha}_{\beta\delta} - \partial_{\delta}\gamma^{\alpha}_{\beta\bar{\gamma}} + \gamma^{s}_{\beta\delta}\, \gamma^{\alpha}_{s\bar{\gamma}} - \gamma^{s}_{\beta\bar{\gamma}}\, \gamma^{\alpha}_{s\delta}$$

$$= \partial_{\bar{\gamma}} \gamma^{\alpha}_{\beta\delta} - 0 + 0 - 0$$

$$= \partial_{\bar{\gamma}} \gamma^{\alpha}_{\delta\beta} \quad .$$

The other components are obtained similarly.

<u>VI.27: THEOREM:</u> On a Kählerian manifold,

$$R_{\alpha\beta cd} = R_{\bar{\alpha}\bar{\beta} cd} = 0 \ ,$$

$$R_{ab\gamma\delta} = R_{ab\bar{\gamma}\bar{\delta}} = 0 \ ,$$

$$R_{\bar{\alpha}\beta\bar{\gamma}\delta} = g_{\bar{\alpha}\mu}\, \partial_{\bar{\gamma}}\, \gamma^{\mu}_{\delta\beta} \ ,$$

and

$$R_{\alpha\bar{\beta}\gamma\bar{\delta}} = g_{\alpha\bar{\mu}}\, \partial_{\gamma}\, \gamma^{\bar{\mu}}_{\bar{\delta}\bar{\beta}} \ .$$

<u>Proof:</u> Follows by direct computation after setting $R_{abcd} = g_{am} R^{m}{}_{bcd}$. (Note that other components of R_{abcd} are obtained by consulting the symmetries of Lemma VI.25, e.g., $R_{\beta\bar{\alpha}\bar{\gamma}\delta} = - g_{\bar{\alpha}\mu}\, \partial_{\bar{\gamma}}\, \gamma^{\mu}_{\delta\beta}$.)

<u>VI.28: THEOREM:</u> On a Kählerian manifold,

$$R_{\bar{\alpha}\beta\bar{\gamma}\delta} = \partial_{\bar{\alpha}} \partial_{\beta} \partial_{\bar{\gamma}} \partial_{\delta} K \ -$$

$$- \ g^{\bar{\mu}\tau} (\partial_{\bar{\mu}} \partial_{\beta} \partial_{\delta} K)\, (\partial_{\tau}\, \partial_{\bar{\alpha}}\, \partial_{\bar{\gamma}} K) \ ,$$

where

$$g_{\alpha\bar{\beta}} = \partial_{\alpha}\, \partial_{\bar{\beta}} K \ ,$$

as in Theorem VI.23.

<u>Proof:</u> $R_{\alpha\beta\bar{\gamma}\delta} = g_{\bar{\alpha}\mu}\, \partial_{\bar{\gamma}}\, \gamma^{\mu}_{\delta\beta}$

$= g_{\bar{\alpha}\mu}\, \partial_{\bar{\gamma}}\, (g^{\mu\bar{\nu}}\, \partial_{\delta}\, g_{\bar{\nu}\beta})$

$= g_{\bar{\alpha}\mu} (\partial_{\bar{\gamma}}\, g^{\mu\bar{\nu}})\, (\partial_{\delta}\, g_{\bar{\nu}\beta})$

$\quad + g_{\bar{\alpha}\mu}\, g^{\mu\bar{\nu}}\, \partial_{\bar{\gamma}}\, \partial_{\delta}\, g_{\bar{\nu}\beta}$

$= - g^{\mu\nu} (\partial_{\bar{\gamma}}\, g_{\bar{\alpha}\mu})\, (\partial_{\delta}\, g_{\beta\bar{\nu}})$

$\quad + \partial_{\bar{\gamma}}\, \partial_{\delta}\, g_{\bar{\alpha}\beta}$

$= \partial_{\bar{\gamma}}\, \partial_{\delta}\, \partial_{\bar{\alpha}}\, \partial_{\beta}\, K$

$\quad - g^{\mu\bar{\nu}} (\partial_{\bar{\gamma}}\, \partial_{\bar{\alpha}}\, \partial_{\mu}\, K)\, (\partial_{\delta}\, \partial_{\beta}\, \partial_{\bar{\nu}}\, K)$

$= \partial_{\bar{\alpha}}\, \partial_{\beta}\, \partial_{\bar{\gamma}}\, \partial_{\delta}\, K$

$\quad - g^{\bar{\mu}\tau} (\partial_{\bar{\mu}}\, \partial_{\beta}\, \partial_{\delta}\, K)\, (\partial_{\tau}\, \partial_{\bar{\alpha}}\, \partial_{\bar{\gamma}}\, K)$, as desired.

<u>VI.29: DEFINITION:</u> The Ricci curvature tensor R_{ab} is given by

$$R_{ab} \equiv R^{s}{}_{abs} \ .$$

<u>VI.30: LEMMA:</u> The Ricci tensor satisfies

$$R_{ab} = R_{ba} \ .$$

<u>Proof:</u> $R_{ab} = R^{s}{}_{abs}$

$= g^{sm}\, R_{mabs}$

$= g^{sm}\, R_{bsma}$ by Lemma VI.25

$= g^{sm}\, R_{sbam}$ by Lemma VI.25

$= R^{m}{}_{bam}$

$= R_{ba}$, as desired.

<u>VI.31: THEOREM:</u> On a Kählerian manifold,

$$R_{\alpha\beta} = R_{\bar{\alpha}\bar{\beta}} = 0 ,$$

$$R_{\alpha\bar{\beta}} = \partial_{\alpha} \partial_{\bar{\beta}} (\ln g) ,$$

where $g = \det(g_{\alpha\bar{\beta}})$.

<u>Proof:</u> $R_{\alpha\beta} = R^{s}{}_{\alpha\beta s}$

$= R^{\sigma}{}_{\alpha\beta\sigma} + R^{\bar{\sigma}}{}_{\alpha\beta\bar{\sigma}}$

$= 0$ by Theorem VI.26;

$R_{\bar{\alpha}\bar{\beta}} = 0$ similarly;

$R_{\alpha\bar{\beta}} = R^{s}{}_{\alpha\bar{\beta}s}$

$= R^{\sigma}{}_{\alpha\bar{\beta}\sigma} + R^{\bar{\sigma}}{}_{\alpha\bar{\beta}\bar{\sigma}}$

$= R^{\sigma}{}_{\alpha\bar{\beta}\sigma} + 0$ by Theorem VI.26

$= \partial_{\bar{\beta}} \gamma^{\sigma}{}_{\sigma\alpha}$ by Theorem VI.26

$= \partial_{\bar{\beta}} (g^{-1} \partial_{\alpha} g)$ by Theorem VI.21

$= \partial_{\bar{\beta}} \partial_{\alpha} (\ln g)$

$= \partial_{\alpha} \partial_{\bar{\beta}} (\ln g)$, as desired.

VI.32: DEFINITION: The Weyl conformal curvature tensor $C^a{}_{bcd}$ is given by

$$C^a{}_{bcd} = R^a{}_{bcd} + \frac{1}{n-2}\left(g_{bd} R^a{}_c - g_{bc} R^a{}_d\right) + \frac{1}{n-2}\left(R_{bd}\, \delta^a{}_c - R_{bc}\, \delta^a{}_d\right) - \frac{R}{(n-1)(n-2)}\left(g_{bd}\, \delta^a{}_c - g_{bc}\, \delta^a{}_d\right) ,$$

where n is the real dimension of the manifold and

$$R \equiv R^a{}_a .$$

VI.33: LEMMA: (i) C_{abcd} has all of the algebraic symmetries of the Riemann tensor (see Lemma VI.25), and in addition satisfies

$$C^s{}_{bcs} = 0 .$$

(ii) On a Kählerian manifold,

$$C_{\alpha\beta\bar{\gamma}\delta} = C_{\overline{\alpha\beta\gamma\delta}} = 0 ,$$

$$C_{\bar{\alpha}\beta\gamma\delta} = C_{\bar{\alpha}\beta\overline{\gamma\delta}} = 0 ,$$

and $$C_{\alpha\beta\gamma\delta} = C_{\overline{\alpha\beta\gamma\delta}} = 0 .$$

(However, $C_{\alpha\beta\overline{\gamma\delta}}$ and $C_{\overline{\alpha\beta}\gamma\delta}$ do not, in general, vanish.)

Proof: These results follow by direct computation starting with Definition VI.32 and using our previous results.

In applications to relativity, the following theorem, whose proof is omitted, will play an important role:

VI.34: THEOREM: $\hat{g}_{ab}$ is conformally related to g_{ab}, so that

$$\hat{g}_{ab} = \Omega^2 g_{ab} ,$$

if and only if

$$\hat{C}^a{}_{bcd} = C^a{}_{bcd} .$$

Proof: Consult any standard text on Riemannian geometry, for example, Eisenhart [16].

VI.35: REMARK: On a Kählerian manifold, the tensors R_{abcd}, R_{ab}, R, and C_{abcd} are all real, i.e.,

$$\overline{R_{abcd}} = R_{\overline{abcd}} ,$$

$$\overline{R_{ab}} = R_{\overline{ab}} ,$$

$$\overline{R} = R ,$$

and

$$\overline{C_{abcd}} = C_{\overline{abcd}} .$$

Recapitulation

To conclude this chapter, it will be helpful to review the various "structures" which we have introduced and to display the interrelationships between them. This is done in the diagram on the following page.

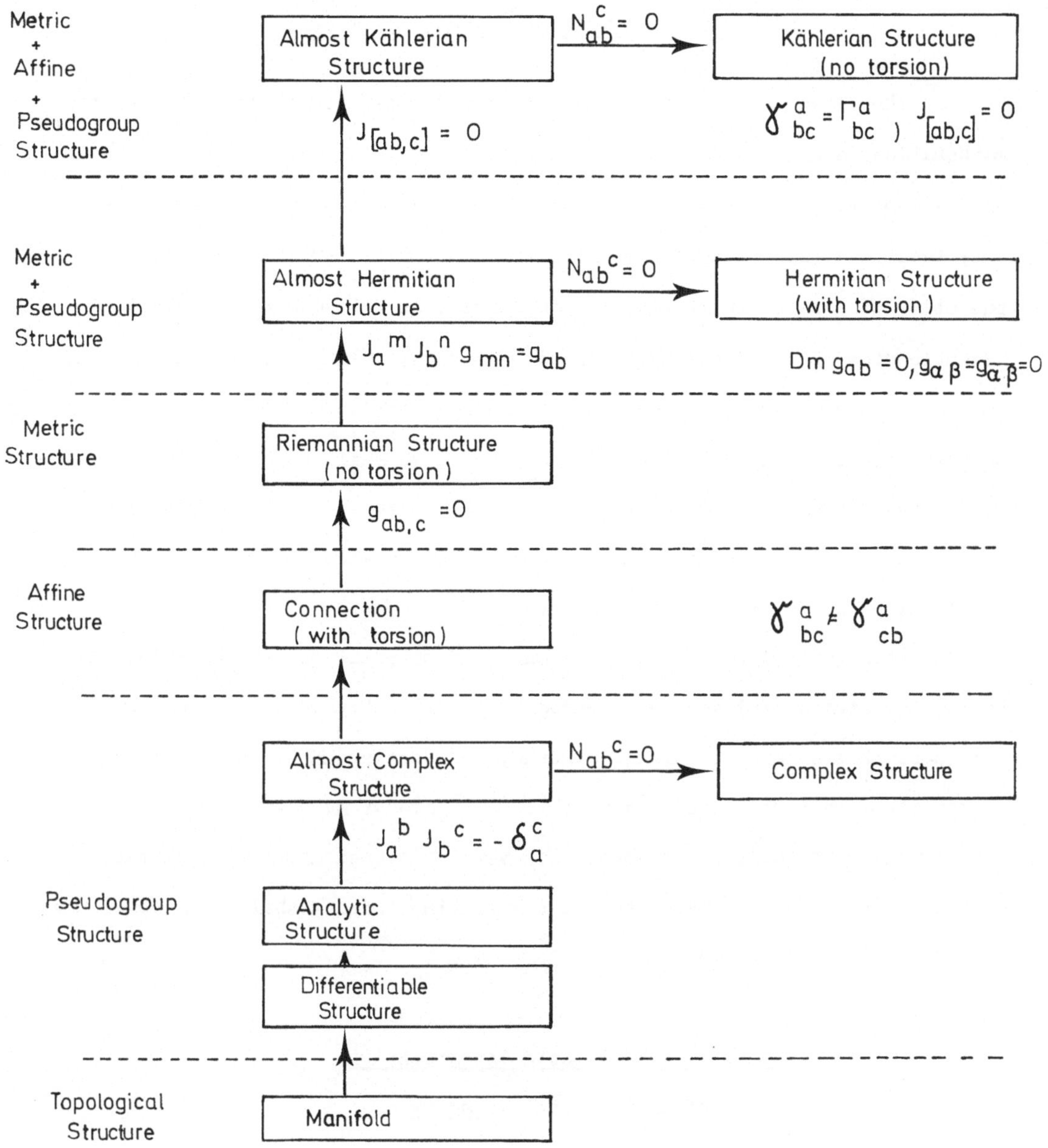
Metric
+
Affine
+
Pseudogroup
Structure
Almost Kählerian Structure
$N_{ab}{}^{c} = 0$
Kählerian Structure (no torsion)
$\gamma^{a}_{bc} = \Gamma^{a}_{bc}$) $J_{[ab,c]} = 0$
$J_{[ab,c]} = 0$
Metric
+
Pseudogroup
Structure
Almost Hermitian Structure
$N_{ab}{}^{c} = 0$
Hermitian Structure (with torsion)
$J_a{}^m J_b{}^n g_{mn} = g_{ab}$
$D_m g_{ab} = 0, g_{\alpha\beta} = g_{\bar{\alpha}\bar{\beta}} = 0$
Metric
Structure
Riemannian Structure (no torsion)
$g_{ab,c} = 0$
Affine
Structure
Connection (with torsion)
$\gamma^{a}_{bc} \neq \gamma^{a}_{cb}$
Almost Complex Structure
$N_{ab}{}^{c} = 0$
Complex Structure
$J_a{}^b J_b{}^c = -\delta^c_a$
Pseudogroup
Structure
Analytic Structure
Differentiable Structure
Topological
Structure
Manifold

CHAPTER VII: REVIEW OF GENERAL RELATIVITY

The first half of this work has been devoted to an exposition of the mathematical apparatus of complex manifold theory. At this point we have in hand all of the ideas necessary to begin the second half of the work, namely, a systematic application of the mathematical results (suitably modified) to the study of general relativity. Accordingly, in this chapter we shall assemble some of the fundamental facts of general relativity as well as some more specialized results which will be needed later on. Our approach will be both eclectic and selective: In the sequel we shall move back and forth freely between various formulations (e.g., NP formalism and coordinate treatments), and topics not of importance later on will only be briefly -- if at all -- discussed in this chapter.

As a general reference for background material on general relativity, Schild's lectures [48] are recommended. On a slightly less advanced level, the thorough treatment of Misner, Thorne, and Wheeler [29], and the more conventional text of Landau and Lifshitz [28] are of much value.

The mathematical conventions not yet introduced above will be those of Schild (with the exception that our spacetime indices shall range over the values $0, \bar{0}, 1, \bar{1}$ rather than 1, 2, 3, 4).

General Relativistic Interpreation of Differential Geometry

The foundation of general relativity theory is the equivalence principle, which implies that the events of spacetime correspond to the points of a four-dimensional pseudo-Riemannian manifold, generally curved, with a metric of signature (+---). For then there exist everywhere local coordinate systems (Fermi coordinate systems) such that the laws of physics expressed in these coordinates are formally identical to the laws of physics in special relativity, expressed in rectangular Cartesian coordinates. This "locally Minkowskian" character of spacetime is reflected in the "locally Euclidean" character of a differentiable manifold (compare Definition III.1), and in the fact that the metric of spacetime is pointwise reducible, by a coordinate transformation, to the form of the Minkowski metric of special relativistic spacetime:

$$ds^2 = dt^2 - dx^2 - dy^2 - dz^2 \quad . \qquad \text{(VII.1)}$$

(In fact, this reduction can even be accomplished along a timelike line in spacetime.) The metric of curved space is thus a generalization of the flat metric VII.1:

$$ds^2 = g_{ab}\, dx^a\, dx^b \quad . \qquad \text{(VII.2)}$$

There are three points to the physical interpretation of the metric VII.2. **Firstly**, the timelike geodesics (see below) with respect to the pseudo-Riemannian connection

$$\Gamma_{ab}{}^{c} = \frac{1}{2}\, g^{cs}\, (g_{as,b} + g_{bs,a} - g_{ab,s}) \qquad \text{(VII.3)}$$

are interpreted as the paths of massive test particles, for which $ds^2 > 0$. Secondly, the element of length ds along the path of such a test particle corresponds to the increment of proper time measured by an observer on the particle itself. Thirdly, the propagation of light and other zero restmass particles is described by the null geodesics of the spacetime, and thus $ds^2 = 0$ along the path of such a particle.

The relationship between the geometry of spacetime and the matter which it contains is summarized by the Einstein gravitational field equations:

$$G_{ab} = -8\pi T_{ab} \tag{VII.4}$$

Here we are using units such that $c = G = 1$. The tensor G_{ab} is the Einstein tensor, defined by

$$G_{ab} \equiv R_{ab} - \frac{1}{2} g_{ab} R , \tag{VII.5}$$

where R_{ab} is the Ricci curvature and R is the scalar curvature. The tensor T_{ab} is the stress-energy tensor.

We shall be especially interested in two types of spacetime. The first is vacuum spacetime, for which T_{ab} is zero. It is easy to show that in this case the Einstein equation VII.4 reduces to

$$R_{ab} = 0 . \tag{VII.6}$$

The second type of field is that for which the stress-energy tensor arises from a source-free electromagnetic field F_{ab} (electrified spacetime). The field F_{ab} satisfies the source-free Maxwell equations

$$\overset{+}{F}_{[ab,c]} = 0 , \tag{VII.7}$$

where $\overset{+}{F}_{ab}$ is the self-dual part of the Maxwell tensor, defined by:

$$\begin{aligned} \overset{+}{F}_{ab} &= F_{ab} + i \overset{*}{F}_{ab} \\ &= F_{ab} + i \frac{1}{2} \eta_{abrs} F^{rs} . \end{aligned} \tag{VII.8}$$

For such an Einstein-Maxwell spacetime, the stress-energy tensor is given by:

$$T_{ab} = \frac{1}{4\pi}(+ F_{as} F^{s}{}_{b} + \frac{1}{4} g_{ab} F^{rs} F_{rs}) \quad . \qquad \text{(VII.9)}$$

In the source-free case T_{ab} is traceless, $T_s{}^s = 0$, and from this it is easy to show that $R = 0$, so that the Einstein field equations reduce to

$$R_{ab} = - 8\pi T_{ab} \quad . \qquad \text{(VII.10)}$$

Newman-Penrose Formalism

The Newman-Penrose (NP) formalism [37], [53], [26], is a particular example of a tetrad formalism which is amazingly useful in many applications. It is very closely related to the two-index spinor formulation of general relativity, which we shall only briefly touch upon later.

At each point of spacetime a tetrad, or set of four vectors, $\{\ell^a, m^a, \bar{m}^a, n^a\}$ is introduced. Two of these vectors, ℓ^a and n^a are real and null, so that $\ell^a \ell_a = n^a n_a = 0$; the further condition is imposed that ℓ^a and n^a satisfy $\ell^a n_a = +1$. The other two legs of the tetrad, m^a and $\bar{m}^a$, are complex null vectors, $m^a m_a = \bar{m}^a \bar{m}_a = 0$, with $\bar{m}^a$ the complex conjugate of m^a and with normalization $m^a \bar{m}_a = -1$. Furthermore, it is required that $\ell^a m_a = \ell^a \bar{m}_a = n^a m_a = n^a \bar{m}_a = 0$. So to summarize, we have

$$\ell^a n_a = - m^a \bar{m}_a = +1 \quad , \qquad \text{(VII.11)}$$

all other products zero.

It is always possible to introduce such a "null tetrad" locally. Globally, the existence of a null tetrad is equivalent to the existence of a global orthonormal tetrad. And, in turn, Geroch [18], [19] has shown that such a global tetrad exists if and only if the spacetime (assumed non-compact) admits a spinor structure. Since there is reason to believe that a spinor structure is physically meaningful (see [2], [18]), it seems reasonable to demand that a spacetime does in fact admit a globally defined normed null tetrad field. (The purpose of this digression is to elevate some of our results below to the status of global theorems.)

The normalizations VII.11 chosen for the tetrad legs imply that the metric of spacetime must be expressible as

$$g_{ab} = \ell_a n_b + n_a \ell_b - m_a \bar{m}_b - \bar{m}_a m_b \ . \qquad \text{(VII.12)}$$

Given the coordinate components of any tensor, we can form the tetrad components of the tensor by contracting with the appropriate tetrad legs, e.g., $T^{a\ldots}{}_{b\ldots} \longrightarrow \ell_a \ldots n^b \ldots T^{a\ldots}{}_{b\ldots}$, and so on. Conversely, given the tetrad components of a tensor, we can recover the coordinate components. To see this we note that VII.12 implies

$$\delta_a{}^b = \ell_a n^b + n_a \ell^b - m_a \bar{m}^b - \bar{m}_a m^b \ . \qquad \text{(VII.13)}$$

Then, for example, we can write

$$\begin{aligned} \xi^b &= \delta_a{}^b \xi^a \\ &= n^b(\ell_a \xi^a) + \ell^b(n_a \xi^a) \\ &\qquad - \bar{m}^b(m_a \xi^a) - m^b(\bar{m}_a \xi^a) \ , \end{aligned}$$

and similarly to recover the other coordinate components.

In a tetrad formalism, the role of the Christoffel symbols is taken over by the Ricci rotation coefficients. In the NP formalism, these are called the spin coefficients, and they are defined as follows:

$$
\begin{aligned}
\alpha &= \tfrac{1}{2}(\ell_{a;b}\, n^a \bar{m}^b - m_{a;b}\, \bar{m}^a \bar{m}^b) \\
\beta &= \tfrac{1}{2}(\ell_{a;b}\, n^a m^b - m_{a;b}\, \bar{m}^a m^b) \\
\gamma &= \tfrac{1}{2}(\ell_{a;b}\, n^a n^b - m_{a;b}\, \bar{m}^a n^b) \\
\epsilon &= \tfrac{1}{2}(\ell_{a;b}\, n^a \ell^b - m_{a;b}\, \bar{m}^a \ell^b) \\
\kappa &= \ell_{a;b}\, m^a \ell^b \\
\lambda &= -\, n_{a;b}\, \bar{m}^a \bar{m}^b \\
\mu &= -\, n_{a;b}\, \bar{m}^a m^b \\
\nu &= -\, n_{a;b}\, \bar{m}^a n^b \\
\pi &= -\, n_{a;b}\, \bar{m}^a \ell^b \\
\rho &= \ell_{a;b}\, m^a \bar{m}^b \\
\sigma &= \ell_{a;b}\, m^a m^b \\
\tau &= \ell_{a;b}\, m^a n^b
\end{aligned}
\qquad \text{(VII.14)}
$$

Making use of the spin coefficients allows one to express tetrad components of covariant derivatives in terms of the spin coefficients themselves and the intrinsic derivative operators, defined by:

$$
\begin{aligned}
D &= \ell^a\, \partial/\partial x^a \\
\Delta &= n^a\, \partial/\partial x^a \\
\delta &= m^a\, \partial/\partial x^a \\
\bar{\delta} &= \bar{m}^a\, \partial/\partial x^a \ .
\end{aligned}
\qquad \text{(VII.15)}
$$

Then we have, for example,

$$\begin{aligned} \ell^a m^b \xi_{a;b} &= m^b(\ell^a \xi_a)_{;b} - m^b \xi_a \ell^a{}_{;b} \\ &= \delta(\ell^a \xi_a) - \ell_{a;b} \xi^a m^b \\ &= \delta(\ell^a \xi_a) - (\bar{\alpha} + \beta)(\ell^a \xi_a) \\ &\quad - 0(n^a \xi_a) - \bar{\rho}(m^a \xi_a) \\ &\quad - \sigma(\bar{m}^a \xi_a) \quad , \end{aligned}$$

where bars on the spin coefficients denote the complex conjugate. (Although at this point the formalism looks vastly more complicated than ordinary methods, there will be great simplifications later on.)

Next, the tetrad components of the various curvature quantities are defined as follows:

$$\begin{aligned} \Psi_0 &= - C_{abcd} \ell^a m^b \ell^c m^d \\ \Psi_1 &= - C_{abcd} \ell^a n^b \ell^c m^d \\ \Psi_2 &= - C_{abcd} \bar{m}^a n^b \ell^c m^d \\ \Psi_3 &= - C_{abcd} \bar{m}^a n^b \ell^c n^d \\ \Psi_4 &= - C_{abcd} \bar{m}^a n^b \bar{m}^c n^d \\ \\ \Phi_{00} &= - \tfrac{1}{2} R_{ab} \ell^a \ell^b \\ \Phi_{01} &= - \tfrac{1}{2} R_{ab} \ell^a m^b \\ \Phi_{02} &= - \tfrac{1}{2} R_{ab} m^a m^b \end{aligned} \qquad \text{(VII.16)}$$

$$\Phi_{10} = -\frac{1}{2} R_{ab} \ell^a \bar{m}^b$$

$$\Phi_{11} = -\frac{1}{4} R_{ab} (\ell^a n^b + m^a \bar{m}^b)$$

$$\Phi_{12} = -\frac{1}{2} R_{ab} n^a m^b$$

$$\Phi_{20} = -\frac{1}{2} R_{ab} \bar{m}^a \bar{m}^b$$

$$\Phi_{21} = -\frac{1}{2} R_{ab} n^a \bar{m}^b$$

$$\Phi_{22} = -\frac{1}{2} R_{ab} n^a n^b$$

$$\Lambda = \frac{1}{24} R \quad .$$

Likewise, the tetrad components of the electromagnetic field are given by:

$$\Phi_0 = F_{ab} \ell^a m^b$$

$$\Phi_1 = \frac{1}{2} F_{ab} (\ell^a n^b + \bar{m}^a m^b) \qquad \text{(VII.17)}$$

$$\Phi_2 = F_{ab} \bar{m}^a n^b$$

If electromagnetism is the only source of stress-energy present, then

$$\Phi_{mn} = \Phi_m \bar{\Phi}_n \quad . \qquad \text{(VII.18)}$$

The NP version of Maxwell's equations becomes:

$$D\Phi_1 - \bar{\delta}\Phi_0 = (\pi - 2\alpha)\Phi_0 + 2\rho\Phi_1 - \kappa\Phi_2$$

$$D\Phi_2 - \bar{\delta}\Phi_1 = -\lambda\Phi_0 + 2\pi\Phi_1 + (\rho - 2\epsilon)\Phi_2$$

(VII.19)

$$\delta\Phi_1 - \Delta\Phi_0 = (\mu - 2\gamma)\Phi_0 + 2\tau\Phi_1 - \sigma\Phi_2$$

$$\delta\Phi_2 - \Delta\Phi_1 = -\nu\Phi_0 + 2\mu\Phi_1 + (\tau - 2\beta)\Phi_2 \ .$$

Now, the so-called NP equations arise from the commutation relations for covariant differentiation, i.e., from the definition of the Riemann tensor in terms of the spin coefficients. The NP equations are:

$$D\rho - \bar{\delta}\kappa = \rho^2 + \sigma\bar{\sigma} + (\epsilon + \bar{\epsilon})\rho - \bar{\kappa}\tau$$
$$- \kappa(3\alpha + \bar{\beta} - \pi) + \Phi_{00}$$

$$D\sigma - \delta\kappa = (\rho + \bar{\rho})\sigma + (3\epsilon - \bar{\epsilon})\sigma$$
$$- (\tau - \bar{\pi} + \bar{\alpha} + 3\beta)\kappa + \Psi_0$$

$$D\tau - \Delta\kappa = (\tau + \bar{\pi})\rho + (\bar{\tau} + \pi)\sigma$$
$$+ (\epsilon - \bar{\epsilon})\tau - (3\gamma + \bar{\gamma})\kappa + \Psi_1 + \Phi_{01}$$

$$D\alpha - \bar{\delta}\epsilon = (\rho + \bar{\epsilon} - 2\epsilon)\alpha + \beta\bar{\sigma} - \bar{\beta}\epsilon$$
$$- \kappa\lambda - \bar{\kappa}\gamma + (\epsilon + \rho)\pi + \Phi_{10}$$

$$D\beta - \delta\epsilon = (\alpha + \pi)\sigma + (\bar{\rho} - \bar{\epsilon})\beta$$
$$- (\mu + \gamma)\kappa - (\bar{\alpha} - \bar{\pi})\epsilon + \Psi_1$$

$$D\gamma - \Delta\epsilon = (\tau + \bar{\pi})\alpha + (\bar{\tau} + \pi)\beta - (\epsilon + \bar{\epsilon})\gamma$$
$$- (\gamma + \bar{\gamma})\epsilon + \tau\pi - \nu\kappa + \Psi_2 - \Lambda + \Phi_{11}$$

$$
\begin{aligned}
D\lambda - \bar{\delta}\pi &= \rho\lambda + \bar{\sigma}\mu + \pi^2 + (\alpha - \bar{\beta})\pi \\
&\qquad - \nu\bar{\kappa} - (3\epsilon - \bar{\epsilon})\lambda + \Phi_{20} \\[1em]
D\mu - \delta\pi &= \bar{\rho}\mu + \sigma\lambda + \pi\bar{\pi} - (\epsilon + \bar{\epsilon})\mu \\
&\qquad - \pi(\bar{\alpha} - \beta) - \nu\kappa + \Psi_2 + 2\Lambda \\[1em]
D\nu - \Delta\pi &= (\pi + \bar{\tau})\mu + (\bar{\pi} + \tau)\lambda + (\gamma - \bar{\gamma})\pi \\
&\qquad - (3\epsilon + \bar{\epsilon})\nu + \Psi_3 + \Phi_{21} \\[1em]
\Delta\lambda - \bar{\delta}\nu &= -(\mu + \bar{\mu})\lambda - (3\gamma - \bar{\gamma})\lambda \\
&\qquad + (3\alpha + \bar{\beta} + \pi - \bar{\tau})\nu - \Psi_4 \\[1em]
\delta\rho - \bar{\delta}\sigma &= \rho(\bar{\alpha} + \beta) - \sigma(3\alpha - \bar{\beta}) + (\rho - \bar{\rho})\tau \\
&\qquad + (\mu - \bar{\mu})\kappa - \Psi_1 + \Phi_{01} \\[1em]
\delta\alpha - \bar{\delta}\beta &= \mu\rho - \lambda\sigma + \alpha\bar{\alpha} + \beta\bar{\beta} - 2\alpha\beta \\
&\qquad + \gamma(\rho - \bar{\rho}) + \epsilon(\mu - \bar{\mu}) - \Psi_2 + \Lambda + \Phi_{11} \\[1em]
\delta\lambda - \bar{\delta}\mu &= (\rho - \bar{\rho})\nu + (\mu - \bar{\mu})\pi \\
&\qquad + \mu(\alpha + \bar{\beta}) + \lambda(\bar{\alpha} - 3\beta) - \Psi_3 + \Phi_{21} \\[1em]
\delta\nu - \Delta\mu &= \mu^2 + \lambda\bar{\lambda} + (\gamma + \bar{\gamma})\mu - \bar{\nu}\pi \\
&\qquad + (\tau - 3\beta - \bar{\alpha})\nu + \Phi_{22}
\end{aligned}
\tag{VII.20}
$$

$$\delta\gamma - \Delta\beta = (\tau - \bar{\alpha} - \beta)\gamma + \mu\tau - \sigma\nu - \epsilon\bar{\nu}$$

$$- \beta(\gamma - \bar{\gamma} - \mu) + \alpha\bar{\lambda} + \Phi_{12}$$

$$\delta\tau - \Delta\sigma = \mu\sigma + \bar{\lambda}\rho + (\tau + \beta - \bar{\alpha})\tau$$

$$- (3\gamma - \bar{\gamma})\sigma - \kappa\bar{\nu} + \Phi_{02}$$

$$\Delta\rho - \bar{\delta}\tau = -\rho\bar{\mu} - \sigma\lambda + (\bar{\beta} - \alpha - \bar{\tau})\tau$$

$$+ (\gamma + \bar{\gamma})\rho + \nu\kappa - \Psi_2 - 2\Lambda$$

$$\Delta\alpha - \bar{\delta}\gamma = (\rho + \epsilon)\nu - (\tau + \beta)\lambda$$

$$+ (\bar{\gamma} - \bar{\mu})\alpha + (\bar{\beta} - \bar{\tau})\gamma - \Psi_3 \; .$$

These NP equations are equivalent to the Einstein equations VII.4 when the proper expression for the stress-energy is inserted. For example, for vacuum we set $\Phi_{ab} = \Lambda = 0$, and for electrified $\Phi_{ab} = \Phi_a \bar{\Phi}_b$, $\Lambda = 0$.

The Bianchi identities,

$$R_{ab[cd;e]} = 0 \; , \qquad \text{(VII.21)}$$

take the following form in the NP formalism:

$$\bar{\delta}\Psi_0 - D\Psi_1 + D\Phi_{01} - \delta\Phi_{00} = (4\alpha - \pi)\Psi_0 - 2(2\rho + \epsilon)\Psi_1$$

$$+ 3\kappa\Psi_2 + (\bar{\pi} - 2\bar{\alpha} - 2\beta)\Phi_{00} + 2(\epsilon + \bar{\rho})\Phi_{01} + 2\sigma\Phi_{10}$$

$$- 2\kappa\Phi_{11} - \bar{\kappa}\Phi_{02}$$

$$\begin{aligned}\Delta\Psi_0 - \delta\Psi_1 + D\Phi_{02} - \delta\Phi_{01} &= (4\gamma - \mu)\Psi_0 - 2(2\tau + \beta)\Psi_1 \\ &+ 3\sigma\Psi_2 - \bar{\lambda}\Phi_{00} + 2(\bar{\pi} - \beta)\Phi_{01} + 2\sigma\Phi_{11} \\ &+ (2\epsilon - 2\bar{\epsilon} + \bar{\rho})\Phi_{02} - 2\kappa\Phi_{12}\end{aligned}$$

$$\begin{aligned}3(\bar{\delta}\Psi_1 - D\Psi_2) + 2(D\Phi_{11} - \delta\Phi_{10}) + \bar{\delta}\Phi_{01} - \Delta\Phi_{00} &= 3\lambda\Psi_0 - 9\rho\Psi_2 \\ &+ 6(\alpha - \pi)\Psi_1 + 6\kappa\Psi_3 + (\bar{\mu} - 2\mu - 2\gamma - 2\bar{\gamma})\Phi_{00} + (2\alpha + 2\pi + 2\bar{\tau})\Phi_{01} \\ &+ 2(\tau - 2\bar{\alpha} + \bar{\pi})\Phi_{10} + 2(2\bar{\rho} - \rho)\Phi_{11} + 2\sigma\Phi_{20} - \bar{\sigma}\Phi_{02} - 2\bar{\kappa}\Phi_{12} - 2\kappa\Phi_{21}\end{aligned}$$

$$\begin{aligned}3(\Delta\Psi_1 - \delta\Psi_2) + 2(D\Phi_{12} - \delta\Phi_{11}) + (\bar{\delta}\Phi_{02} - \Delta\Phi_{01}) &= 3\nu\Psi_0 + 6(\gamma - \mu)\Psi_1 \\ &- 9\tau\Psi_2 + 6\sigma\Psi_3 - \bar{\nu}\Phi_{00} + 2(\bar{\mu} - \mu - \gamma)\Phi_{01} - 2\bar{\lambda}\Phi_{10} + 2(\tau + 2\bar{\pi})\Phi_{11} \\ &+ (2\alpha + 2\pi + \bar{\tau} - 2\bar{\beta})\Phi_{02} + (2\bar{\rho} - 2\rho - 4\bar{\epsilon})\Phi_{12} + 2\sigma\Phi_{21} - 2\kappa\Phi_{22}\end{aligned}$$

$$\begin{aligned}3(\bar{\delta}\Psi_2 - D\Psi_3) + D\Phi_{21} - \delta\Phi_{20} + 2(\bar{\delta}\Phi_{11} - \Delta\Phi_{10}) &= 6\lambda\Psi_1 - 9\pi\Psi_2 \\ &+ 6(\epsilon - \rho)\Psi_3 + 3\kappa\Psi_4 - 2\nu\Phi_{00} + 2\lambda\Phi_{01} + 2(\bar{\mu} - \mu - 2\bar{\gamma})\Phi_{10} \\ &+ (2\pi + 4\bar{\tau})\Phi_{11} + (2\beta + 2\tau + \bar{\pi} - 2\bar{\alpha})\Phi_{20} - 2\bar{\sigma}\Phi_{12} + 2(\bar{\rho} - \rho - \epsilon)\Phi_{21} - \bar{\kappa}\Phi_{22}\end{aligned}$$

(VII.22)

$$\begin{aligned}3(\Delta\Psi_2 - \delta\Psi_3) + D\Phi_{22} - \delta\Phi_{21} + 2(\bar{\delta}\Phi_{12} - \Delta\Phi_{11}) &= 6\nu\Psi_1 - 9\mu\Psi_2 \\ &+ 6(\beta - \tau)\Psi_3 + 3\sigma\Psi_4 - 2\nu\Phi_{01} - 2\bar{\nu}\Phi_{10} + 2(2\bar{\mu} - \mu)\Phi_{11} + 2\lambda\Phi_{02} - \bar{\lambda}\Phi_{20} \\ &+ 2(\pi + \bar{\tau} - 2\bar{\beta})\Phi_{12} + 2(\beta + \tau + \bar{\pi})\Phi_{21} + (\bar{\rho} - 2\epsilon - 2\bar{\epsilon} - 2\rho)\Phi_{22}\end{aligned}$$

$$\bar{\delta}\Psi_3 - D\Psi_4 + \bar{\delta}\Phi_{21} - \Delta\Phi_{20} = 3\lambda\Psi_2 - 2(\alpha + 2\pi)\Psi_3 + (4\epsilon - \rho)\Psi_4 - 2\nu\Phi_{10}$$
$$+ 2\lambda\Phi_{11} + (2\gamma - 2\bar{\gamma} + \bar{\mu})\Phi_{20} + 2(\bar{\tau} - \alpha)\Phi_{21} - \bar{\sigma}\Phi_{22}$$

$$\Delta\Psi_3 - \delta\Psi_4 + \bar{\delta}\Phi_{22} - \Delta\Phi_{21} = 3\nu\Psi_2 - 2(\gamma + 2\mu)\Psi_3 + (4\beta - \tau)\Psi_4$$
$$- 2\nu\Phi_{11} - \bar{\nu}\Phi_{20} + 2\lambda\Phi_{12} + 2(\gamma + \bar{\mu})\Phi_{21} + (\bar{\tau} - 2\bar{\beta} - 2\alpha)\Phi_{22}$$

$$D\Phi_{11} - \delta\Phi_{10} - \bar{\delta}\Phi_{01} + \Delta\Phi_{00} + 3D\Lambda = (2\gamma - \mu + 2\bar{\gamma} - \bar{\mu})\Phi_{00} + (\pi - 2\alpha - 2\bar{\tau})\Phi_{01}$$
$$+ (\bar{\pi} - 2\bar{\alpha} - 2\tau)\Phi_{10} + 2(\rho + \bar{\rho})\Phi_{11} + \bar{\sigma}\Phi_{02} + \sigma\Phi_{20} - \bar{\kappa}\Phi_{12} - \kappa\Phi_{21}$$

$$D\Phi_{12} - \delta\Phi_{11} - \bar{\delta}\Phi_{02} + \Delta\Phi_{01} + 3\delta\Lambda = (2\gamma - \mu - 2\bar{\mu})\Phi_{01} + \bar{\nu}\Phi_{00} - \bar{\lambda}\Phi_{10}$$
$$+ 2(\bar{\pi} - \tau)\Phi_{11} + (\pi + 2\bar{\beta} - 2\alpha - \bar{\tau})\Phi_{02} + (2\rho + \bar{\rho} - 2\bar{\epsilon})\Phi_{12} + \sigma\Phi_{21} - \kappa\Phi_{22}$$

$$D\Phi_{22} - \delta\Phi_{21} - \bar{\delta}\Phi_{12} + \Delta\Phi_{11} + 3\Delta\Lambda = \nu\Phi_{01} + \bar{\nu}\Phi_{10} - 2(\mu + \bar{\mu})\Phi_{11} - \lambda\Phi_{02}$$
$$- \bar{\lambda}\Phi_{20} + (2\pi - \bar{\tau} + 2\beta)\Phi_{12} + (2\beta - \tau + 2\bar{\pi})\Phi_{21} + (\rho + \bar{\rho} - 2\epsilon - 2\bar{\epsilon})\Phi_{22}$$

Finally, we require the commutators of successive intrinsic derivatives acting on scalars:

$$(\Delta D - D\Delta) = (\gamma + \bar{\gamma})D + (\epsilon + \bar{\epsilon})\Delta - (\tau + \bar{\pi})\bar{\delta} - (\bar{\tau} + \pi)\delta$$
$$(\delta D - D\delta) = (\bar{\alpha} + \beta - \bar{\pi})D + \kappa\Delta - \sigma\bar{\delta} - (\bar{\rho} + \epsilon - \bar{\epsilon})\delta$$
$$(\delta\Delta - \Delta\delta) = -\bar{\nu}D + (\tau - \bar{\alpha} - \beta)\Delta + \bar{\lambda}\bar{\delta} + (\mu - \gamma + \bar{\gamma})\delta \qquad \text{(VII.23)}$$
$$(\bar{\delta}\delta - \delta\bar{\delta}) = (\bar{\mu} - \mu)D + (\bar{\rho} - \rho)\Delta - (\bar{\alpha} - \beta)\bar{\delta} - (\bar{\beta} - \alpha)\delta$$

$$(\bar{\delta} D - D\bar{\delta}) = (\alpha + \bar{\beta} - \pi)D + \bar{\kappa}\Delta - (\rho + \bar{\epsilon} - \epsilon)\bar{\delta} - \bar{\sigma}\,\delta$$

$$(\bar{\delta}\Delta - \Delta\bar{\delta}) = -\nu D + (\bar{\tau} - \alpha - \bar{\beta})\Delta + (\bar{\mu} - \bar{\gamma} + \gamma)\bar{\delta} + \lambda\,\delta$$

Lorentz Transformations

The reality conditions and the normalization conditions VII.11 do not uniquely characterize a null tetrad. In fact, there is a six-parameter group of transformations which preserves these conditions, and this group is just the homogeneous Lorentz group acting in the tangent space.

In NP language the proper Lorentz transformations are grouped into three sets as follows:

$$\begin{aligned} \ell^a \rightarrow \hat{\ell}^a &= \ell^a \\ m^a \rightarrow \hat{m}^a &= m^a + a\ell^a \\ \bar{m}^a \rightarrow \hat{\bar{m}}^a &= \bar{m}^a + \bar{a}\ell^a \\ n^a \rightarrow \hat{n}^a &= n^a + a\bar{m}^a + \bar{a}m^a + a\bar{a}\ell^a \quad , \end{aligned} \tag{VII.24}$$

$$\begin{aligned} \ell^a \rightarrow \hat{\ell}^a &= A^{-1}\,\ell^a \\ m^a \rightarrow \hat{m}^a &= e^{i\phi}\,m^a \\ \bar{m}^a \rightarrow \hat{\bar{m}}^a &= e^{-i\phi}\,\bar{m}^a \\ n^a \rightarrow \hat{n}^a &= A\,n^a \quad , \end{aligned} \tag{VII.25}$$

$$\begin{aligned} \ell^a \rightarrow \ell^a &= \ell^a + b\bar{m}^a + \bar{b}m^a + b\bar{b}\,n^a \\ m^a \rightarrow \hat{m}^a &= m^a + b\,n^a \end{aligned}$$

$$\bar{m}^a \longrightarrow \hat{\bar{m}}^a = \bar{m}^a + \bar{b}\, n^a \qquad \text{(VII.26)}$$

$$n^a \longrightarrow \hat{n}^a = n^a \quad .$$

The transformation VII.24 is called a null rotation about ℓ^a; in it, a is an arbitrary complex function. The transformation VII.25 represents a "boost" in the ℓ^a - n^a plane and a rotation in the m^a - $\bar{m}^a$ plane; A and ϕ are arbitrary real functions. Finally, the transformation VII.25 is a null rotation about n^a, with b an arbitrary complex function. It is a straightforward matter to check that these transformations preserve the reality conditions of the tetrad and the normalization VII.11.

It is now necessary to determine how the various NP quantities transform under these Lorentz transformation. This has been done by Kinnersley [26], and the results are as follows.

Under VII.24, we have:

$$\Psi_0 \rightarrow \Psi_0 = \Psi_0$$

$$\Psi_1 \longrightarrow \hat{\Psi}_1 = \Psi_1 + \bar{a}\,\Psi_0$$

$$\Psi_2 \longrightarrow \hat{\Psi}_2 = \Psi_2 + 2\bar{a}\,\Psi_1 + \bar{a}^2\,\Psi_0$$

$$\Psi_3 \longrightarrow \hat{\Psi}_3 = \Psi_3 + 3\bar{a}\,\Psi_2 + 3\bar{a}^2\,\Psi_1 + \bar{a}^3\,\Psi_0$$

$$\Psi_4 \longrightarrow \hat{\Psi}_4 = \Psi_4 + 4\bar{a}\,\Psi_3 + 6\bar{a}^2\,\Psi_2 + 4\bar{a}^3\,\Psi_1 + \bar{a}^4\,\Psi_0$$

$$\Phi_0 \longrightarrow \hat{\Phi}_0 = \Phi_0$$

$$\Phi_1 \longrightarrow \hat{\Phi}_1 = \Phi_1 + \bar{a}\,\Phi_0$$

$$\Phi_2 \longrightarrow \hat{\Phi}_2 = \Phi_2 + 2\bar{a}\,\Phi_1 + \bar{a}^2\,\Phi_0$$

$$
\begin{aligned}
D \longrightarrow \hat{D} &= D \\
\delta \longrightarrow \hat{\delta} &= \delta + a\, D \\
\bar{\delta} \longrightarrow \hat{\bar{\delta}} &= \bar{\delta} + \bar{a}\, D \\
\Delta \longrightarrow \hat{\Delta} &= \Delta + a\,\bar{\delta} + \bar{a}\,\delta + a\bar{a}\, D
\end{aligned}
\qquad \text{(VII.27)}
$$

$$
\begin{aligned}
\alpha \longrightarrow \hat{\alpha} &= \alpha + \bar{a}(\rho + \epsilon) + \bar{a}^2 \kappa \\
\beta \longrightarrow \hat{\beta} &= \beta + a\,\bar{\epsilon} + \bar{a}\sigma + a\bar{a}\,\kappa \\
\gamma \longrightarrow \hat{\gamma} &= \gamma + a\alpha + \bar{a}(\beta + \tau) + a\bar{a}(\rho + \epsilon) \\
&\quad + \bar{a}^2\sigma + a\bar{a}^2\kappa \\
\epsilon \longrightarrow \hat{\epsilon} &= \epsilon + \bar{a}\,\kappa \\
\kappa \longrightarrow \hat{\kappa} &= \kappa \\
\lambda \longrightarrow \hat{\lambda} &= \lambda + \bar{a}(\pi + 2\alpha) + \bar{a}^2(\rho + 2\epsilon) \\
&\quad + \bar{a}^3\kappa + \overline{\delta a} + \bar{a}\, D\,\bar{a} \\
\mu \longrightarrow \hat{\mu} &= \mu + a\pi + 2\bar{a}\beta + 2a\bar{a}\,\epsilon \\
&\quad + \bar{a}^2\sigma + a\bar{a}^2\kappa + \delta\bar{a} + a\, D\,\bar{a} \\
\nu \longrightarrow \hat{\nu} &= \nu + a\lambda + \bar{a}(\mu + 2\gamma) + a\bar{a}(2\alpha + \pi) \\
&\quad + \bar{a}^2(\tau + 2\beta) + a\bar{a}^2(\rho + 2\epsilon) + \bar{a}^3\sigma \\
&\quad + a\bar{a}^3\kappa + \Delta\,\bar{a} + a\,\bar{\delta}\,\bar{a} + \bar{a}\,\delta\,\bar{a} \\
&\quad + a\bar{a}\, D\, a
\end{aligned}
$$

$$\pi \longrightarrow \hat{\pi} = \pi + 2\bar{a}\epsilon + \bar{a}^2\kappa + D\,\bar{a}$$

$$\rho \longrightarrow \hat{\rho} = \rho + \bar{a}\,\kappa$$

$$\sigma \longrightarrow \hat{\sigma} = \sigma + a\,\kappa$$

$$\tau \longrightarrow \hat{\tau} = \tau + a\rho + \bar{a}\sigma + a\bar{a}\,\kappa$$

Under the transformation VII.25, the so-called "spin" $(\phi \neq 0)$ and "boost" $(A \neq 0)$ transformations, we have:

$$\Psi_0 \longrightarrow \hat{\Psi}_0 = A^{-2}\, e^{2i\phi}\Psi_0$$

$$\Psi_1 \longrightarrow \hat{\Psi}_1 = A^{-1}\, e^{i\phi}\Psi_1$$

$$\Psi_2 \longrightarrow \hat{\Psi}_2 = \Psi_2$$

$$\Psi_3 \longrightarrow \hat{\Psi}_3 = A\, e^{-i\phi}\Psi_3$$

$$\Psi_4 \longrightarrow \hat{\Psi}_4 = A^2\, e^{-2i\phi}\Psi_4$$

$$\Phi_0 \longrightarrow \hat{\Phi}_0 = A^{-1}\, e^{i\phi}\,\Phi_0$$

$$\Phi_1 \longrightarrow \hat{\Phi}_1 = \Phi_1$$

$$\Phi_2 \longrightarrow \hat{\Phi}_2 = A\, e^{-i\phi}\,\Phi_2$$

$$D \longrightarrow \hat{D} = A^{-1}\, D$$

$$\delta \longrightarrow \hat{\delta} = e^{i\phi}\,\delta \qquad \text{(VII.28)}$$

$$\bar{\delta} \longrightarrow \hat{\bar{\delta}} = e^{-i\phi}\,\bar{\delta}$$

$$\Delta \longrightarrow \hat{\Delta} = A\,\Delta$$

$$\alpha \longrightarrow \hat{\alpha} = e^{-i\phi}\alpha - \frac{1}{2} A^{-1} e^{-i\phi} \bar{\delta} A + \frac{1}{2} i e^{-i\phi} \bar{\delta}\phi$$

$$\beta \longrightarrow \hat{\beta} = e^{i\phi}\beta - \frac{1}{2} A^{-1} e^{i\phi} \delta A + \frac{1}{2} i e^{i\phi} \delta\phi$$

$$\gamma \longrightarrow \hat{\gamma} = A\gamma - \frac{1}{2} \Delta A + \frac{1}{2} i A\Delta\phi$$

$$\epsilon \longrightarrow \hat{\epsilon} = A^{-1}\epsilon - \frac{1}{2} A^{-2} D A + \frac{1}{2} i A^{-1} D\phi$$

$$\kappa \longrightarrow \hat{\kappa} = A^{-2} e^{i\phi} \kappa$$

$$\lambda \longrightarrow \hat{\lambda} = A e^{-2i\phi} \lambda$$

$$\mu \longrightarrow \hat{\mu} = A\mu$$

$$\nu \longrightarrow \hat{\nu} = A^{2} e^{-i\phi} \nu$$

$$\pi \longrightarrow \hat{\pi} = e^{-i\phi} \pi$$

$$\rho \longrightarrow \hat{\rho} = A^{-1} \rho$$

$$\sigma \longrightarrow \hat{\sigma} = A^{-1} e^{2i\phi} \sigma$$

$$\tau \longrightarrow \hat{\tau} = e^{i\phi} \tau$$

And finally, under VII.26, we have:

$$\Psi_0 \longrightarrow \hat{\Psi}_0 = \Psi_0 + 4b\Psi_1 + 6b^2\Psi_2 + 4b^3\Psi_3 + b^4\Psi_4$$

$$\Psi_1 \longrightarrow \hat{\Psi}_1 = \Psi_1 + 3b\Psi_2 + 3b^2\Psi_3 + b^3\Psi_4$$

$$\Psi_2 \longrightarrow \hat{\Psi}_2 = \Psi_2 + 2b\Psi_3 + b^2\Psi_4$$

$$\Psi_3 \longrightarrow \hat{\Psi}_3 = \Psi_3 + b\Psi_4$$

$$\Psi_4 \longrightarrow \hat{\Psi}_4 = \Psi_4$$

$$\Phi_0 \rightarrow \hat{\Phi}_0 = \Phi_0 + 2b\Phi_1 + b^2\Phi_2$$

$$\Phi_1 \rightarrow \hat{\Phi}_1 = \Phi_1 + b\Phi_2$$

$$\Phi_2 \rightarrow \hat{\Phi}_2 = \Phi_2$$

$$D \rightarrow \hat{D} = D + b\bar{\delta} + \bar{b}\delta + b\bar{b}\Delta$$

$$\delta \rightarrow \hat{\delta} = \delta + b\Delta \qquad \text{(VII.29)}$$

$$\bar{\delta} \rightarrow \hat{\bar{\delta}} = \bar{\delta} + \bar{b}\Delta$$

$$\Delta \rightarrow \hat{\Delta} = \Delta$$

$$\alpha \rightarrow \hat{\alpha} = \alpha + \bar{b}\gamma + b\lambda + b\bar{b}\nu$$

$$\beta \rightarrow \hat{\beta} = \beta + b(\mu + \gamma) + b^2\nu$$

$$\gamma \rightarrow \hat{\gamma} = \gamma + b\nu$$

$$\epsilon \rightarrow \hat{\epsilon} = \epsilon + \bar{b}\beta + b(\alpha + \pi) + b\bar{b}(\mu + \gamma)$$
$$+ b^2\lambda + b^2\bar{b}\nu$$

$$\kappa \rightarrow \hat{\kappa} = \kappa + \bar{b}\sigma + b(\rho + 2\epsilon) + b\bar{b}(2\beta + \tau)$$
$$+ b^2(\pi + 2\alpha) + b^2\bar{b}(\mu + 2\gamma)$$
$$+ b^3\lambda + b^3\bar{b}\nu - D\,b - \bar{b}\delta\,b$$
$$- b\bar{\delta}\,b - b\bar{b}\Delta\,b$$

$$\lambda \rightarrow \hat{\lambda} = \lambda + \bar{b}\nu$$

$$\mu \rightarrow \hat{\mu} = \mu + b\nu$$

$$\nu \rightarrow \hat{\nu} = \nu$$

$$\pi \rightarrow \hat{\pi} = \pi + \bar{b}\mu + b\lambda + b\bar{b}\nu$$

$$\rho \rightarrow \hat{\rho} = \rho + \bar{b}\tau + 2b\alpha + 2b\bar{b}\gamma + b^2\lambda + b^2\bar{b}\nu - \bar{\delta} b - \bar{b}\Delta b$$

$$\sigma \rightarrow \hat{\sigma} = \sigma + b(\tau + 2\beta) + b^2(\mu + 2\gamma) + b^3\nu - \delta b - b\Delta b$$

$$\tau \rightarrow \hat{\tau} = \tau + 2b\gamma + b^2\nu - \Delta b$$

It can be shown that the NP equations VII.20, the Bianchi identities VII.22, the commutators VII.23, and Maxwell's equations VII.19 are all covariant with respect to these proper Lorentz transformations (i.e., they all go into linear combinations of themselves under such a transformation).

In addition to these proper Lorentz transformations, there are three interesting "improper" transformations under which the NP equations are covariant. The first two of them are real improper Lorentz transformations (reflections in the $\ell^a - n^a$ plane and the $m^a - \bar{m}^a$ plane, respectively), while the third (which was first noted by Sachs [20]) will be shown later to be an improper complex Lorentz transformation.

These transformations are:

$$\ell^a \rightarrow \hat{\ell}^a = \ell^a$$

$$m^a \rightarrow \hat{m}^a = \bar{m}^a$$
$$\bar{m}^a \rightarrow \hat{\bar{m}}^a = m^a \qquad \text{(VII.30)}$$
$$n^a \rightarrow \hat{n}^a = n^a$$

$$\ell^a \rightarrow \hat{\ell}^a = n^a$$
$$m^a \rightarrow \hat{m}^a = m^a \qquad \text{(VII.31)}$$
$$\bar{m}^a \rightarrow \hat{\bar{m}}^a = \bar{m}^a$$
$$n^a \rightarrow \hat{n}^a = \ell^a$$

$$\ell^a \rightarrow \hat{\ell}^a = m^a$$
$$m^a \rightarrow \hat{m}^a = -\ell^a \qquad \text{(VII.32)}$$
$$\bar{m}^a \rightarrow \hat{\bar{m}}^a = n^a$$
$$n^a \rightarrow \hat{n}^a = -\bar{m}^a$$

Under VII.30 we have:

$$\Psi_0 \rightarrow \hat{\Psi}_0 = \bar{\Psi}_0$$
$$\Psi_1 \rightarrow \hat{\bar{\Psi}}_1 = \bar{\Psi}_1 \qquad \text{(VII.33)}$$
$$\ldots \qquad ,$$

i.e., all NP variables go into their complex conjugates.

Under VII.31 the transformations are:

$$
\begin{aligned}
\Psi_0 &\rightarrow \hat{\Psi}_0 = \bar{\Psi}_4 \\
\Psi_1 &\rightarrow \hat{\Psi}_1 = \bar{\Psi}_3 \\
\Psi_2 &\rightarrow \hat{\Psi}_2 = \bar{\Psi}_2 \\
\Psi_3 &\rightarrow \hat{\Psi}_3 = \bar{\Psi}_1 \\
\Psi_4 &\rightarrow \hat{\Psi}_4 = \bar{\Psi}_0 \\
\Phi_0 &\rightarrow \hat{\Phi}_0 = \overline{-\Phi_2} \\
\Phi_1 &\rightarrow \hat{\Phi}_1 = \overline{-\Phi_1} \\
\Phi_2 &\rightarrow \hat{\Phi}_2 = \overline{-\Phi_0} \\
D &\rightarrow \hat{D} = \Delta \\
\delta &\rightarrow \hat{\delta} = \delta \\
\bar{\delta} &\rightarrow \hat{\bar{\delta}} = \bar{\delta} \\
\Delta &\rightarrow \hat{\Delta} = D \\
\alpha &\rightarrow \hat{\alpha} = -\bar{\beta} \\
\beta &\rightarrow \hat{\beta} = -\bar{\alpha} \\
\gamma &\rightarrow \hat{\gamma} = -\bar{\epsilon} \\
\epsilon &\rightarrow \hat{\epsilon} = -\bar{\gamma} \\
\kappa &\rightarrow \hat{\kappa} = -\bar{\nu}
\end{aligned}
\tag{VII.34}
$$

$$\lambda \rightarrow \hat{\lambda} = -\bar{\sigma}$$

$$\mu \rightarrow \hat{\mu} = -\bar{\rho}$$

$$\nu \rightarrow \hat{\nu} = -\bar{\kappa}$$

$$\pi \rightarrow \hat{\pi} = -\bar{\tau}$$

$$\rho \rightarrow \hat{\rho} = -\bar{\mu}$$

$$\sigma \rightarrow \hat{\sigma} = -\bar{\lambda}$$

$$\tau \rightarrow \hat{\tau} = -\bar{\pi}$$

Under the Sachs transformation VII.32, the reality conditions of the null tetrad are violated. Therefore, it is necessary to separately list the transformation of the complex conjugate quantities. We have

$$\Psi_0 \rightarrow \hat{\Psi}_0 = \Psi_0 \qquad \bar{\Psi}_0 \rightarrow \hat{\bar{\Psi}}_0 = \bar{\Psi}_4$$

$$\Psi_1 \rightarrow \hat{\Psi}_1 = \Psi_1 \qquad \bar{\Psi}_1 \rightarrow \hat{\bar{\Psi}}_1 = -\bar{\Psi}_3$$

$$\Psi_2 \rightarrow \hat{\Psi}_2 = \Psi_2 \qquad \bar{\Psi}_2 \rightarrow \hat{\bar{\Psi}}_2 = \bar{\Psi}_2$$

$$\Psi_3 \rightarrow \hat{\Psi}_3 = \Psi_3 \qquad \bar{\Psi}_3 \rightarrow \hat{\bar{\Psi}}_3 = -\bar{\Psi}_1$$

$$\Psi_4 \rightarrow \hat{\Psi}_4 = \Psi_4 \qquad \bar{\Psi}_4 \rightarrow \hat{\bar{\Psi}}_4 = \bar{\Psi}_0$$

$$\Phi_0 \rightarrow \hat{\Phi}_0 = \Phi_0 \qquad \bar{\Phi}_0 \rightarrow \hat{\bar{\Phi}}_0 = \bar{\Phi}_2$$

$$\Phi_1 \rightarrow \hat{\Phi}_1 = \Phi_1 \qquad \bar{\Phi}_1 \rightarrow \hat{\bar{\Phi}}_1 = -\bar{\Phi}_1$$

$$\Phi_2 \rightarrow \hat{\Phi}_2 = \Phi_2 \qquad \bar{\Phi}_2 \rightarrow \hat{\bar{\Phi}}_2 = \bar{\Phi}_0$$

$$D \longrightarrow \hat{D} = \delta$$

$$\delta \longrightarrow \hat{\delta} = -D$$

$$\bar{\delta} \longrightarrow \bar{\delta} = \Delta \qquad \text{(VII.35)}$$

$$\Delta \longrightarrow \hat{\Delta} = -\bar{\delta}$$

$$\alpha \longrightarrow \hat{\alpha} = \gamma \qquad \bar{\alpha} \longrightarrow \hat{\bar{\alpha}} = \bar{\epsilon}$$

$$\beta \longrightarrow \hat{\beta} = -\epsilon \qquad \bar{\beta} \longrightarrow \hat{\bar{\beta}} = -\bar{\gamma}$$

$$\gamma \longrightarrow \hat{\gamma} = -\alpha \qquad \bar{\gamma} \longrightarrow \hat{\bar{\gamma}} = \bar{\beta}$$

$$\epsilon \longrightarrow \hat{\epsilon} = \beta \qquad \bar{\epsilon} \longrightarrow \hat{\bar{\epsilon}} = -\bar{\alpha}$$

$$\kappa \longrightarrow \hat{\kappa} = \sigma \qquad \bar{\kappa} \longrightarrow \hat{\bar{\kappa}} = \bar{\lambda}$$

$$\lambda \longrightarrow \hat{\lambda} = \nu \qquad \bar{\lambda} \longrightarrow \hat{\bar{\lambda}} = -\bar{\kappa}$$

$$\mu \longrightarrow \hat{\mu} = -\pi \qquad \bar{\mu} \longrightarrow \hat{\bar{\mu}} = \bar{\tau}$$

$$\nu \longrightarrow \hat{\nu} = -\lambda \qquad \bar{\nu} \longrightarrow \hat{\bar{\nu}} = -\bar{\sigma}$$

$$\pi \longrightarrow \pi = \mu \qquad \bar{\pi} \longrightarrow \hat{\bar{\pi}} = \bar{\rho}$$

$$\rho \longrightarrow \hat{\rho} = \tau \qquad \bar{\rho} \longrightarrow \hat{\bar{\rho}} = -\bar{\pi}$$

$$\sigma \longrightarrow \hat{\sigma} = -\kappa \qquad \bar{\sigma} \longrightarrow \hat{\bar{\sigma}} = \bar{\nu}$$

$$\tau \longrightarrow \hat{\tau} = -\rho \qquad \bar{\tau} \longrightarrow \hat{\bar{\tau}} = -\bar{\mu}$$

Geometrical Meaning of the Spin Coefficients

Most of the spin coefficients can be given a useful geometrical interpretation. For example, a straightforward calculation shows that

$$\ell_{a;b}\,\ell^{b} = -\kappa\,\bar{m}_a - \bar{\kappa}\,m_a + (\epsilon + \bar{\epsilon})\,\ell_a \quad . \qquad \text{(VII.36)}$$

Then if $\kappa = 0$, we have $\ell_{a;b}\,\ell^{b} \propto \ell_a$, so that the curves of the ℓ^a-congruence (curves with ℓ^a as tangent vector) are geodesics. Furthermore, there always exists a tetrad transformation of the form VII.25 which takes $\epsilon + \bar{\epsilon} \rightarrow \hat{\epsilon} + \hat{\bar{\epsilon}} = 0$ and $\kappa = 0 \rightarrow \hat{\kappa} = 0$ (compare equations VII.28). In that case we would have $\hat{\ell}_{a;b}\,\hat{\ell}^{b} = 0$, so that then the ℓ^a-congruence would be affinely parametrized, that is, the parameter y for which $\hat{\ell}^a = \frac{dx^a}{dy}$ would be an affine parameter.

Newman and Penrose [37] have shown that, when $\kappa = \epsilon + \bar{\epsilon} = 0$, the spin coefficients ρ and σ are related to the optical parameters defined by Sachs [47] in the following way:

$$\rho = \frac{1}{2}\left[-\,\mathrm{div}\,\ell^a + i\,\mathrm{curl}\,\ell_a\right]$$

$$\sigma\bar{\sigma} = |\,\mathrm{shear}\,\ell^a|^2 \quad , \qquad \text{(VII.37)}$$

where

$$\mathrm{div}\,\ell^a = \ell^a{}_{;a}$$

$$\mathrm{curl}\,\ell^a = \left(\ell_{[a;b]}\,\ell^{a;b}\right)^{1/2} \qquad \text{(VII.38)}$$

$$|\,\mathrm{shear}\,\ell^a|^2 = \frac{1}{2}\left[\ell_{(a;b)}\,\ell^{a;b} - \frac{1}{2}\,(\ell^a{}_{;a})^2\right] .$$

To see the geometrical meaning of these quantities, we write $m^a = a^a + ib^a$, where a^a and b^a are two spacelike orthogonal unit vectors. Then we have

$$\begin{aligned} \rho &= \ell_{a;b}\, m^a\, \bar{m}^b \\ &= \ell_{a;b}\,(a^a + i\, b^a)\,(a^b - i\, b^b) \qquad \text{(VII.39)} \\ &= \ell_{a;b}\,(a^a\, a^b + b^a\, b^b) + i\,\ell_{a;b}\,(b^a\, a^b - a^a\, b^b)\ . \end{aligned}$$

We interpret the real part of ρ as "the change in ℓ^a moving along a^b, projected onto a^a, plus the change in ℓ^a moving along b^b, projected onto b^a." Similarly, we interpret the imaginary part of ρ as "the change in ℓ^a moving along a^b, projected onto b^a, plus the change in ℓ^a moving along b^b, projected onto a^a." These interpretations are conveniently represented by the following diagram:

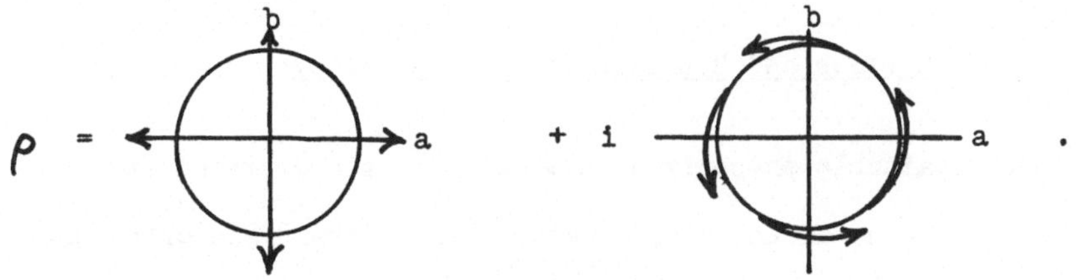

Diagram VII.1

In the diagram, the plane of the paper is the $a^a - b^a$ plane, and the ℓ^a-congruence should be thought of as coming out of the page. Even when κ and $\epsilon + \bar{\epsilon}$ are not equal to zero, we refer to $\rho + \bar{\rho}$ as the divergence (expansion, minus the convergence) of the ℓ^a congruence and $\rho - \bar{\rho}$ as the twist (curl, rotation, drill) of ℓ^a.

In a similar manner we have the following diagram for σ :

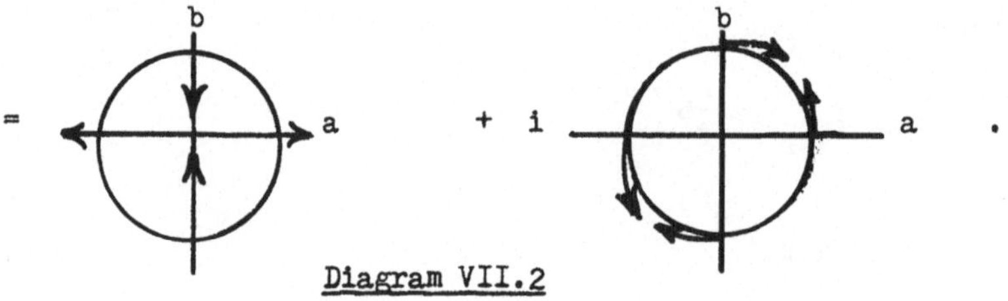

Diagram VII.2

Thus σ (even when κ, $\epsilon + \bar{\epsilon} \neq 0$) measures two independent shearing modes of the ℓ^a-congruence. We refer to σ as the shear of ℓ^a.

Lastly, the spin coefficient $\tau = \ell_{a;b}\, m^a\, n^b$ measures "the change in ℓ^a moving along n^b, projected onto m^a."

All of this has been concerning the ℓ^a-congruence; the quantities which describe the corresponding properties of the n^a congruence are ν, $\gamma + \bar{\gamma}$, $-\mu$, $-\lambda$, and π (corresponding to κ, $\epsilon + \bar{\epsilon}$, ρ, σ, and τ).

Surface and Hypersurface Orthogonality

Another valuable property of some of the spin coefficients is that they indicate when one of the tetrad vectors is hypersurface orthogonal. The Frobenius theorem [50] states that a vector ξ_a is hypersurface orthogonal if and only if $\xi_{[a;b}\xi_{c]} = 0$. Taking the tetrad components of this expression, with ξ_a one of the tetrad legs, yields the desired conditions. In fact, we have:

$$\left.\begin{array}{l} \ell^a \\ m^a \\ \bar{m}^a \\ n^a \end{array}\right\} \text{ is hypersurface orthogonal iff } \left\{\begin{array}{l} \kappa = \rho - \bar{\rho} = 0 \\ \sigma = -\bar{\lambda} = \bar{\pi} + \tau = 0 \\ -\bar{\sigma} = \lambda = \pi + \bar{\tau} = 0 \\ \nu = \mu - \bar{\mu} = 0 \end{array}\right. \qquad \text{(VII.40a)}$$

For the stronger requirement that ξ_a is equal to a gradient, we must impose the condition $\xi_{[a;b]} = 0$. We obtain:

$$\left.\begin{array}{l} \ell^a \\ \\ m^a \\ \\ \bar{m}^a \\ \\ n^a \end{array}\right\} \text{ is equal to a gradient iff } \left\{\begin{array}{l} \epsilon + \bar{\epsilon} = \bar{\alpha} + \beta - \tau = \kappa \\ \qquad = \rho - \bar{\rho} = 0 \\ \bar{\pi} + \tau = \sigma = -\rho + \epsilon - \bar{\epsilon} \\ \qquad = \bar{\alpha} - \beta = \lambda = \gamma - \bar{\gamma} + \bar{\mu} = 0 \\ \pi + \bar{\tau} = \sigma = \bar{\rho} + \epsilon - \bar{\epsilon} \\ \qquad = \alpha - \bar{\beta} = \lambda = \bar{\gamma} - \gamma + \mu = 0 \\ \gamma + \bar{\gamma} = \bar{\alpha} + \beta + \bar{\pi} = \nu \\ \qquad = \mu - \bar{\mu} = 0 \end{array}\right. \qquad \text{(VII.40b)}$$

Note that if m^a or $\bar{m}^a$ is hypersurface orthogonal, the hypersurface must be a "complex" hypersurface, and if one of them is equal to the gradient of a scalar field, this scalar field must be complex. We do not try to give a precise physical meaning to these concepts just yet, but for the time being we agree to use this nomenclature in the obvious formal sense.

In a similar manner, using the Frobenius theorem, it is easy to establish the criteria for surface-orthogonality of the various bivectors formed out of the tetrad legs. In this case the Frobenius Theorem states that a bivector B_{ab} is surface orthogonal if and only if $B_{[ab,c}\, B_{d]e} = 0$. From this, we have:

$$\left.\begin{array}{l} \ell_{[a}n_{b]} \\ m_{[a}\bar{m}_{b]} \\ \ell_{[a}m_{b]} \\ n_{[a}m_{b]} \\ \ell_{[a}\bar{m}_{b]} \\ n_{[a}\bar{m}_{b]} \end{array}\right\} \text{is surface orthogonal iff} \left\{\begin{array}{l} \rho - \bar{\rho} = \mu - \bar{\mu} = 0 \\ \pi + \bar{\tau} = \bar{\pi} + \tau = 0 \\ \kappa = \sigma = 0 \\ \bar{\nu} = \bar{\lambda} = 0 \\ \bar{\kappa} = \bar{\sigma} = 0 \\ \nu = \lambda = 0 \end{array}\right. \qquad \text{(VII.41)}$$

Again, for the complex bivectors, we agree to use the nomenclature "surface orthogonal" in at least a formal sense.

Algebraic Classification of Fields

An extremely useful classification of zero rest mass fields is that developed by Petrov [45], Pirani [46], Debever [10], and Penrose [41]. Although the procedure is usually presented in an explicit spinor formalism [41], it is equally transparent in the NP tetrad formalism we are using [24], at least in the special cases of gravitation and electromagnetism, the only cases we shall treat here.

Consider, then, an electromagnetic field described by the variables Φ_i of VII.17. If we perform a transformation of the form VII.26, we will have, according to VII.29:

$$\Phi_0 \rightarrow \hat{\Phi}_0 = \Phi_0 + 2b\Phi_1 + b^2\Phi_2 \qquad \text{(VII.42)}$$

If we set $\hat{\Phi}_0 = 0$ and solve for b, then in general we will find two roots b_1 and b_2 which will define two null directions $\hat{\ell}^a_{(1)}$ and $\hat{\ell}^a_{(2)}$ by virtue of the transformation VII.26 with $b = b_1$ and $b = b_2$ respectively.

The two null directions so defined are called the principal null directions of the electromagnetic field. If $\hat{\Phi}_0 = 0$ has a double root, $b_1 = b_2$, then only one principal null direction is defined by the field. In the case of two distinct roots the field is said to be algebraically general, or of type $\{1,1\}$, while in the case of a double root the field is said to be algebraically special, or of type $\{2\}$, or null.

In a completely analogous fashion, under a transformation VII.26 we will have, by VII.29,

$$\Psi_0 \rightarrow \hat{\Psi}_0 = \Psi_0 + 4b\Psi_1 + 6b^2\Psi_2 + 4b^3\Psi_3 + b^4\Psi_4 \qquad \text{(VII.43)}$$

Now when we set $\hat{\Psi}_0 = 0$, we will in general have four principal null directions defined by the roots b_1, b_2, b_3, b_4. If all of the roots are distinct, the field is algebraically general, or of type $\{1111\}$, while if some of the roots coincide we have an algebraically special field. The various possibilities are described in the table below:

Roots	Description
all distinct	type $\{1111\}$ or type I
two coincident	type $\{211\}$ or type II
three coincident	type $\{31\}$ or type III
four coincident	type $\{4\}$ or type N (null)
coincident in pairs	type $\{22\}$ or type D

Table VII.1

It should be noted from VII.43 and VII.29 that if $\hat{\Psi}_0 = 0$ has a repeated root, then the transformation defined by this root makes $\hat{\Psi}_1$ as well as $\hat{\Psi}_0$ vanish. Therefore an algebraically special Weyl tensor field is characterized by $\hat{\Psi}_0 = \hat{\Psi}_1 = 0$ for a suitable tetrad.

One reason for the importance of the algebraic classification of the Weyl tensor field is contained in the following theorem [21]:

VII.1: THEOREM (GOLDBERG-SACHS): In a vacuum spacetime, the Weyl tensor is algebraically special if and only if there exists a geodesic and shearfree null congruence ℓ^a (i.e., an ℓ^a such that $\kappa = \sigma = 0$). If such an ℓ^a exists, it is a repeated principal null direction of the Weyl tensor.

Proof: The proof is given by Newman and Penrose in [37].

It should be noted that if there are two repeated principal null directions ℓ^a and n^a for the Weyl tensor, then each must satisfy the Goldberg-Sachs theorem, so that $\kappa = \sigma = \nu = \lambda = 0$ in a suitable tetrad, and Ψ_2 is the only non-vanishing component of the Weyl tensor.

For electromagnetism, there is a result analogous to the Goldberg-Sachs theorem [26]:

VII.2: THEOREM (MARIOT-ROBINSON): In a charge-free region of spacetime, the repeated principal null direction of a null electromagnetic field is geodesic and shearfree.

Proof: See [26].

We notice that the Mariot-Robinson theorem is weaker than the Goldberg-Sachs theorem in that its converse is not true. For example, the Coulomb field in Minkowski space has two geodesic and shearfree principal null directions, but the field is algebraically general.

Type D Spacetimes

Among the most important solutions to Einstein's equations are the type D vacuum and electrified solutions. For the type D vacuum spaces, the Goldberg-Sachs theorem applies to give

$$\kappa = \sigma = \nu = \lambda = 0 \tag{VII.44}$$

in any tetrad for which ℓ^a and n^a point in the repeated principal null directions of the Weyl tensor. Kinnersley [26] has shown that VII.44 also holds in the electrified case in which the principal null directions of the (algebraically general) electromagnetic field coincide with those of the Weyl tensor ("electrified" solution). For such a tetrad, the only non-vanishing component of the Weyl tensor for these spacetimes is Ψ_2. These spacetimes enjoy many desirable properties. They all admit at least two Killing vectors [26] and a conformal Killing tensor [23]. In addition, several of these metrics have physically interesting interpretations (for example, the Kerr-Newman solution describes the spacetime of a rotating charged black hole). And as we shall see later, all of these spacetimes admit a rich integrable structure which is a modification of an integrable almost Hermitian structure.

It is a fortunate circumstance that all of type D vacuum and electrified spacetimes have been explicitly presented by Kinnersley [26],[27].

In fact, from his work it is possible to specify a particular type D spacetime almost uniquely in terms of the four spin coefficients ρ, μ, π, and τ, referred to a tetrad in which ℓ^a and n^a point along the two repeated principal null directions (keep in mind that $\kappa = \sigma = \nu = \lambda = 0$ for all of these metrics in such a tetrad). Such a description in terms of ρ, μ, π, and τ is given for the vacuum metrics in the following table (the electrified versions can be described in the same way). In addition, the form of Ψ_2 for each of these metrics is also presented.

In the table, the quantities m, ℓ, and a are real constants, and the quantity x which appears in the last two metrics is a spacelike coordinate used by Kinnersley. The "related metrics" are metrics which differ from the given metric but have the same characterization in terms of ρ, μ, π, and τ (these related metrics differ in the choice of sign for constants of integration).

Here is the table:

Spacetime	Specification		Ψ_2
Schwarzschild (plus two related metrics)	$\rho \neq 0$	$\rho = \bar{\rho}$	$\Psi_2 = m\rho^3$
	$\mu \neq 0$	$\mu = \bar{\mu}$	
	$\pi = 0$		
	$\tau = 0$		
NUT (plus two related metrics)	$\rho \neq 0$	$\rho \neq \bar{\rho}$	$\Psi_2 = (m + i\ell)\rho^3$
	$\mu \neq 0$	$\mu \neq \bar{\mu}$	
	$\pi = 0$		
	$\tau = 0$		

Spacetime	Specification		Ψ_2
Kerr-NUT (plus five related metrics)	$\rho \neq 0$	$\rho \neq \bar{\rho}$	$\Psi_2 = (m + i\ell)\rho^3$
	$\mu \neq 0$	$\mu \neq \bar{\mu}$	
	$\pi \neq 0$	$\pi + \bar{\tau} \neq 0$	
	$\tau \neq 0$		
"C"	$\rho \neq 0$	$\rho = \bar{\rho}$	$\Psi_2 = m\rho^3$
	$\mu \neq 0$	$\mu = \bar{\mu}$	
	$\pi \neq 0$	$\pi + \bar{\tau} = 0$	
	$\tau \neq 0$		
"Twisting C"	$\rho \neq 0$	$\rho \neq \bar{\rho}$	$\Psi_2 = (m + i\ell)(\mathrm{dn}\, x$ $\frac{1}{2} i \sqrt{2}\, \mathrm{sn}\, x)^3 \rho^3$
	$\mu \neq 0$	$\mu \neq \bar{\mu}$	
	$\pi \neq 0$	$\pi + \bar{\tau} \neq 0$	
	$\tau \neq 0$		
"B" (plus two related metrics)	$\rho = 0$		$\Psi_2 = m\, x^{-3}$
	$\mu = 0$		
	$\pi \neq 0$	$\pi + \bar{\tau} = 0$	
	$\tau \neq 0$		
"Rotating B" (plus two related metrics)	$\rho = 0$		$\Psi_2 = (m + i\ell)(x + ia)^{-3}$
	$\mu = 0$		
	$\pi \neq 0$	$\pi + \bar{\tau} \neq 0$	
	$\tau \neq 0$		

Table VII.2

The first thing to check is that these specifications are invariant under tetrad transformations which have the directions of ℓ^a and n^a unchanged. These are just the "spin" and "boost" transformations VII.25, and it is easy to see from VII.28 that these transformations do indeed preserve the specifications $\rho = 0$ or $\rho \neq 0$, $\rho = \bar{\rho}$ or $\rho \neq \bar{\rho}$, and so on. This shows that our specifications are really properties of the geometry itself.

The most striking feature of the table is the fact that the specification of μ parallels the specification of ρ in every case (i.e., $\rho = 0 \Leftrightarrow \mu = 0$, $\rho = \bar{\rho} \Leftrightarrow \mu = \bar{\mu}$). The reason for this is not known.

It should also be noted that the specification serves to uniquely determine a given spacetime (up to "related metrics"), with one exception: The "twisting C" metric is not distinguished from the Kerr-NUT metrics. The reason for this anomaly is not clear. In the tetrads chosen by Kinnersley, the two are distinguished by $\tau = -\bar{\tau}$ (Kerr-NUT) or $\tau \neq -\bar{\tau}$ ("twisting C") but the condition $\tau = -\bar{\tau}$ is not preserved under spin and boost transformations.

Finally, we should examine the limits in which some of these metrics reduce to others. The various relationships are summarized in Diagram VII. below. The way in which the limits arise depends, of course, on the choice of coordinates. If we agree to use the coordinate systems of reference [26], then two types of limit arise. Those limits which are singular, i.e., under which the metric becomes singular, are indicated by dotted lines; the "regular" limits, for which this behavior does not occur in the given coordinates, are indicated by solid lines. The latter type of limit always arises by setting equal to zero one or more of the constants of integration occurring in the metric. The main point of interest in the diagram is the fact that any of the metrics can be reduced to either Schwarzschild or "B" or "C" by taking an appropriate chain of limits of the second type. We shall have reason to return to these matters later when we examine the so-called "complex coordinate transformations" of Newman.

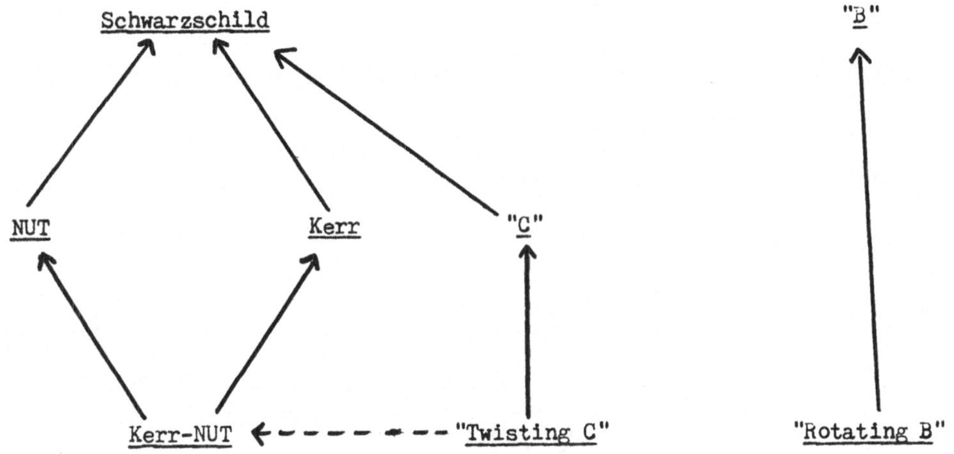

Diagram VII.3

Note: The "related metrics" of Table VII.2 reduce either to the metrics in this diagram, or to one of the corresponding "related metrics."

Kerr-Schild Spacetimes

As another example of algebraically special spacetimes, we shall very briefly look at the Kerr-Schild vacuum spacetimes; for more information, consult [11] and [25].

The Kerr-Schild line element is given by

$$g_{ab} = \eta_{ab} + 2H\, \ell_a\, \ell_b$$
$$g^{ab} = \eta^{ab} - 2H\, \ell^a\, \ell^b \quad , \tag{VII.45}$$

where η_{ab} is the Minkowski metric (choosing signature +2),

$$\eta_{ab} = 2\,du\, dv + 2\,d\zeta\, d\bar{\zeta} \quad , \tag{VII.46}$$

and ℓ^a is a null vector with respect to either η_{ab} or g_{ab}. When the vacuum field equations are imposed, ℓ^a is found to be a repeated principal null direction and thus geodesic and shear-free.

All of the Kerr-Schild spacetimes admit at least one Killing vector. When this Killing vector is timelike, the metric can be presented as

$$ds^2 = 2\,du\, dv + 2\,d\zeta\, d\bar{\zeta}$$
$$- 4\sqrt{2}\;\; m \;\; \mathrm{Re}\left(\frac{1}{F_{,Y}}\right)\left[\frac{du + Y\, d\bar{\zeta} + \bar{Y}\, d\zeta - Y\bar{Y}\, dv}{1 + Y\bar{Y}}\right]^2 \tag{VII.47}$$

where m is a real constant, Y is a complex variable, and $F_{,Y} = \partial F/\partial Y$, where

$$F(Y, \zeta, \bar{\zeta}, u + v) = \phi(Y) + \left[Y^2\, \bar{\zeta} - \zeta + (u + v)Y\right] \tag{VII.48}$$

where $\phi(Y)$ is an arbitrary analytic function of Y. In terms of the coordinates u, v, ζ, and $\bar{\zeta}$, Y is given implicitly by

$$\phi(Y) = - Y^2\, \bar{\zeta} + \zeta - (u + v)Y \quad . \tag{VII.49}$$

Similar results hold in the case of a spacelike or null Killing vector, and electrified versions can also be obtained.

The only type D vacuum solutions which are known to be in the Kerr-Schild class are Schwarzschild and Kerr. The Schwarzschild metric is generated by choosing $\phi(Y) = 0$, and Kerr is generated by choosing $\phi(Y) = -\sqrt{2}\, ia\, Y$.

Conformal Geometry

A conformal rescaling (not to be confused with an element of the group of conformal transformations [44]) is a replacement of the metric g_{ab} of spacetime by a "conformally related" metric $\hat{g}_{ab}$, where

$$\hat{g}_{ab} = \Omega^2 g_{ab} ,$$
$$\hat{g}^{ab} = \Omega^{-2} g^{ab} . \qquad \text{(VII.50)}$$

Here Ω is a function which is ordinarily taken to be real and positive.

A set of equations is said to be conformally invariant if for every variable ϕ in the equations, a transformation of the form $\phi \rightarrow \hat{\phi} = \Omega^{\rho} \phi$ for some ρ together with the transformation VII.50 leaves the equations invariant in form. The zero rest mass free field equations for any spin (including Maxwell's equations and the linearized Einstein equations) are in fact conformally invariant, but the full Einstein equations are not.

We recall Theorem VI.34, which states that the condition for two metrics g_{ab} and $\hat{g}_{ab}$ to be conformally related is $C^a{}_{bcd} = \hat{C}^a{}_{bcd}$, where $C^a{}_{bcd}$ is the Weyl conformal curvature tensor. In particular, $\hat{g}_{ab}$ is conformally related to a flat metric η_{ab} (is "conformally flat") if and only if $\hat{C}^a{}_{bcd} = 0$.

We now consider how the NP variables are transformed under VII.50. Several choices could be made for the scaling of the tetrad legs, but we shall choose the following:

$$
\begin{aligned}
\ell^a &\rightarrow \hat{\ell}^a = \Omega^{-1}\ell^a & \qquad \ell_a &\rightarrow \hat{\ell}_a = \Omega\,\ell_a \\
m^a &\rightarrow \hat{m}^a = \Omega^{-1}\,m^a & \qquad m_a &\rightarrow \hat{m}_a = \Omega\,m_a \\
\bar{m}^a &\rightarrow \hat{\bar{m}}^a = \Omega^{-1}\,\bar{m}^a & \qquad \bar{m}_a &\rightarrow \hat{\bar{m}}_a = \Omega\,\bar{m}_a \\
n^a &\rightarrow \hat{n}^a = \Omega^{-1}\,n^a & \qquad n_a &\rightarrow \hat{n}_a = \Omega\,n_a
\end{aligned}
\qquad \text{(VII.51)}
$$

Then a straightforward computation shows that the rest of the NP variables transform as follows:

$$
\begin{aligned}
&\Psi_i \rightarrow \hat{\Psi}_i = \Omega^{-2}\Psi_i\,, \qquad i = 0, 1, 2, 3, 4 \\
&\Phi_j \rightarrow \hat{\Phi}_j = \Omega^{-2}\,\Phi_j\,, \qquad j = 0, 1, 2 \\
&D \rightarrow \hat{D} = \Omega^{-1}\,D \\
&\delta \rightarrow \hat{\delta} = \Omega^{-1}\,\delta \\
&\bar{\delta} \rightarrow \hat{\bar{\delta}} = \Omega^{-1}\,\hat{\delta} \\
&\Delta \rightarrow \hat{\Delta} = \Omega^{-1}\,\Delta \\
&\alpha \rightarrow \hat{\alpha} = \Omega^{-1}\alpha - \tfrac{1}{2}\,\Omega^{-2}\,\bar{\delta}\Omega \\
&\beta \rightarrow \hat{\beta} = \Omega^{-1}\,\beta + \tfrac{1}{2}\,\Omega^{-2}\,\delta\,\Omega \\
&\gamma \rightarrow \hat{\gamma} = \Omega^{-1}\gamma - \tfrac{1}{2}\,\Omega^{-2}\,\Delta\Omega
\end{aligned}
\qquad \text{(VII.52)}
$$

$$\epsilon \to \hat{\epsilon} = \Omega^{-1}\epsilon + \frac{1}{2}\Omega^{-2} D\Omega$$

$$\kappa \to \hat{\kappa} = \Omega^{-1}\kappa$$

$$\lambda \to \hat{\lambda} = \Omega^{-1}\lambda$$

$$\mu \to \mu = \Omega^{-1}\mu + \Omega^{-2}\Delta\Omega$$

$$\nu \to \hat{\nu} = \Omega^{-1}\nu$$

$$\pi \to \hat{\pi} = \Omega^{-1}\pi + \Omega^{-2}\bar{\delta}\Omega$$

$$\rho \to \hat{\rho} = \Omega^{-1}\rho - \Omega^{-2} D\Omega$$

$$\sigma \to \hat{\sigma} = \Omega^{-1}\sigma$$

$$\tau \to \hat{\tau} = \Omega^{-1}\tau - \Omega^{-2}\delta\Omega$$

The transformation laws for Φ_{ij} and Λ are very complicated in their NP form. It will be much more helpful to consider instead the transformation of R_{ab} and R. This is given by Penrose [43] as follows:

$$\hat{S}^a{}_b = \Omega^2 S^a{}_b + 2\Omega\Omega_;{}^a{}_b - \delta^a{}_b\,\Omega_{;s}\,\Omega_;{}^s \qquad \text{(VII.53)}$$

where $S^a{}_b$ is defined by

$$S_{ab} = \frac{1}{6} R\, g_{ab} - R_{ab} \ . \qquad \text{(VII.54)}$$

Constants of Motion

As mentioned above, the paths of test particles in a spacetime are given by the (timelike or null) geodesics of the spacetime. If $k^a = dx^a/dy$

is the tangent vector to a geodesic $x^a = x^a(y)$ with affine parameter y, then the geodesic equation reads

$$k^b k_{a;b} = 0 . \tag{VII.55}$$

A constant of motion along geodesics is a scalar ϕ such that

$$k^b \phi_{;b} = 0 , \tag{VII.56}$$

i.e., ϕ is constant along the geodesic.

A Killing tensor [49] is a completely symmetric tensor $K_{ab\ldots d} = K_{(ab\ldots d)}$ which satisfies the Killing equation

$$K_{(ab\ldots d;e)} = 0 . \tag{VII.57}$$

Such a tensor gives rise to the constant of motion $\phi = k^a k^b \ldots k^d K_{ab\ldots d}$, because then we have

$$\begin{aligned} k^s \phi_{;s} &= k^s(k^a k^b \ldots k^d K_{ab\ldots d})_{;s} \\ &= k^s k^a{}_{;s} k^b \ldots k^d K_{ab\ldots d} + \ldots \\ &\quad + k^s k^a k^b \ldots k^d{}_{;s} K_{ab\ldots d} \\ &\quad + k^s k^a k^b \ldots k^d K_{ab\ldots d;s} \\ &= 0 + \ldots + 0 + k^s k^a k^b \ldots k^d K_{(ab\ldots d;s)} \\ &= 0 . \end{aligned}$$

A symmetric tensor $K_{ab\ldots d}$ which satisfies

$$K_{(ab\ldots d;s)} = g_{(as} Q_{b\ldots d)} \tag{VII.58}$$

for some symmetric tensor $Q_{b\ldots d}$ is said to be a conformal Killing tensor.

It can be shown [49] that a Killing tensor associated with a metric g_{ab} becomes a conformal Killing tensor for metrics $\hat{g}_{ab}$ conformally related to g_{ab} ; more generally, the property of being a conformal Killing tensor is invariant under the scaling VII.45 if we set $\hat{K}^{ab\ldots d} = K^{ab\ldots d}$ (note the position of the indices).

A conformal Killing tensor gives rise to constants of motion for null geodesics, because for a null geodesic ℓ^a , we have

$$\ell^s(\ell^a \ell^b \ldots \ell^d K_{ab\ldots d})_{;s}$$

$$= 0 + \ldots + 0 + \ell^s \ell^a \ell^b \ldots \ell^d K_{(ab\ldots d;s)}$$

$$= \ell^s \ell^a \ell^b \ldots \ell^d \; g_{(as} Q_{b\ldots d)}$$

$$= 0$$

because ℓ^a is null.

For more information on Killing tensors, see reference [49].

Spinors

The use of spinor methods in general relativity was explored and popularized by Penrose [41]. The NP formalism outlined above can be thought of equally well as a "spinor dyad" formalism rather than a "vector null tetrad" formalism. This is due basically to the fact, first demonstrated by Geroch [18], that a (non-compact) spacetime manifold admits a spinor structure if and only if it admits a global normed null tetrad. However, for some applications, notably twistor theory, the use of spinors is almost inescapable, and hence we present a brief review here.

At the heart of the spinor approach lies the fact that the group SL(2,$\mathbb{C}$) is two-to-one homomorphic to $\mathcal{L}_+^\uparrow$, the group of restricted (proper and orthochronous) Lorentz transformations. To see this, suppose we construct a complex vector bundle S of complex dimension 2 over a spacetime M (see Appendix C). This can always be done uniquely by demanding that S be <u>trivial</u>, i.e.,

$$S = M \times \mathbb{C}^2 .$$

(In the language of Appendix C, a trivial bundle is one for which the transition functions G_{UV} are equal to the identity matrix $I \in GL(2,\mathbb{C})$ for all intersecting charts U, V of M.)

For such a bundle S, the sections

$$o^A(p) = (1,0) \qquad \text{for all } p \in M$$

$$\iota^A(p) = (0,1) \qquad \text{for all } p \in M$$

are global smooth sections of S, and

$$\epsilon^{AB}(p) \equiv \begin{pmatrix} 0 & 1 \\ -1 & 0 \end{pmatrix} \qquad \text{for all } p \in M$$

is a global smooth section of the tensor product bundle $S \otimes S$. Notice that

$$\epsilon^{AB} = o^A \iota^B - \iota^A o^B . \tag{VII.59}$$

The pair (o^A, ι^A) is a <u>spinor dyad</u> for S, since it spans the fibre $\mathbb{C}^2$ at each point $p \in M$. The group of complex linear transformations which preserves Equation VII.59 at a point is easily seen to be $SL(2,\mathbb{C})$. That is, if

$$\begin{pmatrix} o^A \\ \iota^A \end{pmatrix} \longrightarrow \begin{pmatrix} a & b \\ c & d \end{pmatrix} \begin{pmatrix} o^A \\ \iota^A \end{pmatrix} = \begin{pmatrix} a o^A + b \iota^A \\ c o^A + d \iota^A \end{pmatrix} , \tag{VII.60}$$

then

$$o^A \iota^B - \iota^A o^B \rightarrow \begin{pmatrix} 0 & 1 \\ -1 & 0 \end{pmatrix} = \epsilon^{AB} \quad \text{iff} \quad ad - bc = 1 \; .$$

Thus $SL(2,\mathbf{C})$ is a sort of "gauge group" for spinor dyads, which preserves the "normalization condition" Equation VII.59.

Given S , we can immediately construct the dual bundle S^* , the complex conjugate bundle $\bar{S}$, and the dual conjugate bundle $\overline{S^*} = \bar{S}^*$. Defining a section of $S^* \otimes S^*$ by

$$\epsilon_{AB} \equiv \begin{pmatrix} 0 & 0 \\ -1 & 0 \end{pmatrix} ,$$

we have a canonical map $\mu : S \longrightarrow S^*$ given by $\xi^A \rightarrow \xi^A \epsilon_{AB} \equiv \xi_B$. It is crucial to note that $\xi^A \epsilon_{AB} = - \epsilon_{AB} \xi^B$, because of the anti-symmetry of ϵ_{AB} . We have

$$\begin{aligned} \mu(o^A) &\equiv o_A = (0,1) \\ \mu(\iota^A) &\equiv \iota_A = (-1,0) \end{aligned} \qquad \text{(VII.61)}$$

whence

$$o_A \iota_B - \iota_A o_B = \begin{pmatrix} 0 & 1 \\ -1 & 0 \end{pmatrix} \equiv \epsilon_{AB} \; . \qquad \text{(VII.62)}$$

The "normalization condition" VII.59 can now be expressed as

$$\begin{aligned} o^A \iota^B \epsilon_{AB} &\equiv o_B \iota^B = +1 \\ \iota^A o^B \epsilon_{AB} &\equiv \iota_B o^B = -1 \\ o^A o^B \epsilon_{AB} &= o_B o^B = 0 \\ \iota^A \iota^B \epsilon_{AB} &= \iota_B \iota^B = 0 \; , \end{aligned} \qquad \text{(VII.63)}$$

which follow from the antisymmetry of ϵ_{AB} and the relation $\epsilon_{AB}\epsilon^{AB} = +2$. (Notice that for <u>any</u> spinors ξ^A, η^A , we have $\xi^A \eta_A = -\eta^A \xi_A$ and hence $\xi^A \xi_A = 0$.)

We also have canonical maps $\nu : S \longrightarrow \bar{S}$ and $\mu \circ \nu : S \longrightarrow \bar{S}^*$ given by

$$\begin{aligned} \nu(\xi^A) &= \overline{\xi^A} \equiv \bar{\xi}^{A'} \\ \mu\circ\nu(\xi^A) &= \overline{\xi^A \epsilon_{AB}} \equiv \bar{\xi}_{B'} \ , \end{aligned} \tag{VII.64}$$

the primes on the indices serving to denote that $\xi^{A'}$ "lives in" $\bar{S}$ rather than S , and $\xi_{A'}$ "lives in" $\bar{S}^*$. Under these maps we have

$$\begin{aligned} \nu(o^A) &\equiv \bar{o}^{A'} = (1,0) \\ \nu(\iota^A) &\equiv \bar{\iota}^{A'} = (0,1) \\ \mu\circ\nu(o^A) &\equiv \bar{o}_{A'} = (0,1) \\ \mu\circ\nu(\iota^A) &\equiv \bar{\iota}_{A'} = (-1,0) \ . \end{aligned} \tag{VII.65}$$

A general spinor field (= "spin tensor field") is now obtained as a section of one of the tensor product bundles

$$T = S \otimes \ldots \otimes S \otimes \bar{S} \otimes \ldots \otimes \bar{S} \otimes S^* \otimes \ldots \otimes S^* \otimes \bar{S}^* \otimes \ldots \otimes \bar{S}^* .$$

There is an obvious map $\nu_T : T \longrightarrow \bar{T}$, where $\bar{T}$ is the complex conjugate bundle of T (this generalizes the map ν above). If $\xi^{A\ldots B'\ldots}_{C\ldots D'\ldots}(p)$ is an element of T , then $\nu_T\left(\xi^{A\ldots B'}_{C\ldots D'}(p)\right) \equiv \overline{\xi^{A\ldots B'\ldots}_{C\ldots D'\ldots}(p)}$

is denoted $\bar{\xi}^{A'\ldots B\ldots}{}_{C'\ldots D\ldots}(p)$. The only exception to this notation is that $\bar{\epsilon}^{A'B'}$ and $\bar{\epsilon}_{A'B'}$ are usually written as $\epsilon^{A'B'}$ and $\epsilon_{A'B'}$. The complex conjugate of a section is now defined in the obvious way. If $\xi^{A\ldots B'}{}_{C\ldots D'\ldots}$ has the same number of primed and unprimed indices, both "upstairs" and "downstairs", so that we can write it as $\xi^{AB'\ldots}{}_{CD'\ldots}$, then ξ is said to be <u>Hermitian</u> if

$$\xi^{AB'\ldots}{}_{CD'\ldots}(p) = \bar{\xi}^{A'B\ldots}{}_{C'D\ldots}(p) \quad \text{for all } p . \qquad \text{(VII.66)}$$

For example, the spinor field $\xi^{AA'}(p)$ is Hermitian if and only if $\xi^{00'}(p)$ and $\xi^{11'}(p)$ are real for all p, and $\xi^{01'}(p) = \overline{\xi^{10'}(p)}$ for all p. (Note that spinor indices are always thought of as ranging over the values 0 and 1, and primed indices can be freely "commuted" with unprimed, and vice versa.)

With the spinors ϵ^{AB}, ϵ_{AB}, $\epsilon^{A'B'}$, and $\epsilon_{A'B'}$ in hand, the indices of a general spinor $\xi^{A\ldots B'\ldots}{}_{C\ldots D'\ldots}$ can be raised or lowered, this operation corresponding to isomorphisms between the various tensor product bundles involved. The general rule is that an upper index is lowered by contraction on the <u>first</u> index of ϵ_{AB} or $\epsilon_{A'B'}$, while a lower index is raised by contraction on the <u>second</u> index of ϵ^{AB} or $\epsilon^{A'B'}$. For example,

$$\xi^{AB'}{}_{C} \equiv \epsilon^{AD}\,\epsilon^{B'E'}\,\xi_{DE'}{}^{F}\,\epsilon_{FC} .$$

So far, everything we have done is applicable to any spacetime manifold M. We now have to make a connection between the action of the spinor dyad gauge group $SL(2,\mathbb{C})$, as discussed above, and the restricted

Lorentz group $\mathcal{L}_+^\uparrow$, of which SL(2,$\mathbb{C}$) is a two-to-one covering. Following Geroch [18], we define a spinor structure for M , as follows. We assume that M is space and time orientable, and that a space and time orientation has been chosen (see [18]). We construct the bundle B of oriented orthonormal tetrads on M . Thus B is a principal fibre bundle (see Steenrod [76]) with $\mathcal{L}_+^\uparrow$ as the group of the bundle; a point of B is an orthonormal tetrad (T^a, X^a, Y^a, Z^a) at a point of M . A spinor structure for M is then defined to be a second principal fibre bundle $\tilde{B}$ over M , with group SL(2,$\mathbb{C}$) , such that there exists a map $\varphi : \tilde{B} \rightarrow B$ with the following properties:

(i) φ is a two-to-one map;

(ii) for each $p \in M$, φ maps the fibre of $\tilde{B}$ over p to the fibre of B over p ;

(iii) if $\Lambda : SL(2,\mathbb{C}) \rightarrow \mathcal{L}_+^\uparrow$ is the two-to-one covering map of $\mathcal{L}_+^\uparrow$ by SL(2,$\mathbb{C}$) and $G \in SL(2,\mathbb{C})$, then for each $\tilde{b} \in \tilde{B}$ we have $\varphi(G(\tilde{b})) = \Lambda(G)(\varphi(\tilde{b}))$. (This condition will be interpreted in a moment.)

If a spinor structure $\tilde{B}$ exists for a given M , it is not hard to show that $\tilde{B}$ can be thought of as the bundle of spinor dyads (o^A, ι^A) satisfying Equation (VII.59). Then φ is a map which sends pairs $[(o^A, \iota^A), (-o^A, -\iota^A)]$ of spinor dyads into orthonormal tetrads (T^a, X^a, Y^a, Z^a). To each such orthonormal tetrad there corresponds a unique normed null tetrad $(\ell^a, n^a, m^a, \bar{m}^a)$ defined by

$$\ell^a = T^a + Z^a$$
$$n^a = T^a - Z^a$$
$$m^a = X^a + iY^a$$
$$\bar{m}^a = X^a - iY^a ,$$

and φ is most easily thought of as a map taking $[(o^A, \iota^A), (-o^A, -\iota^A)]$ into $(\ell^a, n^a, m^a, \bar{m}^a)$. Then the action of φ is described by the van der Waerden symbols $\sigma^a{}_{AA'}$. These are sections of the complex vector bundle $\mathbb{C}TM \otimes S^* \otimes \bar{S}^*$ with the property $\overline{\sigma^a{}_{AA'}} = \sigma^a{}_{A'A}$ (for each value of a, $\sigma^a{}_{AA'}$ is a two-by-two Hermitian matrix). Then $\varphi([(o^A, \iota^A), (-o^A, -\iota^A)]) = (\ell^a, n^a, m^a, \bar{m}^a)$ if and only if

$$\begin{aligned}
\sigma^a{}_{AA'}\, o^A \bar{o}^{A'} &= \ell^a \\
\sigma^a{}_{AA'}\, o^A \bar{\iota}^{A'} &= m^a \\
\sigma^a{}_{AA'}\, \iota^A \bar{o}^{A'} &= \bar{m}^a \\
\sigma^a{}_{AA'}\, \iota^A \bar{\iota}^{A'} &= n^a \quad .
\end{aligned} \qquad \text{(VII.67)}$$

The nature of the homomorphism $\Lambda : SL(2,C) \longrightarrow \mathcal{L}_+^\uparrow$ can now be exhibited as follows. Under the dyad transformations

$$\begin{pmatrix} o^A \\ \iota^A \end{pmatrix} \rightarrow \pm \begin{pmatrix} 1 & 0 \\ \bar{a} & 1 \end{pmatrix} \begin{pmatrix} o^A \\ \iota^A \end{pmatrix} \quad \text{and} \quad \begin{pmatrix} \bar{o}^{A'} \\ \bar{\iota}^{A'} \end{pmatrix} \rightarrow \pm \begin{pmatrix} 1 & 0 \\ a & 1 \end{pmatrix} \begin{pmatrix} \bar{o}^{A'} \\ \bar{\iota}^{A'} \end{pmatrix} , \qquad \text{(VII.68)}$$

$$\begin{pmatrix} o^A \\ \iota^A \end{pmatrix} \rightarrow \pm \begin{pmatrix} A^{-1} e^{i\phi} & 0 \\ 0 & A e^{-i\phi} \end{pmatrix} \begin{pmatrix} o^A \\ \iota^A \end{pmatrix} \quad \text{and} \quad \begin{pmatrix} \bar{o}^{A'} \\ \bar{\iota}^{A'} \end{pmatrix} \rightarrow \pm \begin{pmatrix} A^{-1} e^{-i\phi} & 0 \\ 0 & A e^{i\phi} \end{pmatrix} \begin{pmatrix} \bar{o}^{A'} \\ \bar{\iota}^{A'} \end{pmatrix} , \qquad \text{(VII.69)}$$

$$\begin{pmatrix} o^A \\ \iota^A \end{pmatrix} \rightarrow \pm \begin{pmatrix} 1 & b \\ 0 & 1 \end{pmatrix} \begin{pmatrix} o^A \\ \iota^A \end{pmatrix} \quad \text{and} \quad \begin{pmatrix} \bar{o}^{A'} \\ \bar{\iota}^{A'} \end{pmatrix} \rightarrow \pm \begin{pmatrix} 1 & \bar{b} \\ 0 & 1 \end{pmatrix} \begin{pmatrix} \bar{o}^{A'} \\ \bar{\iota}^{A'} \end{pmatrix} , \qquad \text{(VII.70)}$$

the null tetrad $(\ell^a, n^a, m^a, \bar{m}^a)$ undergoes the transformations VII.24, VII.25, and VII.26, respectively. Note that for a real Lorentz transformation, the conjugate dyad $(\bar{o}^{A'}, \bar{\iota}^{A'})$ is transformed by the complex conjugate $SL(2,\mathbb{C})$ matrix.

The two-to-one nature of φ is expressed by the fact that $\sigma^a{}_{AA'}$ maps (o^A, ι^A) and $(-o^A, -\iota^A)$ into the same null tetrad. Indices on $\sigma^a_{AA'}$ may be raised and lowered with g_{ab}, ϵ^{AB}, and $\epsilon^{A'B'}$ to produce maps between the various complexified tensor bundles on M and the corresponding spinor tensor-product bundles. For example, for a complexified tensor $T^{a\ldots}{}_{b\ldots}$, we can write

$$T^{AA'\ldots}{}_{BB'\ldots} = T^{a\ldots}{}_{b\ldots} \quad \sigma_a{}^{AA'\ldots} \quad \sigma^b{}_{BB'\ldots} \quad .$$

From the Hermiticity of the $\sigma_a{}^{AA'}$, it follows that a complexified tensor $T^{a\ldots}{}_{b\ldots}$ is real if and only if the corresponding spinor $T^{AA'\ldots}{}_{BB'\ldots}$ is Hermitian as defined above.

Since o^A, ι^A, and $\sigma^a{}_{AA'}$ are global fields, it follows from Equation VII.67 that M admits a global null tetrad. Geroch has shown ([18],[19]) that the existence of a global null tetrad (equivalently, a global orthonormal tetrad) is in fact sufficient as well as necessary for a (non-compact) spacetime to admit a spinor structure.

We now turn to the question of defining a connection on the spinor bundles which is in some sense compatible with the pseudo-Riemannian connection on M. For the bundle S, a connection ∇_m is defined by writing

$$\nabla_m \xi^A = \xi^A{}_{,m} + \Gamma^A_{Bm} \xi^B \quad . \qquad \text{(VII.71)}$$

Similarly, for $\bar{S}$, we write

$$\nabla_m \bar{\xi}^{A'} = \bar{\xi}^{A'}{}_{,m} + \bar{\Gamma}^{A'}{}_{B'm} \bar{\xi}^{B'} . \qquad \text{(VII.72)}$$

(In the general case, no relationship between the $\Gamma^A{}_{Bm}$ and the $\bar{\Gamma}^{A'}{}_{B'm}$ need exist.)

The coefficients $\Gamma^A{}_{Bm}$, $\bar{\Gamma}^{A'}{}_{B'm}$, together with the Christoffel symbols $\Gamma^a{}_{bm}$, allow us to define connections for all bundles formed as tensor products of $S, \bar{S}, S^*$, $\bar{S}^*$, $\mathbb{C}TM$, and $\mathbb{C}TM^*$. For example, for a mixed tensor-spinor $T^{aBc'}{}_{dEF'}$, we can write:

$$\begin{aligned}\nabla_m T^{aBC'}{}_{dEF'} = T^{aBC'}{}_{dEF',m} &+ \Gamma^a{}_{ms} T^{sBC'}{}_{dEF'} - \Gamma^s{}_{md} T^{aBC'}{}_{SEF'} \\ &+ \Gamma^B{}_{Sm} T^{aSC'}{}_{dEF'} - \Gamma^S{}_{Em} T^{aBC'}{}_{dSF'} \\ &+ \bar{\Gamma}^{C'}{}_{S'm} T^{aBS'}{}_{dEF'} - \bar{\Gamma}^{S'}{}_{F'm} T^{aBC'}{}_{dES'} . \qquad \text{(VII.73)}\end{aligned}$$

The affine geometry of the spinor bundles is now "locked in" to the Riemannian geometry of M by requiring $\Gamma^A{}_{Bm}$ and $\bar{\Gamma}^{A'}{}_{B'm}$ to be chosen so that

(i) $\nabla_m \epsilon_{AB} = \nabla_m \epsilon_{A'B'} = 0$

(ii) $\nabla_m \sigma^a{}_{AA'} = 0$. (VII.74)

These two requirements serve to determine the spinor connections uniquely. Provided the metric g_{ab} is real, they imply that

$$\bar{\Gamma}^{A'}{}_{B'm} = \overline{\Gamma^A{}_{Bm}} . \qquad \text{(VII.75)}$$

Defining $\Gamma_{ABm} = \Gamma^{C}{}_{Bm}\, \epsilon_{CA}$ and $\bar{\Gamma}_{A'B'm} = \bar{\Gamma}^{C'}{}_{B'm}\, \epsilon_{C'A'}$, condition (i) is equivalent to

$$\text{(iii)}\quad \Gamma_{ABm} = \Gamma_{BAm}\,,$$
$$\bar{\Gamma}_{A'B'm} = \bar{\Gamma}_{B'A'm}\,. \tag{VII.76}$$

In Minkowski space, in a Cartesian coordinate system, condition (ii) insures that the $\sigma^{a}{}_{AA'}$ can be chosen to be four constant Hermitian matrices. In a curved spacetime, such a choice is no longer possible.

Because of VII.74(ii) and VII.67, the NP spin coefficients defined above can be expressed very elegantly in spinor language. First note, from Equations VII.12, VII.67, and the relations

$$\epsilon_{AB} = o_A \iota_B - \iota_A o_B$$
$$\epsilon_{A'B'} = \bar{o}_{A'} \bar{\iota}_{B'} - \bar{\iota}_{A'} \bar{o}_{B'}$$

it follows that

$$g_{ab} = \sigma_a{}^{AA'}\, \sigma_b{}^{BB'}\, \epsilon_{AB}\, \epsilon_{A'B'}\,,$$

and

$$\epsilon_{AB}\, \epsilon_{A'B'} = g_{ab}\, \sigma^{a}{}_{AA'}\, \sigma^{b}{}_{BB'}$$
$$\Leftrightarrow \sigma_{aAA'}\, \sigma^{a}{}_{BB'} = \epsilon_{AB}\, \epsilon_{A'B'}\,. \tag{VII.77}$$

Then using VII.74(ii) and VII.67, we have, for example,

$$
\begin{aligned}
\kappa \;&\equiv\; \ell_{a;b}\, m^a \ell^b \\
&=\; \ell^b \nabla_b (\sigma_a{}^{AA'} o_A \bar{o}_{A'})\; \sigma^{aBB'} o_B \bar{\iota}_{B'} \\
&=\; \ell^b \nabla_b (o_A \bar{o}_{A'})\; o^A \bar{\iota}^{A'} \\
&=\; \ell^b [o_A \nabla_b\; \bar{o}_{A'} + \bar{o}_{A'} \nabla_b\, o_A]\; o^A \bar{\iota}^{A'} \\
&=\; o^A \ell^b \nabla_b\, o_A \,,
\end{aligned}
$$

$$\text{since } o_A o^A = 0 \quad \text{and} \quad \bar{o}_A \bar{\iota}^{A'} = 1 \,.$$

The other spin coefficients have correspondingly simple spinor expressions, Defining $\nabla_{AA'} \equiv \sigma^a{}_{AA'} \nabla_a$, the information is usually summarized in a table of the following sort:

$\nabla =$	$o^A \nabla o_A$ or $m^a \nabla \ell_a$	$o^A \nabla \iota_A = \iota^A \nabla o_A$ or $\frac{1}{2}(n^a \nabla \ell_a + m^a \nabla \bar{m}_a)$	$\iota^A \nabla \iota_A$ or $n^a \nabla \bar{m}_b$
$\ell^b \nabla_b = o^B \bar{o}^{B'} \nabla_{BB'}$	κ	ϵ	π
$\bar{m}^b \nabla_b = \iota^B \bar{o}^{B'} \nabla_{BB'}$	ρ	α	λ
$m^b \nabla_b = o^B \bar{\iota}^{B'} \nabla_{BB'}$	σ	β	μ
$n^b \nabla_b = \iota^b \bar{\iota}^{B'} \nabla_{BB'}$	τ	γ	ν

Table VII.3

The complex conjugate spin coefficients are defined by conjugating the corresponding spinor expression, e.g.,

$$\bar{\kappa} = \overline{o^A o^B \bar{o}^{B'} \nabla_{BB'} o_A}$$

$$= \bar{o}^{A'} \bar{o}^{B'} o^B \nabla_{BB'} \bar{o}_{A'} \ .$$

(Note that $\nabla_{BB'}$ is an Hermitian operator.)

The spinor transcription of the Riemann curvature tensor is (following the conventions of Newman and Penrose [37])

$$R_{AE'BF'CG'DH'} = \sigma^a{}_{AE'}\, \sigma^b{}_{BF'}\, \sigma^c{}_{CG'}\, \sigma^d{}_{DH'}\ R_{abcd},$$

where the symmetries of R_{abcd} imply that $R_{AE'BF'CG'DH'}$ can be written as

$$\begin{aligned}-R_{AE'BF'CG'DH'} = {} & \Psi_{ABCD}\, \epsilon_{E'F'}\, \epsilon_{G'H'} + \epsilon_{AB}\, \epsilon_{CD}\, \bar{\Psi}_{E'F'G'H'} \\ & + \epsilon_{AB}\, \Phi_{CDE'F'}\, \epsilon_{G'H'} + \epsilon_{CD}\, \Phi_{ABG'H'}\, \epsilon_{E'F'} \\ & + 2\Lambda(\epsilon_{AC}\, \epsilon_{BD}\, \epsilon_{E'F'}\, \epsilon_{G'H'} + \epsilon_{AB}\, \epsilon_{CD}\, \epsilon_{E'H'}\, \epsilon_{F'G'})\end{aligned} \tag{VII.78}$$

where the spinors on the right-hand side have the properties

$$\Psi_{ABCD} = \Psi_{(ABCD)}$$

$$\Phi_{ABC'D'} = \Phi_{(AB)(C'D')} = \bar{\Phi}_{A'B'CD}$$

$$\Lambda = \frac{1}{24} R = \bar{\Lambda}. \tag{VII.79}$$

The field $C_{AE'BF'CG'DH'}$ defined by

$$-C_{AE'BF'CG'DH'} \equiv \Psi_{ABCD}\,\epsilon_{E'F'}\,\epsilon_{G'H'} + \epsilon_{AB}\,\epsilon_{CD}\,\overline{\Psi}_{E'F'G'H'} \tag{VII.80}$$

turns out to be just the spinor transcription of the Weyl tensor, Definition VI.32. (See Schild [48] for the details of this decomposition.)

In terms of the "curvature spinors" Ψ_{ABCD} and $\Phi_{ABC'D'}$, the NP curvature scalars can be expressed as follows:

$$\Psi_0 = \Psi_{ABCD}\,o^A o^B o^C o^D \qquad \Phi_{00} = \Phi_{ABC'D'}\,o^A o^B \bar{o}^{C'} \bar{o}^{D'}$$

$$\Psi_1 = \Psi_{ABCD}\,o^A o^B o^C \iota^D \qquad \Phi_{01} = \Phi_{ABC'D'}\,o^A o^B \bar{o}^{C'} \bar{\iota}^{D'}$$

$$\Psi_2 = \Psi_{ABCD}\,o^A o^B \iota^C \iota^D \qquad \Phi_{02} = \Phi_{ABC'D'}\,o^A o^B \bar{\iota}^{C'} \bar{\iota}^{D'}$$

$$\Psi_3 = \Psi_{ABCD}\,o^A \iota^B \iota^C \iota^D \qquad \Phi_{10} = \Phi_{ABC'D'}\,o^A \iota^B \bar{o}^{C'} \bar{o}^{D'}$$

$$\Psi_4 = \Psi_{ABCD}\,\iota^A \iota^B \iota^C \iota^D \qquad \Phi_{11} = \Phi_{ABC'D'}\,o^A \iota^B \bar{o}^{C'} \bar{\iota}^{D'}$$

$$\Phi_{12} = \Phi_{ABC'D'}\,o^A \iota^B \bar{\iota}^{C'} \bar{\iota}^{D'}$$

$$\Phi_{20} = \Phi_{ABC'D'}\,\iota^A \iota^B \bar{o}^{C'} \bar{o}^{D'}$$

$$\Phi_{21} = \Phi_{ABC'D'}\,\iota^A \iota^B \bar{o}^{C'} \bar{\iota}^{D'}$$

$$\Phi_{22} = \Phi_{ABC'D'}\,\iota^A \iota^B \bar{\iota}^{C'} \bar{\iota}^{D'} \tag{VII.81}$$

Finally, we have the "anticommutation relations",

$$\nabla_{(A}{}^{M'}\,\nabla_{B)M'}\,\xi_C = -\Psi_{ABCD}\,\xi^D + 2\Lambda\,\xi_{(A}\,\epsilon_{B)C}\,,$$

$$\nabla_{M(P'}\,\nabla^{M}{}_{Q')}\,\xi_A = \Phi_{ABP'Q'}\,\xi^B\,, \tag{VII.82}$$

and the spinor form of the Bianchi identities:

$$\nabla^{D}{}_{G'} \Psi_{ABCD} = \nabla_{(C}{}^{H'} \Phi_{AB)G'H'} ,$$

$$\nabla^{AG'} \Phi_{ABG'H'} = -3 \nabla_{BH'} \Lambda . \qquad \text{(VII.83)}$$

From the fact that $\xi^A \xi_A = 0$ for any spinor ξ^A, it follows that $k_a = \sigma_a{}^{AA'} \xi_A \eta_{A'}$ is a <u>complex null vector</u> for any spinors ξ_A and $\eta_{A'}$ (compare, for example, Equation VII.67). If $\eta_{A'} = \bar{\xi}_{A'}$, then k_a is also real and future-pointing. Penrose has pointed out [44] that a spinor field ξ_A defines two pieces of information. These are the real future-pointing null vector field $\sigma_a{}^{AA'} \xi_A \bar{\xi}_{A'}$ ("the flagpole field") and the real bivector field $\sigma_a{}^{AA'} \sigma_b{}^{BB'} (\epsilon_{AB} \bar{\xi}_{A'} \bar{\xi}_{B'} + \xi_A \xi_B \epsilon_{A'B'})$ ("the flag-plane field").

Finally, we mention the fact that any spinor field $\phi_{AB\ldots L} = \phi_{(AB\ldots L)}$ which is completely symmetric in all of its indices can always be decomposed into the form

$$\phi_{AB\ldots L} = \phi \, \alpha_{(A} \beta_{B} \cdots \lambda_{L)} , \qquad \text{(VII.84)}$$

where ϕ is a scalar, and the one-spinors $\alpha_A, \beta_B, \ldots, \lambda_L$ are unique up to rescalings, and are called the <u>principal spinors</u> of $\phi_{AB\cdots L}$. When $\phi_{AB\cdots L}$ has four indices and represents a curvature spinor $\Psi_{ABCD} = \Psi_{(ABCD)}$, then $\Psi_{(ABCD)} = \Psi \, \alpha_{(A} \beta_B \gamma_C \delta_{D)}$, and the null directions $\sigma_a{}^{AA'} \alpha_A \bar{\alpha}_{A'}, \ldots, \sigma_a{}^{AA'} \delta_A \bar{\delta}_{A'}$ are just the principal null directions of the corresponding Weyl tensor. Viewed in this light, the algebraic classification of fields discussed above is equivalent to classifying symmetric spinor fields according to coincidences among their principal spinors. For example, a null Maxwell field corresponds to a spinor of the form $\Phi_{AB} = \phi\, \alpha_{(A} \beta_{B)}$, a type D Weyl tensor corresponds to a spinor of the form $\Psi_{ABCD} = \psi\, \alpha_{(A} \alpha_B \beta_C \beta_{D)}$, and so forth. (See Penrose [41] or Schild [48] for details on these assertions.)

CHAPTER VIII: STRUCTURES ON SPACETIME MANIFOLDS

We now have the necessary mathematical and physical background to begin applying complex manifold theory to the study of general relativity. We will find that in order to get meaningful results, some modifications of complex manifold theory will be necessary. As mentioned at the beginning of Chapter VI, the necessity for these modifications arises from the fact that the signature of a spacetime metric is indefinite: The "Hermitian" metric should be a generalization of the Minkowski metric, $ds^2 = du\,dv - d\zeta\,d\bar{\zeta}$ (u and v real, ζ complex), rather than the positive definite form $ds^2 = dw\,d\bar{w} + dz\,d\bar{z}$ (w and z complex).

After discarding a "genuine" almost complex structure for spacetime, we shall find a modified almost Hermitian structure which is integrable for spacetimes admitting two geodesic and shearfree null congruences, in particular, for the type D vacuum and electrified spaces. We shall show that these spaces are not in general Kählerian, but that in the vacuum and electrified cases they are conformally related by a complex conformal factor to spacetimes which are Kählerian. And finally, we shall show that spacetimes admitting only one geodesic and shearfree null congruence -- including all the algebraically special vacuum spaces -- admit an almost Hermitian structure which is "halfway integrable" in an appropriate sense.

Throughout what follows we shall assume that there is given a normed null tetrad, so that the metric can be written as in Equation VII.12 as

$$g_{ab} = \ell_a n_b + n_a \ell_b - m_a \bar{m}_b - \bar{m}_a m_b \quad . \qquad \text{(VIII.1)}$$

All of our results will be <u>local</u>, in the sense that we will be dealing with only one coordinate patch at a time. But many of our results will be valid *globally*, provided that a global normed null tetrad exists (compare the discussion following Equation VII.11). We shall not concern ourselves with any specific global questions.

A Genuine Almost Complex Structure for Spacetimes

We now present a genuine example of a family of almost complex structures for spacetimes (the adjective "genuine" is used to distinguish the structures from the "modified" structures which appear later). One member of the family is determined by each choice of normed null tetrad satisfying VII.11 and hence VIII.1:

VIII.1: THEOREM: Given a spacetime with a normed null tetrad, an almost complex structure for the spacetime is given (locally) by

$$\begin{aligned} J_a{}^b = & - \ell_a \ell^b + n_a n^b \\ & - i\, m_a \bar{m}^b + i\, \bar{m}_a m^b \; . \end{aligned} \tag{VIII.2}$$

Proof: Given the reality character of the tetrad, $J_a{}^b$ is manifestly real, and it is trivial to verify that

$$\begin{aligned} J_a{}^s J_s{}^b &= - \ell_a n^b - n_a \ell^b + m_a \bar{m}^b + \bar{m}_a m^b \\ &= - \delta_a{}^b \; . \end{aligned}$$

Equivalently (compare Chapter V), the almost complex structure is defined by its forms of type (1,0), which are given by linear combinations of the forms $\omega^{(b)} = (J_a{}^b + i\,\delta_a{}^b)dx^a$. If we take instead the equivalent linear combinations $\ell_b \omega^{(b)}$, $n_b \omega^{(b)}$, $m_b \omega^{(b)}$, and $\bar{m}_b \omega^{(b)}$, we find directly that the type (1,0) forms are given by linear combinations of

$$(n_a + i\,\ell_a)dx^a \quad \text{and} \quad m_a\,dx^a \ . \tag{VIII.3}$$

The type (0,1) forms are given by linear combinations of the complex conjugates of these, namely,

$$(n_a - i\,\ell_a)dx^a \quad \text{and} \quad \bar{m}_a\,dx^a \ . \tag{VIII.4}$$

The integrability conditions for the almost complex structure are given by the vanishing of the Nijenhuis tensor, $N_{ab}{}^c = 0$ (see Chapter V). (Since we are working locally, there is no problem with differentiability requirements: Within one coordinate patch, the tetrad field can be chosen to be (real) analytic.) Since $J_a{}^b$ is given in terms of the null tetrad, it is easier to evaluate the tetrad components of the Nijenhuis tensor rather than $N_{ab}{}^c$ itself. In fact, it is easier still to apply the Frobenius Theorem directly by demanding the "surface orthogonality" of the bivector $(n_a + i\,\ell_a)m_b\,dx^a \wedge dx^b$ (and of its complex conjugate). In any case, it is found that the integrability condition $N_{ab}{}^c = 0$ is equivalent to the following relations among the spin coefficients formed from the given null tetrad:

$$\begin{aligned} \ell^a m^b N_{ab}{}^c \ell_c &= -2i(\bar{\alpha} + \beta) + \kappa - \bar{\nu} + i(\bar{\pi} + \tau) = 0 \ , \\ \ell^a m^b N_{ab}{}^c m_c &= -2i\,\bar{\lambda} + 2\sigma = 0 \ . \end{aligned} \tag{VIII.5}$$

(The other tetrad components of $N_{ab}{}^c$ either vanish or are linear combinations of these and their complex conjugates.)

In looking for spacetimes with tetrads for which these integrability conditions are satisfied, the presence of the spin coefficients κ, σ, ν, and λ in Equations VIII.5 leads us to consider the type D vacuum and electrified solutions, for which tetrads exist with $\kappa = \sigma = \nu = \lambda = 0$ (see Equation VII.44). In this case, the integrability conditions reduce to the single requirement

$$2(\bar{\alpha} + \beta) - (\bar{\pi} + \tau) = 0 \quad . \qquad \text{(VIII.6)}$$

If we choose the tetrads used by Kinnersley [26], we find that the only spacetimes which satisfy Equation VIII.6 are NUT space together with its two related metrics (see Table VII.2). (Note that Schwarzschild space is the special case of NUT space for which the NUT parameter ℓ vanishes. Furthermore, a tetrad for flat space is obtained from this Schwarzschild tetrad by setting the mass parameter m equal to zero: This is the "Coulomb tetrad" for flat space.)

It will be instructive to look at the actual process of integration which produces the complex coordinates. We shall do this for the case of NUT spacetime, using the tetrad given by Kinnersley [26]. The covariant components of this tetrad, in the (real) coordinate system $\{u, r, \theta, \phi\}$, are given by:

$$\ell_a = (1, 0, 0, -2\ell\cos\theta)$$

$$n_a = \left(\frac{1}{2}\,\frac{r^2 - 2mr - \ell^2}{r^2 + \ell^2}, 1, 0, -\frac{\ell\cos\theta(r^2 - 2mr - \ell^2)}{r^2 + \ell^2} \right) \qquad \text{(VIII.7)}$$

$$m_a = -\frac{r^2 + \ell^2}{\sqrt{2}(r - i\ell)}\,(0, 0, 1, i\sin\theta)$$

The complex coordinates z^0 and z^1 must be such that dz^0 and dz^1 are linearly independent linear combinations of the type (1,0) forms, $(n_a + i\,\ell_a)dx^a$ and $m_a\,dx^a$; then we will have

$$z^0 = \int dz^0$$

and $$z^1 = \int dz^1 .$$

It is not hard to find such linear combinations which are integrable. We have:

$$\frac{2(r^2+\ell^2)}{r^2 - 2mr - \ell^2 + 2i(r^2+\ell^2)}\,(n_a + i\,\ell_a) \equiv \omega_a^{(1)} = \left(1, \frac{2(r^2+\ell^2)}{r^2 - 2mr - \ell^2 + 2i(r^2+\ell^2)}, 0, -2\ell\cos\theta\right), \qquad \text{(VIII.8)}$$

$$-\frac{\sqrt{2}\,(r - i\ell)}{r^2+\ell^2}\,m_a \equiv \omega_a^{(2)} = (0, 0, 1, i\sin\theta) .$$

From these we form:

$$dz^0 = \omega_a^{(1)}\,dx^a - 2i\ell\cot\theta\;\omega_a^{(2)}\,dx^a$$

$$= \left(1, \frac{2(r^2+\ell^2)}{r^2 - 2mr - \ell^2 + 2i(r^2+\ell^2)}, -2i\ell\cot\theta, 0\right),$$

(VIII.9)

$$dz^1 = \frac{1}{\sin\theta}\, \omega_a^{(2)}\, dx^a$$

$$= (0,\ 0,\ \frac{1}{\sin\theta},\ i) \quad .$$

So finally we obtain the following complex coordinates:

$$z^0 = \int dz^0 = u + R(r) - 2i\,\ell\,\ln\sin\theta \quad ,$$

$$z^1 = \ln\tan\theta/2 + i\,\phi \quad , \qquad \text{(VIII.10)}$$

where

$$R(r) \equiv \int \frac{2(r^2 + \ell^2)}{r^2 - 2mr - \ell^2 + 2i(r^2 + \ell^2)}\, dr \quad .$$

(The constants of integration can, of course, be set equal to zero, as we have done here. Furthermore, we have ignored the question of coordinate singularities, in accordance with the local nature of our treatment.) It is a straightforward matter to verify that the coordinate transformation $\{u, r, \theta, \phi\} \rightarrow \{z^0, z^1, \bar{z}^0, \bar{z}^1\}$ has a non-vanishing Jacobean (at least away from any coordinate singularities) and is therefore a "valid" coordinate transformation, even at the level where "i" is treated purely formally as a square root of minus one.

Summarizing our results, we have:

<u>VIII.2: THEOREM:</u> The almost complex structure given by Equation VIII.2 is integrable if and only if

$$-2i(\bar{\alpha} + \beta) + \varkappa - \bar{\nu} + i(\bar{\pi} + \tau) = 0 \quad ,$$

$$-2i\,\bar{\lambda} + 2\sigma = 0 \quad .$$

For a type D vacuum or electrified spacetime, in

the tetrad used by Kinnersley, the almost complex structure is integrable if and only if

$$2(\bar{\alpha} + \beta) - (\bar{\pi} + \tau) = 0 \quad ,$$

hence, if and only if the space is NUT (or one of the two related spaces), with Schwarzschild space as a special case.

Proof: See above.

Necessity for Complex Valued
Almost Hermitian Structure Tensor

The almost complex structure of Equation VIII.2 has some seriously undesirable features. The first of these is that this $J_a{}^b$ is not invariant under the spin and boost transformations of the tetrad (Equations VII.25). Under such a transformation, we have

$$J_a{}^b \longrightarrow - A^{-2} \ell_a \ell^b + A^2 n_a n^b - i m_a \bar{m}^b + i \bar{m}_a m^b \quad . \qquad \text{(VIII.11)}$$

This is undesirable for treating the type D vacuum spaces, where the geometry singles out only two preferred directions, the principal null directions. Since these directions are preferred by the geometry, we would not expect physically meaningful entities based on these directions to be invariant under Lorentz tetrad rotations which change the directions of ℓ and n, when these are chosen to be the principal null directions. However, we would expect physically meaningful entities to be invariant under the spin and boost transformations, which leave the directions of ℓ and n unchanged. Equation VIII.11 shows that this is not the case for our almost complex structure tensor.

Another disturbing feature of our almost complex structure is that it is not Hermitian. Consulting Definition VI.5 and Equation VIII.2, we have:

$$\begin{aligned} J_a{}^m J_b{}^n g_{mn} &= -\ell_a n_b - n_a \ell_b - m_a \bar{m}_b - \bar{m}_a m_b \\ &\neq g_{ab} \quad . \end{aligned} \tag{VIII.12}$$

We can describe this situation by saying that $J_a{}^b$ splits the metric g_{ab} into an "Hermitian part,"

$$\begin{aligned} h_{ab} &= \frac{1}{2}(g_{ab} + J_a{}^m J_b{}^n g_{mn}) \\ &= -m_a \bar{m}_b - \bar{m}_a m_b \quad , \end{aligned} \tag{VIII.13}$$

and an "anti-Hermitian part,"

$$\begin{aligned} a_{ab} &= \frac{1}{2}(g_{ab} - J_a{}^m J_b{}^n g_{mn}) \\ &= \ell_a n_b + n_a \ell_b \quad . \end{aligned} \tag{VIII.14}$$

(The fact that the coordinate $z^1 = \ell n \tan \theta/2$ of Equation VIII.10 is a very useful coordinate (compare Aronson, et al. [3]) is precisely because it arises from the "Hermitian part" of the NUT metric.)

The most serious failing of our almost complex structure is, quite simply, that it is not very interesting or useful. Working locally, as we are, and in the absence of the Hermitian compatibility requirement, we are open to the following objection: Why not simply discard the almost complex structure tensor and use instead the structure for which the complex coordinates are $z^0 = u + ir$, $z^1 = \theta + i\phi$, or some similar choice? In the absence of the Hermitian property, there is really no valid answer to this objection.

Furthermore our $J_a{}^b$ has been shown to be integrable only for Schwarzschild and NUT, among the type D vacuum spaces. In particular, it is not integrable for the Kerr spacetime (at least not for any of the natural choices of tetrad; whether a more exotic tetrad exists for Kerr in which $J_a{}^b$ is integrable is not known).

All of this suggests that we discard this choice of almost complex structure and search for one that is Hermitian. However, we are then immediately faced with the following problem:

<u>VIII.3: THEOREM</u>: A manifold with a metric of Lorentz signature cannot admit an almost Hermitian structure.

<u>Proof</u>: Suppose an almost Hermitian structure $J_a{}^b$ does exist. At any point of the manifold, choose coordinates such that g_{ab} takes the form $_{ab}$ at that point. Then by the Hermitian property, we will have $J_a{}^m J_b{}^n {}_{mn} = {}_{ab}$. Therefore $J_a{}^b$ defines a Lorentz transformation at the given point. But then we have a <u>real</u> Lorentz transformation whose square $J_a{}^s J_s{}^b$ is equal to minus the identity. We shall now show that this leads to a contradiction. Since $J_a{}^b$ is a Lorentz transformation, it is non-singular and hence has four linearly independent eigenvectors with non-zero eigenvalues. Since $J^2 = -1$, we know that the eigenvalues +i, +i, -i, -i. Furthermore, the eigenvectors are null due to the antisymmetry of J_{ab}:

$$J_a{}^b v^a = \quad v^b$$

$$\Rightarrow \quad v_b J_a{}^b v^a = \quad v_b v^b$$

$$\Rightarrow \qquad 0 = v_b v^b , \qquad \text{because } \lambda \neq 0 .$$

Lastly, from the reality of $J_a{}^b$ we know that the eigenvectors come in complex conjugate pairs. What we have then is four linearly independent null vectors in complex conjugate pairs. (None of these null vectors can be real because the eigenvalues are not real.) But this can be the case only if the signature of the metric is (++++), (++--), or (----). Hence we have a contradiction and the stated result follows.

It is important to note that this theorem does not depend on any global properties of the space: It is a purely local (in fact, even pointwise) result.

In the face of this depressing result, a clue as to where to proceed from here is provided by the observation that there do exist complex Lorentz transformations whose square is minus one; in fact, in coordinates for which $\eta_{ab} = \mathrm{diag}(1, -1, -1, -1)$, one such complex Lorentz transformation is given by

$$\Lambda_a{}^b = \begin{pmatrix} 0 & i & 0 & 0 \\ i & 0 & 0 & 0 \\ 0 & 0 & 0 & 1 \\ 0 & 0 & -1 & 0 \end{pmatrix} . \qquad \text{(VIII.15)}$$

It is straightforward to verify that $\Lambda_a{}^m \Lambda_b{}^n \eta_{mn} = \eta_{ab}$. In view of our use of the null tetrad formalism, it is helpful to transform to null coordinates in which η_{ab} takes the form

$$\eta_{ab} = \begin{pmatrix} 0 & 1 & 0 & 0 \\ 1 & 0 & 0 & 0 \\ 0 & 0 & 0 & -1 \\ 0 & 0 & -1 & 0 \end{pmatrix} . \qquad \text{(VIII.16)}$$

In these coordinates, a suitable complex $\Lambda_a{}^b$ is given by:

$$\Lambda_a{}^b = \begin{pmatrix} i & 0 & 0 & 0 \\ 0 & -i & 0 & 0 \\ 0 & 0 & -i & 0 \\ 0 & 0 & 0 & i \end{pmatrix} ,$$

or, for the covariant components,

$$\Lambda_{ab} = \begin{pmatrix} 0 & i & 0 & 0 \\ -i & 0 & 0 & 0 \\ 0 & 0 & 0 & i \\ 0 & 0 & -i & 0 \end{pmatrix} . \qquad \text{(VIII.17)}$$

It is these observations, together with the general desirability of the Hermitian compatibility property, that motivate the discarding of the almost complex structure of Equation VIII.2 and the examination of the possibility of complex-valued almost complex structure tensors possessing the Hermitian property.

The Modified Almost Hermitian Structure

In accordance with the motivation above, we now present a modified almost Hermitian structure for spacetime manifolds. Here and in the sequel we shall always use the adjective "modified" to denote a structure whose structure tensor is complex-valued. We shall sometimes abbreviate this by writing "almost m-Hermitian", "m-Kählerian", and so on.

VIII.4: DEFINITION: Given a spacetime and a normed null tetrad, the modified almost Hermitian structure for the spacetime is given by the structure tensor

$$J_a{}^b = i\,\ell_a n^b - i\, n_a \ell^b - i\, m_a \bar{m}^b + i\, \bar{m}_a m^b \quad . \qquad \text{(VIII.18)}$$

Since we want to treat this $J_a{}^b$ at least formally as an "ordinary" almost Hermitian structure, it is essential to have the following facts:

VIII.5: THEOREM: For the $J_a{}^b$ of Definition VIII.4 we have:

(i) $J_a{}^m J_m{}^b = -\delta_a{}^b$

(ii) $J_a{}^m J_b{}^n g_{mn} = g_{ab}$

(iii) $J_a{}^b \ell_b = i\,\ell_a$, $J_a{}^b m_b = i\, m_a$,

$J_a{}^b n_b = -i\, n_a$, $J_a{}^b \bar{m}_b = -i\, \bar{m}_a$.

Proof: Each of these results follows from a simple computation.

It turns out that the three properties of the above theorem are precisely the properties needed to treat $J_a{}^b$ formally as an almost Hermitian structure. Note that property (iii) of the theorem shows that $J_a{}^b$ has eigenvalues +i, +i, -i, -i like a real almost complex structure. This turns out to be the crucial property when the integrability of $J_a{}^b$ is examined. (L. Shepley has shown that properties (i) and (ii) imply the local existence of a null tetrad for which property (iii) holds.)

We also have the following facts:

<u>VIII.6: THEOREM:</u> The forms of type (1,0) defined by $J_a{}^b$ are linear combinations of $\ell_a\, dx^a$ and $m_a\, dx^a$. The forms of type (0,1) are linear combinations of $n_a\, dx^a$ and $\bar{m}_a\, dx^a$.

<u>Proof</u>: Follows directly from property (iii) of Theorem VIII.5.

It is noteworthy that the forms of type (0,1) are <u>not</u> the complex conjugates of the forms of type (1,0). This is due, of course, to the complex nature of $J_a{}^b$. Essentially, $J_a{}^b$ treats n as the formal complex conjugate of ℓ.

<u>VIII.7: REMARK:</u> We note that there are three trivially different almost Hermitian structures that could have been chosen instead of Definition VIII.4. There is the structure formally conjugate to $J_a{}^b$ (compare Chapter III) with type (1,0) froms given by linear combinations of $n_a\, dx^a$ and $\bar{m}_a\, dx^a$ and type (0,1) forms formed from $\ell_a\, dx^a$ and $m_a\, dx^a$. There is also the structure conjugate to $J_a{}^b$ in the usual sense, with type (1,0) forms composed from $\ell_a\, dx^a$ and $\bar{m}_a\, dx^a$ and type (0,1) forms from $n_a\, dx^a$ and $m_a\, dx^a$. Finally, there is the formal conjugate of this latter structure, with type (1,0) forms built out of $n_a\, dx^a$ and $m_a\, dx^a$ and type (0,1) forms arising from $\ell_a\, dx^a$ and $\bar{m}_a\, dx^a$. It is clear that the choice of any one of these structures gives analogous results to those that hold for our choice of $J_a{}^b$. If any one of these structures is integrable, the other three also are, provided the metric is real (see below).

VIII.8: REMARK: Since our $J_a{}^b$ is already complex-valued, we could just as well choose to work with the structure tensor $K_a{}^b = i\, J_a{}^b$. Then we would have $K_a{}^m K_m{}^b = +\,\delta_a{}^b$, which is the analogue of the condition for an almost product structure (see Appendix B). However, then we would have $K_a{}^m K_b{}^n g_{mn} = -\,g_{ab}$, rather than the condition $K_a{}^m K_b{}^n g_{mn} = +\,g_{ab}$ for a locally product Riemannian space (Appendix B). Hence we choose to work with $J_a{}^b$ with its Hermiticity property.

VIII.9: REMARK: It is easy to verify that our $J_a{}^b$ remains invariant under the spin and boost transformations of Equations VII.25. This answers the objection raised about our previous genuine almost complex structure with regard to treating type D spacetimes: For these spacetimes, the modified almost Hermitian structure depends only on the directions of ℓ and n.

Modified Integrability Considerations

What we want to do now is use the modified almost complex structure $J_a{}^b$ to set up complex coordinates on a spacetime manifold. This suggests trying to prove an analogue of the integrability theorems of Chapter V. In fact, a perusal of these theorems shows that, assuming our $J_a{}^b$ is analytic in the real sense, the proofs do indeed go through (with one proviso; see below). This is true essentially because the properties of Theorem VIII.5 hold. For example, the fact that our $J_a{}^b$ has eigenvalues +i, +i, -i, -i implies that $J_a{}^b + i\,\delta_a{}^b$ still has rank m (for real dimension 2m) as in the case of a genuine almost complex structure, so that the Frobenius Theorem can still be invoked (assuming analyticity) to produce two complex coordinates z^0 and z^1 such that dz^0 and dz^1 are type (1,0) forms. (It is this fact which disallows a modified almost complex structure such as $J_a{}^b = i\,\delta_a{}^b$, which has eigenvalues +i, +i, +i, +i.)

There is, however, one important distinction to be made in the case of a complex $J_a{}^b$, which arises from the fact that $J_a{}^b - i\,\delta_a{}^b$ is no longer the complex conjugate of $J_a{}^b + i\,\delta_a{}^b$. This means that the integrability of the type (1,0) forms no longer implies the integrability of the type (0,1) forms. In particular, the structure coordinates $\overline{z^0}$ and $\overline{z^1}$ will no longer be obtainable as the complex conjugates of some structure coordinates z^0 and z^1. Again, this is ultimately a consequence of the indefinite character of the spacetime metric. (In spite of this, the integrability condition $N_{ab}{}^c = 0$ still "contains" the integrability condition for the type (0,1) forms as well as the type (1,0) forms: Compare the proof given in Chapter III.)

Even though the structure coordinates will no longer come in complex conjugate pairs, it will still be true that $\{z^0, z^1, \overline{z^0}, \overline{z^1}\}$ will form a valid coordinate system (because $J_a{}^b$ is nonsingular). That is to say, the Jacobean for the transformation from real coordinates to "structure coordinates" will still be nonsingular.

To summarize, we have the following results:

<u>VIII.10: DEFINITION</u>: The modified almost Hermitian structure of Definition VIII.4 is said to be integrable if coordinates, generally complex, exist such that $J_a{}^b$ has the components

$$J_a{}^b = \begin{pmatrix} +i & 0 & 0 & 0 \\ 0 & +i & 0 & 0 \\ 0 & 0 & -i & 0 \\ 0 & 0 & 0 & -i \end{pmatrix}.$$

<u>VIII.11: THEOREM:</u> The modified almost Hermitian structure $J_a{}^b$ is integrable

(i) if and only if the Nijenhuis tensor formed from $J_a{}^b$ vanishes,

(ii) if and only if the systems of equations

$$(J_a{}^b + i\,\delta_a{}^b)\,dx^a = 0$$

$$(J_a{}^b - i\,\delta_a{}^b)\,dx^a = 0$$

are both completely integrable.

<u>Proof:</u> Consult Theorem VIII.5, the theorems of Chapter V, and the comments above.

If $J_a{}^b$ is integrable it defines, as in the genuine case, a pseudo-group structure for the manifold (see Appendix A). More precisely:

<u>VIII.12: THEOREM:</u> Let $J_a{}^b$ be an analytic integrable modified almost Hermitian structure. Let w^α, $\overline{w^\alpha}$ and z^α, $\overline{z^\alpha}$ (not in general complex conjugates of each other) be structure coordinates for intersecting charts. Then in the intersection, the w^α are analytic functions of the z^α, and the $\overline{w^\alpha}$ are analytic functions of the $\overline{z^\alpha}$ (in the sense that power series expansions exist).

<u>Proof:</u> Again, apply Theorem VIII.5, the theorems of Chapter V, and the comments above.

The following observation will be useful later on:

<u>VIII.13: THEOREM:</u> The forms of type (1,0) and of type (0,1) for our $J_a{}^b$ are null vectors.

<u>Proof:</u> A typical form of type (1,0) is given by $a\,\ell_a + b\,m_a$, so we have $(a\,\ell_a + b\,m_a)(a\,\ell^a + b\,m^a) = 0$, and similarly for the forms of type (0,1), given by $a\,n_a + b\,\bar{m}_a$.

<u>VIII.14: THEOREM:</u> The structure coordinates obtained from an integrable modified $J_a{}^b$ are (generally complex) null coordinates.

<u>Proof:</u> If z is one of the structure coordinates, then dz is a form of type (1,0) or type (0,1); hence Theorem VIII.12 applies and dz is null.

Generalized Reality Conditions and the Product Manifold as the Complex Extension

The structure coordinates will generally no longer come in complex conjugate pairs for our modified structure, and therefore we shall now begin to write $\widetilde{z^\alpha}$ instead of $\overline{z^\alpha}$ for the coordinates arising from the type (0,1) forms. So the bar over a symbol shall still be taken to mean "genuine" complex conjugation. We note that in general,

$$\overline{z^\alpha} \neq \widetilde{z^\alpha} ,$$

$\widetilde{\overline{z^\alpha}}$ is undefined,

and $\widetilde{\widetilde{z^\alpha}}$ is undefined,

but $\overline{\widetilde{z^\alpha}}$ is defined, being the complex conjugate of $\widetilde{z^\alpha}$.

Although it is no longer generally true that $\widetilde{z}^{\alpha} = \overline{z^{\alpha}}$, the structure coordinates nevertheless do continue to obey "generalized" reality conditions. In the four real dimensions of spacetime, we should expect two complex conditions (or possibly two real and one complex condition) which generalize the standard conditions $\widetilde{z}^{0} = \overline{z^{0}}$ and $\widetilde{z}^{1} = \overline{z^{1}}$. This is as it should be, because we must expect two complex or four real conditions on four complex or eight real coordinates, in order to have two complex or four real independent numbers to describe a point in a four-real-dimensional manifold.

The algorithm for obtaining the generalized reality conditions on the structure coordinates is as follows: Solve for real coordinates x^a in terms of structure coordinates z^{α}, $\widetilde{z}^{\alpha}$, and then set $x^a = \overline{x^a}$.

As an example, suppose the structure coordinates are given in flat space by

$$z^0 = u \ ,$$

$$\widetilde{z}^0 = u + 2r \ ,$$

$$z^1 = \ln \tan \theta/2 + i\phi \ ,$$

$$\widetilde{z}^1 = \ln \tan \theta/2 - i\phi \ ,$$

with u, r, θ, ϕ real coordinates. Solving for the latter, we find:

$$u = z^0$$

$$2r = \widetilde{z}^0 - z^0$$

$$2 \ln \tan \theta/2 = z^1 + \widetilde{z}^1$$

$$2\phi = -i\,(z^1 - \widetilde{z}^1) \ .$$

Then the reality conditions are:

$$u = \bar{u} \Longleftrightarrow z^0 = \overline{z^0}$$

$$r = \bar{r} \Longleftrightarrow \overline{\tilde{z}^0} - \overline{z^0} = \tilde{z}^0 - z^0$$

$$\theta = \bar{\theta} \Longleftrightarrow \overline{z^1} + \overline{\tilde{z}^1} = z^1 + \tilde{z}^1$$

$$\phi = \bar{\phi} \Longleftrightarrow i(\overline{z^1} - \overline{\tilde{z}^1}) = -i(z^1 - \tilde{z}^1) \quad .$$

These in turn are equivalent to:

$$z^0 = \overline{z^0}$$

$$\overline{\tilde{z}^0} = \tilde{z}^0$$

$$\tilde{z}^1 = \overline{z^1} \quad ,$$

that is, z^0 and $\tilde{z}^0$ are real and z^1 and $\tilde{z}^1$ are complex conjugate (which in this particular case was obvious from the start!). Notice that we obtained two real and one complex condition, as is appropriate for four real dimensions.

As in the case of a genuine complex manifold, a <u>product manifold</u> is obtained, roughly speaking, by "violating the reality conditions." The procedure is a straightforward generalization of the ordinary case discussed in Chapter III. Given the manifold M, we construct the <u>conjugate</u> manifold $\overline{M}$ (here "conjugate" is used in the <u>ordinary</u> sense) with coordinates $z^{\tilde{\alpha}}, \tilde{z}^{\tilde{\alpha}}$ and a map $*: M \rightarrow \overline{M}$ such that

$$z^{\tilde{\alpha}}\big(*(p)\big) = \overline{z^{\alpha}}(p)$$

and $$\tilde{z}^{\tilde{\alpha}}\big(*(p)\big) = \overline{\tilde{z}^{\alpha}}(p) \quad .$$

(There is a technicality here which must be mentioned. In order to define the product manifold properly (see below) we must define the conjugate manifold in terms of coordinates z^{α} and $\widetilde{z^{\alpha}}$ on M which are genuinely complex. The only cases in which trouble might arise are those for which $\ell_a\, dx^a$ (or $n_a\, dx^a$) is itself integrable, so that one of the coordinates z^0, z^1 (or $\widetilde{z^0}$, $\widetilde{z^1}$) is real. If we start with such a coordinate, we must first make a suitable transformation to coordinates which are all genuinely complex.)

Then, as before, the product manifold is obtained as M x $\bar{M}$ (see Chapter III), with complex coordinates z^{α}, $\widetilde{z^{\alpha}}$, $z^{\hat{\alpha}}$, $\widetilde{z^{\hat{\alpha}}}$. For the four real dimensions of spacetime, the product manifold will have eight real dimensions. This corresponds to the fact that four independent complex numbers (eight real) are determined by the eight complex coordinates z^0, z^1, $\widetilde{z^0}$, $\widetilde{z^1}$, $z^{\hat{0}}$, $z^{\hat{1}}$, $\widetilde{z^{\hat{0}}}$, and $\widetilde{z^{\hat{1}}}$, after two complex reality conditions have been applied in M and two more have been applied in $\bar{M}$ (four complex conditions on eight complex variables).

This definition of the product manifold actually coincides with Ehresmann's procedure [58] for endowing a neighborhood of the diagonal of any product manifold M x M with a (genuine) complex structure, where M is assumed to be a real analytic manifold. One simply takes the real charts $\{(U, x^a), (U', x^{a'}),\dots\}$ of M , with real analytic transition functions $x^{a'} = x^{a'}(x^b),\dots$, and extends these transition functions to complex values of their arguments to obtain complex analytic transition functions $z^{a'} = z^{a'}(z^b)$. Since these functions are guaranteed to be non-singular for real values of z^b , we conclude that they remain non-singular in a finite domain away from real values of z^b . (For example, $x' = \frac{1}{1+x^2}$ is non-singular for all real values of x , and hence $z' = \frac{1}{1+z^2}$ is non-singular for a finite neighborhood of the real axis, which does not extend out to the values $z = \pm i$, where the function does become singular.) Thus we obtain an atlas $\{(\tilde{U}, z^a), (\tilde{U}', z^{a'}), \ldots\}$ where $\tilde{U}, \tilde{U}', \ldots$ can be thought of as neighborhoods in $M \times M$ which contain $U, U' \subseteq \Delta(M \times M)$, where $\Delta(M \times M) \cong M$ is the diagonal of $M \times M$, i.e., if a point of $M \times M$ is given by an ordered pair (p,q) , where $p,q \in M$, then $\Delta(M \times M)$ is given by all pairs of the form (p,p) , and the map $\mu: (p,p) \longrightarrow p$ defines an analytic diffeomorphism between $\Delta(M \times M)$ and M . Then it is easy to see that the atlas above defines a (genuine) complex manifold $\mathbb{C}M$ which is complex analytically diffeomorphic to a neighborhood $\mathcal{N}(\Delta(M \times M))$, this diffeomorphism being given by $\nu: \mathbb{C}M \longrightarrow \mathcal{N}(\Delta(M \times M)): z^a = x^a + i y^a \longrightarrow (x^a + y^a, x^a - y^a)$, where y^a is "sufficiently small". We shall refer to $\mathbb{C}M \cong \mathcal{N}(\Delta(M \times M))$ as the complex extension of M .

It is intuitively obvious that this process works, and it was presented by Ehresmann [58] as self-evident. A detailed proof, however, has been given by Shutrick [73].

The product manifold $M \times \bar{M}$, as we have defined it, is now seen to be locally identical to the manifold $\mathbb{C}M$ arising from Ehresmann's construction. That is, we can think of it as being obtained by "violating the reality conditions" of the structure coordinates $z^\alpha, \widetilde{z^\alpha}$ or equally well by violating the reality conditions of the real coordinates x^a . (The term "liberation of the coordinates" has been suggested for this procedure.) In either case the transition functions are "extended out until singularities develop". Note that our product manifold is a genuine complex manifold by virtue of this construction, and that both the "liberated" structure coordinates z^a, $z^{\tilde{a}}$ and the "liberated" real coordinates x^a become complex analytic coordinates in the complex extension $\mathbb{C}M \subseteq M \times \bar{M}$. Since we have agreed to limit ourselves to local considerations, we shall look upon $M \times \bar{M}$ as being defined only on a neighborhood of the diagonal, so that in fact it is identical to $\mathbb{C}M$.

The analytic continuation of functions and tensors from M to $\mathbb{C}M$ is now defined precisely as in Chapter IV. Notice that the analytic continuation of a tensor on M is a <u>holomorphic tensor</u> on $\mathbb{C}M$, that is, a section of one of the holomorphic tensor product bundles $T_hM \otimes \ldots \otimes T_hM \otimes T_hM^* \otimes \ldots \otimes T_hM^*$. Again, we stress the fact that the analytic continuation of zero onto the product manifold must give zero (see Remark IV.15).

Modified Complex Analysis

It is (to our knowledge) an open question as to how much of standard complex analysis carries over to the case of our modified complex manifolds with their generalized reality conditions. Certainly, some fundamental compromises have to be made in constructing a "modified complex analysis." For example, a function $f(z, \bar{z})$ for which $\partial f/\partial \bar{z} = 0$ is analytic in z, provided $\bar{z}$ is the complex conjugate of z; but this result certainly does not hold if z and $\bar{z}$ are replaced by real variables x and y. This seems to indicate that -- at the very least -- we must deal exclusively with functions analytic in all their arguments right from the very start. We must console ourselves with the fact that non-analytic functions can be approximated arbitrarily closely by analytic ones.

The modifications of other results from complex analysis must be dealt with, it is felt, one at a time as they arise. We maintain that everything used in the remainder of this work "carries through," provided that we always look at things locally and deal only with manifestly analytic functions. The generalization of results such as the Dolbeault-Grothendieck Lemma (Chapter IV) and the Newlander-Nirenberg Theorem (Chapter V) comprises, we feel, a substantial task for future mathematical research.

The Integrability Conditions $\kappa = \sigma = \nu = \lambda = 0$

We return now to the matter of the integrability conditions of our modified almost Hermitian structure, Equation VIII.18. Once again, we can either compute directly the tetrad components of the Nijenhuis tensor constructed from $J_a{}^b$, or we can apply the Frobenius Theorem directly to check the conditions for the (complex-) surface orthogonality of the bivector

$\ell_a m_b dx^a \wedge dx^b$ arising from the type (1,0) forms and the bivector $n_a \bar{m}_b dx^a \wedge dx^b$ arising from the type (0,1) forms. In fact, we have already found the latter conditions in Chapter VII, and thus we have:

<u>VIII.15: THEOREM:</u> The modified almost Hermitian structure of Equation VIII.18 is integrable if and only if

$$\kappa = \sigma = \nu = \lambda = 0 \ .$$

<u>Proof:</u> Consult Equation VII.41 and the comments above.

Thus the structure is integrable if and only if the spacetime admits two geodesic and shearfree null congruences, with the tetrad chosen so that ℓ and n are respectively tangent to these two congruences. We will refer to this property so often in the sequel that it is desirable to introduce the following terminology:

<u>VIII.16: DEFINITION:</u> A <u>2GSF</u> <u>spacetime</u> is a spacetime which admits two geodesic and shearfree null congruences.

We shall always assume that we are working in a tetrad for which $\kappa = \sigma = \nu = \lambda = 0$ when we are dealing with a 2GSF spacetime.

By virtue of the Goldberg-Sachs Theorem and its generalization to type D electrified spaces we have at once:

<u>VIII.17: THEOREM:</u> Among the vacuum and electrified spacetimes, the modified almost Hermitian structure of Equation VIII.18 is integrable if and only if the spacetime is of type D.

Proof: Apply Theorem VIII.15, Theorem VII.1, and Kinnersley's generalization of Theorem VII.1 (see Equation VII.44 and the paragraph following it).

Example: Structure Coordinates for Schwarzschild and Kerr Spacetimes

In the tetrad and coordinates of Kinnersley [26], we have the following null tetrad for Kerr spacetime (ℓ and n point along the principal null directions):

$$\ell_a = (1, 0, 0, -a\sin^2\theta) ,$$

$$n_a = \left(\frac{1}{2}\frac{r^2 - 2mr + a^2}{r^2 + a^2\cos^2\theta}, 1, 0, -\frac{1}{2} a \sin^2\theta \frac{r^2 - 2mr + a^2}{r^2 + a^2\cos^2\theta} \right) , \qquad \text{(VIII.19)}$$

$$m_a = \frac{1}{\sqrt{2}\,(r + ia\cos\theta)} \left(i a \sin\theta, 0, -(r^2 + a^2\cos^2\theta), -i\sin\theta\,(r^2 + a^2) \right) .$$

A set of structure coordinates is obtained from the forms $a\,\ell_a + b\,m_a$ of type (1,0) and the forms $a\,n_a + b\,\bar{m}_a$ of type (0,1), as follows:

$$z^0 = \int \frac{(r^2 + a^2)\ell_a + \sqrt{2}\,(r + ia\cos\theta)\, i a \sin\theta\, m_a}{r^2 + a^2\cos^2\theta}\, dx^a$$

$$= u + i a \cos\theta ,$$

$$z^1 = \int \frac{i\, a \sin\theta\, \ell_a - \sqrt{2}\,(r + i\, a \cos\theta)\, m_a}{\sin\theta\,(r^2 + a^2 \cos^2\theta)}\, dx^a$$

$$= \ln \tan \theta/2 + i\,\phi \quad ,$$

$$\tilde{z}^0 = \int \left(\frac{2(r^2 + a^2)}{r^2 - 2mr + a^2}\, n_a - \frac{\sqrt{2}\; i\, a \sin\theta}{r + i\, a \cos\theta}\, \bar{m}_a \right) dx^a$$

$$= u + 2r + 2m \ln (r^2 - 2mr + a^2)$$

$$- \frac{4m^2}{\sqrt{m^2 - a^2}} \tanh^{-1} \left(\frac{r - m}{\sqrt{m^2 - a^2}} \right)$$

$$- i\, a \cos\theta \quad , \qquad \text{(VIII.20)}$$

$$\tilde{z}^1 = \int \left(\frac{-2\, i\, a \sin\theta\, (r^2 + a^2 \cos^2\theta)}{r^2 - 2mr + a^2}\, n_a \right.$$

$$\left. - \sqrt{2}\; (r - i\, a \cos\theta)\, \bar{m}_a \right) \frac{dx^a}{\sin\theta\, (r^2 + a^2 \cos^2\theta)}$$

$$= \ln \tan \theta/2 - i\,\phi + \frac{2ia}{\sqrt{m^2 - a^2}} \tanh^{-1} \frac{r - m}{\sqrt{m^2 - a^2}} \quad .$$

Structure coordinates for Schwarzschild are obtained by setting the angular momentum parameter a equal to zero:

$$z^0 = u \quad ,$$

$$z^1 = \ln \tan \theta/2 + i\,\phi \quad ,$$

$$\tilde{z}^0 = u + 2r + 2m \ln (r^2 - 2mr) - 4m \tanh^{-1} \left(\frac{r-m}{m} \right)$$

$$= u + 2r + 4m \ln(r-2m) + \text{constant}, \qquad \text{(VIII.21)}$$

$$\tilde{z}^1 = \ln \tan \theta/2 - i\,\phi \ .$$

Modified Kählerian Spacetimes

We have seen in Chapter VI that the "natural" differential geometry of an Hermitian manifold coincides with the ordinary Riemannian geometry if and only if the manifold is Kählerian. It therefore behooves us to examine the Kähler condition $d(J_{ab}\, dx^a \wedge dx^b) = 0 \Leftrightarrow J_{ab;c} = 0$ for the case of our modified Hermitian structure. We have the following result:

VIII.18: THEOREM: The modified Hermitian structure $J_a{}^b$ satisfies the Kählerian condition $d(J_{ab}\, dx^a \wedge dx^b) = 0 \Leftrightarrow J_{ab;c} = 0$ if and only if, in addition to

$$\kappa = \sigma = \nu = \lambda = 0 \ ,$$

we have

$$\rho = \mu = \pi = \tau = 0 \qquad \text{(VIII.22)}$$

for the given tetrad.

Proof: A straightforward computation of the tetrad components of $J_{ab;c} = 0$ gives

$$\kappa = \sigma = \nu = \lambda = \rho = \mu = \pi = \tau = 0 \ .$$

Consulting Table VII.2, we find that none of the Type D vacuum spacetimes satisfy the additional conditions of Equation VIII.22. The same is true for the electrified Type D spacetimes. To find 2GSF spacetimes which do satisfy the Kählerian condition, we have a clue in the form of the following theorem:

<u>VIII.19: THEOREM:</u> Under the conformal rescaling of Equation VII.51, the modified almost Hermitian structure is invariant, but the Kählerian condition is not invariant.

<u>Proof:</u> Consulting Equations VII.51, we see that under the conformal rescaling VII.50, we have

$$J_a{}^b \rightarrow J_a{}^b \quad .$$

On the other hand, for the Kähler form J_{ab}, we have

$$J_{ab} \rightarrow \Omega^2 J_{ab} \quad ,$$

and $d(J_{ab}\, dx^a \wedge dx^b) = 0$ is not equivalent to $d(\Omega^2 J_{ab}\, dx^a \wedge dx^b) = 0$.

This result suggests that to find 2GSF spacetimes which satisfy the Kählerian condition, we should look for spacetimes conformally related to the Type D vacuum (and electrified) spacetimes. The conformally related spacetime will carry the same modified complex structure, but it will have a different Kähler form, given by $\Omega^2 J_{ab}$; our task, then, is to find a conformal factor for which $\Omega^2 J_{ab}$ is a closed form.

We shall show below that the Kähler form J_{ab} of a modified almost Hermitian structure is a <u>self-dual</u> two-form (this may be verified by direct computation). Hence $\Omega^2 J_{ab}$ is also self-dual, so that in looking for an Ω^2 such that $d(\Omega^2 J_{ab}\, dx^a \wedge dx^b) = 0$, we are looking for a closed self-dual two-form, which is to say, for a (test) <u>solution to Maxwell's equations</u> on the spacetime. Because of the eigenvector structure of $J_a{}^b$, Theorem VIII.5 Equation (iii), this Maxwell field is algebraically general, with principal null directions given by ℓ_a and n_a. And in fact we find the following result:

<u>VIII.20: THEOREM:</u> Every type D vacuum spacetime admits a Maxwell field with principal null directions coaligned with the two repeated principal null directions of the Weyl tensor. The self-dual Maxwell tensor is given by $F_{ab} = \Omega^2 J_{ab}$, with

$$\Omega^2 = (\Psi_2)^{2/3} .$$

<u>Proof:</u> The tetrad components of such an F_{ab} are given by (see Equation VII.17):

$$\Phi_0 = 0 ,$$

$$\Phi_1 = 2i\Omega^2 ,$$

$$\Phi_2 = 0 .$$

Since the space is type D vacuum, all of the curvature quantities must vanish except for Ψ_2, and in addition, $\kappa = \sigma = \nu = \lambda = 0$. Inserting this information into the Bianchi Identities, Equations VII.22, we find:

$$0 = 0$$

$$0 = 0$$

$$-3D\Psi_2 = -9\rho\Psi_2$$

$$-3\delta\Psi_2 = -9\tau\Psi_2$$

$$3\bar{\delta}\Psi_2 = -9\pi\Psi_2$$

$$3\Delta\Psi_2 = -9\mu\Psi_2$$

$$
\begin{aligned}
0 &= 0 \\
0 &= 0 \\
0 &= 0 \\
0 &= 0 \\
0 &= 0
\end{aligned}
\qquad \text{(VIII.23)}
$$

And Maxwell's Equations, Equations VII.19, become:

$$
\begin{aligned}
D\,\Phi_1 &= 2\rho\,\Phi_1 \\
-\delta\,\Phi_1 &= 2\pi\,\Phi_1 \\
\delta\,\Phi_1 &= 2\tau\,\Phi_1 \\
-\Delta\Phi_1 &= 2\mu\Phi_1 \quad .
\end{aligned}
$$

Then, we find by using Equations VIII.23 that

$$\Phi_1 = (\Psi_2)^{2/3}$$

solves Maxwell's Equations, whence, we obtain

$$2i\,\Omega^2 = (\Psi_2)^{2/3} \quad ,$$

Of course, we can drop the factor of 2i and take

$$\Omega^2 = (\Psi_2)^{2/3} \quad ,$$

as desired. (In the electrified case, the result follows in a similar manner after Φ_{11} has been found; for details, see Kinnersley [26], especially Equations V.1 - V.3, V.8, and V.14.

See also Hughston and Sommers [23].)

There is one important "catch" to all of this: The correct conformal factor $\Omega^2 = (\Psi_2)^{2/3}$ will in general be complex (see Table VII.2). This means that our conformally related Kählerian spacetime will have a complex metric (in a real coordinate system). Because the metric is complex, the spin coefficients $\bar{\alpha}$, $\bar{\beta}$,... are no longer given by the complex conjugates of α, β,... . In particular, while our construction here succeeds in reducing ρ, μ, π, and τ to zero, it does not follow that $\bar{\rho}$, $\bar{\mu}$, $\bar{\pi}$, and $\bar{\tau}$ will also vanish, unless the conformal factor Ω^2 is real (this will be the case only for Schwarzschild, "B", and "C" spacetimes). Clearly, a careful examination of complex metrics, complex conformal factors, and complex tetrads is called for at this point; this will be done at the beginning of the next chapter. In the meantime, we can summarize our results as follows:

VIII.21: THEOREM: A type D vacuum spacetime is conformally related, by a (generally complex) conformal vector $\Omega^2 = (\Psi_{2(\text{vac})})^{2/3}$, to a space with a complex metric for which

$$\kappa = \sigma = \nu = \lambda = \rho = \mu = \pi = \tau = 0$$

(with a similar result holding for the type D electrified spaces).

Proof: See above.

Half-Hermitian Spacetimes

The 2GSF spacetimes for which our modified almost Hermitian structure is integrable form a very interesting but also a very small set of spacetimes (e.g., only type D among vacuum spaces). It would be nice to obtain results for a larger set of spacetimes. In fact, we shall now show that we can get some of the properties of an Hermitian spacetime for a class which includes all of the algebraically special vacuum spacetimes.

We have mentioned above that in the case of our complex-valued almost Hermitian structure tensor, the integrability of the type (1,0) forms $(J_a{}^b + i\delta_a{}^b)dx^a$ does not imply the integrability of the type (0,1) forms $(J_a{}^b - i\delta_a{}^b)dx^a$. Hence it is possible to have one-half of the integrability conditions satisfied without the other half being satisfied. Namely, $\ell_a m_b\, dx^a \wedge dx^b$ may be complex-surface-orthogonal while $n_a \bar{m}_b\, dx^a \wedge dx^b$ is not, so that structure coordinates z^0, z^1 can be obtained while their formal complex conjugates $\widetilde{z^0}$, $\widetilde{z^1}$ are not defined. It is easy to see that we have the following state of affairs:

<u>VIII.22: DEFINITION:</u> A 1GSF spacetime is a spacetime which admits at least one geodesic and shearfree null congruence.

<u>VIII.23: DEFINITION:</u> A modified almost Hermitian structure is said to be <u>half-integrable</u> if and only if

$$\kappa = \sigma = 0$$

$$\text{or} \qquad \nu = \lambda = 0 \; .$$

<u>VIII.24: THEOREM:</u> A modified almost Hermitian structure is half-integrable if and only if the spacetime is a 1GSF spacetime. Such a spacetime admits structure coordinates z^0, z^1 or $\widetilde{z^0}$, $\widetilde{z^1}$, but not necessarily both.

<u>Proof:</u> See above.

<u>VIII.25: COROLLARY:</u> The modified almost Hermitian structure of an algebraically special vacuum spacetime is half-integrable.

<u>Proof:</u> Apply the Goldberg-Sachs Theorem.

Duality and Almost m-Hermitian Structure

We recall that the duality operation "*" operating on the vector space B of bivectors over a point of spacetime is defined as follows (see Schild [48]):

$$T_{ab}{}^{*} = \frac{1}{2}\,\eta_{abcd}\,T^{cd} \,, \qquad \text{(VIII.24)}$$

where

$$\eta_{abcd} = \sqrt{-g}\;\epsilon_{abcd} \,,$$

$$\epsilon_{0123} = -1 \,.$$

Because of the Lorentz signature of the spacetime metric, the duality operator obeys the following property:

$$(T_{ab}{}^{*})^{*} = -\,T_{ab} \,. \qquad \text{(VIII.25)}$$

A self-dual bivector $T_{ab}{}^{+}$ is a complex bivector (element of the complexification B^{c} of B) for which

$$(T_{ab}{}^{+})^{*} = i\,T_{ab}{}^{+} \,. \qquad \text{(VIII.26)}$$

Given any bivector T_{ab}, the bivector

$$T_{ab}^{+} \equiv T_{ab} + i\, T_{ab}^{*} \qquad \text{(VIII.27)}$$

is self-dual.

Now, "*" is an endomorphism of B into itself, and by virtue of Equation VIII.25 , the square of this endomorphism is minus the identity. Therefore "*" is a vector space complex structure on B (see Chapter II).

In a recent preprint, Brans [6] has identified "*" as a vector space complex structure, and has shown that Einstein's vacuum field equations are equivalent to the condition that "*" commutes with the curvature two-form.

We shall now show that "*" can be described remarkably naturally in terms of our almost m-Hermitian structure of Equation VIII.18. From Equation VIII.26 it is seen that the type (1,0) elements arising from the complex structure "*" are precisely the complex self-dual bivectors. It can be shown (see Sachs [47]) that a basis for the self-dual bivectors is given, in terms of a null tetrad ℓ_a, n_a, m_a, $\bar{m}_a$, by

$$V_1 = \ell_a m_b \, dx^a \wedge dx^b$$

$$V_2 = (\ell_a n_b - m_a \bar{m}_b) \, dx^a \wedge dx^b \qquad \text{(VIII.28)}$$

$$V_3 = n_a \bar{m}_b \, dx^a \wedge dx^b \; .$$

Rewriting this as

$$V_1 = \ell_a \, dx^a \wedge m_b \, dx^b$$

$$V_2 = - i \, J_{ab} \, dx^a \wedge dx^b \qquad \text{(VIII.29)}$$

$$V_3 = n_a \, dx^a \wedge \bar{m}_b \, dx^b \; ,$$

it is seen that the self-dual bivectors (bivectors of type (1,0) with respect to "*") are given by the forms of type (2,0) with respect to $J_a{}^b$, the forms of type (0,2), and the forms proportional to $J_{ab} \, dx^a \wedge dx^b$, the Kähler form. If $J_a{}^b$ is integrable, then a basis for the self-dual bivectors is given by

$$W_1 = dz^0 \wedge dz^1$$

$$W_2 = J_{ab} \, dz^a \wedge dz^b \qquad \text{(VIII.30)}$$

$$W_3 = d\tilde{z}^0 \wedge d\tilde{z}^1 \; .$$

We see that $dW_1 = dW_3 = 0$ always, and $dW_2 = 0$ if and only if $J_a{}^b$ is (m-left-) Kählerian (see below).

CHAPTER IX: KÄHLERIAN SPACETIMES

In this chapter we shall use the NP formalism to solve the Einstein field equations assuming that the metric is real and that a null tetrad exists for which the almost m-Hermitian structure is integrable and the m-Kählerian condition is satisfied. In this way we shall obtain the most general (real) m-Kählerian spacetime that can be obtained by modifying standard complex manifold theory in this particular fashion. We shall find that these m-Kählerian spacetimes possess a much stronger structure than might be supposed to start with: they admit a locally product Riemannian structure (see Appendix B) which together with the m-Kählerian structure forms what we shall call a "crossed" structure for the spacetime. This "crossed" structure determines the coordinates up to a gauge-like sort of freedom. We shall also find that these m-Kählerian spacetimes admit (in addition to the Kähler **fore** J_{ab}) a covariantly constant tensor which is closely related to the conformal Killing tensors of Hughston and Sommers [23].

It might be argued that the rich structure of our m-Kählerian spacetimes merely reflects how very strong a restriction the m-Kählerian conditions are, and that precisely because these conditions are so restrictive, the m-Kählerian spacetimes are rather uninteresting. This objection cannot be raised, however, in the case of spaces with a <u>complex</u> metric and tetrad satisfying the m-Kählerian conditions. For reasons discussed below, we shall call such a space a modified left-Kählerian spacetime. There is a large class of interesting m-left-Kählerian spacetimes, namely, those whose complex extensions are the class of (left) <u>Heavenly spacetimes</u> recently discovered by Newman and Penrose. We shall discuss Heavenly spacetimes and their relation to m-left-Kählerian spacetimes in Chapter XI.

We begin this chapter with a discussion of complex Lorentz transformations, complex null tetrads, and complex metrics, as these become essential in what follows.

Complex Lorentz Transformations, Complex Null Tetrads, and Complex Metrics

A formulation of a physical theory which is invariant under real Lorentz transformations is also invariant under complex Lorentz transformations, except for reality conditions on quantities appearing in the formulation. Here, by complex Lorentz transformation, we mean a transformation represented by a complex $\Lambda_a{}^b$ such that $\Lambda_a{}^m \Lambda_b{}^n \eta_{mn} = \eta_{ab}$, where $\eta_{ab} = \mathrm{diag}(+1, -1, -1, -1)$.

In the Newman-Penrose null tetrad formulation of general relativity, Lorentz transformations act on the tetrad legs satisfying the normalization conditions of Equation VII.11: In this context, a real Lorentz tetrad transformation is one which preserves the conditions of Equation VII.11 as well as the reality conditions of the tetrad, i.e., ℓ_a and n_a real, m_a and $\bar{m}_a$ complex conjugates. A complex Lorentz tetrad transformation violates these reality conditions, but still preserves Equation VII.11: The new tetrad will satisfy generalized reality conditions different from the standard ones. Thus the NP formulation of general relativity is invariant under complex tetrad rotations (i.e., the NP equations go into linear combinations of themselves) except for the fact that the reality conditions on the null tetrad are generally changed (hence reality conditions on other variables are also changed).

In the Newman-Penrose formalism, the (proper) real Lorentz transformations of the tetrad are given by Equations VII.24 through VII.26, with A and ϕ real and $\bar{a}, \bar{b}$ the complex conjugates of a, b. The (proper) complex Lorentz transformations are obtained by letting A and ϕ become complex and letting $\bar{a}, \bar{b}$ assume values different from the complex conjugates of a, b. It is easy to verify directly that such transformations preserve the null character of the tetrad legs and the normalization conditions of Equation VII.11, and furthermore that the NP equations remain invariant (go into linear combinations of themselves) under these transformations. We shall refer to a null tetrad for which the ordinary reality conditions are violated as a "complex null tetrad", and a null tetrad for which the ordinary reality conditions do hold, a "real null tetrad". For a complex null tetrad, we shall write $\tilde{m}^a$ instead of $\bar{m}^a$.

As an example of an improper complex Lorentz transformation, we have the Sachs transformation of Equation VII.32.

When dealing with a complex null tetrad, care must be exercised in the use of the NP formalism. A general rule of thumb is the following: an NP quantity which is real for a real null tetrad should never appear with a bar over it even when the tetrad is complex, and an NP quantity which appears with a bar over it for a real null tetrad should be written with a "twiddle" over it when the tetrad is complex, to indicate that it is no longer the complex conjugate of the unbarred quantity. For example, the spin coefficient $\tilde{\sigma}$ is given by $\tilde{\sigma} = \ell_{a;b}\, \tilde{m}^a \tilde{m}^b$ ($\ell_{a;b}$ does not appear with a bar over it, even though ℓ_a is now complex); and is not equal to the complex conjugate of $\sigma = \ell_{a;b}\, m^a m^b$.

Similarly, the curvature quantities $\widetilde{\Psi}_0$, $\widetilde{\Psi}_1, \ldots$ will no longer be the complex conjugates of $\Psi_0, \Psi_1, \ldots$ and Φ_{00} will not be a real funciton, Φ_{01} and Φ_{10} will not be complex conjugates, and so forth. For example,

$$\widetilde{\Psi}_2 \equiv -C_{abcd}\, m^a n^b \ell^c \tilde{m}^d \neq \overline{\Psi_2},$$

$$\Phi_{00} \equiv -\tfrac{1}{2} R_{ab} \ell^a \ell^b \neq \overline{\Phi_{00}},$$

$$\Phi_{01} \equiv -\tfrac{1}{2} R_{ab} \ell^a m^b \neq \overline{\Phi_{10}}.$$

As an example of the proper transcription of the NP equations, consider the fourth of Equations VII.20 and its complex conjugate:

$$D\alpha - \bar{\delta}\epsilon = (\rho + \bar{\epsilon} - 2\epsilon)\alpha + \beta\bar{\sigma} - \bar{\beta}\epsilon - \mu\lambda - \bar{\mu}\gamma + (\epsilon + \rho)\pi + \Phi_{10},$$

$$D\bar{\alpha} - \delta\bar{\epsilon} = (\bar{\rho} + \epsilon - 2\bar{\epsilon})\bar{\alpha} + \bar{\beta}\sigma - \beta\bar{\epsilon} - \bar{\mu}\bar{\lambda} - \mu\bar{\gamma} + (\bar{\epsilon} + \bar{\rho})\bar{\pi} + \Phi_{01}.$$

For a complex null tetrad, these two equations are properly rewritten as:

$$D\alpha - \hat{\delta}\epsilon = (\rho + \hat{\epsilon} - 2\epsilon)\alpha + \beta\hat{\sigma} - \hat{\beta}\epsilon - \mu\lambda - \hat{\mu}\gamma + (\epsilon + \rho)\pi + \Phi_{10},$$

$$D\hat{\alpha} - \delta\hat{\epsilon} = (\hat{\rho} + \epsilon - 2\hat{\epsilon})\hat{\alpha} + \hat{\beta}\sigma - \beta\hat{\epsilon} - \hat{\mu}\hat{\lambda} - \mu\hat{\gamma} + (\hat{\epsilon} + \hat{\rho})\hat{\pi} + \Phi_{01}.$$

From the spinor point of view, a complex null tetrad corresponds to choosing a spinor dyad $(\tilde{o}^{A'}, \tilde{\iota}^{A'})$ which is not the complex conjugate of the dyad (o^A, ι^A), and a complex Lorentz transformation of the tetrad corresponds to transforming (o^A, ι^A) and $(\tilde{o}^{A'}, \tilde{\iota}^{A'})$ by <u>independent</u> elements of $SL(2, \mathbb{C})$.

Up to this point, we have been implicitly assuming that the underlying metric g_{ab} of spacetime is a <u>real</u> tensor. But in view of Theorem VIII.21 (and in order to treat Heavenly spaces) we must now consider what happens when the metric is allowed to be complex, i.e., when g_{ab} is allowed to be a <u>complexified tensor</u>. (Note, however, that we are still considering the underlying manifold M to be a <u>four-real-dimensional</u> manifold with real coordinates x^0, x^1, x^2, x^3, or possibly with structure coordinates $z^0, z^1, z^{\bar{0}}, z^{\bar{1}}$ satisfying generalized reality conditions.) For such a complex-valued g_{ab}, we can still construct tetrads such that

$g_{ab} = l_a n_b + n_a l_b - m_a \tilde{m}_b - \tilde{m}_a m_b$ with the normalizations

$l^a n_a = -m^a \tilde{m}_a = 1$ holding, but the standard reality conditions on $l^a, n^a, m^a, \tilde{m}^a$ will now generally be impossible to fulfill. This means that if we consider the complex Lorentz group as the "tetrad gauge group", there is nothing left in the geometry to single out the real Lorentz group as a subgroup which preserves anything over and above the normalization conditions.

Suppose, then, that we apply the NP formalism to a spacetime with such a complex metric g_{ab}. There will generally be <u>no</u> choice of tetrad for which $\tilde{\alpha}, \tilde{\beta}, \ldots, \tilde{\Psi}_0, \ldots$ are the complex conjugates of $\alpha, \beta, \ldots, \Psi_0, \ldots$. In terms of spinors, the $\sigma^a{}_{AA'}$ cannot be chosen to be Hermitian, and the $\Gamma^{A'}{}_{B'm}$ will not be the complex conjugates of the $\Gamma^{A}{}_{Bm}$ (so we will write them as $\tilde{\Gamma}^{A'}{}_{B'm}$) . Hence, even though we can still choose a dyad $(\tilde{o}^{A'}, \tilde{\iota}^{A'})$ which is complex conjugate to any choice of dyad (o^{A}, ι^{A}), such a choice will no longer produce a real tetrad or NP quantities $\tilde{\alpha}, \tilde{\beta}, \ldots, \tilde{\Psi}_{0}, \ldots$ which, are complex conjugates of $\alpha, \beta, \ldots, \Psi_{0}, \ldots,$ because $\nabla_{AA'} = \sigma^{a}{}_{AA'} \nabla_{a}$ is no longer an Hermitian operator.

We have already been led to consider complex metrics in Chapter VIII when we sought to find spaces satisfying the m-Kählerian condition $dJ = 0$. We found that this condition was expressible as

$$\kappa = \sigma = \nu = \lambda = \rho = \mu = \pi = \tau = 0,$$

and was satisfied in spaces with complex metrics of the form

$$g_{ab} = \Omega^{2}\, g_{ab(vac)}$$

where $g_{ab(vac)}$ is the (real) metric of a Type D vacuum or electrified spacetime and $\Omega^{2} = (\Psi_{2(vac)})^{2/3}$ is a generally complex conformal factor. It was mentioned at the time that $\tilde{\kappa}, \tilde{\sigma}, \tilde{\nu}, \tilde{\lambda}, \tilde{\rho}, \tilde{\mu}, \tilde{\pi}$ and $\tilde{\tau}$ will <u>not</u> generally vanish for these spaces, even though their formal conjugates are zero.

For two complex metrics related by a complex conformal factor as in Equation VII.50, and with the choice of Equations VII.51 for the scaling of the tetrad, the transformation of the NP equations is still given by Equations VII.52, with the proviso that "bars go to twiddles" and "no $\bar{\Omega}$'s may appear". For example,

$$\hat{\alpha} = \Omega^{-1}\alpha - \tfrac{1}{2}\,\Omega^{-2}\,\tilde{\delta}\Omega$$

$$\hat{\tilde{\alpha}} = \Omega^{-1}\tilde{\alpha} - \tfrac{1}{2}\,\Omega^{-2}\,\delta\Omega \quad , \text{etc.}$$

At this point a recapitulation of terminology is in order.

IX.1: **DEFINITION**: A complex spacetime is a four-real-dimensional real manifold with a complex metric $g_{ab} \in \Gamma(\mathbb{C}TM^* \otimes \mathbb{C}TM^*)$ which is everywhere non-singular and of signature (+---). If g_{ab} is real ($g_{ab} \in \Gamma(TM^* \otimes TM^*)$), the spacetime will be called a real spacetime.

Since there should be little danger of confusion at this point, we shall henceforth drop the adjective "modified" or its abbreviation "m-" when referring to the various structures defined in terms of our tensor $J_a{}^b$ of Equation VIII.11.

IX.2: **DEFINITION**: A complex spacetime is said to be left-Hermitian iff there exists a tetrad for which $J_a{}^b$ is integrable , or equivalently, for which $\kappa = \sigma = \nu = \lambda = 0$; it is said to be full-Hermitian if for the same tetrad we also have $\tilde{\kappa} = \tilde{\sigma} = \tilde{\nu} = \tilde{\lambda} = 0$.

IX.3: DEFINITION: A complex spacetime is said to be left-Kählerian iff there exists a tetrad for which $J_a{}^b$ integrable and $J_a{}^b$ is closed, or equivalently, for which $\kappa = \sigma = \nu = \lambda = \rho = \mu = \pi = \tau = 0$; it is said to be full-Kählerian if for the same tetrad we also have $\tilde{\kappa} = \tilde{\sigma} = \tilde{\nu} = \tilde{\lambda} = \tilde{\rho} = \tilde{\mu} = \tilde{\pi} = \tilde{\tau} = 0$.

We could also define right-Hermitian and right-Kählerian spacetimes by switching the roles of the "twiddled" and "untwiddled" spin coefficients in the above definitions. These terms would refer to the

the properties of the structure tensor $\overline{J_a{}^b}$ (genuine complex conjugate!) The "left-right" terminology has been introduced by Newman and Penrose in a different but analogous context (see Chapter XI).

<u>IX.4: THEOREM</u>: If a real spacetime is left-Hermitian, it is full-Hermitian; and if it is left-Kählerian, it is full-Kählerian.

<u>Proof</u>: Suppose $(\ell^a, n^a, m^a, \tilde{m}^a)$ is a complex null tetrad which makes the spacetime left-Hermitian. Since the metric is real, the $\sigma^a{}_{AA'}$ can be chosen to be Hermitian; make such a choice and let (o^A, ι^A), $(\tilde{o}^{A'}, \tilde{\iota}^{A'})$ be spinor dyads corresponding to (ℓ^a, n^a, m^a, m^a). Perform a complex Lorentz transformaition which leaves (o^A, ι^A) unchanged, but sends $(\tilde{o}^{A'}, \hat{\iota}^{A'})$ into $(\bar{o}^{A'}, \bar{\iota}^{A'})$ the complex conjugate of (o^A, ι^A). This will leave the "untwiddled" spin coefficients unchanged but will transform the "twiddled" ones into the complex conjugates of the "untwiddled" (since $\nabla_{AA'}$ is Hermitian). <u>Hence</u> $\kappa_{new} = \kappa_{old} = 0 \Rightarrow \hat{\kappa}_{new} = \overline{\kappa_{new}} = \overline{\kappa_{old}} = 0$, and so forth.

In view of this theorem, the designations "full-Hermitian" and "full-Kählerian" are redundant in the case of a <u>real</u> spacetime, so we shall henceforth simply refer to "real Hermitian spacetimes" and "real Kählerian spacetimes". But we must be careful to retain the distinction between "left-" and "full-" properties in the case of complex spacetimes.

Derivation of the Real Kählerian Spacetimes

We are now in a position to derive an explicit form of the metric for the most general real Kählerian spacetime. In view of theorem IX.4, we can assume we are given a real null tetrad for which

$$\kappa = \sigma = \nu = \lambda = \rho = \mu = \pi = \tau = 0 ,$$

$$\bar{\kappa} = \bar{\sigma} = \bar{\nu} = \bar{\lambda} = \bar{\rho} = \bar{\mu} = \bar{\pi} = \bar{\tau} = 0 . \qquad \text{(IX.1)}$$

From these it follows, of course, that

$$\kappa = \rho - \bar{\rho} = 0$$

$$\sigma = - \lambda = \bar{\pi} + \tau = 0$$

$$-\bar{\sigma} = \lambda = \pi + \bar{\tau} = 0 \qquad \text{(IX.2)}$$

and $$\nu = \mu - \bar{\mu} = 0 .$$

But these are precisely the conditions of Equations VII.40a for the vectors l, m, $\bar{m}$, and n to be hypersurface-orthogonal. (Recall that we have agreed to use the term "hypersurface orthogonal" in a purely formal sense when it refers to a complexified vector; the precise analytical meaning of such a statement is expressed by Equations IX.3 below.) We can therefore write:

$$dz^0 = a\, l_a dx^a$$

$$dz^1 = b\, m_a dx^a$$

$$d\tilde{z}^0 = e\, n_a dx^a$$

$$d\bar{z}^1 = \bar{b}\, \bar{m}_a dx^a \qquad \text{(IX.3)}$$

for some real coordinate system x^a on M and some functions z^0, z^1, $\tilde{z}^0$, a, b, and e of these coordinates. From the reality of ℓ and n, we conclude that a, e, z^0, and $\tilde{z}^0$ can be chosen to be real functions of x^a.

Since the tetrad legs are linearly independent, the functions z^0, z^1, $\tilde{z}^0$, and $\tilde{z}^1 \equiv \overline{z^1}$ can be taken as coordinates. Because of Equations IX.3, they are in fact structure coordinates for the integrable $J_a{}^b$ of the spacetime (i.e., dz^0 and dz^1 are linear combinations of the type (1,0) forms and $d\tilde{z}^0$ and $d\tilde{z}^1$ are linear combinations of the type (0,1) forms). We have already obtained the generalized reality conditions for these coordinates; they are simply that z^0 and $\tilde{z}^0$ are real and $\tilde{z}^1 = \overline{z^1}$.

In these coordinates, the null tetrad becomes:

$$
\begin{aligned}
\ell_a &= a^{-1}\, \delta_a{}^0 \\
m_a &= b^{-1}\, \delta_a{}^1 \\
n_a &= e^{-1}\, \delta_a{}^{\tilde{0}} \\
\bar{m}_a &= \bar{b}^{-1}\, \delta_a{}^{\tilde{1}}
\end{aligned}
\qquad , \qquad \text{(IX.4)}
$$

where the indices $\tilde{0}$, $\tilde{1}$ refer to the coordinates $\tilde{z}^0$, $\tilde{z}^1$. The corresponding contravariant components follow easily from the normalization conditions for a null tetrad:

$$
\begin{aligned}
\ell^a &= e\, \delta_{\tilde{0}}{}^a \\
m^a &= -\bar{b}\, \delta_{\tilde{1}}{}^a \\
n^a &= a\, \delta_0{}^a \\
\bar{m}^a &= -b\, \delta_1{}^a \, .
\end{aligned}
\qquad \text{(IX.5)}
$$

Constructing the metric in these coordinates according to Equation VII.12 we find:

$$ds^2 = 2\,A\,dz^0\,dz^{\widetilde{0}} - 2\,B\,dz^1\,dz^{\widetilde{1}}\,, \qquad \text{(IX.6)}$$

where $A = a^{-1}\,e^{-1}$ and $B = b^{-1}\,\bar{b}^{-1}$. So we have:

$$g_{0\widetilde{0}} = A \qquad g_{1\widetilde{1}} = -\,B\,,$$

$$g^{0\widetilde{0}} = A^{-1} \qquad g^{1\widetilde{1}} = -\,B^{-1}\,, \qquad \text{(IX.7)}$$

all other components zero.

We must now return and make further use of the conditions of Equations IX.1. These are now used to obtain conditions on the functions A and B of Equation IX.6. For example, from the condition $\rho = 0$ we obtain:

$$\begin{aligned} 0 = \rho &= \ell_{a;b}\, m^a\, \bar{m}^b \\ &= \ell_{\widetilde{1};1}\,(-\bar{b})\,(-b) \\ &= B^{-1}\,(\ell_{\widetilde{1},1} - \Gamma^m_{\widetilde{1}1}\,\ell_m) \\ &= -B^{-1}\,\Gamma^0_{\widetilde{1}1}\,a^{-1} \end{aligned} \qquad \text{(IX.8a)}$$

$$\Rightarrow \Gamma^0_{11} = 0 = \tfrac{1}{2}\,A^{-1}\,B_{,\widetilde{0}}$$

$$\Rightarrow B_{,\widetilde{0}} = 0\,.$$

In a similar manner, we find from the rest of Equations IX.1 that

$$B_{,0} = 0$$

$$A_{,1} = 0$$

and $$A_{,\tilde{1}} = 0 . \qquad \text{(IX.8b)}$$

We have now arrived at the following theorem:

<u>IX.2: THEOREM</u>: The most general real Kählerian spacetime which arises from the modified almost Hermitian structure of Equation VIII.18 is given by

$$ds^2 = 2\,A\,(z^0, z^{\tilde{0}})\,dz^0\,dz^{\tilde{0}} - 2\,B\,(z^1, z^{\tilde{1}})\,dz^1\,dz^{\tilde{1}} \;, \qquad \text{(IX.9)}$$

where $z^0, z^{\tilde{0}}, z^1, z^{\tilde{1}}$ are structure coordinates for the Hermitian structure, with reality conditions

$$z^0 = \overline{z^0}$$
$$z^{\tilde{0}} = \overline{z^{\tilde{0}}}$$
$$z^{\tilde{1}} = \overline{z^1} \;,$$

and the functions A and B take on real values when their arguments satisfy these reality conditions.

The only non-vanishing Christoffel symbols for the metric IX.9 are given by:

$$\Gamma^{0}_{00} = A^{-1} A_{,0}$$

$$\Gamma^{\tilde{0}}_{\tilde{0}\tilde{0}} = A^{-1} A_{,\tilde{0}}$$

$$\Gamma^{1}_{11} = B^{-1} B_{,1}$$

$$\Gamma^{\tilde{1}}_{\tilde{1}\tilde{1}} = B^{-1} B_{,\tilde{1}} , \qquad \text{(IX.10)}$$

all others zero.

We now want to obtain the curvature information for our real Kählerian spacetimes in terms of the NP quantities Ψ_i, Φ_{ij}, and Λ. It turns out to be easiest to first perform a tetrad rotation taking us from the tetrad of Equations IX.4 and IX.5 to the following complex tetrad:

$$l_a = \delta_a{}^0$$

$$m_a = \delta_a{}^1$$

$$n_a = A\, \delta_a{}^{\tilde{0}}$$

$$\tilde{m}_a = B\, \delta_a{}^{\tilde{1}}$$

$$l^a = A^{-1}\, \delta_{\tilde{0}}{}^a$$

$$m^a = -\, B^{-1}\, \delta_{\tilde{1}}{}^a$$

$$n^a = \delta_0{}^a$$

$$\tilde{m}^a = -\, \delta_1{}^a . \qquad \text{(IX.11)}$$

This transformation preserves Equations IX.1. At this point it is easiest to proceed by computing the remaining spin coefficients α, β, γ, ϵ,

$\tilde{\alpha}$, $\hat{\beta}$, $\tilde{\gamma}$, and $\tilde{\epsilon}$ and then inserting the results into the NP Equations VII.20. We have:

$$\alpha = \frac{1}{2} B^{-1} B_{,1} \qquad \beta = 0$$

$$\tilde{\alpha} = 0 \qquad \hat{\beta} = -\frac{1}{2} B^{-1} B_{,1}$$

$$\gamma = -\frac{1}{2} A^{-1} A_{,0} \qquad \epsilon = 0$$

$$\tilde{\gamma} = -\frac{1}{2} A^{-1} A_{,0} \qquad \tilde{\epsilon} = 0 \; . \tag{IX.12}$$

Then the "untwiddled" NP equations give the following information:

$$0 = \Phi_{00}$$

$$0 = \Psi_0$$

$$0 = \Psi_1 + \Phi_{01}$$

$$0 = \Phi_{10}$$

$$0 = \Psi_1$$

$$-\frac{1}{2} A^{-1} (A^{-1} A_{,0})_{,\tilde{0}} = \Psi_2 + \Phi_{11} - \Lambda$$

$$0 = \Phi_{20}$$

$$0 = \Psi_2 + 2\Lambda$$

$$0 = \Psi_3 + \Phi_{21}$$

$$0 = -\Psi_4$$

$$0 = -\Psi_1 + \Phi_{01} \tag{IX.13}$$

$$-\tfrac{1}{2}B^{-1}(B^{-1}B_{,1})_{,\tilde{1}} = -\Psi_2 + \Phi_{11} + \Lambda$$

$$0 = -\Psi_3 + \Phi_{21}$$

$$0 = \Phi_{22}$$

$$0 = \Phi_{12}$$

$$0 = \Phi_{02}$$

$$0 = -\Psi_2 - 2\Lambda$$

$$0 = -\Psi_3 \quad .$$

From these we find directly that the only non-vanishing "untwiddled" curvature quantities are given by:

$$24\,\Psi_2 = -4A^{-1}(A^{-1}A_{,0})_{,\tilde{0}} + 4B^{-1}(B^{-1}B_{,1})_{,\tilde{1}}$$

$$24\,\Phi_{11} = -6A^{-1}(A^{-1}A_{,0})_{,\tilde{0}} - 6B^{-1}(B^{-1}B_{,1})_{,\tilde{1}}$$

$$24\,\Lambda = 2A^{-1}(A^{-1}A_{,0})_{,\tilde{0}} - 2B^{-1}(B^{-1}B_{,1})_{,\tilde{1}} \quad . \qquad \text{(IX.14)}$$

The only quantities left to be obtained are the $\widetilde{\Psi}_i$. Rather than solve the "twiddled" NP equations for these, we shall transform to a real null tetrad, where we will then have $\widetilde{\Psi}_i = \overline{\Psi}_i$. The new tetrad is:

$$l^a = A^{-1}\,\delta_{\tilde{0}}{}^a$$

$$m^a = -B^{-1/2}\,\delta_{\tilde{1}}{}^a$$

$$n^a = \delta_0{}^a$$

$$\bar{m}^a = -B^{-1/2}\,\delta_1{}^a \quad . \qquad \text{(IX.15)}$$

This is obtained from the tetrad IX.11 by a complex "spin" transformation, of the form Equation VII.25 with $e^{i\phi} = B^{1/2}$ and $A = 1$ (since B is real, choose $B^{1/2} = +\sqrt{B}$). The transformed curvature quantities are now easily found to be

$$24\,\Psi_2 = -4A^{-1}(A^{-1}A_{,0})_{,\hat{0}} + 4B^{-1}(B^{-1}B_{,1})_{,\hat{1}}$$

$$24\,\Phi_{11} = -6A^{-1}(A^{-1}A_{,0})_{,\tilde{0}} - 6B^{-1}(B^{-1}B_{,1})_{,\tilde{1}}$$

$$24\,\Lambda = 2A^{-1}(A^{-1}A_{,0})_{,\tilde{0}} - 2B^{-1}(B^{-1}B_{,1})_{,\tilde{1}}$$

$$\tilde{\Psi}_2 = \overline{\Psi_2} \qquad \text{(IX.16)}$$

all others zero.

From these we have at once the following interesting results:

<u>IX.3: THEOREM</u>: If a real Kählerian spacetime is Ricci flat, then it is flat.

<u>Proof</u>: From Equations IX.16, $R_{ab} = 0 \Leftrightarrow \Phi_{11} = \Lambda = 0$. But $\Lambda = -2\,\Psi_2$, so $R_{ab} = 0 \Rightarrow R_{abcd} = 0$.

<u>IX.4: THEOREM</u>: A real Kählerian spacetime is of type D, with repeated principal null directions dz^0 and $d\tilde{z}^0$.

<u>Proof</u>: $\Psi_0 = \Psi_1 = \Psi_3 = \Psi_4 = 0$ in the real null tetrad of Equations IX.15. Note the result does <u>not</u> follow from the Goldberg-Sachs theorem, because the Kählerian spacetimes are not vacuum (unless they are flat).

<u>IX.5: THEOREM</u>: Among the type D vacuum spacetimes, Schwarzschild, "B", and "C" are conformally related to real Kählerian spacetimes.

Proof: Recall from Theorem VIII.21 and Definition IX.2 that the complex spaces $g_{ab} = (\Psi_{2(vac)})^{2/3} g_{ab(vac)}$ are left-Kählerian, where $g_{ab(vac)}$ is any of the type D vacuum spaces. Then g_{ab} will be real Kählerian if and only if $\Psi_{2(vac)}$ is real, and from Table VII.2, this occurs only for Schwarzschild, "B", and "C".

IX.6: THEOREM: Minkowski space with rectangular null corrdinates

$$\sqrt{2}\,u = t + z$$
$$\sqrt{2}\,v = t - z$$
$$\sqrt{2}\,\zeta = x + iy$$
$$\sqrt{2}\,\bar{\zeta} = x - iy \;,$$

for which

$$ds^2 = 2\,dudv - 2d\zeta\, d\bar{\zeta} \quad ,$$

is a real Kählerian spacetime.

Proof: Put $A = B = 1$ in Equation IX.9.

Structure of the Real Kählerian Spacetimes

The real Kählerian spacetimes which we have found possess a very rich structure even beyond the Kählerian structure. In addition to the latter, they admit a locally product structure which arises from the integrability of the almost product structure given by

$$K_a{}^b = \ell_a n^b + n_a \ell^b + m_a \bar{m}^b + \bar{m}_a m^b \;. \qquad \text{(IX.17)}$$

The integrability conditions for this almost product structure are given by

$$\begin{aligned} \rho - \bar{\rho} &= 0 \\ \mu - \bar{\mu} &= 0 \\ \pi + \bar{\tau} &= 0 \\ \bar{\pi} + \tau &= 0 \quad , \end{aligned} \tag{IX.18}$$

which are satisfied for our Kählerian spacetime in the given tetrads, by virtue of Equation IX.1.

The locally product structure is even richer than this, It is, in fact, a locally decomposable Riemannian structure. This means that coordinates x^A and $x^{\tilde{A}}$ exist for which the metric takes the form

$$\begin{aligned} ds^2 = {} & g_{AB}\,(x^M)\, dx^A\, dx^B \\ & + g_{\tilde{A}\tilde{B}}\,(x^{\tilde{M}})\, dx^{\tilde{A}}\, dx^{\tilde{B}} \quad , \end{aligned} \tag{IX.19}$$

in any system of structure coordinates. In our case, from Equation IX.11, we see that product structure coordinates are given by

$$\begin{aligned} x^A &= z^0, z^{\tilde{0}} \\ x^{\tilde{A}} &= z^1, z^{\tilde{1}} \end{aligned} \tag{IX.20}$$

For more details on product structures, Appendix B should be consulted.

The product structure together with the Kählerian structure of our spacetimes comprise what we shall call a "crossed" structure; this terminology refers to the following diagram:

z^0	$z^{\tilde{0}}$
z^1	$z^{\tilde{1}}$

DIAGRAM IX.1

The Kählerian structure limits coordinate transformations to those which mix z^0, z^1 and $\widetilde{z^0}$, $\widetilde{z^1}$ (vertical line in diagram), and the locally decomposable Riemannian structure limits coordinate transformations to those which mix z^0, $\widetilde{z^0}$ and z^1, $\widetilde{z^1}$ (horizontal line in diagram). Thus the two structures taken together define a "crossed" structure for which the only allowable coordinate transformations are the "gauge-like" transformations

$$\begin{aligned} z^0 &\to f(z^0) \\ \widetilde{z^0} &\to g(\widetilde{z^0}) \\ z^1 &\to h(z^1) \\ \widetilde{z^1} &\to e(\widetilde{z^1}) \ . \end{aligned} \qquad \text{(IX.21)}$$

Symmetries in Real Kählerian Spacetimes

Given the rich structure discussed above for our modified Kählerian spacetimes, it is perhaps not too surprising to find that these spacetimes admit symmetries described by a Killing tensor.

First, we have the following result:

IX.7: THEOREM: For the modified Kählerian spacetimes of Equation IX.9, and the tetrad of Equations IX.11, the tensor

$$K_{ab} = \ell_a n_b + n_a \ell_b + m_a \bar{m}_b + \bar{m}_a m_b \qquad \text{(IX.22)}$$

is covariantly constant, and hence is a Killing tensor.

<u>Proof</u>: A computation of the tetrad components of $K_{ab;c} = 0$ gives precisely the conditions of Equations IX.1. Then $K_{ab;c} = 0 \Rightarrow K_{(ab;c)} = 0$.

We notice that the Killing tensor K_{ab} is precisely the (covariant form of the) almost product structure tensor of Equation IX.17. However, the integrability conditions for the almost product structure are <u>not</u> equivalent to the Killing equation for K_{ab} .

In accordance with the discussion in Chapter VII, the Killing tensor K_{ab} gives rise to constants of the motion $K_{ab}\, v^a\, v^b$ for geodesics, where $v^a = dz^a/d\lambda$ and $z^a(\lambda)$ is an affinely parametrized geodesic. In fact, by a direct examination of the geodesic equations,

$$\frac{d\, v^a}{d\, \lambda} + \Gamma^a_{bc}\, v^b\, v^c = 0$$

we find, using Equations IX.10, that these equations "separate" (by virtue of the existence of the locally decomposable Riemannian structure) to give two constants of the motion

$$C = A\, v^0\, v^0$$

$$\text{and} \qquad D = B\, v^1\, v^1 \; . \tag{IX.23}$$

It is then easy to see that the constant of motion $(C + D)$ is that which arises from the Killing tensor K_{ab}, and the constant of motion $(C - D)$ is that which arises from the metric Killing tensor g_{ab} . (Note in this context that K_{ab} considered as a metric tensor is projectively related to g_{ab} , i.e., they both have the same Christoffel symbols, given by Equations IX.10.)

We now limit our attention to real Kählerian spacetimes which are conformally related to Schwarzschild, "B", and "C" spacetimes (see Theorem IX.5). In these cases, K_{ab} is precisely the conformal Killing tensor for type D spacetimes found by Hughston and Sommers (see [23]). Thus K_{ab} gives (compare Chapter VII) constants of motion $(C + D)$ for null geodesics in Schwarzschild, "B", and "C". (The constant of motion $(C - D)$ will be zero for these geodesics.)

Left-Kählerian Spacetimes

We now want to consider the more general situation of a left-Kählerian spacetime, i.e., a complex spacetime with a complex null tetrad satisfying the Kählerian conditions $\kappa = \sigma = \nu = \lambda = \rho = \mu = \pi = \tau = 0$, <u>without</u> necessarily having $\tilde{\kappa} = \tilde{\sigma} = \tilde{\nu} = \tilde{\lambda} = \tilde{\rho} = \tilde{\mu} = \tilde{\pi} = \tilde{\tau} = 0$. We shall now derive the most general form of the metric for such a spacetime.

First we note that, since the Frobenius Theorem has nothing to do with whether the metric is real or not, Equations VII.41 continue to hold, with bars replaced by "twiddles", in the case of a complex null tetrad. Hence the integrability conditions $\kappa = \sigma = \nu = \lambda = 0$ continue to guarantee the existence of four complex functions z^0, z^1, $\tilde{z}^0$, and $\tilde{z}^1$ such that

$$dz^0 = A' \ell_a dx^a + B' m_a dx^a$$

$$dz^1 = C' \ell_a dx^a + D' m_a dx^a$$

$$d\tilde{z}^0 = E' n_a dx^a + F' \tilde{m}_a dx^a$$

$$d\tilde{z}^1 = G' n_a dx^a + H' \tilde{m}_a dx^a , \qquad \text{(IX.24)}$$

for some functions $A', \ldots, H'$, with $A'D' - B'C' \neq 0$ and $E'H' - F'G' \neq 0$. (Here the x^a are some unspecified real coordinates on the underlying manifold.) Now dz^0, dz^1, $d\widetilde{z^0}$, and $d\widetilde{z^1}$ are linearly independent, so we can invert the two pairs of these equations to obtain

$$\begin{aligned} \ell_a dx^a &= A''\, dz^0 + B''\, dz^1 \\ m_a dx^a &= C''\, dz^0 + D''\, dz^1 \\ n_a dx^a &= E''\, d\widetilde{z^0} + F''\, d\widetilde{z^1} \\ \widetilde{m}_a dx^a &= G''\, d\widetilde{z^0} + H''\, d\widetilde{z^1} \end{aligned} \tag{IX.25}$$

for some functions $A'', \ldots, H''$, with

$$\begin{aligned} &A''D'' - B''C'' \neq 0 \\ \text{and} \quad &E''H'' - F''G'' \neq 0 \, . \end{aligned} \tag{IX.26}$$

Because dz^0, dz^1, $d\widetilde{z^0}$, and $d\widetilde{z^1}$ are linearly independent, we can now take z^0, z^1, $\widetilde{z^0}$, $\widetilde{z^1}$ as structure coordinates for a patch of the manifold. They will satisfy some definite but unascertainable generalized reality conditions which reduce them to describing the four real degrees of freedom required from the four-real-dimensionality of the underlying manifold. (The best we can do toward specifying these reality conditions is to point out that they must be equivalent to $\overline{x^a} = x^a$.) It will be convenient to rename these structure coordinates as follows:

$$\begin{aligned} u &\equiv z^0 \\ \zeta &\equiv z^1 \\ v &\equiv \widetilde{z^0} \\ \widetilde{\zeta} &\equiv \widetilde{z^1} \, . \end{aligned} \tag{IX.27}$$

When enumerating the components of tensors, we shall use the indices 0, 1, $\tilde{0}$, $\tilde{1}$ - always in that order - to denote the u-, ζ-, v-, and $\tilde{\zeta}$-components, respectively. For example,

$$\ell_a = (\ell_0, \ell_1, \ell_{\tilde{0}}, \ell_{\tilde{1}}) \quad ,$$

$$g_{ab} = \begin{pmatrix} g_{00} & g_{01} & g_{0\tilde{0}} & g_{0\tilde{1}} \\ g_{10} & g_{11} & g_{1\tilde{0}} & g_{1\tilde{1}} \\ g_{\tilde{0}0} & g_{\tilde{0}1} & g_{\tilde{0}\tilde{0}} & g_{\tilde{0}\tilde{1}} \\ g_{\tilde{1}0} & g_{\tilde{1}1} & g_{\tilde{1}\tilde{0}} & g_{\tilde{1}\tilde{1}} \end{pmatrix} , \qquad \text{(IX.28)}$$

and so forth.

Using Equations IX.25, IX.26, and VII.12, we can now write the metric of our left-Kahlerian spacetime as

$$ds^2 = 2\,A\,du\,dv + 2\,B\,d\zeta\,d\tilde{\zeta} + 2\,C\,du\,d\tilde{\zeta} + 2\,D\,d\zeta\,dv \qquad \text{(IX.29)}$$

where

$$\begin{aligned} A &= A''E'' - C''G'' \\ B &= B''F'' - D''H'' \\ C &= A''F'' - C''H'' \\ D &= B''E'' - D''G'' . \end{aligned} \qquad \text{(IX.30)}$$

We also write this as (see Equation IX.28)

$$g_{ab} = \begin{pmatrix} 0 & 0 & A & C \\ 0 & 0 & D & B \\ A & D & 0 & 0 \\ C & B & 0 & 0 \end{pmatrix} \qquad \text{(IX.31)}$$

or

$$g_{\alpha\tilde{\beta}} = \begin{pmatrix} A & C \\ D & B \end{pmatrix} . \qquad \text{(IX.32)}$$

The contravariant metric is found to be:

$$g^{ab} = \begin{pmatrix} 0 & 0 & B\Delta^{-1} & -D\Delta^{-1} \\ 0 & 0 & -C\Delta^{-1} & A\Delta^{-1} \\ B\Delta^{-1} & -C\Delta^{-1} & 0 & 0 \\ -D\Delta^{-1} & A\Delta^{-1} & 0 & 0 \end{pmatrix} \tag{IX.33}$$

or

$$g^{\hat{\beta}\gamma} = \begin{pmatrix} B\Delta^{-1} & -C\Delta^{-1} \\ -D\Delta^{-1} & A\Delta^{-1} \end{pmatrix} \tag{IX.34}$$

where $\Delta \equiv AB - CD$. (IX.35)

Notice the two-by-two blocks of zeroes on the diagonals of g_{ab} and g^{ab} , which signal the Hermitian character of the metric. From Equation IX.30 we find that $\Delta = AB - CD = (A''D'' - B''C'')(F''G'' - E''H'')$; this is always nonzero by virtue of Equations IX.26, so g_{ab} is always non-singular:

$$\det(g_{ab}) = -[\det(g_{\alpha\hat{\beta}})]^2 = -\Delta^2 \neq 0 . \tag{IX.36}$$

We have not yet made use of the remaining conditions $\rho = \mu = \pi = \tau = 0$. We can "use up" the information contained in these conditions by recalling that they are equivalent to the condition that the Kähler form be closed. From Equation IX.32 and Theorem VI.9, we find that the Kähler form is given by

$$J = 2i\,(A du \wedge dv + B d\zeta \wedge d\hat{\zeta} + C\, du \wedge d\hat{\zeta} + D d\zeta \wedge dv) . \tag{IX.37}$$

Therefore

$$dJ = 2i\,[A_{,1}\, d\zeta \wedge du \wedge dv + A_{,\tilde{1}}\, d\tilde{\zeta} \wedge du \wedge dv$$
$$+ B_{,0}\, du \wedge d\zeta \wedge d\tilde{\zeta} + B_{,\tilde{0}}\, dv \wedge d\zeta \wedge d\tilde{\zeta}$$
$$+ C_{,1}\, d\zeta \wedge du \wedge d\tilde{\zeta} + C_{,\tilde{0}}\, dv \wedge du \wedge d\tilde{\zeta}$$
$$+ D_{,0}\, du \wedge d\zeta \wedge dv + D_{,\tilde{1}}\, d\tilde{\zeta} \wedge d\zeta \wedge dv\,] \qquad \text{(IX.38)}$$

and

$$dJ = 0 \qquad \begin{aligned} A_{,1} &= D_{,0} \\ A_{,\tilde{1}} &= C_{,\tilde{0}} \\ B_{,0} &= C_{,1} \\ B_{,\tilde{0}} &= D_{,\tilde{1}} \end{aligned} \quad . \qquad \text{(IX.39)}$$

The information contained in the left-Kähler conditions has now been exhausted. It will be noted that if the metric is derivable from a (complex-valued) "Kähler scaler" K, so that

$$g_{\alpha\tilde{\beta}} = \begin{pmatrix} A & C \\ D & B \end{pmatrix} = \begin{pmatrix} K_{,0\tilde{0}} & K_{,0\tilde{1}} \\ K_{,1\tilde{0}} & K_{,1\tilde{1}} \end{pmatrix}, \qquad \text{(IX.40)}$$

then the conditions IX.39 will be automatically satisfied. Hence we would like to invoke an analogue of Theorem VI.23, which guarantees the local existence of a Kähler scalar in the standard case of a positive definite real Kählerian metric. But the proof of this theorem depends ultimately on the Dolbeault-Grothendieck Lemma (Theorem IV.33), and a re-examination of the proof of the latter may leave one reluctant to assert that a similar proof will go through for our modified complex manifolds. In any case, we will be able to prove the existence of a Kähler scalar by more direct means, so we leave it as a conjecture that a "modified Dolbeault-Grothendieck Lemma" can in fact be proven.

To show that a local, complex-valued Kähler scalar always exists for our left-Kählerian spacetimes, it suffices to realize that Poincare's Lemma (Theorem IV.21) holds for complexified forms (complex-valued forms) as well as real forms. (If $J = \alpha + i\beta$, where α and β are real k-forms, then $dJ = 0 \Rightarrow d\alpha + i\, d\beta = 0 \Rightarrow d\alpha = d\beta = 0 \Rightarrow \alpha = d\gamma$ and $\beta = d\epsilon$ locally $\Rightarrow J = d(\gamma + i\epsilon)$ locally.) The only restrictive assumption we will have to make is that all our functions are analytic (in the power series sense); but this restriction has been a general feature of our modified complex analysis all along (compare the remarks on page 198).

Now to the construction of the Kähler scalar. Since $DJ = 0$, we can write (locally)

$$\begin{aligned} J &= d\,(E\,du + F\,dv + G\,d\zeta + H\,d\hat{\zeta}) \\ &= (F,_0 - E,_{\hat{0}})\,du \wedge dv + (G,_0 - E,_1)\,du \wedge d\zeta \\ &\quad + (H,_0 - E,_{\hat{1}})\,du \wedge d\hat{\zeta} + (G,_{\hat{0}} - F,_1)\,dv \wedge d\zeta \\ &\quad + (H,_{\hat{0}} - F,_{\hat{1}})\,dv \wedge d\hat{\zeta} + (H,_1 - G,_{\hat{1}})\,d\zeta \wedge d\hat{\zeta} \end{aligned} \qquad \text{(IX.40)}$$

Comparing Equations IX.40 and IX.37, we conclude that

$$G,_0 - E,_1 = 0$$

$$\text{and} \qquad H,_{\hat{0}} - F,_{\hat{1}} = 0 \ . \qquad \text{(IX.41)}$$

Now go to the product manifold, by "liberating" u, v, ζ , $\tilde{\zeta}$ from their generalized reality conditions. Take a v = const, $\tilde{\zeta}$ = const slice of the product manifold: on this submanifold, call it N , the first of Equations IX.41 states that $E\,du + G\,d\zeta\ /_N$ is a closed one-form. Hence Poincare's Lemma applies to give a local scalar function $\hat{\psi}$ on N for which $(E\,du + G\,d\zeta)/_N = (\hat{\psi}_{,0}\,du + \hat{\psi}_{,1}\,d\zeta)/_N$. $\hat{\psi}$ will depend analytically on v, $\tilde{\zeta}$ as N varies. Hence in a finite ball in the product manifold, we have

$$E\,du + G\,d\zeta \;=\; \psi_{,0}\,du + \psi_{,1}\,d\zeta \qquad \text{(IX.42)}$$

for some $\psi(u, \zeta, v, \tilde{\zeta})$. Similarly, obtain a scalar χ on an open ball in the product manifold for which

$$F\,dv + H\,d\tilde{\zeta} \;=\; \chi_{,\tilde{0}}\,dv + \chi_{,\tilde{1}}\,d\tilde{\zeta}\;. \qquad \text{(IX.43)}$$

The intersection I of the two balls can be chosen to be nonzero. Apply the generalized reality conditions to project I down to a finite region of the spacetime manifold, on which Equations IX.42 and IX.43 continue to hold. Insert Equations IX.42 and IX.43 into IX.40 to obtain:

$$\begin{aligned} J = {} & (\chi_{,\tilde{0}0} - \psi_{,0\tilde{0}})\,du\wedge dv + (\psi_{,10} - \psi_{,01})\,du\wedge d\zeta \\ & + (\chi_{,\tilde{1}0} - \psi_{,0\tilde{1}})\,du\wedge d\tilde{\zeta} + (\psi_{,1\tilde{0}} - \chi_{,\tilde{0}1})\,dv\wedge d\zeta \\ & + (\chi_{,\tilde{1}\tilde{0}} - \chi_{,\tilde{0}\tilde{1}})\,dv\wedge d\tilde{\zeta} + (\chi_{,\tilde{1}1} - \psi_{,1\tilde{1}})\,d\zeta\wedge d\tilde{\zeta}\,. \end{aligned} \qquad \text{(IX.44)}$$

Finally, defining $2iK = \chi - \psi$ and comparing Equations IX.44 with IX.37, we obtain the desired result (in a suitably small region):

$$\begin{aligned} A &= g_{0\tilde{0}} = K_{,0\tilde{0}} \\ B &= g_{1\tilde{1}} = K_{,1\tilde{1}} \\ C &= g_{0\tilde{1}} = K_{,0\tilde{1}} \\ D &= g_{1\tilde{0}} = K_{,1\tilde{0}}\;. \end{aligned} \qquad \text{(IX.45)}$$

We have arrived at the following:

IX.8: THEOREM: The most general analytic left-Kählerian spacetime is given locally by

$$ds^2 = 2\,A\,du\,dv + 2\,B\,d\zeta\,d\hat{\zeta} + 2\,C\,du\,d\hat{\zeta} + 2\,D\,d\zeta\,dv \tag{IX.46}$$

where $u, v, \zeta, \hat{\zeta}$ satisfy generalized reality conditions, and A, B, C, D are functions satisfying Equations IX.39 and hence are derivable from a complex-valued analytic Kähler scalar $K(u, v, \zeta, \hat{\zeta})$ as in Equations IX.45.

IX.9: THEOREM: All of the type D vacuum and electrified spacetimes are conformally related to left-Kählerian spacetimes of the above form, with a generally complex conformal factor.

Proof: Restatement of Theorem VIII.21 in updated forms.

We now would like to obtain the curvature information for our left-Kählerian manifolds, using the NP formalism. We choose the following null tetrad, in which the Kählerian conditions are satisfied (see Equation IX.28):

$$\begin{aligned}
\ell_a &= (1 \quad 0 \quad 0 \quad 0)\\
m_a &= (0 \quad 1 \quad 0 \quad 0)\\
\tilde{m}_a &= (0 \quad 0 \quad -D \quad -B)\\
n_a &= (0 \quad 0 \quad A \quad C)\\
\ell^a &= (0 \quad 0 \quad B/\Delta \quad -D/\Delta)\\
m^a &= (0 \quad 0 \quad -C/\Delta \quad A/\Delta)\\
\tilde{m}^a &= (0 \quad -1 \quad 0 \quad 0)\\
n^a &= (1 \quad 0 \quad 0 \quad 0)\ .
\end{aligned} \tag{IX.47}$$

The Christoffel symbols are readily calculated with the help of Theorem VI.17 (and Corollary VI.20):

$$
\begin{aligned}
\Gamma^{0}_{00} &= g^{0\hat{\mu}}\, g_{0\hat{\mu},0} = \frac{B}{\Delta} A_{,0} - \frac{D}{\Delta} C_{,0} \\
\Gamma^{0}_{01} &= g^{0\hat{\mu}}\, g_{0\hat{\mu},1} = \frac{B}{\Delta} A_{,1} - \frac{D}{\Delta} C_{,1} \\
\Gamma^{0}_{11} &= g^{0\hat{\mu}}\, g_{1\hat{\mu},1} = \frac{B}{\Delta} D_{,1} - \frac{D}{\Delta} B_{,1} \\
\\
\Gamma^{1}_{00} &= g^{1\hat{\mu}}\, g_{0\hat{\mu},0} = \frac{-C}{\Delta} A_{,0} + \frac{A}{\Delta} C_{,0} \\
\Gamma^{1}_{10} &= g^{1\hat{\mu}}\, g_{1\hat{\mu},0} = \frac{-C}{\Delta} D_{,0} + \frac{A}{\Delta} B_{,0} \\
\Gamma^{1}_{11} &= g^{1\hat{\mu}}\, g_{1\hat{\mu},1} = \frac{-C}{\Delta} D_{,1} + \frac{A}{\Delta} B_{,1} \\
\\
\Gamma^{\tilde{0}}_{\tilde{0}\tilde{0}} &= g^{\tilde{0}\mu}\, g_{\tilde{0}\mu,\tilde{0}} = \frac{B}{\Delta} A_{,\tilde{0}} - \frac{C}{\Delta} D_{,\tilde{0}} \\
\Gamma^{\tilde{0}}_{\tilde{0}\tilde{1}} &= g^{\tilde{0}\mu}\, g_{\tilde{0}\mu,\tilde{1}} = \frac{B}{\Delta} A_{,\tilde{1}} - \frac{C}{\Delta} D_{,\tilde{1}} \\
\Gamma^{\tilde{0}}_{\tilde{1}\tilde{1}} &= g^{\tilde{0}\mu}\, g_{\tilde{1}\mu,\tilde{1}} = \frac{B}{\Delta} C_{,\tilde{1}} - \frac{C}{\Delta} B_{,\tilde{1}} \\
\\
\Gamma^{\tilde{1}}_{\tilde{0}\tilde{0}} &= g^{\tilde{1}\mu}\, g_{\tilde{0}\mu,\tilde{0}} = \frac{-D}{\Delta} A_{,\tilde{0}} + \frac{A}{\Delta} D_{,\tilde{0}} \\
\Gamma^{\tilde{1}}_{\tilde{0}\tilde{1}} &= g^{\tilde{1}\mu}\, g_{\tilde{0}\mu,\tilde{1}} = \frac{-D}{\Delta} A_{,\tilde{1}} + \frac{A}{\Delta} D_{,\tilde{1}} \\
\Gamma^{\tilde{1}}_{\tilde{1}\tilde{1}} &= g^{\tilde{1}\mu}\, g_{\tilde{1}\mu,\tilde{1}} = \frac{-D}{\Delta} C_{,\tilde{1}} + \frac{A}{\Delta} B_{,\tilde{1}}
\end{aligned}
\tag{IX.48}
$$

all others zero.

From these, the spin coefficients for the tetrad IX.47 are found to be:

$$
\begin{aligned}
\alpha &= \tfrac{1}{2}\,\Delta^{-1}\,\Delta_{,1} \\
\beta &= 0 \\
\gamma &= -\tfrac{1}{2}\,\Delta^{-1}\,\Delta_{,0} \\
\epsilon &= 0 \\
\kappa &= \lambda = \mu = \nu = \pi = \rho = \sigma = \tau = 0 \\
\tilde{\alpha} &= 0 \\
\tilde{\beta} &= \tfrac{1}{2}\,\Delta^{-1}\,(B\,A_{,1} - A\,B_{,1} + C\,D_{,1} - D\,C_{,1}) \\
\tilde{\gamma} &= \tfrac{1}{2}\,\Delta^{-1}\,(-B\,A_{,0} + A\,B_{,0} - C\,D_{,0} + D\,C_{,0}) \\
\tilde{\epsilon} &= 0 \\
\tilde{\kappa} &= 0 \\
\tilde{\lambda} &= 0 \\
\tilde{\mu} &= \Delta^{-1}\,(A\,B_{,0} - C\,D_{,0}) \\
\tilde{\nu} &= \Delta^{-1}\,(C\,A_{,0} - A\,C_{,0}) \\
\tilde{\pi} &= 0 \\
\tilde{\rho} &= 0 \\
\tilde{\sigma} &= \Delta^{-1}\,(D\,B_{,1} - B\,D_{,1}) \\
\tilde{\tau} &= \Delta^{-1}\,(B\,A_{,1} - D\,C_{,1}) \quad .
\end{aligned}
\tag{IX.49}
$$

It is now a lengthy but straightforward matter to insert these spin coefficients into the NP Equations VII.20, and their "twiddled" versions, to obtain the curvature quantities Ψ_i, $\tilde{\Psi}_i$, Φ_{ij}, and Λ. We shall merely quote the results here:

$$
\begin{aligned}
\Psi_2 &= \tfrac{1}{3}(\mathcal{D}\gamma - \delta\alpha) \\
\tilde{\Psi}_0 &= \mathcal{D}\tilde{\sigma} \\
\tilde{\Psi}_1 &= \mathcal{D}\tilde{\beta} \\
\tilde{\Psi}_2 &= \mathcal{D}\tilde{\gamma} - \tfrac{1}{3}\delta\alpha - \tfrac{2}{3}\mathcal{D}\gamma \\
\tilde{\Psi}_3 &= \delta\tilde{\gamma} \\
\tilde{\Psi}_4 &= \delta\tilde{\nu} \\
\Phi_{10} &= \mathcal{D}\alpha \\
\Phi_{11} &= \tfrac{1}{2}(\delta\alpha + \mathcal{D}\gamma) \\
\Phi_{12} &= \delta\gamma \\
\Lambda &= \tfrac{1}{6}(\delta\alpha - \mathcal{D}\gamma),
\end{aligned}
\qquad \text{(IX.50)}
$$

where $\mathcal{D} \equiv \Delta^{-1} B\, \partial_{\tilde{0}} - \Delta^{-1} D\, \partial_{\tilde{1}}$

and $\delta \equiv -\Delta^{-1} C\, \partial_{\tilde{0}} + \Delta^{-1} A\, \partial_{\tilde{1}}$.

Zero Rest Mass Free Fields

We now want to consider (test) solutions to the zero rest mass free field equations on the background of a left-Kählerian complex space-time. The most straightforward version of these equations is the spinor form:

$$\nabla^{A'A} \varphi_{AB\ldots L} = 0 \qquad \text{(IX.51)}$$

$$\tilde{\nabla}^{AA'} \tilde{\varphi}_{A'B'\ldots L'} = 0 \quad . \qquad \text{(IX.52)}$$

Here $\varphi_{AB\ldots L}$ and $\tilde{\varphi}_{A'B'\ldots L'}$ are the field spinors and $\nabla^{A'A} = \epsilon^{A'B'} \epsilon^{AB} \sigma^{a}{}_{BB'} \nabla_a$ is the spinor covariant derivative. $\varphi_{AB\ldots L}$ and $\tilde{\varphi}_{A'B'\ldots L'}$ are both completely symmetric in all of their indices, and both have n indices for a field of spin $= \frac{1}{2}n$. If n is even, then Equations IX.51 are equivalent to

$$\nabla_{[a} K_{bc]de\ldots \ell m} = 0 \quad , \qquad \text{(IX.53)}$$

where $K_{bcde\ldots \ell m}$ is the (complexified) tensor corresponding to the spinor

$$\varphi_{BCDE\ldots LM}\, \epsilon_{B'C'}\, \epsilon_{D'E'} \cdots \epsilon_{L'M'} +$$

$$\epsilon_{BC}\, \epsilon_{DE} \cdots \epsilon_{LM}\, \tilde{\varphi}_{B'C'D'E'\ldots L'M'} \quad . \qquad \text{(IX.54)}$$

(See Penrose and MacCallum [44] for more details on these matters.)

In a <u>real</u> spacetime, a physically realistic <u>classical</u> field is required to satisfy the Hermitian property

$$\tilde{\varphi}_{A'B'\ldots L'} = \bar{\varphi}_{A'B'\ldots L'} . \qquad \text{(IX.55)}$$

(For even spin, this means that $K_{bcde\ldots \ell m}$ is a real tensor.) In a complex spacetime, where $\tilde{\Gamma}^{A'}{}_{B'm} \neq \bar{\Gamma}^{A'}{}_{B'm}$, this requirement loses its significance. Furthermore, the field equations IX.51 and IX.52 are now completely independent, in the sense that the first set involve only the "untwiddled" spin coefficients, and the second set only the "twiddled" spin coefficients. In view of Equations IX.49, it is easy to foresee that Equations IX.51 are greatly simplified by the left-Kählerian conditions, while Equations IX.52 remain quite untractable.

However, there is some motivation for considering solutions to Equations IX.51 alone and ignoring Equations IX.52. The zero rest mass free field equations are <u>conformally invariant</u> if we take the scaling to be

$$\begin{aligned} \varphi_{AB\ldots L} &\rightarrow \Omega^{-s-1}\, \varphi_{AB\ldots L} \\ \tilde{\varphi}_{A'B'\ldots L'} &\rightarrow \Omega^{-s-1}\, \tilde{\varphi}_{A'B'\ldots L'} \end{aligned} \qquad \text{(IX.56)}$$

when $g_{ab} \rightarrow \Omega^2 g_{ab}$ and $\sigma^a{}_{AA'} \rightarrow \Omega^{-1} \sigma^a{}_{AA'}$, $\epsilon_{AB} \rightarrow \epsilon_{AB}$ (see Penrose [ ]). Because of this conformal invariance, we can make use of Theorem IX.8 to obtain Hermitian zero rest mass fields on Type D vacuum and electrified spacetimes, by solving Equations IX.51 in the corresponding left-Kählerian spacetime. Specifically, if $\varphi_{AB\ldots L}$ is a solution of

IX.51 in the left-Kählerian space, then $\Omega^{s+1}\ \varphi_{AB\ldots L}$ is a solution in the vacuum or electrified spacetimes. But the latter are real spacetimes, so $\overline{\Omega^{s+1}\ \varphi_{AB\ldots L}}$ will automatically satisfy Equations IX.52. Thus the desired Hermitian fields are obtained.

When one is interested in a quantum treatment of zero rest mass test fields, there is even more justification for considering Equations IX.51 alone. As pointed out by Penrose [64], positive energy solutions of IX.51 and IX.52 with $\tilde{\varphi}_{A'B'\ldots L'} = 0$ represent right-handed solutions, i.e., states of helicity $+\hbar$, while positive energy solutions with $\varphi_{AB\ldots L} = 0$ represent left-handed solutions, i.e., states of helicity $-\hbar$. Thus positive energy solutions to Equations IX.51 alone may be thought of as given right-handed states in our left-Kählerian spacetimes.

Accordingly, we now examine Equations IX.51 in left-Kählerian spacetimes. (In what follows, we shall ignore the so-called Buchdahl conditions [78], which are algebraic constraints on the test fields for spin $s \geq 3/2$ in the presence of a curved background. These constraints couple curvature quantities algebraically to the $\varphi_{AB\ldots L}$ and $\tilde{\varphi}_{A'B'\ldots L'}$. However, in the cases of greatest physical interest, Heavenly left-Kählerian spacetimes, the Buchdahl conditions are vacuous for right-handed solutions.) It is a straightforward but tedious exercise to reexpress Equations IX.51 for any spin $s = \frac{1}{2}n$ in terms of $D, \Delta, \delta, \tilde{\delta}$, the spin coefficients, and the dyad components of $\varphi_{AB\ldots L}$ defined by

$$\varphi_p = \varphi_{AB\ldots L}\, \iota^A \cdots \iota^D o^E \cdots o^L , \qquad \text{(IX.57)}$$

where p iotas and $n-p$ omicrons appear, as p takes on the values $0, 1, 2, \ldots, n$. (For the Maxwell field, $n = 2$, $\varphi_0 = \Phi_0$, $\varphi_1 = \Phi_1$, and $\varphi_2 = \Phi_2$.) The resulting expressions for the field equations are generally quite involved. However, for a left-Kählerian spacetime with the tetrad IX.47, the field equations are considerably simplified. After a moderately long calculation we find that they are given by

$$D\,\varphi_r - \hat{\delta}\,\varphi_{r-1} + (n - 2r + 2)\,\alpha\,\varphi_{r-1} = 0$$
$$\delta\,\varphi_r - \Delta\,\varphi_{r-1} + (n - 2r + 2)\,\gamma\,\varphi_{r-1} = 0 \, , \qquad \text{(IX.57)}$$

$$r = 1, 2, \ldots, n = 2s \, ,$$

$$\text{where} \quad \alpha = \tfrac{1}{2}\Delta^{-1}\Delta_{,1} \quad \text{and} \quad \gamma = -\tfrac{1}{2}\Delta^{-1}\Delta_{,0} \, .$$

Solutions on Left-Kählerian Spacetimes

Suppose we look for solutions to these equations with the property that φ_p is the only non-vanishing component of the field, for some value of p between 0 and n. This would mean that the field has 0_A as a p-fold repeated principal spinor and ι_A as an (n-p)-fold repeated principal spinor (see Penrose [41]. (Note that for a non-Hermitian field, the principal spinors of $\tilde{\varphi}_{A'B'\ldots L'}$ need not be the complex conjugates of the principal spinors of $\varphi_{AB\ldots L}$.) We must distinguish four cases: $p = n$, $p = \frac{1}{2}n$ if n is even, $p = 0$, and $p \neq n, \frac{1}{2}, 0$.

For the case $p = n$, the field equations IX.57 reduce to

$$D\,\varphi_n = 0.$$
$$\delta\,\varphi_n = 0. \qquad \text{(IX.58)}$$

This is a "null" field with the n-fold repeated principal spinor o_A .

Since $D = \Delta^{-1} (B\,\partial_{\tilde{0}} - D\,\partial_{\tilde{1}})$ and $\delta = \Delta^{-1} (-C\,\partial_{\tilde{0}} + A\,\partial_{\tilde{1}})$ (see Equations IX.47), we must have that φ_n (assumed analytic in all of the coordinates) is independent of the structure coordinates $\tilde{z}^{\tilde{0}}$ and $\tilde{z}^{\tilde{1}}$. We thus have the interesting situation that <u>null solutions to the zero rest mass free field equations are given by analytic functions of the structure coordinates z^0 and z^1</u> .

For the case $p = \frac{1}{2}n$, if n is even, the field Equations IX.57 reduce to

$$\begin{aligned} D\,\varphi_{\frac{1}{2}n} &= 0 \\ \delta\,\varphi_{\frac{1}{2}n} &= 0 \\ \tilde{\delta}\,\varphi_{\frac{1}{2}n} &= 0 \\ \Delta\,\varphi_{\frac{1}{2}n} &= 0 \ , \end{aligned} \qquad \text{(IX.59)}$$

so that $\varphi_{\frac{1}{2}n}$ must be a constant. This is a "type D" solution, with o_A and ι_A as the two $\frac{1}{2}$ n-fold repeated principal spinors.

Next we consider the case $p = 0$. The field equations become

$$\begin{aligned} \tilde{\delta}\,\varphi_0 &= n\,\kappa\,\varphi_0 \\ \Delta\,\varphi_0 &= n\,\gamma\,\varphi_0 \ . \end{aligned} \qquad \text{(IX.60)}$$

Reexpressing $\tilde{\delta}$, Δ and α, γ for a left-Kählerian spacetime in terms of Equations IX.47 and IX.49 these equations can be written as

$$- \frac{1}{\varphi_0} \frac{\partial \varphi_0}{\partial z^1} = \frac{1}{2} n \Delta^{-1} \Delta_{,1} \qquad \text{(IX.61)}$$

$$- \frac{1}{\varphi_0} \frac{\partial \varphi_0}{\partial z^0} = \frac{1}{2} n \Delta^{-1} \Delta_{,0}$$

The solutions are quite easily found to be

$$\varphi_0 = \Delta^{-\frac{1}{2}n} \, C(\tilde{z}^0, \tilde{z}^1) , \qquad \text{(IX.62)}$$

where C is an arbitrary function of its arguments. These are the null solutions with ι_A as n-fold repeated principal spinor. The factor $\Delta^{-\frac{1}{2}n}$ appears because of our choice of tetrad.

Finally we consider the case of a general p, $p \neq n, \frac{1}{2}n, 0$. Here the field equations become:

$$D \varphi_p = 0$$

$$\delta \varphi_p = 0$$

$$\tilde{\delta} \varphi_p = (n - 2p) \alpha \varphi_p$$

$$\Delta \varphi_p = (n - 2p) \gamma \varphi_p .$$

The first two of these imply that φ_p is an analytic function of z^0 and z^1 only, while the second two equations imply (compare Equation IX.62) imply that

$$\varphi_p = \Delta^{-\frac{1}{2}n+p} \, C(\tilde{z}^0, \tilde{z}^1) .$$

Thus we have a contradiction, and we conclude that there are no solutions in general for this case.

Let us summarize. We have found that a left-Kählerian spacetime admits three families of solutions to Equations IX.57 having the form $\varphi_p \neq 0$, other $\varphi_i = 0$ for some value of $p = 0, 1, \ldots, n$:

(i) null fields with principal spinor o_A described by $\varphi_n = C(z^0, z^1)$ where C is an analytic function of its arguments;

(ii) "type D" fields with principal spinors o_a, ι_A described by $\varphi_{\frac{1}{2}n}$ = constant; these exist only when n is even;

(iii) null fields with principal spinor ι_A, described by $\varphi_0 = \Delta^{-\frac{1}{2}n}\, C(\tilde{z}^0, \tilde{z}^1)$ where C is an analytic function of its arguments.

We should remark that since z^0, z^1 and $\tilde{z}^0$, $\tilde{z}^1$ are <u>null</u> coordinates, the null solutions (i) and (iii) give physically realistic null fields when transformed to the real type D vacuum or electrified spacetimes.

It is also noteworthy that null EM fields $(n = 2)$ of the form (i) can be constructed even when the almost Hermitian structure is only half-integrable, i.e., for lGSF spacetimes (see Chapter VIII). This is because all we really need to construct such solutions is the null coordinates z^0 and z^1 and the geodesic and shearfree conditions on ℓ_a. Then, a null electromagnetic field would be given by

$$\varphi_2 = F^{+}_{ab}\, dx^a \wedge dx^b = C\,(z^0, z^1)\, dz^0 \wedge dz^1 \;, \qquad \text{(IX.63)}$$

where C is analytic in its arguments. This is a closed self-dual two-form (see Equations VIII.30), hence a (complex) Maxwell field. Furthermore, it is not hard to show that any null EM field must have its repeated principal null direction geodesic and shearfree (actually, this is true for a null zero rest mass field of any spin $s > 0$). So for any null EM field, structure coordinates must exist for which $F^{+} \propto l_{[\mu} m_{\nu]}\, dz^0 \wedge dz^1$, where l is the repeated principal null direction (compare Equations VII.17 with $\varphi_1 = \varphi_2 = 0$ and Equations VIII.28, and note that $F_{ab}\, l^a m^b = F^{+}_{ab}\, l^a m^b$). Hence our solution IX.63 is really the most general null EM field. The fact that null EM fields can exist only in 1GSF spacetimes and are determined by one arbitrary function of two complex variables was first obtained by Robinson [71]. Our result for left-Kählerian spacetimes generalizes Robinson's result to any spin $s > 0$; let us now state this formally:

<u>IX.10: THEOREM</u>: In any spacetime, a null solution to Equations IX.51 for any spin $s > 0$ has a repeated principal spinor for which $\mu = \sigma = 0$.

<u>Proof</u>: If the field is null, it can be written (see Chapter VII) as

$$\varphi_{AB\ldots L} = \phi\, o_A\, o_B \ldots o_L \;,$$ where o_A is the repeated principal spinor, and then Equations IX.51 become

$$\nabla^{A'A} (\phi\, o_A\, o_B \ldots o_L) = 0 \;.$$

Now choose any spinor ι^A such that $\iota^A o_A = 1$, so that (o^A, ι^A) can be used as a spinor dyad. Taking the dyad components of the above equations and consulting Table VII.3 yields eventually:

$$D\phi + (n\epsilon - \rho) = 0$$
$$\delta\phi + (n\beta - \tau) = 0$$
$$\kappa\phi = 0$$
$$\sigma\phi = 0 .$$

So if the field does not vanish, we must have $\kappa = \sigma = 0$.

<u>IX.11: THEOREM</u>: In a left-Kähler ian spacetime, the most general null solution to Equations IX.51 for any spin $s > 0$ takes one of the forms (i) or (iii) above, and hence is determined by an arbitrary analytic function of the form $C(z^0, z^1)$ or $C(\widetilde{z^0}, \widetilde{z^1})$.

<u>Proof</u>: In a complex spacetime, the Goldberg-Sachs Theorem applies separately to Ψ_{ABCD} and $\widetilde{\Psi}_{A'B'C'D'}$ (compare the proof given in [37]). Hence, Ψ_{ABCD} can have <u>at most</u> two independent repeated principal spinors. In our left-Kahlerian spacetimes, then, any null field must have either o_A or ι_A as a principal spinor. Hence the solutions (i) and (iii) are the most general null solutions.

Besides generalizing Robinson's theorem, these results are reminiscent of the twistor description of null zero rest mass fields in Minkowski space (see [79] and Chapter XI.)

Geroch-Held-Penrose Formalism

A remarkably concise formulation of the field Equations IX.57 for left-Kählerian spacetimes can be obtained by using the formalism developed by Geroch, Held, and Penrose in reference [20].

We recall that the complex spin and boost transformations for a null tetrad are given by

$$\begin{aligned} l^a &\to A^{-1}\, l^a \\ m^a &\to e^{i\phi}\, m^a \\ \tilde{m}^a &\to e^{-i\phi}\, \tilde{m}^a \\ n^a &\to A\, n^a \quad . \end{aligned} \qquad \text{(IX.64)}$$

This is equivalent to a spinor dyad transformation of the form

$$\begin{aligned} o^A &\to \lambda\, o^A \\ \iota^A &\to \lambda^{-1}\, \iota^A \\ \tilde{o}^{A'} &\to \tilde{\lambda}\, \tilde{o}^{A'} \\ \tilde{\iota}^{A'} &\to \tilde{\lambda}^{-1}\, \tilde{\iota}^{A'} \quad , \end{aligned} \qquad \text{(IX.65)}$$

where we can identify $A^{-1} = \lambda\tilde{\lambda}$, $e^{i\phi} = \lambda\tilde{\lambda}^{-1}$, $e^{-i\phi} = \tilde{\lambda}\lambda^{-1}$, and $A = \lambda^{-1}\tilde{\lambda}^{-1}$.

A scalar quantity η is now said to be of type $\{k,j\}$ (curly brackets to distinguish from the type (i,j) designation with respect to an almost complex structure) if it transforms in the following way under Equation IX.64 or IX.65:

$$\eta \to \lambda^k\, \tilde{\lambda}^j\, \eta \, . \qquad \text{(IX.66)}$$

The derivations $D, \delta, \tilde{\delta}, \Delta$ are replaced by derivations þ, $\eth$, $\eth'$, $\text{þ}'$ which are defined to act on type $\{k,j\}$ scalars η as follows:

$$\begin{aligned}
\text{þ}\eta &= (D - k\epsilon - j\hat{\epsilon})\eta \\
\eth\eta &= (\delta - k\beta - j\tilde{\alpha})\eta \\
\eth'\eta &= (\tilde{\delta} - k\alpha - j\hat{\beta})\eta \\
\text{þ}'\eta &= (\Delta - k\gamma - j\tilde{\gamma})\eta .
\end{aligned} \qquad \text{(IX.67)}$$

Now we consider again the field Equations IX.57 for left-Kählerian spacetimes. For our choice of tetrad, the operators of Equations IX.67 reduce to

$$\begin{aligned}
\text{þ}\eta &= D\eta \\
\eth\eta &= \delta\eta \\
\eth'\eta &= (\tilde{\delta} - k\alpha - j\hat{\beta})\eta \\
\text{þ}'\eta &= (\Delta - k\gamma - j\tilde{\gamma})\eta .
\end{aligned} \qquad \text{(IX.68)}$$

Next we look at the spin and boost transformation law for the φ_p under Equations IX.65:

$$\varphi_p = \varphi_{AB\ldots L} \underbrace{\iota^A \ldots \iota^D}_{p} \underbrace{o^E \ldots o^L}_{n-p}$$

$$\Rightarrow \varphi_p \rightarrow (\lambda^{-1})^p (\lambda)^{n-p} \varphi_p$$

$$= \lambda^{n-2p} \varphi_p . \qquad \text{(IX.69)}$$

Therefore φ_p is of type $\{n-2p, 0\}$, and we have:

$$
\begin{aligned}
\text{þ}\varphi_r &= D\varphi_r \\
\eth\varphi_r &= \delta\varphi_r \qquad\qquad (\text{IX.70}) \\
\eth'\varphi_{r-1} &= \tilde{\delta}\varphi_{r-1} + (2r - n - 2)\alpha\varphi_{r-1} \\
\text{þ}'\varphi_{r-1} &= \Delta\varphi_{r-1} + (2r - n - 2)\gamma\varphi_{r-1} \quad .
\end{aligned}
$$

From these we see at once that the zero rest mass free field Equations IX.57 for Kählerian spacetimes are given in the Geroch-Held-Penrose formalism by the remarkably simple expressions

$$
\begin{aligned}
\text{þ}\varphi_r - \eth'\varphi_{r-1} &= 0 \\
\eth\varphi_r - \text{þ}'\varphi_{r-1} &= 0 \quad , \\
r = 1, 2, \ldots, n \quad . & \qquad\qquad (\text{IX.71}).
\end{aligned}
$$

CHAPTER X: COMPLEX TRANSFORMATIONS

The subject matter of this chapter is the so-called "complex coordinate transformations" discovered by Newman and others ([36], [35], [13]). The most famous of these is the transformation which takes Schwarzschild spacetime to Kerr spacetime. These transformations provided the original motivation for this study of complex manifold theory in connection with general relativity.

We begin by looking at the generation of new solutions to linear field equations from known solutions by a complexification procedure, all in flat Minkowski space. This is followed by a discussion of the full curved-space transformations of Newman, and the "explanations" thereof which have been offered in the literature. We shall make a mild contribution to this literature by showing that Schwarzschild and Kerr (as well as other pairs of Kerr-Schild spacetimes) can be obtained as different "real slices" of a four-complex-dimensional complex manifold with an Hermitian metric.

All of the "explanations" of this type suffer from the defect of being rather remote from any direct physical interpretation. However, Newman has recently shown that the Schwarzschild-to-Kerr transformation is reflected in the "Heaven picture" as an imaginary displacement of a complex center of mass line in complex Minkowski space, and that such an imaginary displacement appears to have a general physical interpretation as the addition of intrinsic spin to a physical system. We will end this chapter with a discussion of these complex center of mass lines in complex Minkowski space, while their relation to the Schwarzschild-to-Kerr transformation and the "Heaven picture" will be discussed in Chapter XI.

Generation of Twist

The term "generation of twist" is used to express the fact that certain complexification prodecures produce fields with twisting principal null directions ($\rho - \bar{\rho} \neq 0$) from fields with twistfree principal null directions ($\rho - \bar{\rho} = 0$), e.g., Schwarzschild-to-Kerr. It is a remarkable fact that for the type D vacuum spaces, the two repeated principal null directions are either both twisting or both twistfree, so that no ambiguity arises from referring to the solutions themselves as being either twisting or twistfree. (It is a trivial matter to show that the property of having twist is a property of the direction of a null vector only and is invariant under spin and boost transformations of the tetrad.)

Generation of Twist in Minkowski Space

The first reference to the generation of twisting fields from twistfree fields by means of some sort of complexification procedure is a paper of A. Trautman's [52] (in which some unpublished results of I. Robinson are quoted). The title of the paper is "Analytic solutions of Lorentz-invariant linear equations," and the basic idea is quite straightforward, and can be described by the following algorithm.

We take a real analytic solution $\phi(x^a)$ to any linear real-Lorentz-invariant field equations in Minkowski space. Then, as we have remarked already, the field equations are also invariant under complex Lorentz transformations, modulo reality conditions. Now we analytically continue the solution ϕ for complex values of the coordinates x^a, obtaining a complex analytic function $\phi(z^a)$, i.e., we go to the complex extension of Minkowski space.

Now we perform a complex, inhomogeneous Lorentz transformation of the complex coordinates z^a :

$$z^a \longrightarrow \hat{z}^a = \Lambda^a{}_b \, z^b + \lambda^a ,$$

under which $\phi\,(z^a)$ will transform as a scalar, a spinor, a tensor, or whatever the case may be. So we obtain the solution $\hat{\phi}\,(\hat{z}^a)$ in the new coordinates. Now if we restrict the $\hat{z}^a$ to real values $\hat{x}^a$, we will obtain a new solution $\hat{\phi}\,(\hat{x}^a)$. This solution will generally be complex valued, and it is precisely here that the linearity of the field equations plays a crucial role, because the linearity implies that the real and imaginary parts of $\hat{\phi}\,(\hat{x}^a)$ will be new real solutions to the field equations. Since $z^a \longrightarrow \hat{z}^a$ is a complex Lorentz transformation, the metric will again be Minkowskian on the "new real slice" defined by $\hat{z}^a = \overline{\hat{z}^a}$.

As an example of this procedure which produces a twisting solution from a twistfree solution, Trautman presents the following null self-dual Maxwell tensor:

$$F_{ab}{}^{+} = A(\xi, \eta)\ \xi_{,[a}\,\eta_{,b]} \quad , \tag{X.1}$$

where

$$\begin{aligned} \xi &= -\eta_{ab}\, x^a x^b / 2\, n_a x^a \\ \eta &= m_a x^a / n_a x^a \quad , \end{aligned} \tag{X.2}$$

where ℓ_a, m_a, n_a form a constant real normed null tetrad in Minkowski space. Since ξ is real, $\xi_{,a}$ is a principal null vector, and the twist of $\xi_{,a}$ vanishes, because it is a gradient.

Now Trautman lets x^a go to z^a and performs the complex Lorentz transformation

$$z^a \longrightarrow \hat{z}^a = z^a + i\beta \ell^a , \tag{X.3}$$

where β is a constant. Then we have:

$$\xi(z^a) \longrightarrow \hat{\xi}(\hat{z}^a) = \xi(\hat{z}^a - i\beta\ell^a) ,$$

$$\eta(z^a) \longrightarrow \hat{\eta}(z^a) = \eta(\hat{z}^a - i\beta\ell^a) .$$

Then letting $\hat{z}^a$ go to $\hat{x}^a$, we obtain the new solution

$$\hat{F}_{ab}{}^{+} = A(\hat{\xi},\hat{\eta}) \ \hat{\xi}_{,[a}\hat{\eta}_{,b]} \tag{X.4}$$

(notice that self-duality is invariant under complex Lorentz transformations, so there is no trouble with reality conditions). Now we have

$$\hat{\xi} = (-\eta_{ab}\hat{x}^a\hat{x}^b + 2i\beta\ell_a\hat{x}^a)/(n_a\hat{x}^a - i\beta) ,$$

$$\hat{\eta} = (m_a\hat{x}^a)/(n_a\hat{x}^a - i\beta) . \tag{X.5}$$

The principal null direction is now a linear combination $E\hat{\xi}_{,a} + F\hat{\eta}_{,a} = k_a$ such that k_a is real. It is found after quite a lengthy calculation that k_a possesses a non-vanishing twist (Robinson, quoted in Trautman [52]).

All of this has a familiar ring to it for us. The solutions X.1 and X.4 are precisely of the form of Equation IX.63 for the most general null EM field. Hence we know that the principal null directions $\xi_{,a}$ and $k_a = E\hat{\xi}_{,a} + F\hat{\eta}_{,a}$ must be geodesic and shearfree, and furthermore that ξ, η and $\hat{\xi}, \hat{\eta}$ must be structure coordinates for some half-integrable almost Hermitian structures on Minkowski space.

It is also easy to see why the procedure generates a twisting solution from a twistfree solution, because there is nothing to guarantee that $k_a = E\hat{\xi}_{,a} + F\hat{\eta}_{,a}$ should be hypersurface-orthognal: the only condition on E,F is that they must reduce to a real vector on the new "real slice" $\hat{z}^a = \overline{\hat{z}^a}$. (Note that in general there is only one real direction among the type (1,0) forms, namely, that of λ_a.)

The Trautmann procedure provides one of the inspirations for twistor theory. We shall have much more to say on this in Chapter XI, but a few preliminary remarks are in order here.

Recall that the twistfree condition means that the principal null directions ℓ_a are orthogonal to real null hypersurfaces L = real constant, where $\ell_a \propto L_{,a}$. For the field X.1, these are just the null hypersurfaces $\xi(x^a)$ = real constant. When we go to the complex extension $\mathbb{C}$M of Minkowski space, the analytic continuation $\xi(z^a)$ of the function $\xi(x^a)$ defines three-complex-dimensional hypersurfaces $\xi(z^a)$ = complex constant, because now ξ has been "liberated" from the constraint of being real-valued.

But, in the complex extension $\mathbb{C}$M, the functions $\xi(z^a)$ and $\eta(z^a)$ stand on a completely equal footing, because there are no reality conditions left to single out ξ. Therefore, the only objects which retain intrinsic significance in the extended manifold CM are the two-complex-dimensional complex surfaces defined by $\xi(z^a)$ = complex constant, $\eta(z^a)$ = complex constant. In Chapter XI we shall show that these complex surfaces are in fact twistor surfaces, i.e., totally geodesic, null, two-complex dimensional surfaces, in a sense to be defined later. The point to be made here is that the analytic continuation of ξ, η onto CM is the same as the analytic continuation of $\hat{\xi}, \hat{\eta}$, i.e., the EM fields on the two "real slices" <u>define the same twistor surfaces on the complex extension</u>.

Consider one of these twistor surfaces T in $\mathbb{C}M$, given by

$$\xi(z^a) = A = \text{complex constant} = \hat{\xi}(\hat{z}^a)$$

$$\eta(z^a) = B = \text{complex constant} = \hat{\eta}(\hat{z}^a).$$

Let M denote the "real slice" of $\mathbb{C}M$ defined by $z^a = \overline{z^a}$, and M' the "real slice" defined by $\hat{z}^{a'} = \overline{\hat{z}^{a'}}$. Coordinatize M with the structure coordinates ξ, η and two other arbitrary coordinates χ, ψ; these four coordinates will satisfy some generalized reality conditions. Similarly M' with $\hat{\xi}, \hat{\eta}, \hat{\chi}, \hat{\psi}$ satisfying some generalized reality conditions. Now consider the intersection $T \cap M$ of the twistor surface T with the "real slice" M. It is not hard to show directly that this intersection is empty <u>unless</u> values of χ, ψ exist which make $\xi = A$, $\eta = B$, χ, ψ satisfy the generalized reality conditions for M. When this is the case, $T \cap M$ is a <u>single null line</u> with tangent vector $l_a = \xi_{,a}$. We call the twistor surface a <u>null twistor surface</u> when values of χ, ψ exist such that $\xi = A$, $\eta = B$, χ, ψ satisfy the generalized reality conditions for M. In a similar fashion, we can show that $T \cap M'$ is empty unless values of $\hat{\chi}$, $\hat{\psi}$ exist for which $\hat{\xi} = A$, $\hat{\eta} = B$, $\hat{\chi}$, $\hat{\psi}$ satisfy the generalized reality conditions for M', in which case $T \cap M'$ is a single null line with tangent vector $k_a = E\hat{\xi}_{,a} + F\hat{\eta}_{,a}$.

<u>Newman's Complex Transformations</u>

It is generally agreed that the generation of twisting fields on Minkowski space as described above is a mathematical rigorous and well-understood procedure. We now move to more mysterious matters and consider the so-called "complex coordinate transformations" discovered by Newman and others [36], [35], [13].

We start with Schwarzschild spacetime in coordinates for which

$$ds^2 = (1 - \frac{2m}{r})\, du^2 + 2\, du\, dr$$
$$- r^2 (d\theta^2 + \sin^2\theta\, d\phi^2) . \qquad (X.6)$$

A real null tetrad for this metric is provided by:

$$\begin{aligned} \ell^a &= (0, 1, 0, 0) , \\ n^a &= (1, -\frac{1}{2}(1 - \frac{2m}{r}), 0, 0) , \\ m^a &= \frac{1}{\sqrt{2}r} (0, 0, 1, i \csc\theta) , \\ \bar{m}^a &= \frac{1}{\sqrt{2}r} (0, 0, 1, - i \csc\theta) . \end{aligned} \qquad (X.7)$$

The coordinate r is now allowed to "take on complex values" (this is sort of a limited complex extension of the spacetime manifold), and the tetrad is rewritten as follows:

$$\begin{aligned} {}^a &= (0, 1, 0, 0) , \\ n^a &= (1, -\frac{1}{2}[1 - m (\frac{1}{r} + \frac{1}{\bar{r}})], 0, 0) , \\ m^a &= \frac{1}{\sqrt{2}\bar{r}} (0, 0, 1, i \csc\theta) , \\ \bar{m}^a &= \frac{1}{\sqrt{2}r} (0, 0, 1, - i \csc\theta) . \end{aligned} \qquad (X.8)$$

The purpose of this step is to insure that the tetrad continues to satisfy the usual reality conditions even for complex values of r . Why this should be necessary is not clear. This step of the algorithm is by no means unique; for example, $2/r$ could have been replaced by $4/(r+\bar{r})$ instead of by $1/r + 1/\bar{r}$ in n^a . In fact, different choices will generally lead to "wrong" results (i.e., a new vacuum spacetime will not be obtained).

The next step is to let all of the coordinates "take on complex values" and then to formally perform the complex coordinate transformation given by

$$\begin{aligned} u' &= u - i\,a\cos\theta \\ r' &= r + i\,a\cos\theta \\ \theta' &= \theta \\ \phi' &= \phi \ , \end{aligned} \tag{X.9}$$

where a = real constant. Under this transformation, the tetrad is transformed by the tensor transformation law:

$$\begin{aligned} \ell^{a'} &= (0, 1, 0, 0) \ , \\ n^{a'} &= \left(1, -\frac{1}{2}\left[1 - \frac{2mr'}{r'^2 + a^2\cos^2\theta'}\right], 0, 0\right) \ , \\ m^{a'} &= \frac{1}{\sqrt{2}\,(r' + i\,a\cos\theta')}\left(i\,a\sin\theta', -i\,a\sin\theta', 1, i\csc\theta'\right), \\ \bar{m}^{a'} &= \frac{1}{\sqrt{2}\,(r' - i\,a\cos\theta')}\left(-i\,a\sin\theta', +i\,a\sin\theta', 1, -\csc\theta'\right) . \end{aligned} \tag{X.10}$$

When the new coordinates u', r', θ', ϕ' are restricted to real values, it is found that the tetrad of Equation X.5 becomes a real tetrad describing Kerr spacetime.

It should be noted that a "clue" as to the proper complex coordinate transformation is provided by the fact that, in these coordinates and tetrads, the Weyl tensor for Schwarzschild is given by $\Psi_2 = - m/r^3$, while for Kerr it is given by $\Psi_2 = - m/(r' - i\ a \cos \theta')^3$.

Newman and his coworkers have found similar complex transformations which produce Kerr-Newman ("charged Kerr") spacetime from the Reissner-Nordstrom spacetime, and a combined Kerr-NUT solution from Schwarzschild. The transformations all take type D vacuum or electrified spacetimes into type D vacuum or electrified spacetimes. The transformed $\ell^{a'}$ continues to be a principal null direction for the spacetime; however, the transformed $n^{a'}$ is no longer a principal null direction for the new spacetime. The transformation applied directly to the metric tensor g_{ab} rather than to the tetrad legs does not yield the desired results.

In a recent paper [12], Demianski has solved the vacuum field equations to find the most general vacuum metric which can be generated from Schwarzschild by the process above with a coordinate transformation of the form

$$\begin{aligned} u' &= u + i\ G(\theta, \phi) \\ r' &= r + i\ F(\theta, \phi) \\ \theta' &= \theta \\ \phi' &= \phi \ . \end{aligned} \qquad \text{(X.11)}$$

The solution he obtains has four parameters, the Schwarzschild mass m , the Kerr angular momentum parameter a , the NUT constant b , and a new parameter c . When $c \neq 0$, the solution is, surprisingly, type II instead of type D . (This solution provides a good illustration of the ambiguity of the term "explanation"; Where does "explaining" stop and "solving the vacuum field equations" begin? See next section.)

"Explanations" of the Newman Transformations

It is natural to ask for an "explanation" of these complex transformations. However, then we are immediately faced with the problems of deciding what constitutes an "explanation." It is because of this indeterminancy that we always put quotes on the word "explanation." The "explanations" offered in the past range from "luck" to the postulation of some additional structure of physical significance (the "Heavenly explanation" discussed below). Intermediate between these two extremes is the rather popular opinion that whatever procedure is "sanctioned by the vacuum field equations" should be accepted as such at face value, although it seems to us that this is a rather empty assertion. We shall now review some of the other "explanations" that have been given.

To begin with, everyone seems to be in agreement that the generation of twist in Minkowski space, for linear Lorentz-invariant theories, is well understood and mathematically straightforward (see above).

As for the Schwarzschild-to-Kerr transformation, it was shown by Kerr [36], and later in more detail by Talbot [51], that the Kerr-Schild vacuum field equations reveal that any transformation of the form of Equations X.6 can be used as an artifice to obtain a possibly new Kerr-Schild solution from any given Kerr-Schild solution. We refer to [51] for the details of this, but the general idea is as follows.

We recall from Chapter VII that a Kerr-Schild solution is specified by an arbitrary analytic function $\phi(Y)$ of a complex variable Y, where Y is given in terms of the coordinates $u, v, \zeta, \bar{\zeta}$ by Equation VII.49:

$$\phi(Y) = -Y^2\bar{\zeta} + \zeta - (u+v)Y \ . \qquad \text{(X.12)}$$

Suppose a particular $\Phi(Y)$ is chosen, for example, $\Phi(Y) = 0$ for Schwarzschild spacetime. Now we take a new choice for $\Phi(Y)$ given by

$$\hat{\Phi}(Y) = \Phi(Y) - \sqrt{2}\, i\, a\, Y . \tag{X.13}$$

Then if we insert $\hat{\Phi}(Y)$ into Equation X.7 we find that:

$$\begin{aligned} \Phi(Y) - \sqrt{2}\, i\, a\, Y &= -Y^2\bar{\zeta} + \zeta - (u+v)Y \\ \Longrightarrow \quad \Phi(Y) &= -Y^2\bar{\zeta} + \zeta - (u+v)Y + \sqrt{2}\, i\, a\, Y \\ &= -Y^2\bar{\zeta} + \zeta - \left[(u + \sqrt{2}\, i\, a) + v\right] Y . \end{aligned} \tag{X.14}$$

That is, the new solution corresponding to $\hat{\Phi}(Y)$ is obtained formally by performing the "complex transformation"

$$u \longrightarrow u + \sqrt{2}\, i\, a \tag{X.15}$$

in Equation X.12. In particular, Kerr is "generated" from Schwarzschild in this way. It should be pointed out that iterating this procedure does not necessarily lead to more new solutions; for example, in the Schwarzschild-to-Kerr case, iteration just keeps increasing the Kerr parameter by units of a.

This is the primary example of a "sanctioned-by-the-field-equations" explanation. It is believed by many to be due to the striking similarity between the Kerr-Schild vacuum field equations and the linearized gravitational field equations in Minkowski space (see Debney, Kerr, and Schild [11]). This is related to the fact that a Kerr-Schild spacetime is defined in terms of an auxiliary Minkowski space (see Chapter VII).

Recently, Newman has given a new "explanation" for the Schwarzschild-to-Kerr transformation [34]. The method depends crucially on the existence of auxiliary Minkowski spaces associated with the Kerr-Schild nature of these space times. Thus Schwarzschild is written in the Kerr-Schild form

$$g^{ab} = \eta^{ab} - \lambda \, l^a l^b \, , \qquad (X.16)$$

in coordinates for which the components of the repeated principal null vector l^a are given by $l^a = (0, 1, 0, 0)$ and for which

$$\eta_{ab} \, dx^a \, dx^b = du^2 + 2 \, du \, de - r^2 (d\theta^2 + \sin^2\theta \, d\phi^2) \, . \qquad (X.17)$$

Then the coordinates are allowed to "go complex" by going to the complex extension, and the following complex coordinate transformation, considered as acting on the complexified Minkowski space, is performed:

$$\begin{aligned} u &= u' + i \, a \cos \theta' \\ r &= r' - i \, a \cos \theta' \\ \cos \theta &= \frac{r' \cos \theta' - i \, a}{r' - i \, a \cos \theta'} \\ \cos 2(\phi - \phi') &= \frac{r'^2 - a^2}{r'^2 + a^2} \, . \end{aligned} \qquad (X.18)$$

The transformed Minkowski metric becomes:

$$\begin{aligned} ds^2 = du'^2 &+ 2 \, du' \, dr' - 2 \, a \sin^2 \theta' \, dr' \, d\phi' \\ &- (r'^2 + a^2 \cos^2 \theta') (d\theta'^2 + \sin^2 \theta' \, d\phi'^2) \\ &- a^2 \sin^4 \theta' \, d\phi'^2 \, , \end{aligned} \qquad (X.19)$$

and then the coordinates u', r', θ', ϕ' are restricted to real values.

Next a tetrad for the Minkowski metric of Equation X.12 is presented:

$$\begin{aligned} l^a &= (0, 1, 0, 0) \, , \\ n^a &= (1, -\tfrac{1}{2}, 0, 0) \, , \\ m^a &= \frac{1}{\sqrt{2} \, r} (0, 0, 1, i \csc \theta) \, , \\ m^a &= \frac{1}{\sqrt{2} \, r} (0, 0, 1, - i \csc \theta) , \end{aligned} \qquad (X.20)$$

where ℓ^a is one of the Schwarzschild principal null vectors (ℓ^a is null with respect to the auxiliary Minkowski metric as well as with respect to the Schwarzschild metric).

Likewise, a tetrad for the Minkowski metric of Equation X.14 is presented:

$$\hat{\ell}^{a'} = (0, 1, 0, 0) \quad ,$$

$$\hat{n}^{a'} = (1, -\tfrac{1}{2}, 0, 0) \quad ,$$

$$\hat{m}^{a'} = \frac{1}{\sqrt{2}\,(r' + i\, a \cos\theta')} \left(i\, a \sin\theta', -\, i\, a \sin\theta' , 1, i \csc\theta'\right) \quad ,$$

$$\hat{\bar{m}}^{a'} = \frac{1}{\sqrt{2}\,(r' - i\, a \cos\theta')} \left(-\, i\, a \sin\theta', +\, i\, a \sin\theta', 1, -\, i \csc\theta'\right) \quad . \qquad \text{(X.21)}$$

It is shown that the tetrad of Equations X.16 is obtainable from that of Equations X.15 by a complex Lorentz transformation of the tetrad legs. Also, it is obvious that the first tetrad is a real null tetrad when u, r, θ, ϕ are restricted to real values, and the second tetrad is likewise real when u', r', θ', ϕ' are real.

Now, the key to the "explanation" lies in interpreting the Weyl tensor $\overset{+}{C}_{abcd}$ as a field defined on the auxiliary complexified Minkowski space. The field is chosen so that its tetrad components with respect to the tetrad of Equation X.20 give the Schwarzschild values, i.e.,

$$\Psi_2 = -\, m/r^3 \qquad \text{others zero} \quad , \qquad \text{(X.22)}$$

when u, r, θ, ϕ are real. It is then found that the same field, re-expressed in terms of the coordinates u', r', θ', ϕ' and the tetrad of Equation X.21, gives the Kerr values for Ψ_i when u', r', θ', ϕ' are restricted to real values.

Finally, the function λ appearing in Equation X.11 is given for Schwarzschild by

$$\lambda = 2m/r \tag{X.23}$$

in the coordinates u, r, θ, ϕ. If this is rewritten as

$$\lambda = m\,(1/r + 1/\bar{r}) \tag{X.24}$$

and the coordinate transformation of Equation X.18 is applied, we get

$$\lambda \rightarrow \frac{2mr'}{r'^2 + a^2 \cos^2 \theta'} , \tag{X.25}$$

which is the correct value for the Kerr solution in the form of Equation X.16 for real values of u', r', θ', ϕ'.

The most serious drawback of this "explanation" is the necessity of arbitrarily rewriting functions of r as functions of r and $\bar{r}$. This means that the continuation of functions to the complex extension is not being done uniquely. For example, the only unique continuation of the function $\lambda = 2m/r$ is the analytic continuation $\tilde{\lambda} = 2m/r$, where now r is complex. If we write instead $\tilde{\lambda} = m(1/r + 1/\bar{r})$, there is nothing to favor this choice over others, such as $\tilde{\lambda} = 2m/r + \bar{r}$, etc. This was a problem with the original form of the Schwarzschild-to-Kerr transformation: recall that the tetrad X.7 had to be arbitrarily rewritten as X.8, in which $\bar{r}$ appears. Even worse, under the coordinate transformation X.9, $\bar{r}$ was transformed into $r' + i\, a \cos\theta'$ instead of $\bar{r}' + i\, a \cos\bar{\theta}'$.

There are similar problems with the Kerr-Talbot "explanation." If Schwarzschild is written in the coordinates $u, v, \zeta, \tilde{\zeta}$ and the coordinate transformation X.15 ($u' = u + \sqrt{2}\, i\, a$) is performed on the complex extended manifold, Kerr is not obtained on the "new real slice" $u' = \bar{u}'$, as can be verified directly. The problem is that while X.15 correctly takes $Y(x^a)$ for Schwarzschild into $Y(x^{a'})$ for Kerr, it "messes up" the transformation of $\bar{Y}(x^a)$. Again, the problem is the non-unique nature of the continuation of $\bar{Y}$ onto the complex extension: to get Kerr, we must arbitrarily rewrite $\bar{Y}(u)$ as $\bar{Y}(\bar{u})$, and this is a non-analytic continuation.

Analytic Complex Transformations

Faced with this situation, it is natural to ask if the procedure can be modified in such a way that functions of r do not have to be arbitrarily rewritten as functions of r and $\bar{r}$. If this were the case, we would have the situation depicted in Diagram X.1. Referring to this diagram, the point is the following: if g_{ab} satisfies the vacuum field equations $R_{ab}(x) = 0$, then we would have $R_{ab}(z) = 0$ because "the analytic continuation of zero is zero." Then after the complex analytic coordinate transformation we would have $R_{a'b'}(z') = 0$, and hence $R_{a'b'}(x') = 0$ on the new "real slice" $z^{a'} = \overline{z^{a'}}$. So the requirement that $g_{ab}(z)$ be obtained from $g_{ab}(x)$ by analytic continuation would automatically imply that $g_{a'b'}(x')$ was again a solution to the vacuum field equations on any new "real slice." (Of course, it would still be a non-trivial matter to find a new "real slice" on which $g_{a'b'}(z')$ reduced to a real tensor $g_{a'b'}(x')$.)

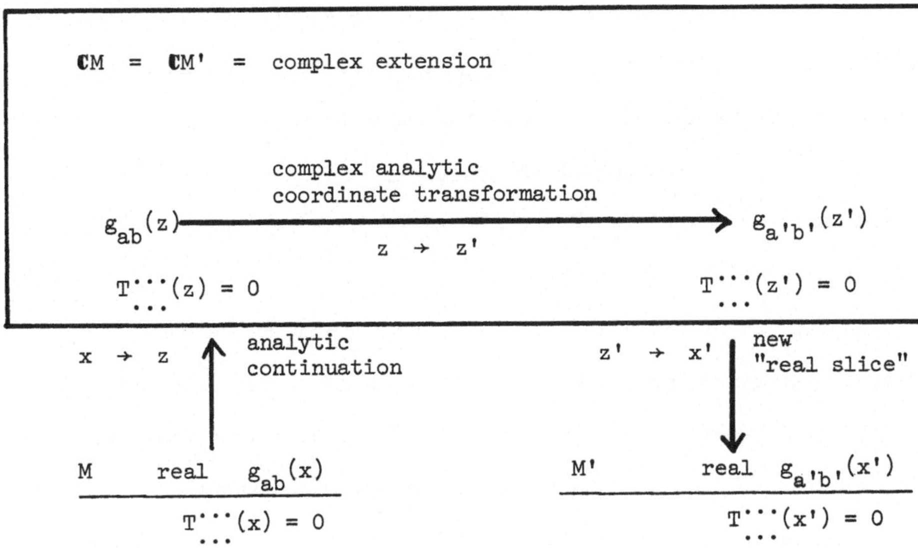
ℂM = ℂM' = complex extension
complex analytic
coordinate transformation
$g_{ab}(z)$
$g_{a'b'}(z')$
z → z'
$T^{\cdots}_{\cdots}(z) = 0$
$T^{\cdots}_{\cdots}(z') = 0$
x → z
analytic
continuation
z' → x'
new
"real slice"
M real $g_{ab}(x)$
$T^{\cdots}_{\cdots}(x) = 0$
M' real $g_{a'b'}(x')$
$T^{\cdots}_{\cdots}(x') = 0$

Diagram X.1

We have already seen one example of this algorithm in Trautman's procedure: in that case, $g_{ab}(x) = \eta_{ab}(x)$, and the complex analytic coordinate transformation $z \to z'$ was a complex Lorentz transformation. The fact that $F_{ab}(x)$ was analytically continued to the complex extension guaranteed that the new field $F_{a'b'}(x')$ on the new "real slice" was again a solution to Maxwell's equations.

An example of the algorithm in which M and M' are non-flat real vacuum spacetimes was found by Ehlers and Kundt [57]. The algorithm transforms the Schwarzschild metric into the "B" metric (see Table VII.2 and Diagram VII.3). The transformation is especially transparent when the Schwarzschild metric is written in the following (standard) coordinates:

$$ds^2 = (1 - 2m/r)\, dt^2 - (1 - 2m/r)^{-1}\, dr^2 - r^2 \left(d\theta^2 + \sin^2\theta\, d\phi^2\right). \tag{X.26}$$

Now go to the complex extension by allowing t, r, θ, ϕ to take on complex values, and perform the complex analytic coordinate transformation

$$\begin{aligned} t' &= -\, i\phi \\ r' &= r \\ \theta' &= \theta \\ \phi' &= -\, i\, t \ . \end{aligned} \tag{X.27}$$

The metric becomes

$$ds'^2 = r'^2 \sin^2\theta'\, dt'^2 - (1 - 2m/r')^{-1}\, dr'^2 - r'^2\, d\theta'^2 - (1 - 2m/r')\, d\phi'^2 \tag{X.28}$$

and on the new "real slice" given by t', r', θ', ϕ' real, this is just the "B" metric (the "B1" metric in Ehlers and Kundt [57]; in this reference two similar transformations were given relating Schwarzschild to the "A2" and "B2" metrics).

Since Schwarzschild and "B" are both type D vacuum spacetimes, they both admit structure coordinates arising from an integrable almost m-Hermitian structure. It is instructive to look at the analytic transformation X.27 in terms of these structure coordinates. To do this, it is convenient to transform first to the coordinates used by Kinnersley [26]:

$$ds^2 \text{ (Schwarzschild)} = (1 - 2m/r)\, du^2 + 2\, du\, dr$$

$$- r^2 (d\theta^2 + \sin^2\theta\, d\phi^2) \qquad (X.29)$$

$$ds'^2 \text{ ("B")} = \frac{r'^2}{x'^2}\, du'^2 + du'\, dr' - 4\frac{r}{x'}\, du'\, dx'$$

$$+ (1 - 2m/x')^{-1}\, dx'^2 - \frac{1}{4}(1 - 2m/x')\, dy'^2 \qquad (X.30)$$

The relevant tetrads are:

$$l_a = (l_u, l_r, l_\theta, l_\phi) = (1, 0, 0, 0)$$

$$n_a = (\tfrac{1}{2}(1 - 2m/r), 1, 0, 0)$$

$$m_a = \frac{r}{\sqrt{2}}(0, 0, -1, +i \sin\theta)$$

$$\bar{m}_a = r/\sqrt{2}\,(0, 0, -1, -i \sin\theta) \qquad (X.31)$$

for Schwarzschild and

$$\ell_{a'} = (\ell_{u'}, \ell_{r'}, \ell_{x'}, \ell_{y'}) = (\frac{r'^2}{x'^2}, 1, -\frac{2r'}{x'}, 0)$$

$$n_{a'} = (1, 0, 0, 0)$$

$$m_{a'} = (0, 0, -1/2\,\xi', +i\,\xi'/2)$$

$$\bar{m}_{a'} = (0, 0, -1/2\,\xi', -i\,\xi'/2) \qquad (X.32)$$

for "B" , where $\xi' \equiv \sqrt{m/x' - 1}$. The respective structure coordinates are found to be:

$$z^0 = u$$

$$z^1 = \ln(\tan\,\theta/2) + i\phi$$

$$\widetilde{z^0} = u + 2r + 4m\,\ln\,(r - 2m)$$

$$\widetilde{z^1} = \ln(\tan\,\theta/2) - i\phi \qquad (X.33)$$

for Schwarzschild, and

$$z^{0'} = \frac{1}{2}u' - \frac{x'^2}{r'}$$

$$z^{1'} = x' + 2m\,\ln\,(m - \frac{1}{2}x') + \frac{1}{2}i\,y' \qquad (X.34)$$

$$\widetilde{z^{0'}} = u'$$

$$\widetilde{z^{1'}} = x' + 2m\,\ln\,(m - \frac{1}{2}x') - \frac{1}{2}i\,y'$$

for "B". Finally, the transformation X.27 becomes, in the new coordinates,

$$
\begin{aligned}
u' &= e^{-i\phi} \tan \theta/2 \\
r' &= r^2 \sin\theta \, e^{i\phi} \\
x' &= r \\
y' &= 2i(u + r + 2m \ln(r - 2m)), \qquad \text{(X.35)}
\end{aligned}
$$

and from this it is easy to show that in terms of the structure coordinates the transformation is given by:

$$
\begin{aligned}
z^{0'} &= -\tfrac{1}{2}\, e^{-z^1} \\
z^{1'} &= -z^0 + \text{complex constant} \\
\widetilde{z}^{0'} &= e^{\widetilde{z}^{1}} \\
\widetilde{z}^{1'} &= \widetilde{z}^{0} + \text{complex constant} \qquad \text{(X.36)}
\end{aligned}
$$

The point of all of this is the following. In the complex extension, which contains both Schwarzschild and "B" as different real slices, we can consider twistor surfaces exactly as we did in the case of the complex extension of Minkowski space. Associated with Schwarzschild, we will have twistor surfaces $T(z^a_0)$ given by z^0 = complex constant, z^1 = complex constant, as well as different twistor surfaces $T(\widetilde{z}^a{}_0)$ given by $\widetilde{z}^0$ = complex constant, $\widetilde{z}^1$ = constant. Similarly, the "B slice" gives rise to twistor surfaces $T'(z^{a'}_0)$ given by $z^{0'}$ = complex constant, $z^{1'}$ = complex constant, as well as twistor surfaces $T'(\widetilde{z}^{a'}_0)$ given by $\widetilde{z}^{0'}$ = complex constant, $\widetilde{z}^{1'}$ = complex constant. The interesting fact is that, from Equations X.36, we see that <u>the twistor surfaces T arising from Schwarzschild coincide with the twistor surfaces T' arising from "B"</u>.

Now let S denote the Schwarzschild "real slice" of the complex extension and B the "B real slice." Then proceeding exactly as above in the Minkowski space case, we can show that for an arbitrary twistor surface $T(z^a{}_0) \equiv T'(z_0^{a'})$ or $T(\tilde{z}^a{}_0) \equiv T'(\tilde{z}_0^{a'})$, the intersection $T \cap S$ will be empty unless T satisfies a condition arising from the reality conditions defining S , in which case $T \cap S$ will be a single null line belonging to one of the two geodesic and shearfree principal null congruences of Schwarzschild and similarly for $T \cap B$. In this sense we can "explain" the Schwarzschild-to-"B" transformation by saying that both spacetimes inherit their geometry from one and the same "twistor-surface-geometry" on the complex extension, by the application of two different sets of reality conditions.

Non-Analytic Complex Transformations

Again we have strayed somewhat from the point. Returning to the Schwarzschild-to-Kerr transformation, we would like to know if this transformation can be modified in such a way that our analysis of the Schwarzschild-to-"B" transformation carries through in this case too. As we have already remarked, this would entail finding a way to write the transformation such that complex conjugates of the coordinates do not enter in an arbitrary way. That is, we want $g_{ab}(z)$ to be the unique analytic continuation of $g_{ab}(x)$, and we want $z \to z'$ to be a complex analytic transformation.

Unfortunately, and perhaps somewhat surprisingly, such a transformation apparently does not exist, and Newman and Winicour [63] have given a very simple argument to show this. If an analytic transformation as in Diagram X.1 did exist for Schwarzschild-to-Kerr, then a real Killing vector in Schwarzschild spacetime would be mapped onto a (possibly complex) Killing vector on Kerr. Now Schwarzschild has three $\mathbb{R}$-linearly independent

Killing vectors; if we denote these by T, Θ, and Φ, then under an analytic complex transformation, these would be mapped into three complex Killing vectors $T_1 + i\,T_2$, $\Theta_1 + i\,\Theta_2$, and $\Phi_1 + i\,\Phi_2$ in Kerr. Furthermore, since T, Θ, and Φ are $\mathbb{R}$-linearly independent, it follows that T_1, Θ_1, and Φ_1 must be $\mathbb{R}$-linearly independent in Kerr. Since we know that Kerr has only two independent Killing vectors, we conclude that such an analytic complex transformation cannot exist. (It can be shown that the "B" metric has three independent Killing vectors, so that this argument does not contradict the existence of the analytic Schwarzschild-to-"B" transformation discussed above.)

In the face of this result, the best we can do is to ask how the algorithm of Diagram X.1 can be "weakened" in such a way as to accommodate the Schwarzschild-to-Kerr transformation. Such a weakened algorithm is depicted in Diagram X.2. In this case, $g_{ab}(x)$ is allowed to be "continued non-analytically" to an Hermitian metric $g_{a\bar{b}}(z, \bar{z})$ on CM. (Note that the analytic tensor $g_{ab}(z)$ in the previous algorithm was <u>not</u> a metric in the ordinary sense of complex manifold theory, since it was of the form

$$\begin{pmatrix} g_{ab} & g_{a\bar{b}} \\ g_{\bar{a}b} & g_{\bar{a}\bar{b}} \end{pmatrix} = \begin{pmatrix} g_{ab}(z) & 0 \\ 0 & 0 \end{pmatrix} ;$$

here, however, we want to produce a genuine Hermitian metric on $\mathbb{C}M$, of the form

$$\begin{pmatrix} g_{ab} & g_{a\bar{b}} \\ g_{\bar{a}b} & g_{\bar{a}\bar{b}} \end{pmatrix} = \begin{pmatrix} 0 & g_{a\bar{b}}(z,\bar{z}) \\ g_{\bar{a}b}(z,\bar{z}) & 0 \end{pmatrix} ;$$

note we are using Latin indices to denote the complex coordinates z^a, $z^{\bar{b}}$ arising from the real coordinates x^a.)

The manifold M is now given as the "real slice" on which $z^a = \overline{z^a}$, as before. Now a complex analytic coordinate transformation $z^a \to z^{a'}$ is to be performed, producing $g_{a'\bar{b}'}(z', \bar{z}')$. On the new "real slice" M' given by $z^{a'} = \overline{z^{a'}}$, $g_{a'\bar{b}'}$ is required to reduce to a new real metric tensor $g_{a'b'}(x')$.

Note that the coordinate transformation $z^a \to z^{a'}$ is still required to be complex analytic; however, in the title of this section, we have referred to this second algorithm as a "non-analytic complex transformation" to indicate that the tensor $g_{ab}(x)$ is continued non-analytically to the complex extension. Perhaps more properly we should not speak of "continuing $g_{ab}(x)$" at all, since now only the reverse procedure $g_{ab}(z, \bar{z}) \to g_{ab}(x)$ is uniquely well-defined. Thus in Diagram X.2, we have drawn the left-hand arrow pointing down instead of up as in Diagram X.1.

Note also that in this second algorithm, we are no longer guaranteed to have $R_{a\bar{b}}(z, \bar{z}) = 0$ on $\mathbb{C}M$ just because $R_{ab}(x) = 0$ on M. Again, this is due to the fact that $g_{ab}(x)$ has been "non-analytically continued" onto $\mathbb{C}M$, so we no longer can say that the continuation of $R_{ab} = 0$ is $R_{a\bar{b}} = 0$. This is the real weakness of the second algorithm: the vacuum field equations must be checked "by hand" on each "real slice."

It is not hard to show that the Schwarzschild-to-Kerr transformation can now be cast in the form of the weakened algorithm of Diagram X.2. Consider, then, a four-complex-dimensional genuine complex manifold $\mathbb{C}M$

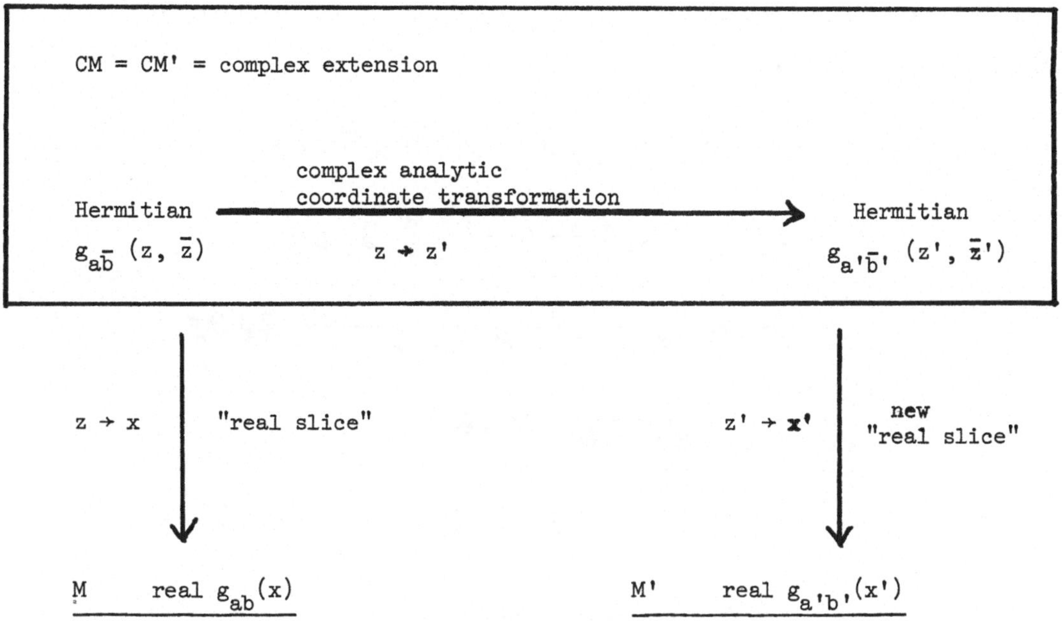
CM = CM' = complex extension
Hermitian
$g_{a\bar{b}}(z, \bar{z})$
complex analytic
coordinate transformation
$z \to z'$
Hermitian
$g_{a'\bar{b}'}(z', \bar{z}')$
$z \to x$
"real slice"
$z' \to x'$
new
"real slice"
M real $g_{ab}(x)$
M' real $g_{a'b'}(x')$

Diagram X.2

with local complex coordinates $z^a = u, r, \theta, \phi$ and an Hermitian metric $g_{a\bar{b}}$ given by

$$(\partial/\partial s)^2 \equiv g^{a\bar{b}} \partial/\partial z^a \partial/\partial \overline{z^b} + g^{\bar{a}b} \partial/\partial \overline{z^a} \partial/\partial z^b \tag{X.37}$$

$$= \partial_r \partial_{\bar{u}} + \partial_u \partial_{\bar{r}} - (1 - \frac{m}{r} - \frac{m}{\bar{r}}) \partial_r \partial_{\bar{r}}$$

$$- \frac{1}{r\bar{r}} \partial_\theta \partial_{\bar{\theta}} - \frac{1}{r \sin\theta \, \bar{r} \sin\bar{\theta}} \partial_\phi \partial_{\bar{\phi}}$$

$$+ \frac{2i}{r\bar{r} (\sin\theta + \sin\bar{\theta})} \partial_\theta \partial_{\bar{\phi}}$$

$$- \frac{2i}{r\bar{r} (\sin\theta + \sin\bar{\theta})} \partial_\phi \partial_{\bar{\theta}} ,$$

m = real constant.

(It turns out to be simpler to use the contravariant components g_{ab}, although the covariant components $g_{a\bar{b}}$ could just as well have been used.)

This metric is manifestly Hermitian $(\overline{g^{a\bar{b}}} = g^{\bar{a}b})$ and can easily be shown to be non-singular. On the "real slice" M given by $u = \bar{u}$, $r = \bar{r}$, $\theta = \bar{\theta}$, $\phi = \bar{\phi}$, the metric reduces to

$$(\partial/\partial s)_M^2 = 2 \partial_u \partial_r - (1 - 2m/r) \partial_r \partial_r \tag{X.38}$$

$$- 1/r^2 \partial_\theta \partial_\theta - \frac{1}{r^2 \sin^2\theta} \partial_\phi \partial_\phi ,$$

which is just the Schwarzschild metric (see [36]) on M .

We now perform the following complex analytic coordinate transformation in $\mathbb{C}M$:

$$u' = u - i\,a\cos\theta \qquad \bar{u}' = \bar{u} + i\,a\cos\bar{\theta}$$

$$r' = r + i\,a\cos\theta \qquad \bar{r}' = \bar{r} - i\,a\cos\bar{\theta}$$

$$\theta' = \theta \qquad \bar{\theta}' = \bar{\theta}$$

$$\phi' = \phi \qquad \bar{\phi}' = \bar{\phi}$$

a = real constant.

In these new coordinates, the Hermitian metric X.37 becomes:

$$\left(\partial/\partial s'\right)^2 = \partial_{u'}\partial_{\bar{r}'} + \partial_{r'}\partial_{\bar{u}'} - \left(1 - \frac{m}{r' - i\,a\cos\theta'} - \frac{m}{\bar{r}' + i\,a\cos\bar{\theta}'}\right)\partial_{r'}\partial_{\bar{r}'}$$

$$- \frac{1}{(r' - i\,a\cos\theta')(\bar{r}' + i\,a\cos\bar{\theta}')}\Big[(\partial_{\theta'} + i\,a\sin\theta'\,\partial_{u'} - i\,a\sin\theta'\,\partial_{r'})$$

$$(\partial_{\bar{\theta}'} - i\,a\sin\bar{\theta}'\,\partial_{\bar{u}'} + i\,a\sin\bar{\theta}'\,\partial_{\bar{r}'})\Big]$$

$$- \frac{1}{(r' - i\,a\cos\theta')\sin\theta'\,(\bar{r}' + i\,a\cos\bar{\theta}')\sin\bar{\theta}'}\,\partial_{\phi'}\partial_{\bar{\phi}'}$$

$$+ \frac{2i}{(r' - i\,a\cos\theta')(\bar{r}' + i\,a\cos\bar{\theta}')(\sin\theta' + \sin\bar{\theta}')}\cdot$$

$$\Big[(\partial_{\theta'} + i\,a\sin\theta'\,\partial_{u'} - i\,a\sin\theta'\,\partial_{r'})\,\partial_{\bar{\phi}'}$$

$$- \partial_{\phi'}(\partial_{\bar{\theta}'} - i\,a\sin\bar{\theta}'\,\partial_{\bar{u}'} + i\,a\sin\bar{\theta}'\,\partial_{\bar{r}'})\Big] \,,$$

(X.40)

and on the new "real slice" M' given by $u' = \bar{u}'$, $r' = \bar{r}'$, $\theta' = \bar{\theta}'$, $\phi' = \bar{\phi}'$, this reduces to:

$$\left(\partial_{s'}\right)^2_{M'} = \frac{-a^2 \sin^2\theta'}{r'^2 + a^2\cos^2\theta'}\,\partial^2_{u'} + \frac{2(r'^2 + a^2)}{r'^2 + a^2\cos^2\theta'}\,\partial_{u'}\partial_{r'}$$
$$- \left(1 - \frac{2mr' - a^2\sin^2\theta'}{r'^2 + a^2\cos^2\theta'}\right)\partial^2_{r'} \tag{X.41}$$
$$- \frac{1}{(r'^2 + a^2\cos^2\theta')}\left[\partial^2_{\theta'} + \frac{1}{\sin^2\theta'}\,\partial^2_{\phi'}\right]$$
$$+ \frac{2a}{(r'^2 + a^2\cos^2\theta')}\left[\partial_{r'}\partial_{\phi'} - \partial_{u'}\partial_{\phi'}\right],$$

and this can be shown to be (see [36]) the Kerr metric on M'.

We can generalize this result to treat the general Kerr-Talbot complex transformation in terms of the algorithm of Diagram X.2. To see this, consider a four-complex-dimensional genuine complex manifold $\mathbb{C}M$ with complex coordinates $z^a = u$, v, ζ, η and an Hermitian metric $g_{a\bar{b}}$ given by

$$ds^2 = d\zeta\, d\bar{\zeta} + d\eta\, d\bar{\eta} + du\, d\bar{v} + dv\, d\bar{u} \tag{X.42}$$
$$+4\sqrt{2}\, m\left(\frac{1}{F_Y} + \frac{1}{\bar{F}_Y}\right)\left[\frac{du + Y d\eta + \bar{Y} d\zeta - Y\bar{Y}\, dv}{1 + Y\bar{Y}}\right]\cdot$$
$$\left[\frac{d\bar{u} + \bar{Y}\, d\bar{\eta} + Y d\bar{\zeta} - Y\bar{Y}\, d\bar{v}}{1 + Y\bar{Y}}\right],$$

m = real constant,

$$F \equiv \phi(Y) + Y^2 \eta - \zeta + (u + v) Y ,$$

where

$$F_Y = \left[\phi'(Y) + 2 Y \eta + (u + v)\right]_{F = 0} , \qquad (X.43)$$

and Y is given implicitly as a function of the coordinates u, v, ζ, η by $F(Y, u, v, \zeta, \eta) = 0$. Here $\phi(Y)$ is an arbitrary analytic function of Y and $\phi'(Y)$ is its first derivative. Also, note that $\bar{Y}$, $\overline{F_Y}$ are meant to be genuine complex conjugates, so that in particular $\bar{Y} = \bar{Y}(\bar{u}, \bar{v}, \bar{\zeta}, \bar{\eta})$ and $\overline{F_Y} = \overline{F_Y}(\bar{u}, \bar{v}, \bar{\zeta}, \bar{\eta})$.

Now consider the four-real-dimensional "real slice" M defined by

$$\begin{aligned} u &= \bar{u} \\ v &= \bar{v} \\ \zeta &= \bar{\eta} \\ \eta &= \bar{\zeta} . \end{aligned} \qquad (X.44)$$

On this slice, the metric X.42 reduces to

$$ds_M^2 = 2d\zeta\, d\bar{\zeta} + 2\, dudv + \left(\frac{1}{F_Y} + \frac{1}{\overline{F_Y}}\right) \left[\frac{du + Yd\bar{\zeta} + \bar{Y}d\zeta - Y\bar{Y}dv}{1 + Y\bar{Y}}\right]^2 \qquad (X.45)$$

where now $\bar{Y} = \bar{Y}(u, v, \bar{\zeta}, \zeta)$ and $\overline{F_Y} = \overline{F_Y}(u, v, \bar{\zeta}, \zeta)$.

But this is precisely of the form of Equation VII.47 for a vacuum Kerr-Schild spacetime with timelike Killing vector (for ease of comparison with reference [11], we have retained Kerr's choice of signature -+++). So on M, $g_{a\bar{b}}$ reduces to the Kerr-Schild metric determined by the analytic function $\phi(Y)$.

Now in $\mathbb{C}M$ we perform the following complex analytic coordinate transformation:

$$\begin{aligned} u' &= u + \sqrt{2}\, i\, a & \bar{u}' &= \bar{u} - \sqrt{2}\, i\, a \\ v' &= v & \bar{v}' &= \bar{v} \\ \zeta' &= \zeta & \bar{\zeta}' &= \bar{\zeta} \\ \eta' &= \eta & \bar{\eta}' &= \bar{\eta} \end{aligned} \qquad \text{(X.46)}$$

a = real constant.

When the metric X.42 is transformed to these new coordinates, and the new "real slice" M' given by

$$\begin{aligned} u' &= \bar{u}' \\ v' &= \bar{v}' \\ \zeta' &= \bar{\eta}' \\ \eta' &= \bar{\zeta}' \end{aligned} \qquad \text{(X.47)}$$

is taken, we find that the induced metric $ds'^2_{M'}$ is again in Kerr-Schild form with

$$\phi'(Y) = \phi(Y) + \sqrt{2}\, i\, a\, Y\,. \qquad \text{(X.48)}$$

(Schwarzschild-to-Kerr is now recovered as the special case $\phi(Y) = 0$ on M, $\phi'(Y) = \sqrt{2}\, i\, a\, Y$ on M'.) It should be stressed that the <u>imaginary</u> nature of the displacement, $u' = u + i a$, is essential for obtaining a really different metric on M' : if instead we perform a real displacement $u' = u + a$, the "real slice" M' is the same as M , and we get the same metric in a different coordinate system (i.e., $u' = u + a$, $v' = v$, $\zeta' = \zeta$).

It can be argued with some justification that our treatment of the Schwarzschild-to-Kerr transformation in terms of our second algorithm (Diagram X.2) is not much more than a restatement of the original derivation [36] in slightly more respectable mathematical terms. In particular, the rather pleasing interpretation of Schwarzschild-to-"B" in terms of twistor surfaces on the complex extension is absent for Schwarzschild-to-Kerr. However, it is an interesting fact that a "flat space limit" of the Schwarzschild-to-Kerr transformation, to be defined precisely in a moment, can be put in terms of the stronger algorithm of Diagram X.1 and be given a twistor surface interpretation.

By the "flat space limit", we mean the limit $m \to 0$ in the tetrad of Equation VIII.19, with $a = 0$ for Schwarzschild and $a \neq 0$ for Kerr. From Table VII.2 it can be seen that $\Psi_2 \to 0$ as $m \to 0$ for both Schwarzschild and Kerr, so that the limit really is Minkowski space in both cases.

For the flat space limit M of Schwarzschild, then, the tetrad VIII.19 with $m = a = 0$ gives the metric

$$ds_M^2 = du^2 + 2\,du\,dr - r^2\,(d\theta^2 + \sin^2\theta\,d\phi^2) \qquad \text{(X.49)}$$

and structure coordinates (see Equations VIII.21)

$$\begin{aligned} z^0 &= u \\ z^1 &= \ln \tan \theta/2 + i\phi \\ \widetilde{z^0} &= u + 2r \\ \widetilde{z^1} &= \ln \tan \theta/2 - i\phi , \end{aligned} \qquad \text{(X.50)}$$

and for the flat space limit M' of Kerr, the tetrad VIII.19 with $m = 0$, $a \neq 0$ gives the metric

$$ds^2_{M'} = du'^2 + 2\, du'\, dr' - (r'^2 + a^2 \sin^2 \theta')\, d\theta'^2 \tag{X.51}$$

$$- (r'^2 + a^2) \sin^2 \theta' d\phi'^2 - 2a \sin^2 \theta'\, dr'\, d\phi'$$

and structure coordinates (see Equations VIII.20)

$$\begin{aligned} z^{0'} &= u' + i\, a \cos\theta' \\ z^{1'} &= \ln \tan \theta'/2 + i\,\phi' \\ \widetilde{z}^{0'} &= u' + 2r' - i\, a \cos\theta' \\ \widetilde{z}^{1'} &= \ln \tan \theta'/2 - i\,\phi' + 2\, i \tan^{-1}(a/r') \, . \end{aligned} \tag{X.52}$$

Newman has given [34] an analytic complex transformation which relates the two metrics X.49 and X.51 according to the strong algorithm of Diagram X.1. Let u, r, θ, ϕ go complex (i.e., extend M to $\mathbb{C}M$) and perform the coordinate transformation

$$\begin{aligned} u &= u' + i\, a \cos\theta' \\ r &= r' - i\, a \cos\theta' \\ r \cos\theta &= r' \cos\theta' - i\, a \\ \cos 2(\phi - \phi') &= \frac{r'^2 - a^2}{r'^2 + a^2} \quad . \end{aligned} \tag{X.53}$$

It can be checked that this transforms the metric X.49 (extended to $\mathbb{C}M$) into the metric X.51 (extended to $\mathbb{C}M$).

Now we re-express this transformation in terms of the liberated structure coordinates X.50 and X.52. We find that X.53 is equivalent to

$$z^0 = z^{0'}$$

$$z^1 = z^{1'}$$

$$\tilde{z}^0 = \tilde{z}^{0'}$$

$$\tilde{z}^1 = \tilde{z}^{1'} \quad , \qquad (X.54)$$

and from this we conclude that the twistor surfaces for the flat space limit of Schwarzschild coincide with the twistor surfaces for the flat space limit of Kerr, on the complex extension $\mathbb{C}M$.

Newman and his co-workers have also found [35] a complex transformation taking Reissner-Nordström (electrified Schwarzschild) spacetime to Kerr-Newman (electrified Kerr) spacetime. Everything we have said about the Schwarzschild-to-Kerr transformation can be generalized straightforwardly to this case, also. In particular, the full curved space transformation can be cast in the form of the weak algorithm of Diagram X.2, and the flat space limit can be put in the form of the strong algorithm of Diagram X.1, with a corresponding twistor surface interpretation.

The "Heavenly Explanation"

We have now discussed several "explanations" and interpretations of the various complex transformations. It must be admitted that everything said so far about these transformations has been quite far removed from any physical considerations. However, it now appears that the recent developments in asymptotic twistor theory and Heaven theory provide a definitive "explanation" of these complex transformations, with at least a tentative interpretation.

The seeds of this physical interpretation lie in some observations of Newman and Winicour [62] concerning special relativistic systems of non-interacting free particles. Such a system can be described classically by a constant momentum vector p^a and a constant angular momentum tensor $M^{ab} = -M^{ba}$, where M^{ba} is taken with respect to some origin 0^a in Minkowski space. Under a change of origin

$$0^a \rightarrow \hat{0}^a = 0^a + x^a \quad , \tag{X.55}$$

we have

$$\begin{aligned} p^a &\rightarrow \hat{p}^a = p^a \\ M^{ab} &\rightarrow \hat{M}^{ab} = M^{ab} + p^a x^b - x^a p^b \quad , \end{aligned} \tag{X.56}$$

(we are working in one rectangular coordinate system throughout; x^a is the displacement vector from 0^a to $\hat{0}^a$).

We want to consider the case in which the momentum p^a of the system is timelike, i.e., $p^a p_a \equiv m^2 > 0$. Newman and Winicour point out that when $p^a p_a > 0$, the angular momentum tensor M^{ab} can be decomposed into an orbital part L^{ab} and a spin (intrinsic) part S^{ab} ,

$$M^{ab} = L^{ab} + S^{ab} \tag{X.57}$$

such that under X.55, we have

$$\begin{aligned} L^{ab} &\rightarrow \hat{L}^{ab} = L^{ab} + p^a x^b - x^a p^b \\ S^{ab} &\rightarrow \hat{S}^{ab} = S^{ab} \ . \end{aligned} \tag{X.58}$$

There is a remarkable symmetry in the definitions of L^{ab} and S^{ab}. To see this, we first define

$$A^a \equiv M^{ab} p_b$$

$$S^a \equiv M^{*ab} p_b \qquad (X.59)$$

(see Equations VIII.24 for the definition of $M^{*ab} = g^{am} g^{bn} M^{*}{}_{ab}$). Note that from these definitions we have

$$A^a p_a = S^a p_a = 0 . \qquad (X.60)$$

In terms of the vectors A^a and S^a, the orbital and spin parts of M^{ab} can now be written as

$$L^{ab} = m^{-2} (A^a p^b - p^a A^b)$$

$$S^{*ab} = m^{-2} (S^a p^b - p^a S^b) . \qquad (X.61)$$

We now ask for changes of origin X.55 which reduce the orbital part of the angular momentum to zero. From Equations X.58 and X.61, we can write for such a change of origin

$$L^{ab} = 0 = m^{-2} (A^a p^b - p^a A^b) + p^a x^b - x^a p^b \qquad (X.62)$$

$$= (m^{-2} A^a - x^a)p^b - p^a(m^{-2} A^b - x^b) .$$

Clearly this can be satisfied whenever $m^{-2} A^a - x^a$ is proportional to p^a :

$$m^{-2} A^a - x^a = - \lambda \ p^a . \qquad (X.63)$$

As λ takes on all real values, this defines a real timelike line in Minkowski space:

$$x^a(\lambda) = m^{-2} A^a + \lambda p^a , \qquad (X.64)$$

which we define to be the <u>center of mass line</u> for the system.

The symmetry between L^{ab} and S^{ab} is even more evident when we consider the self-dual part M^{+ab} of the angular momentum tensor, defined by

$$M^{+ab} = M^{ab} + i\, M^{*ab} . \qquad (X.65)$$

Using X.57, X.61, and the fact that $** = - i\, d$, this can be expressed as

$$\begin{aligned} M^{+ab} &= L^{ab} + S^{ab} + i\, L^{*ab} + i\, S^{*ab} \\ &= (L^{ab} + i\, S^{*ab}) + i\, (L^{ab} + i\, S^{*ab})^* \\ &= m^{-2} [(A^a + i\, S^a)\, p^b - p^a\, (A^b + i\, S^b)] \qquad (X.66) \\ &+ i\, m^{-2} [(A^a + i\, S^a)\, p^b - p^a\, (A^b + i\, S^b)]^* . \end{aligned}$$

Now under a change of origin X.55 we have

$$\begin{aligned} M^{+ab} \rightarrow \hat{M}^{+ab} &= M^{+ab} + p^a x^b - x^a p^b \\ &\quad + i\, (p^a x^b - x^a p^b)^* \qquad (X.67 \\ &= [m^{-2} (A^a + i\, S^a) - x^a]\, p^b - p^a [m^{-2} (A^b + i\, S^b) - x^b] \\ &+ i \left\{ [m^{-2} (A^a + i\, S^a) - x^a]\, p^b - p^a\, [m^{-2} (A^b + i\, S^b) - x^b] \right\}^* . \end{aligned}$$

(X.68)

Newman and Winicour now ask for changes of origin which reduce M^{+ab} to zero. As above, we see that this will happen whenever

$$m^{-2} (A^a + i S^a) - x^a = - \lambda p^a . \qquad (X.69)$$

But now, assuming $S^a \neq 0$, there will be no real values of x^a satisfying this condition. Hence we analytically continue this equation into the complex extension of Minkowski space, and define the <u>complex center of mass line</u> as the set of points

$$z^a (\lambda) = m^{-2} (A^a + i S^a) + \lambda p^a , \qquad (X.70)$$

where now λ ranges over all complex values, there being no reason now to restrict it to real values. Thus the complex center of mass line is a one-complex-dimensional set of points in the complex extension of Minkowski space.

(Notice that the <u>anti-self-dual</u> part $M^{-ab} \equiv M^{ab} - i M^{*ab}$ will not vanish on the line X.70, but rather on the conjugate line $z^a(\lambda) = m^{-2} (A^a - i S^a) + \lambda p^a$. Thus we might more properly speak of left and right complex center of mass lines.)

Newman and Winicour also show that if the system has a net charge e and a magnetic moment, so that a dipole tensor D^{ab} can be defined, then a complex center of charge line can be defined in close analogy to the complex center of mass line. It is then found, rather amazingly, that if the complex center of charge line coincides with the complex center of mass line, then the gyromagnetic ratio (ratio of magnetic moment to spin) takes the Dirac value e/mc in conventional units.

But what does all of this have to do with Schwarzschild-to-Kerr transformation? In the next Chapter we shall discuss the fact that with each asymptotically flat real vacuum or Einstein-Maxwell spacetime there is associated a four-complex-dimensional manifold called "Heaven." For Schwarzschild and Kerr (in fact, for all stationary spacetimes) the associated Heavens are just complex Minkowski space, that is, the complex extension of Minkowski space. A momentum vector p^a and an angular momentum tensor M^{ab} can be defined in Heaven from the curvature data at null infinity in the original spacetime, and from these, a complex center of mass line in Heaven can be associated with the original spacetime by proceeding exactly as above. (This has been shown by Newman and independently by B. D. **Bramson.**) Then it is found that the complex center of mass lines for Schwarzschild and Kerr <u>differ by an imaginary translation</u>. We have already seen (compare Equations X.64 and X.70) that an imaginary displacement of the center of mass line signals the presence of <u>intrinsic spin</u> in the system (i.e., the presence of $i\,S^a$ in Equation X.70). Thus it would appear that the angular momentum of the Kerr solution corresponds to intrinsic spin in the associated Heavenly system.

This interpretation is strengthened when the Heavens associated with Reissner-Nordström and Kerr-Newman spacetimes are considered. In this case complex center of charge lines can also be constructed in the Heavens, and it is found that the Kerr-Newman complex center of charge line differs from that of Reissner-Nordström by an imaginary translation, and both center of charge lines coincide with the respective complex center of mass lines, so that the Dirac gyromagnetic ratio is associated with the Heavenly system arising from Kerr-Newman (the ratio is indeterminate in the case of Reissner-Nordström).

We began this chapter by describing the complex transformations as "generators of twist." We have now seen that in the Heavenly picture, this generation of twist seems to be reflected as a generation of intrinsic spin and magnetic moment. This is a very tentative interpretation, to be sure, but it is the only known interpretation which injects genuine physical concepts into the picture.

CHAPTER XI: TWISTOR THEORY AND HEAVEN

Twistor theory, developed by Roger Penrose and his co-workers, originated as an attempt to give the conformal group of Minkowski space a position of central importance in physics. Twistors are essentially objects which transform covariantly under the action of the group SU(2,2), which is a four-to-one covering of the conformal group. Some of the roots of twistor theory include the work of Robinson and Trautman ([71], [72], [52]) and Debney, Kerr, and Schild [11] on null congruences in Minkowski space. In fact, the twisting EM field which we examined in Chapter X gave the first example of a Robinson congruence, which is a representation of a (non-null) twistor.

Twistor theory has as its ultimate goal the elimination of the continuum in physics and the formulation of physical laws in purely combinational mathematical terms. This is the so-called "twistor program." Penrose has shown that starting from the group SO(3) and the combinatorial rules for non-relativistic angular momentum, Euclidean space can be obtained as a <u>derived</u> concept. Roughly speaking, the goal of the twistor program is to derive relativistic spacetime from combinatorial principles applied to the group SU(2,2). The twistor formulation of fields and scattering processes in terms of contour integrals appears to be an important step towards this goal.

There is no space here to give even a reasonably complete summary of twistor theory, much less a thorough exposition. We will have to content ourselves with a brief sketch of the topic, emphasizing the connections with our own work and referring to the literature for details on the rest. There are three especially important papers on twistor theory. "Twistor

Algebra" (Penrose, [66]) is the original paper on twistors, and discusses the tensor algebra and physical interpretation of twistors. The second paper, "Twistor Quantisation and Curved Space-Time" (Penrose, [67]) discusses the twistor description of zero-rest-mass free fields, global twistors for curved spacetimes, and the passage to quantum theory. Finally, "Twistor Theory: An Approach to the Quantisation of Fields and Space-Time" (Penrose and MacCallum, [44]) is a sort of "1972 state of the art" paper which includes a presentation of twistor diagrams for scattering problems and the first discussion of asymptotic twistor theory.

We shall find two rather surprising connections between twistor theory and our work. The "twistor surfaces" introduced in Chapter X will be seen to be very natural generalizations of a certain description of flat space twistors due to Hansen and Newman. And we will see that "Heavenly spacetimes", which can be thought of as arising from the structure of asymptotic twistor spaces, are closely related to our left-Kählerian spacetimes, being in fact the complex extension of vacuum left-Kählerian spacetimes.

Whenever possible, we shall follow the conventions of MacCallum's paper [44] unless otherwise stated (these differ significantly from earlier conventions, e.g., those of [66]).

Twistors for Minkowski Space

We start by defining twistors for Minkowski space, or so-called flat-space twistors. In the sequel, the unmodified term "twistor" will refer to these flat-space twistors, as opposed to local twistors, asymptotic twistors, and so forth.

XI.1: DEFINITION: Twistor space $\mathbb{T}$ is the complex vector space $\mathbf{C}^4$ with an indefinite Hermitian structure given, in a standard basis for $\mathbf{C}^4$, by

$$H(Z, W) = Z^0 \overline{W^2} + Z^1 \overline{W^3} + Z^2 \overline{W^0} + Z^3 \overline{W^1} .$$

A twistor of valence $\begin{bmatrix}1\\0\end{bmatrix}$ is a vector in $\mathbb{T}$.

(Recall Definition VI.1 for a vector space Hermitian structure; a standard basis for $\mathbf{C}^4$ is in fact a basis of the form $[E_\beta - i E_{\bar{\beta}}]$, $\beta = 0, 1, 2, 3$, with E_β, $E_{\bar{\beta}}$ as in Chapter II. Then a self-conjugate vector is one which has real components with respect to the basis $[E_\beta , E_{\bar{\beta}}]$.)

From our viewpoint, it is somewhat clearer to think of twistors slightly differently. We can think of twistor space $\mathbb{T}$ as the complex manifold $\mathbf{C}^4$ with the canonical complex structure (see page) and the flat indefinite Hermitian metric given by

$$g_{\alpha\bar{\beta}} = \begin{pmatrix} 0 & 0 & 1 & 0 \\ 0 & 0 & 0 & 1 \\ 1 & 0 & 0 & 0 \\ 0 & 1 & 0 & 0 \end{pmatrix} . \qquad \text{(XI.1)}$$

A twistor of valence $\begin{bmatrix}1\\0\end{bmatrix}$ can then be thought of as a real (self-conjugate) vector in the tangent space at the origin of $\mathbb{T}$, in the sense of Chapter IV, i.e., one whose components Z^α, $Z^{\bar{\alpha}}$ satisfy

$$\overline{Z^\alpha} = Z^{\bar{\alpha}} . \qquad \text{(XI.2)}$$

All of this is easily seen to be equivalent to Definition XI.1. This type of identification is familiar from special relativity, where Minkowski space can be though of either as a vector space or as a manifold. Using

XI.1 and XI.2, the Hermitian inner product is now expressible as

$$H(Z, W) = g_{\alpha\bar{\beta}} Z^{\alpha} \overline{W^{\beta}} = g_{\alpha\bar{\beta}} Z^{\alpha} W^{\bar{\beta}} . \qquad (XI.3)$$

We shall refer to $H(Z, W)$ as the twistor inner product of the valence $\begin{bmatrix}1\\0\end{bmatrix}$ twistors Z and W. Notice that the definition of the twistor inner product in terms of Equation (XI.3) involves partial contractions over unbarred indices only and over barred indices only. It is an easy exercise to show that such partial contractions are tensorial with respect to complex analytic coordinate transformations.

XI.2: DEFINITION: The twistor norm of a valence $\begin{bmatrix}1\\0\end{bmatrix}$ twistor is given by

$$H(Z, Z) = g_{\alpha\bar{\beta}} Z^{\alpha} \overline{Z^{\beta}} = g_{\alpha\bar{\beta}} Z^{\alpha} Z^{\bar{\beta}} .$$

Note that the twistor norm is real, $\overline{H(Z, Z)} = H(Z, Z)$, whereas the twistor inner product of two different twistors is not necessarily real.

Because of Equation XI.2, there is no ambiguity in denoting a valence $\begin{bmatrix}1\\0\end{bmatrix}$ twistor by its unbarred components Z^{α}. We now want to define twistors of arbitrary valence.

XI.3: DEFINITION: A twistor of valence $\begin{bmatrix}p\\q\end{bmatrix}$ is a self-conjugate (real) vector in the tensor-product space

$$\underbrace{\mathbb{T}\otimes\cdots\otimes\mathbb{T}}_{\text{p factors}} \otimes \underbrace{\mathbb{T}^*\otimes\cdots\otimes\mathbb{T}^*}_{\text{q-factors}}$$

whose only non-vanishing components in a standard basis are

$$T^{\alpha\cdots\beta}{}_{\gamma\cdots\delta}$$

and

$$T^{\bar{\alpha}\cdots\bar{\beta}}{}_{\bar{\gamma}\cdots\bar{\delta}} \equiv \overline{T^{\alpha\cdots\beta}{}_{\gamma\cdots\delta}} .$$

For example, the alternating twistor $\epsilon_{\alpha\beta\gamma\delta}$ is a twistor of valence $\begin{bmatrix}0\\4\end{bmatrix}$ defined, in a standard basis, by

$$\epsilon_{\alpha\beta\gamma\delta} = \epsilon_{[\alpha\beta\gamma\delta]}$$

$$\epsilon_{0123} = +1$$

$$\epsilon_{\bar{\alpha}\bar{\beta}\bar{\gamma}\bar{\delta}} = \overline{\epsilon_{\alpha\beta\gamma\delta}} \quad . \tag{XI.4}$$

Twistors W_α of valence $\begin{bmatrix}0\\1\end{bmatrix}$ are called dual twistors, since they are self-conjugate vectors in the dual twistor space $\mathbb{T}^*$.

Finally, we need the following definition:

XI.4: DEFINITION: The twistor conjugate of a valence $\begin{bmatrix}p\\q\end{bmatrix}$ twistor $T^{\alpha\beta\cdots\gamma}{}_{\delta\cdots\epsilon}$ is the twistor $\bar{T}^{\delta\cdots\epsilon}{}_{\alpha\beta\cdots\gamma}$ of valence $\begin{bmatrix}q\\p\end{bmatrix}$ defined by

$$\bar{T}^{\delta\cdots\epsilon}{}_{\alpha\beta\cdots\gamma} = g_{\alpha\bar{\kappa}}\, g_{\beta\bar{\lambda}} \cdots g_{\gamma\bar{\mu}}\, g^{\delta\bar{\nu}} \cdots g^{\epsilon\bar{\rho}}\, \overline{T^{\kappa\lambda\cdots\mu}{}_{\nu\cdots\rho}}$$

$$= g_{\alpha\bar{\kappa}}\, g_{\beta\bar{\lambda}} \cdots g_{\gamma\bar{\mu}}\, g^{\delta\bar{\nu}} \cdots g^{\epsilon\bar{\rho}}\, T^{\bar{\kappa}\bar{\lambda}\cdots\bar{\mu}}{}_{\bar{\nu}\cdots\bar{\rho}} \quad .$$

In particular, the twistor conjugate of the valence $\begin{bmatrix}1\\0\end{bmatrix}$ twistor Z^α is the valence $\begin{bmatrix}0\\1\end{bmatrix}$ twistor $\bar{Z}_\alpha \equiv g_{\alpha\bar{\mu}}\, \overline{Z^\mu}$. In terms of components,

$$(\bar{Z}_0, \bar{Z}_1, \bar{Z}_2, \bar{Z}_3) = (\overline{Z^2}, \overline{Z^3}, \overline{Z^0}, \overline{Z^1}), \tag{XI.5}$$

using Equation XI.1 .

Our definitions and conventions are set up in such a way that we need never explicitly consider barred twistor indices. For example, the twistor inner product can now be expressed as

$$H(Z, W) = Z^\alpha \bar{W}_\alpha \quad . \tag{XI.6}$$

More generally, a contraction $T^{\alpha \cdots}_{\cdots} S_{\alpha \cdots}^{\cdots}$ shall always be taken to mean a partial contraction over unbarred indices only.

From Equation XI.1, it can be shown that $g_{\alpha\bar{\beta}}$ has signature (++--). Hence the twistor norm, now expressible as $H(Z, Z) = g_{\alpha\bar{\beta}} Z^{\alpha} \bar{Z}^{\bar{\beta}} = Z^{\alpha}\bar{Z}_{\alpha}$, may be positive, negative, or zero for a non-zero twistor Z^{α}.

<u>XI.5: DEFINITION</u>: The set of <u>right-handed twistors</u> (of valence $\begin{bmatrix}1\\0\end{bmatrix}$) is the set

$$\mathbb{T}^{+} = \{Z^{\alpha} \in \mathbb{T} \mid Z^{\alpha}\bar{Z}_{\alpha} > 0\},$$

The set of <u>left-handed twistors</u> is the set

$$\mathbb{T}^{-} = \{Z^{\alpha} \in \mathbb{T} \mid Z^{\alpha}\bar{Z}_{\alpha} < 0\},$$

and the set of <u>null twistors</u> is the set

$$\mathbb{N} = \{Z^{\alpha} \in \mathbb{T} \mid Z^{\alpha}\bar{Z}_{\alpha} = 0\}.$$

For reasons which will become clear below, it is desirable to consider equivalence classes $[\lambda Z^{\alpha}]$ of proportional twistors, where λ is a non-zero complex number. Notice that if $Y^{\alpha} = \lambda Z^{\alpha}$ is right-handed (resp. left-handed, null) then Z^{α} is also right-handed (resp. left-handed, null). Hence we can define the following sets in terms of equivalence classes $[\lambda Z^{\alpha}]$:

<u>XI.6: DEFINITION</u>: $\mathbb{PT}$ = projective twistor space

$$= \{[\lambda Z^{\alpha}]\};$$

$$\mathbb{PT}^{+} = \{[\lambda Z^{\alpha}] \mid Z^{\alpha}\bar{Z}_{\alpha} > 0\};$$

$$\mathbb{PT}^{-} = \{[\lambda Z^{\alpha}] \mid Z^{\alpha}\bar{Z}_{\alpha} < 0\};$$

$$\mathbb{PN} = \{[\lambda Z^{\alpha}] \mid Z^{\alpha}\bar{Z}_{\alpha} = 0\}.$$

In older references, the sets $\mathbb{PT}^{+}$ and $\mathbb{PT}^{-}$ were denoted by C^{+} and C^{-}, respectively. The structure of projective twistor space is often represented schemetically as in Figure XI.1.

Clearly, the projective twistor space $\mathbb{PT}$ is just the complex manifold $\mathbb{CP}^{3}$, complex projective three-space, discussed in Chapter III. Notice that the vector space nature of $\mathbb{T}$ is lost in the passage to $\mathbb{PT}$, even though the concept of handedness survives.

We shall now give a very brief description of the correspondence between geometrical objects in twistor space $\mathbb{T}$ and geometrical objects in Minkowski space $\mathbb{M}$. For details, [66] or [44] should be consulted.

Choose a standard basis in twistor space and a constant set $\sigma_{a}{}^{AA'}$ of van der Waerden symbols and a fixed rectangular coordinate system for Minkowski space. With respect to these fixed bases, we now set up the following fundamental correspondences: with a twistor Z^{α} of valence $\begin{bmatrix}1\\0\end{bmatrix}$ we associate a pair of constant spinors w^{A} and $\pi_{A'}$ according to the scheme

$$Z^{\alpha} \longleftrightarrow (w^{A}\ ,\ \pi_{A'})$$

$$(Z^{0}, Z^{1}, Z^{2}, Z^{3}) \quad \longleftrightarrow \quad (w^{0}, w^{1}, \pi_{0'}, \pi_{1'}) \ . \qquad \text{(XI.7)}$$

Similarly, with a valence $\begin{bmatrix}0\\1\end{bmatrix}$ twistor we associate a pair of spinors as follows:

$$W_{\alpha} \longleftrightarrow (\mu_{A}\ ,\ \lambda^{A'})$$

$$(W_{0}, W_{1}, W_{2}, W_{3}) \longleftrightarrow (\mu_{0}, \mu_{1}, \lambda^{0'}, \lambda^{1'}) \ . \qquad \text{(XI.8)}$$

We can extend this correspondence to higher valence twistors by taking tensor products, e.g.:

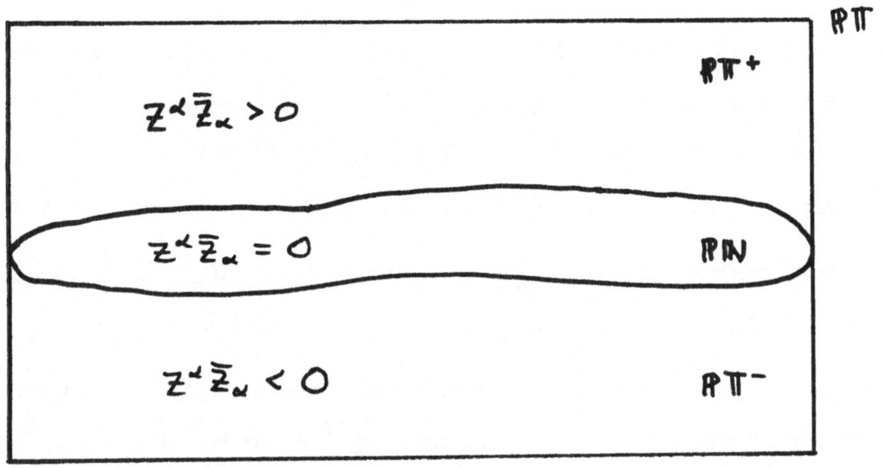
PT
PT⁺
$Z^\alpha \bar{Z}_\alpha > 0$
$Z^\alpha \bar{Z}_\alpha = 0$
PN
$Z^\alpha \bar{Z}_\alpha < 0$
PT⁻

Figure XI.1

$$Z^{\alpha\beta} \leftrightarrow \begin{pmatrix} \alpha^{AB} & \beta^{A}{}_{B'} \\ \gamma_{A'}{}^{B} & \delta_{A'B'} \end{pmatrix} , \qquad \text{(XI.9)}$$

$$Z_{\alpha\beta} \leftrightarrow \begin{pmatrix} \alpha_{AB} & \beta_{A}{}^{B'} \\ \gamma^{A'}{}_{B} & \delta^{A'B'} \end{pmatrix} , \qquad \text{(XI.10)}$$

and so forth. Note that if Z^{α} corresponds to $(w^A, \pi_{A'})$, then the conjugate twistor $\bar{Z}_{\alpha}$ corresponds to $(\bar{\pi}_A, \bar{w}^{A'})$.

The physical interpretation of a twistor $Z^{\alpha} \leftrightarrow (w^A, \pi_{A'})$ is that it represents a zero rest mass particle with momentum $p_a = \sigma_a{}^{AA'}\bar{\pi}_A \pi_{A'}$ and angular momentum $\mu_{AB} = i\, w_{(A}\bar{\pi}_{B)}$, where $M_{ab} = \sigma_a{}^{AA'}\sigma_b{}^{BB'} \cdot (\mu_{AB}\epsilon_{A'B'} + \epsilon_{AB}\bar{\mu}_{A'B'})$. From Equation X.59 we find that $S_a = s\, p_a$ where $2s = w^A\bar{\pi}_A + \pi_{A'}\bar{w}^{A'} = Z^{\alpha}\bar{Z}_{\alpha}$, so that the norm of the twistor is twice the helicity of the particle. Notice that $e^{i\theta}Z^{\alpha} \leftrightarrow (e^{i\theta}w^A, e^{i\theta}\pi_{A'})$ gives the same momentum and angular momentum as $Z^{\alpha} \leftrightarrow (w^A, \pi_{A'})$, and hence represents the same zero rest mass particle.

Now, there is a one-to-one correspondence between proportionality classes $[\lambda Z^{\alpha}]$ of <u>null</u> twistors and null straight lines in (compactified) Minkowski space. (This corresponds to the fact that a unique center of mass line for a zero rest mass particle can be found only if the helicity s vanishes; see [62] or [44]. This is seen as follows. We can describe any null straight line in Minkowski space by giving a point $\underset{0}{x}{}^a$ on the line and a null tangent vector p_a to the line. Let $\underset{0}{x}{}^{AA'}$ and $\bar{\pi}_A \pi_{A'}$ be the corresponding spinor quantities. Then the associated projective twistor is given by $\lambda Z^{\alpha} \leftrightarrow \lambda\,(i\,\underset{0}{x}{}^{AA'}\pi_{A'}, \pi_{A'})$. Any other point on the null line will have a position spinor of the form

$\underset{\circ}{x}^{AA'} + r\ \bar{\pi}_A \pi_{A'}$, r real, but since

$$i(\underset{\circ}{x}^{AA'} + r\ \bar{\pi}_A \pi_{A'}) \pi_{A'} = i\ \underset{\circ}{x}^{AA'} \pi_{A'} ,$$

the choice of point on the null line does not affect the associated twistor. The norm of Z^α is:

$$Z^\alpha \bar{Z}_\alpha = i\ \underset{\circ}{x}^{AA'} \pi_{A'} \bar{\pi}_A - i\ \underset{\circ}{\bar{x}}^{AA'} \bar{\pi}_A \pi_{A'} \tag{XI.13}$$

$$= i\ (\underset{\circ}{x}^{AA'} - \underset{\circ}{\bar{x}}^{AA'})\ \bar{\pi}_A \pi_{A'}$$

$$= 0 ,$$

because the position spinor $\underset{\circ}{x}^{AA'}$ is Hermitian.

The converse of this proposition is slightly more involved (see [66] or [59] for a more detailed discussion). Suppose $Z^\alpha \leftrightarrow (w^A, \pi_{A'})$, and for the moment assume $\pi_{A'} \neq 0$. Then we can always find spinors $x^{AA'}$ such that $w^A = i\ x^{AA'} \pi_{A'}$. If $\underset{\circ}{x}^{AA'}$ is any such spinor, then the most general one can be written as

$$x^{AA'} = \underset{\circ}{x}^{AA'} + \xi^A \pi^{A'} , \tag{XI.14}$$

where ξ^A is completely arbitrary. One particular solution is $\underset{\circ}{x}^{AA'} = i\ w^A \mu^{A'}$, where $\mu^{A'}$ is any spinor for which $\pi^{A'} \mu_{A'} = +1$. Thus Equation (XI.14) can be written as:

$$x^{AA'} = i\ w^A \mu^{A'} + \xi^A \pi^{A'} . \tag{XI.15}$$

As ξ^A takes on values $r\bar{\pi}_A$, r real, the position vector $x^{AA'}$ moves along a null line through the "point" $i\, w^A \mu^{A'}$. But we must show that $i\, w^A \mu^{A'}$ can be chosen to be Hermitian, so that it represents a real point in Minkowski space. Clearly, we have to choose $\mu^{A'}$ so that $i\, w^A = r\bar{\mu}^A$, where r is real. Since $\pi^{A'}\mu_{A'} = 1$, we must for consistency have

$$1 = \pi^{A'}\mu_{A'} = -i\, r^{-1}\, \bar{w}_{A'}\pi^{A'}$$

$$\Rightarrow \quad \bar{w}_{A'}\pi^{A'} \text{ is pure imaginary} \qquad \text{(XI.16)}$$

$$\Rightarrow \quad \bar{w}^{A'}\pi_{A'} + \bar{\pi}_A w^A = 0 \; .$$

But the last condition is simply $Z^\alpha \bar{Z}_\alpha = 0$. Hence we can associate a real null line with Z^α if and only if Z^α is null. If we multiply Z^α by λ , then $\mu^{A'}$ will be multiplied by λ^{-1} so that Equation XI.15 is left unchanged. Thus each member of the equivalence class $[\lambda Z^\alpha]$ will determine the same null line.

In the case $\pi_{A'} = 0$, the twistor $Z^\alpha \leftrightarrow (w^A, 0)$ is automatically null, but it corresponds to no finite line in Minkowski space. Instead, it can be thought of as representing a "line at infinity" in the compactified Minkowski space M_c . We refer the reader to [66] for more information on this case. There is a symmetry here between null twistors $Z^\alpha \leftrightarrow (0, \pi_{A'})$ which represent null lines through the origin and null twistors $Z^\alpha \leftrightarrow (w^A, 0)$ which represent null lines through the "point at infinity."

Non-null twistors can also be represented as geometrical objects in Minkowski space. Suppose $R^{\alpha}\bar{R}_{\alpha} \neq 0$ and consider the set of all projective twistors $[\lambda Z^{\alpha}]$ satisfying

$$R^{\alpha}\bar{Z}_{\alpha} = 0$$

$$Z^{\alpha}\bar{Z}_{\alpha} = 0 . \qquad \text{(XI.17)}$$

Each $[\lambda Z^{\alpha}]$ determines a null line, and the set of all the $[\lambda Z^{\alpha}]$ determines a congruence of null lines in Minkowski space. It turns out (see Kerr's theorem below) that this congruence is geodesic, shearfree, and twisting. This congruence associated with a non-null twistor R^{α} is called a <u>Robinson congruence</u>.

There is also a representation of valence $\begin{bmatrix}1\\0\end{bmatrix}$ twistors in terms of geometrical objects in complex Minkowski space $\mathbb{CM}$ (the complex extension of Minkowski space) which has the advantage of describing null and non-null twistors on an equal footing. This representation is discussed at length by Hansen and Newman [59]. Recall that for an arbitrary twistor $Z^{\alpha} \longleftrightarrow (w^{A}, \pi_{A'})$, the general solution to the equation

$$w^{A} = i\, x^{AA'}\pi_{A'} \qquad \text{(XI.18)}$$

can be written as

$$x^{AA'}(\xi^{A}) = \underset{\circ}{x}^{AA'} + \xi^{A}\pi^{A'} , \qquad \text{(XI.19)}$$

where ξ^{A} is arbitrary and $\underset{\circ}{x}^{AA'}$ is any particular solution, not necessarily Hermitian. Equation XI.18 can be thought of as a two-complex-dimensional surface in complex Minkowski space, the two complex dimensions corresponding to the two complex degrees of freedom in ξ^A . (Again, there are technicalities involving "points at infinity", so that we should really be considering compactified complexified Minkowski space $\mathbb{C}M_c$.)

Clearly, this "twistor surface" is a totally null surface, i.e., every tangent vector to the surface is null with respect to the analytic continuation of the Minkowski metric η_{ab} , because every tangent vector must be of the form $t^A\pi^{A'}$. (To define spinors on the complex extension $\mathbb{C}M$ of a spacetime M which admits a spinor structure, we simply extend the trivial spinor bundle S to all of $\mathbb{C}M$ and relate the resulting bundle to the holomorphic tangent bundle of $\mathbb{C}M$ by the analytic continuation of the van der Waerden symbols: $\sigma^a{}_{AA'}(x) \longrightarrow \sigma^a{}_{AA'}(z)$.) Furthermore, the twistor surface XI.19 is a totally geodesic submanifold of $\mathbb{C}M$; this concept requires some definition, and the matter will be gone into in some detail below.

It is easy to see from Equations XI.18 and XI.19 that λZ^α defines the same twistor surface as Z^α . Also, it is clear from the previous paragraphs that a twistor surface intersects the "real slice" M_c if and only if the corresponding twistor is null, in which case the intersection is a single null straight line (compare our "twistor surfaces" in Chapter X, and see below).

We can now obtain a twistor representation of points in (compactified) complex Minkowski space $\mathbb{C}M_c$. Consider two <u>non-proportional</u> twistors $Z^\alpha \longleftrightarrow (w^A, \pi_{A'})$, $V^\beta \longleftrightarrow (\lambda^A, \mu_{A'})$ and their associated twistor surfaces. Since the twistor surfaces are two-complex- or four-real-dimensional sets in the eight-real-dimensional manifold $\mathbb{C}M_c$, they will intersect in a zero-dimensional set. By ordinary linear algebra, we find that they intersect in precisely one point given by (see [44])

$$x^{AA'} = -i\,(\pi_{M'}\mu^{M'})(w^{A}\mu^{A'} - \lambda^{A}\pi^{A'}) \qquad \text{(XI.20)}$$

(if $\pi_{M'}\mu^{M'} = 0$, the intersection point is "at infinity").

If the intersection point $x^{AA'}$ is Hermitian, then writing $Z^{\alpha} \longleftrightarrow (i\,x^{AA'}\pi_{A'}, \pi_{A'})$, $W^{\beta} \longleftrightarrow (i\,x^{AA'}\mu_{A'}, \mu_{A'})$, it is easily shown that

$$\begin{aligned} Z^{\alpha}\bar{Z}_{\alpha} &= 0 \\ W^{\alpha}\bar{W}_{\alpha} &= 0 \\ Z^{\alpha}\bar{W}_{\alpha} &= 0\ . \end{aligned} \qquad \text{(XI.21)}$$

In this case we can think of Z^{α} and W^{β} as representing real null lines in Minkowski space which intersect at the point $x^{AA'}$. Equations XI.21 are in fact both necessary and sufficient for the intersection point of the twistor surfaces to lie in the "real slice" M_c .

Notice that if the pair (Z^{α}, W^{β}) satisfies XI.21, then so do the pairs $(a\,Z^{\alpha} + b\,W^{\alpha}, W^{\beta})$, $(Z^{\alpha}, c\,Z^{\beta} + d\,W^{\beta})$, and $(a\,Z^{\alpha} + b\,W^{\alpha}, c\,Z^{\beta} + d\,W^{\beta})$, where a, b, c, d are any complex numbers, with $ad-bc \neq 0$ in the last pair. Furthermore, all of these pairs define the same real point $x^{AA'}$ by Equation XI.20. This means that we can think of the real point $x^{AA'}$ as being represented in twistor space by the set $\{a\,Z^{\alpha} + b\,W^{\alpha} \mid a, b, \in \mathbb{C}\} \subseteq N$. Since we are really only interested in proportionality classes of twistors, the real point $x^{AA'}$ is more properly represented by the set $P = \{[\lambda(a\,Z^{\alpha} + b\,W^{\alpha})] \mid \lambda, a, b \in \mathbb{C}\} \subseteq PN$ in projective twistor space. This set P is in fact a <u>holomorphic projective line</u> (one-complex-dimensional) in projective twistor space $P\mathbb{T} = \mathbb{C}P^3$.

If we drop the requirements of Equation XI.21, so that $x^{AA'}$ is now generally a point in (compactified) complex Minkowski space $\mathbb{C}M_c$, it remains true that $(a\,Z^\alpha + b\,W^\alpha, W^\beta)$, $(Z^\alpha, c\,Z^\beta + d\,W^\beta)$, and $(a\,Z^\alpha + b\,W^\alpha, c\,Z^\beta + d\,W^\beta)$ define the same point as (Z^α, W^β). Hence the points of $\mathbb{C}M_c$ can all be represented as holomorphic projective lines $P \subseteq \mathbb{P}\mathbb{T}$, but now P is not necessarily a subset of $\mathbb{P}\mathbb{N}$.

If the twistor pair $Z^\alpha \leftrightarrow (w^A, \pi_{A'}), W^\beta \leftrightarrow (\lambda^B, \mu_{B'})$ represents the point $x^{AA'} \in CM_c$, we can also represent this point by the valence $\begin{bmatrix}2\\0\end{bmatrix}$ twistor

$$p^{\alpha\beta} = Z^\alpha W^\beta - W^\alpha Z^\beta \leftrightarrow (\pi_{M'}, \mu^{M'}) \begin{pmatrix} -\frac{1}{2}\epsilon^{AB} x_{MM'} x^{MM'} & i\,x^A{}_{B'} \\ -i\,x_{A'}{}^B & \epsilon_{A'B'} \end{pmatrix}. \tag{XI.22}$$

Note that any rescaling $p^{\alpha\beta} \to \lambda\, p^{\alpha\beta}$ defines the same point. Defining the dual twistor $p_{\alpha\beta}$ by

$$p_{\alpha\beta} \equiv p^*_{\alpha\beta} = \frac{1}{2}\epsilon_{\alpha\beta\gamma\delta}\, p^{\gamma\delta}, \tag{XI.23}$$

it can be shown that $p^{\alpha\beta}$ represents a point of real (compactified) Minkowski space M_c if and only if

$$p_{\alpha\beta} = \bar{p}_{\alpha\beta}, \tag{XI.24}$$

where $\bar{p}_{\alpha\beta}$ is the twistor conjugate of $p^{\alpha\beta}$, defined by $\bar{p}_{\alpha\beta} = g_{\alpha\bar{\mu}}\, g_{\beta\bar{\nu}}\, \overline{p^{\mu\nu}}$.

If we are interested in doing Poincaré-invariant physics in terms of twistors, rather than conformally-invariant physics, then we must single out a preferred point of compactified Minkowski space, the "point at infinity" (see Penrose in [14]). Since we have been assuming a specific coordinate system and spin frame in Minkowski space with respect to which the correspondences of Equations XI.7 through XI.10 are to hold, we can find an explicit representation for the point at infinity. This can be thought of as being obtained from Equation XI.22 by rescaling $p^{\alpha\beta}$ according to

$$p^{\alpha\beta} \rightarrow \frac{\pi_{M'}\,\mu^{M'}}{-\frac{1}{2}\,x_{MM'}\,x^{MM'}}\,p^{\alpha\beta} \;,$$

and then letting $x^{AA'}$ "recede to infinity" to obtain the <u>infinity twistor</u>:

$$I^{\alpha\beta} \leftrightarrow \begin{pmatrix} \epsilon^{AB} & 0 \\ 0 & 0 \end{pmatrix} . \tag{XI.25}$$

The dual of $I^{\alpha\beta}$ is

$$I_{\alpha\beta} \leftrightarrow \begin{pmatrix} 0 & 0 \\ 0 & \epsilon^{A'B'} \end{pmatrix} , \tag{XI.26}$$

and it is easy to show that

$$I_{\alpha\beta} = \bar{I}_{\alpha\beta} \;, \tag{XI.27}$$

so that $I^{\alpha\beta}$ does indeed represent a point of M_c. $I^{\alpha\beta}$ can be thought of as arising from any two non-proportional twistors of the form $Z^{\alpha} \leftrightarrow (w^A, 0)$, $W^{\beta} \leftrightarrow (\lambda^A, 0)$, which represent "null lines at infinity meeting at the point at infinity."

The invariance groups associated with twistor theory can now be discussed in terms of the "twistor metric tensor" $g_{\alpha\bar{\beta}}$ and the twistors $\epsilon_{\alpha\beta\gamma\delta}$ and $I^{\alpha\beta}$. Since, from Equation XI.1, the signature of $g_{\alpha\bar{\beta}}$ is (++--), the group of holomorphic coordinate transformations in $\mathbb{T}$ (considered as a complex manifold) which preserves the form of $g_{\alpha\bar{\beta}}$

is just $\mathbf{U}(2,2)$. If we demand that $\epsilon_{\alpha\beta\gamma\delta}$ as well as $g_{\alpha\beta}$ be preserved, then the group is reduced to $\mathbf{SU}(2,2)$. Since $\mathbf{SU}(2,2)$ is four-to-one homomorphic to the conformal group of Minkowski space, it is precisely these transformations that we are interested in when examining conformally invariant concepts in twistor terms. Finally, if we demand that the infinity twistor $I^{\alpha\beta}$ be preserved in form along with $g_{\alpha\beta}$ and $\epsilon_{\alpha\beta\gamma\delta}$, the symmetry group is reduced to the Poincare group $\mathcal{P}$. When a homogeneous Lorentz transformation acts on twistor space, the correspondences of Equations XI.7 through XI.10 are preserved after the spinors on the right-hand side have been transformed by an appropriate element of $SL(2,\mathbb{C})$ corresponding to the given homogeneous Lorentz transformation of $\mathbb{T}$. For more information on the group-theoretical aspects of twistor theory, see [66] and [65].

Twistor Physics

Given the correspondences discussed above between geometrical objects in twistor space and geometrical objects in compactified Minkowski space, it is evident that any conformally invariant or Lorents-invariant physical theory can be "translated" into twistor terms - at least in principle (see [66]). However, one of the advantages of a reformulation of a physical theory is that different approaches to unsolved problems may be suggested by the new formalism. In the case of twistor theory, contour integrals of holomorphic functions of twistor variables seem strongly to "suggest themselves" as the most natural way of describing physical fields in twistor terms. In particular, such a contour-integral formulation seems to be leading to a twistor version of quantum theory which is free of divergences, and this approach appears to be the first tangible step toward a realization of the goals of the "twistor program" discussed above.

At the heart of the twistor approach to physical fields is the contour-integral formulation of solutions to the zero-rest-mass free field equations in Minkowski space. The fundamental formula for spin $s = \frac{1}{2}n$ is the following:

$$\underbrace{\phi_{AB\ldots L}}_{n\,=\,2s}(x^m) = \frac{1}{2\pi i}\oint \underbrace{w_A\, w_B \ldots w_L}_{n\,=\,2s}\, f(w_{M},\, -i\, x^{MM'}\, w_{M})\, \Delta W \tag{XI.28}$$

where $\Delta W \equiv \epsilon^{AB}\, w_A\, dw_B$ and $f(W_\alpha)$ is a holomorphic (meromorphic) function of the twistor variable $W_\alpha \longleftrightarrow (w_M,\, -i\, x^{MM'}\, w_M)$ which is homogeneous of degree -n-2. (This homogeneity degree is necessary in order that $\phi_{AB\ldots L}$ have the transformation properties of an n-index spinor; see [67] and [44].) The integral is to be taken around a contour which surrounds the poles of $f(W_\alpha)$ in a suitable fashion. (In twistor theory, the term "holomorphic function" denotes a function which is analytic everywhere except at some isolated poles.)

It is easy to check that $\phi_{AB\ldots L}(x^m)$ satisfies the zero-rest-mass free field Equations IX.51 (when $s = 0$, $\phi(x^m)$ satisfies the scalar wave equation $\Box\phi = 0$, which we have not discussed). By a suitable choice of function $f(W_\alpha)$, algebraically special fields can be easily obtained. Furthermore, by an appropriate choice of function and contour, positive frequency solutions can be ensured. See [67] for details.

Quantum ideas are introduced into twistor theory by the observation that, for a twistor $Z^\alpha \longleftrightarrow (i\, x^{AA'}\, \pi_{A'},\, \pi_{A'})$, we have formally:

$$i\, Z^{\alpha}\, d\bar{Z}_{\alpha} = \bar{\pi}_{A}\, \pi_{A'}\, dx^{AA'} \leftrightarrow p_a\, dx^a$$

$$i\, dZ^{\alpha} \wedge d\bar{Z}_{\alpha} = d(\bar{\pi}_{A}\, \pi_{A'}) \wedge dx^{AA'} \leftrightarrow dp_a \wedge dx^a \qquad \text{(XI.29)}$$

This suggests that $-i\, Z^{\alpha}$ and $\bar{Z}_{\alpha}$ should be thought of as canonically conjugate variables, and that in the passage to quantum mechanics they should become canonically conjugate operators:

$$\underset{\sim}{Z}^{\alpha} \leftrightarrow \partial/\partial\bar{Z}_{\alpha}$$

$$\underset{\sim}{\bar{Z}}_{\alpha} \leftrightarrow -\partial/\partial Z^{\alpha}$$

$$[\underset{\sim}{Z}^{\alpha}, \underset{\sim}{\bar{Z}}_{\beta}] = \delta^{\alpha}_{\beta} \,. \qquad \text{(XI.30)}$$

Now the idea is to think of zero-rest-mass fields as being represented by the functions $f(W_{\alpha})$ of Equation XI.28 (there is a lot of "gauge freedom" involved in this: for example, $f(W_{\alpha})$ and $f(W_{\alpha}) + h\,(W_{\alpha})$, where $h(W_{\alpha})$ has no poles inside the chosen contour, represent the same field). These $f(W_{\alpha})$ are then thought of as state vectors upon which operators like those of Equations XI.30 may act. For example, the spin operator

$$\underset{\sim}{S} \equiv \frac{1}{4}\,(\underset{\sim}{Z}^{\alpha}\underset{\sim}{\bar{Z}}_{\alpha} + \underset{\sim}{\bar{Z}}_{\alpha}\,\underset{\sim}{Z}^{\alpha}) \qquad \text{(XI.31)}$$

acts on functions $f(W_{\alpha})$ to produce $sf(W_{\alpha})$, where s is the spin of the field represented by $f(W_{\alpha})$. Similarly, $\underset{\sim}{Z}^{\alpha}$ and $\underset{\sim}{\bar{Z}}_{\alpha}$ operating on $f(W_{\alpha})$ have the effect of lowering or raising the helicity by one-half.

When the concept of global twistors is introduced (see below), it is possible to treat the scattering of zero-rest-mass particles by weak gravitational or electromagnetic waves in a Hamiltonian formalism which again employs holomorphic functions of twistor variables. The scattering of zero-rest-mass fields can also be handled in the formalism. The calculation of simple processes such as Compton scattering gives agreement with the standard results. Some of the most recent progress in twistor physics involves the description of potentials and massive fields in twistor terms (see [44], [68]).

Analytic Metrics, Twistor Surfaces, and Kerr's Theorem

We must now consider a matter which we have already postponed several times. When complex Minkowski space $\mathbb{CM}$ is obtained as the complex extension of real Minkowski space $\mathbb{M}$, the tensor $\eta_{ab}(z)$ obtained by analytic continuation of $\eta_{ab}(x)$ will be singular when considered as a metric on CM. We have

$$\eta_{ab}(z) = \begin{pmatrix} \eta_{\alpha\beta} & \eta_{\alpha\bar{\beta}} \\ \eta_{\bar{\alpha}\beta} & \eta_{\bar{\alpha}\bar{\beta}} \end{pmatrix} = \begin{pmatrix} \eta_{\alpha\beta} & 0 \\ 0 & 0 \end{pmatrix}. \qquad \text{(XI.32)}$$

The problem is that η_{ab} does not assign a length to type (0,1) vectors $V^a = (0, V^{\bar{\alpha}})$. Nevertheless, it turns out that η_{ab} is the appropriate object to look at for many applications, including Heaven theory. Hence we make the following definition:

XI.7: DEFINITION: An analytic metric on a complex manifold is a tensor of the form

$$g_{ab} = \begin{pmatrix} g_{\alpha\beta} & g_{\alpha\bar{\beta}} \\ g_{\bar{\alpha}\beta} & g_{\bar{\alpha}\bar{\beta}} \end{pmatrix} = \begin{pmatrix} g_{\alpha\beta}(z^{\mu}) & 0 \\ 0 & 0 \end{pmatrix},$$

where $g_{\alpha\beta}$ is a non-singular matrix and each $g_{\alpha\beta}(z^{\mu})$ is an analytic function of the z^{μ}.

It is easy to check that everything in the definition is invariant under complex analytic coordinate transformations. Notice that no signature can be associated with an analytic metric, because, for example, a coordinate transformation of the form $z^\mu \rightarrow i\, z^\mu$ will change the signature. Also note that the complex extension $\mathbb{C}M$ of any Riemannian manifold M admits an analytic metric which is the analytic continuation of the Riemannian metric on M.

XI.8: DEFINITION: Define

$$g^{ab} \equiv \begin{pmatrix} g^{\alpha\beta}(z^\mu) & 0 \\ 0 & 0 \end{pmatrix}$$

and

$$\Gamma^\alpha_{\beta\gamma} \equiv \frac{1}{2} g^{\alpha\mu} \left(g_{\beta\mu,\gamma} + g_{\gamma\mu,\beta} - g_{\beta\gamma,\mu} \right).$$

Then a geodesic with respect to an analytic metric $g_{\alpha\beta}(z^\mu)$ on a complex manifold $\mathbb{C}M$ is a holomorphic curve $\mathcal{C} : \mathbb{C} \to \mathbb{C}M : \lambda \to z^\mu(\lambda)$ such that

$$\frac{d^2 z^\mu}{d\lambda^2} + \Gamma^\mu_{\alpha\beta} \frac{dz^\alpha}{d\lambda} \frac{dz^\beta}{d\lambda} = k \frac{dz^\mu}{d\lambda} .$$

Notice that if $x^\mu(\lambda)$ (λ real) is a geodesic in the ordinary sense on M, then the analytic continuation $z^\mu(\lambda)$ (λ complex) is a geodesic on the complex extension $\mathbb{C}M$ in the above sense.

It is not hard to convince oneself that through each point p of a complex manifold $\mathbb{C}M$ with an analytic metric and for each tangent vector $V^\mu(p)$ of type (1,0) at p, there passes a (locally) unique geodesic $z^\mu(\lambda)$ with $dz^\mu/d\lambda\,|_p = V^\mu(p)$. (The integrability conditions are simply that the $z^\mu(\lambda)$ be complex analytic functions of λ.)

<u>XI.9: DEFINITION</u>: Let S be a complex analytic submanifold of a complex manifold $\mathbb{C}M$ with an analytic metric $g_{\alpha\beta}(z^{\mu})$. Then S is said to be <u>totally geodesic</u> with respect to $g_{\alpha\beta}(z^{\mu})$ if for each point $p \in S$ the geodesics through p and tangent to S at p lie entirely within S .

The point of all of these definitions is that we will now show that the twistor surfaces discussed in Chapter X in connection with integrable almost Hermitian structures, as well as the Hansen-Newman twistor surfaces of Equation XI.19, are in fact totally null totally geodesic submanifolds of the complex manifolds in which they are defined. The totally null condition simply means that every tangent vector to the surface (necessarily of type (1, 0), since the surface is an analytic submanifold) is null with respect to the analytic metric. A formal definition is now in order:

<u>XI.10: DEFINITION</u>: Let $\mathbb{C}M$ be a four-complex-dimensional complex manifold with an analytic metric $g_{\alpha\beta}(z^{\mu})$. A <u>twistor surface</u> S in $\mathbb{C}M$ is a two-complex-dimensional complex analytic submanifold S of $\mathbb{C}M$ which is (i) totally null and (ii) totally geodesic with respect to $g_{\alpha\beta}(z^{\mu})$.

<u>XI.11: THEOREM</u>: Let M be a complex (real) spacetime with a complex (real) metric $g_{ab}(x^m)$. Let $\mathbb{C}M$ be the complex extension of M , with the analytic metric $g_{\alpha\beta}(z^{\mu})$ which is the analytic continuation of $g_{ab}(x^m)$. Let $J_a{}^b$ be an almost Hermitian structure for M which is at least half-integrable, with structure coordinates

z^0, z^1 . Then the surface S in $\mathbb{C}M$ given locally by z^0 = constant, z^1 = constant is a <u>twistor surface</u> with respect to $g_{\alpha\beta}(z^\mu)$.

<u>Proof</u>: The coordinates z^0, z^1 are valid coordinates for M , hence when liberated they are valid complex coordinates for $\mathbb{C}M$. Therefore S is a complex analytic submanifold of $\mathbb{C}M$, obviously of complex dimension two. Let $(\ell_a, n_a, m_a, \tilde{m}_a)$ be the analytic continuation of a tetrad for which $J_a{}^b$ is half-integrable, so that $\kappa = \sigma = 0$ both on M and when analytically continued to $\mathbb{C}M$. Now any normal to S will be a linear combination of dz^0 and dz^1 , hence a linear combination of $\ell_a dz^a$ and $m_a\, dz^a$, the analytic continuations of the type (1,0) forms of $J_a{}^b$. So any vector tangent to S will be orthogonal to ℓ_a and m_a , hence of the general form

$$t^a = A\,\ell^a + Bm^a \quad , \tag{XI.33}$$

because $\ell_a \ell^a = m_a\, m^a = \ell_a m^a = 0$. Therefore we have $t_a t^a = 0$ for any tangent vector t^a , so S is totally null.

It remains to be shown that S is totally geodesic. The key to this is that the totally geodesic condition can be expressed tensorially. At a point $p \in S$, take an arbitrary tangent vector t^a as in Equation XI.33 and propogate it parallelly to itself. In general we will have

$$t^b t_{a;b} = (A\,\ell^b + Bm^b)(A\,\ell_a + Bm_a)_{;b} \tag{XI.34}$$

$$\equiv C\,\ell_a + Dm_a + En_a + F\tilde{m}_a \quad \underline{\text{at } p} .$$

Now, the totally geodesic condition states that t^a remains in the surface S when propagated parallelly to itself. Hence it will be sufficient to show that

$$E = F = 0 \quad \text{at } p . \tag{XI.35}$$

Contracting Equation XI.34 with ℓ^a and then m^a, and using Equations VII.14, we find:

$$\ell^a t^b t_{a;b} = E = -A B \kappa - B^2 \sigma \quad , \tag{XI.36}$$

$$m^a t^b t_{a;b} = -F = A^2 \kappa + A B \sigma \quad \text{at } p .$$

But $\kappa = \sigma = 0$ everywhere by assumption, hence $E = F = 0$ at each $p \in S$, hence S is totally geodesic, and the theorem is proven. (When we contract Equation XI.34 with n^a and $\tilde{m}^a$ and set $C = kA$, $D = kB$ for some function k, we simply get equations for the geodesic propagation of A and B away from p.)

XI.12: COROLLARY: The surface

$$x^{AA'}(\xi^A) = \underset{o}{x}^{AA'} + \xi^A \pi^{A'} \quad , \tag{XI.37}$$

with $\underset{o}{x}^{AA'}$ and $\pi^{A'}$ constant and ξ^A arbitrary, is a twistor surface in complex Minkowski space $\mathbb{CM}$.

Proof: This can be thought of as a trivial special case of the above theorem when the tetrad associated with $J_a{}^b$ is a constant tetrad. Let $\mu^{A'}$ be any constant spinor for which $\mu_{A'}\pi^{A'} = 1$ and consider the tetrad

$$\ell_a = \sigma_{aAA'}\,\bar{\pi}^{A}\,\pi^{A'}$$

$$m_a = \sigma_{aAA'}\,\bar{\mu}^{A}\,\pi^{A'}$$

$$n_a = \sigma_{aAA'}\,\bar{\mu}^{A}\,\mu^{A'}$$

$$\tilde{m}_a = \sigma_{aAA'}\,\bar{\pi}^{A}\,\mu^{A'} \qquad \text{(XI.38)}$$

(Recall that spinors can be defined on **CM** by, roughly speaking, taking the analytic continuation of the $\sigma_{aAA'}$.) This is a constant null tetrad, hence the associated $J_a{}^b$ is integrable. Taking the standard van der Waerden symbols with respect to coordinates for which $ds^2 = 2\,du\,dv - 2\,d\zeta\,d\tilde{\zeta}$, the structure coordinates are easily seen to be

$$z^0 = \sqrt{2}\int \ell_a\,dx^a = x^{AA'}\,\bar{\pi}_A\,\pi_{A'}$$

$$z^1 = \sqrt{2}\int m_a\,dx^a = x^{AA'}\,\bar{\mu}_A\,\pi_{A'}$$

$$z^0 = \sqrt{2}\int n_a\,dx^a = x^{AA'}\,\bar{\mu}_A\,\mu_{A'}$$

$$z^1 = \sqrt{2}\int \tilde{m}_a\,dx^a = x^{AA'}\,\bar{\pi}_A\,\mu_{A'}\,, \qquad \text{(XI.39)}$$

where

$$x^{AA'} = \begin{pmatrix} u & \zeta \\ \tilde{\zeta} & v \end{pmatrix} .$$

Then the surface XI.37 is seen to correspond to the surface

$$z^0 = \underset{o}{x}^{AA'}\,\bar{\pi}_A\,\pi_{A'} = \text{constant}, \quad z^1 = \underset{o}{x}^{AA'}\,\bar{\mu}_A\,\pi_{A'} = \text{constant}.$$

Hence this is a twistor surface. (This single twistor surface is <u>not</u> to be thought as associated with the constant tetrad XI.38; this was merely introduced to facilitate the proof.)

These results provide a remarkable link between twistor theory and the complex transformations discussed in Chapter X. To see this link clearly, it is necessary to consider a result in twistor theory known as Kerr's Theorem, which gives an algorithm for obtaining the most general geodesic and shearfree null congruence (GSF congruence) in Minkowski space.

XI.13: THEOREM (KERR'S THEOREM): Let the metric of Minkowski space $\mathbb{M}$ be written as

$$ds^2 = 2\,du\,dv - 2\,d\zeta\,d\bar{\zeta} + (\ell_a n_b + n_a \ell_b - m_a \bar{m}_b - \bar{m}_a m_b)\,dx^a dx^b ,$$

where $(\ell_a, n_a, m_a, \bar{m}_a)$ is a constant normed null tetrad. The most general analytic GSF congruence ξ_a in M is given by either $\xi_a = n_a$ or by

$$\xi_a = \ell_a + \lambda m_a + \bar{\lambda}\bar{m}_a + \lambda\bar{\lambda} n_a , \qquad \text{(XI.40)}$$

where λ is a complex function of $u, v, \zeta, \bar{\zeta}$ defined implicitly by $F = 0$, where

$$F = F(\lambda, u + \lambda\zeta, \bar{\zeta} + \lambda v) \qquad \text{(XI.41)}$$

is an arbitrary complex analytic function of its three arguments.

Proof: (This proof has been given by D. Cox and the author in [56]; for alternative proofs, see [11] and [66].)

The first step is to recall that the GSF property depends only on the directions of the vectors of the congruence, i.e., $A\xi_a$ is GSF iff ξ_a is GSF. Next, it is easy to show that Equation XI.40 sweeps out every null direction except n_a as λ is varied. (The case $\xi_a = n_a$ can be thought of as the limit in which $\lambda \to \infty$.) We can look upon ξ_a in the form XI.40 as the new tetrad leg $\hat{l}_a$ resulting from a null rotation of the form of Equation VII.26, with $b = \bar{\lambda}$. Then by Equations VII.29, we will have

$$\begin{aligned} \mu \to \hat{\mu} &= -D\bar{\lambda} - \lambda\delta\bar{\lambda} - \bar{\lambda}\bar{\delta}\bar{\lambda} - \lambda\bar{\lambda}\Delta\bar{\lambda} = -\hat{D}\bar{\lambda} \\ \sigma \to \hat{\sigma} &= -\delta\bar{\lambda} - \bar{\lambda}\Delta\bar{\lambda} = -\hat{\delta}\bar{\lambda} \ , \end{aligned} \qquad \text{(XI.42)}$$

since all of the original spin coefficients vanish under our assumption of a constant null tetrad (l_a, n_a, m_a, $\bar{m}_a$) (which can happen only in Minkowski space). There is no loss of generality in choosing this constant tetrad in such a way that in the coordinate system $u, v, \zeta, \bar{\zeta}$ we have

$$\begin{aligned} l^a &= (0, 1, 0, 0) \ , & D &= \partial_v \\ n^a &= (1, 0, 0, 0) \ , & \Delta &= \partial_u \\ m^a &= (0, 0, 0, -1) \ , & \delta &= -\partial_{\bar{\zeta}} \\ \bar{m}^a &= (0, 0, -1, 0) \ , & \bar{\delta} &= -\partial_{\zeta} \ . \end{aligned} \qquad \text{(XI.43)}$$

Then Equations XI.42 become:

$$\hat{\kappa} = -(\partial_v - \bar{\lambda}\partial_\zeta)\bar{\lambda} + \lambda(\partial_{\bar{\zeta}} - \bar{\lambda}\partial_u)\bar{\lambda}$$
$$\hat{\sigma} = (\partial_{\bar{\zeta}} - \bar{\lambda}\partial_u)\bar{\lambda} . \qquad \text{(XI.44)}$$

To apply the GSF conditions, we set $\hat{\kappa} = \hat{\sigma} = 0$. The differential equations which result are equivalent to the following, obtained from XI.44 by taking appropriate linear combinations and complex conjugates:

$$(\partial_v - \lambda\partial_{\bar{\zeta}})\lambda = 0 \ (= \hat{D}\lambda)$$
$$(\partial_\zeta - \lambda\partial_u)\lambda = 0 \ (= \hat{\bar{\delta}}\lambda) . \qquad \text{(XI.45)}$$

Now we consider the system of equations

$$(\partial_v - \lambda\partial_{\bar{\zeta}})X = 0 \ (= \hat{D}X)$$
$$(\partial_\zeta - \lambda\partial_u)X = 0 \ (= \hat{\bar{\delta}}X), \qquad \text{(XI.46)}$$

where for the moment λ is considered to be known. This is a system of linear partial differential equations for X. A necessary and sufficient condition for this system to be completely integrable, i.e., for two independent solutions to exist, is given by Eisenhart ([16], pp. 69-70). This integrability condition is simply that the commutator of the two linear differential operators $(\partial_v - \lambda\partial_{\bar{\zeta}})$ and $(\partial_\zeta - \lambda\partial_u)$ should take the form

$$[(\partial_\zeta - \lambda\partial_u), (\partial_v - \lambda\partial_{\bar\zeta})]\,\emptyset$$

$$= A(\partial_\zeta - \lambda\partial_u)\,\emptyset + B\,(\partial_v - \lambda\partial_{\bar\zeta})\,\emptyset \qquad \text{(XI.47)}$$

for any function $\emptyset$, where A, B are some functions. Working out this commutator, we find:

$$[(\partial_\zeta - \lambda\partial_u), (\partial_v - \lambda\partial_{\bar\zeta})]\,\emptyset$$

$$= (\partial_u\,\emptyset)(\partial_v - \lambda\,\partial_{\bar\zeta})\lambda - (\partial_{\bar\zeta}\,\emptyset)(\partial_\zeta - \lambda\,\partial_u)\lambda .$$

(XI.48)

The only way for this expression to be of the form XI.47 for arbitrary $\emptyset$ is to have

$$(\partial_v - \lambda\partial_{\bar\zeta})\lambda = 0$$

$$(\partial_\zeta - \lambda\partial_u)\lambda = 0 , \qquad \text{(XI.49)}$$

that is to say, λ must satisfy Equations XI.45.

So assume that λ is indeed a solution of XI.45. (Clearly, at least one such solution, λ = constant, exists.) Then the system XI.46 is integrable, and two independent solutions are easily seen to be

$$X_1 = u + \lambda\zeta$$

$$X_2 = \bar\zeta + \lambda v \ . \qquad \text{(XI.50)}$$

Then the most general (analytic) solution must be given by

$$X = f(X_1, X_2) = f(u + \lambda\zeta\ ,\ \bar\zeta + \lambda v)\ , \qquad \text{(XI.51)}$$

where f is an arbitrary analytic function of its arguments. In particular, since λ was assumed to be a solution, it must be expressible in this form:

$$\lambda = g\,(u + \lambda\zeta\ ,\ \bar{\zeta} + \lambda v\,) \tag{XI.52}$$

Then λ is given implicitly in terms of the coordinates by

$$\begin{aligned} &G(\,\lambda,\, u + \lambda\zeta\ ,\ \bar{\zeta} + \lambda v\,) \\ &\quad \equiv \lambda - g(u + \lambda\zeta\ ,\ \bar{\zeta} + \lambda v\,) = 0\ . \end{aligned} \tag{XI.53}$$

Clearly, no generality is lost, and none is gained, by allowing λ to be defined implicitly by <u>any</u> analytic function

$$F(\,\lambda,\, u + \lambda\zeta\ ,\ \bar{\zeta} + \lambda v\,) = 0\ . \tag{XI.54}$$

Thus Kerr's Theorem is proven.

To see the connection of Kerr's Theorem with twistor theory, consider a null twistor $Z_\alpha \longleftrightarrow (\,\pi_{A'}\ ,\ -\,i\,x^{AA'}\,\pi_{A'})$ with $x^{AA'}$ Hermitian. Go to a spin frame such that $\pi_{A'} = (1, \lambda)$ and $x^{AA'} = \begin{pmatrix} u & \zeta \\ \bar{\zeta} & v \end{pmatrix}$. Then $x^{AA'}\pi_{A'} = (u + \lambda\zeta\ ,\bar{\zeta} + \lambda v\,)$. Then we have

$$Z_\alpha \longleftrightarrow \big(1,\ \lambda\ ,\ -\,i\,(u + \lambda\zeta),\ -\,i(\bar{\zeta} + \lambda v\,)\big). \tag{XI.55}$$

Consider the function

$$\begin{aligned} f(Z_\alpha\,) &= (Z_0)^{-n-2}\, F(Z_1/Z_0,\ i\,Z_2/Z_0\ ,\ i\,Z_3/Z_0) \\ &= F\,(\,\lambda,\, u + \lambda\zeta\ ,\ \bar{\zeta} + \lambda v\,)\ . \end{aligned} \tag{XI.56}$$

Then $f(Z_\alpha)$ is a homogeneous function of degree $-n-2$ in Z_α, and $f(Z_\alpha) = 0$ defines a GSF congruence in Minkowski space by Equation XI.54. The corresponding function F can be thought of as defining a complex analytic surface $F = 0$ in the projective twistor space $\mathbb{P}\mathbb{T}$. We have:

<u>XI.14: THEOREM</u>: **(**Twistor statement of Kerr's Theorem**)**: A complex analytic surface $F = 0$ in projective twistor space $\mathbb{P}\mathbb{T}$ corresponds to a GSF congruence in Minkowski space.

<u>Proof</u>: See above; the members of the congruence are the null lines corresponding to projective twistors which are null and lie on the surface $F = 0$. The exceptional case $\xi_a = n_a$ corresponds to the surface $F = 0$ defined by $F(\alpha, \beta, \gamma) = \alpha$.

The twistor statement of Kerr's Theorem is represented schematically in Figure XI.2: the shaded region represents the intersection of $\mathbb{P}\mathbb{N}$ with the complex analytic surface $F = 0$; every twistor lying in this intersection determines one line of the corresponding GSF congruence in Minkowski space.

Since the surfaces $f(W_\alpha) = 0$ or $F([\lambda W_\alpha]) = 0$ are complex analytic functions of the complex coordinates in $\mathbb{T}$ or $\mathbb{P}\mathbb{T}$, Kerr's Theorem can be thought of as giving a geometrical interpretation of the complex structure of twistor space $\mathbb{T}$ and projective twistor space $\mathbb{P}\mathbb{T}$.

The link to our "twistor surfaces" of Chapter X now becomes clear in light of the following two results:

<u>XI.15: THEOREM</u>: Let $\hat{\ell}_a$ be a GSF congruence in Minkowski space, with $J_a{}^b$ the associated half-integrable almost Hermitian structure tensor. Suppose $\hat{\ell}_a$ is generated according to Kerr's Theorem by

$$F(\lambda,\ u + \lambda\zeta\ ,\ \bar{\zeta} + \lambda v_v) = 0.$$

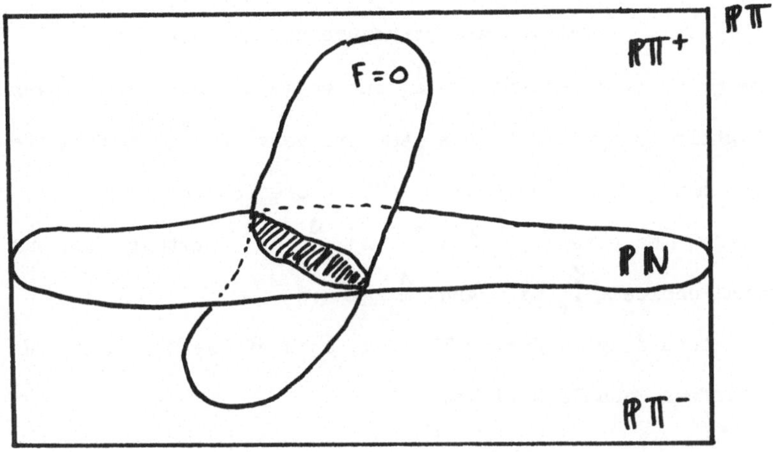

Figure XI.2

Define

$$W^0 = \lambda|_{F=0}$$
$$W^1 = (u + \lambda\zeta)|_{F=0}$$
$$W^2 = (\bar{\zeta} + \lambda v)|_{F=0} . \qquad \text{(XI.57)}$$

Then at least two of the W are not identically zero and are structure coordinates for $J_a{}^b$.

<u>Proof</u>: The proof is straightforward, but we must change our conventions slightly to agree with the standard conventions used in the statement of Kerr's Theorem. Therefore let the type (1,0) forms of $J_a{}^b$ be given by $\hat{\ell}_a dx^a$ and $\hat{\bar{m}}_a dx^a$, rather than our usual choice $\hat{\ell}_a dx^a$ and $\hat{m}_a dx^a$.

According to Kerr's Theorem, each of the functions W satisfies Equations XI.46:

$$\hat{D}W = \hat{\bar{\delta}} W = 0.$$

But this can be written

$$\hat{\ell}^a W_{,a} = \hat{\bar{m}}^a W_{,a} = 0 ,$$

which, because of $\hat{\ell}^a \hat{\ell}_a = \hat{\bar{m}}^a \hat{\bar{m}}_a = \hat{\ell}^a \hat{\bar{m}}_a = 0$, implies that

$$W_{,a} = A \hat{\ell}_a + B\hat{\bar{m}}_a$$

for some A, B . But this means that $dW = (A \hat{\ell}_a + B\hat{\bar{m}}_a) dx^a$ is a form of type (1,0) with respect to $J_a{}^b$, and therefore W is a structure coordinate for $J_a{}^b$, unless W vanishes identically. But neither W^1 nor W^2 is identically zero unless F takes the form $F(\alpha, \beta, \gamma) = \beta$ or $F(\alpha, \beta, \gamma) = \gamma$, respectively. And in these two cases, W^0, W^2 and W^0, W^1, respectively, will be not identically zero. In the general case, W^1 and W^2 were chosen to be independent, and in the two special cases, it is easy to see that W^0, W^2 or W^0, W^1 are also independent. Hence the desired result.

<u>XI.16: THEOREM</u>: In complex Minkowski space $\mathbb{CM}$, every twistor surface S arising from constant values of structure coordinates z^0, z^1 for some half-integrable almost Hermitian structure $J_a{}^b$ is in fact a Hansen-Newman twistor surface. (See the discussion concerning Equation XI.19.)

<u>Proof</u>: Let $\hat{l}_a$ be the GSF congruence leading to the structure coordinates z^0, z^1, and let $\hat{l}_a$ be generated according to Kerr's Theorem by some function $F(\lambda, u + \lambda\zeta, \bar{\zeta} + \lambda v) = 0$. By theorem XI.15, two valid structure coordinates can be chosen from among

$W^0 = \lambda|_{F=0}$, $W^1 = (u + \lambda\zeta)|_{F=0}$, and $W^2 = (\bar{\zeta} + \lambda v)|_{F=0}$.

But two of these constant, together with $F = 0$, immplies the third is constant also. So in any case we have

$$u + \lambda\zeta = \text{const}$$

$$\bar{\zeta} + \lambda v = \text{const}$$

$$\lambda = \text{const}\ . \tag{XI.58}$$

Writing $x^{AA'} = \begin{pmatrix} u & \zeta \\ \bar{\zeta} & v \end{pmatrix}$ and $\pi_{A'} = (1, \lambda)$, this is equivalent to

$$x^{AA'}\pi_{A'} = \text{const} \equiv \underset{0}{x}^{AA'}\pi_{A'}\ , \tag{XI.59}$$

$$\pi_{A'} = \text{const}.$$

But this is precisely the Hansen-Newman twistor surface

$$x^{AA'} = \underset{o}{x}^{AA'} + \xi^A \pi^{A'} ,$$

where ξ^A is arbitrary and $\underset{o}{x}^{AA'}$ is a point on S .

The link to our work in Chapter X is now complete. In the case of complex Minkowski space, the congruences of twistor surfaces defined by constant values of the structure coordinates are in fact equivalent to congruences of Hansen-Newman twistor surfaces, by virtue of Kerr's Theorem and Theorem XI.15. And in the case of Schwarzschild and "B" spacetimes, and other algebraically special spacetimes our curved space twistor surfaces in the complex extended manifold provide a natural generalization of the flat space twistor surfaces, in view of their totally geodesic properties.

Curved-Space Twistors

We have seen that geometrical objects in twistor spaces have a close connection to geometrical objects in compactified Minkowski space M_c. Because of the SU(2,2)-invariance of twistor space, this correspondence is really valid for any conformally flat spacetime. The question arises as to how to generalize twistor concepts so as to be able to treat curved spacetimes in twistor terms. There are several more or less plausible ways of doing this, which lead to several different concepts of twistors for curved spacetimes. We shall very briefly examine some of these ideas now, referring the reader to [44] and [68] for more detailed discussions.

(We should remark at this point that Definition XI.10 of twistor surfaces does not lead to a suitable generalization of twistors to general curved spacetimes, because they exist only for algebraically special spacetimes; and even when they do exist, there are not "enough" of them to form a four-complex-dimensional twistor space. However, see the definition of hypersurface twistors below.)

The simplest curved-space twistor concept is that of a local twistor. Assume the curved spacetime M admits a spinor structure, and let S be the bundle of unprimed contravariant spinors ξ^A . We define a local twistor bundle $T = S \otimes \bar{S}^*$; then a local twistor field of arbitrary valence is defined to be a section of one of the tensor product bundles $T \otimes \ldots \otimes T \otimes T^* \otimes \ldots \otimes T^*$. Thus, for example a valence $\begin{bmatrix}1\\0\end{bmatrix}$ local twistor field corresponds to a pair of spinor fields $(\omega^A(x), \pi_{A'}(x))$ defined at each point of M . (When M is Minkowski space $\mathbb{M}$, a constant local twistor field corresponds to a flat-space twistor of Definition XI.3.)

Local twistor fields are required to have transformation properties under conformal rescalings of the metric which are analogous to the transformation properties of flat-space twistors under conformal rescalings of Minkowski space (see [47]). Then connections for the various twistor bundles $T, T^*, \ldots$ can be defined by requiring that the derivative of a local twistor itself transforms like a local twistor under conformal rescalings. These connections are defined in terms of the Riemannian connection of M and the curvature tensors R_{ab} and R . These connections have torsion, and the curvature forms depend on the Weyl tensor C_{abcd} and some components of its covariant derivative $C_{abcd;e}$. "Twistor transport" of a local twistor along a curve in M is defined by parallel propagation with respect to these connections.

The local twistor concept is undesirable from the viewpoint of the twistor program, which seeks to obtain the points of spacetime as secondary objects, with twistors as the primary objects. The fibre bundle approach, with a twistor space attached to each point of a pre-existing spacetime, is obviously inimical to such a viewpoint. Somewhat more in the spirit of the twistor program is the definition of a global twistor associated with a curved spacetime M . A null global twistor for M is defined to be a null geodesic $\underline{Z}$ in M , together with a spinor $\pi_{A'}$ which is parallelly propagated along $\underline{Z}$ and such that $\bar{\pi}_A \pi_{A'}$ is everywhere tangent to $\underline{Z}$. The analogy to a null flat-space twistor is obvious. The set of null global twistors for M will generally form a seven-real-dimensional manifold—five dimensions to parametrize null geodesics in M and two more dimensions for the complex scaling of $\pi_{A'}$. The non-null global twistors are formally defined by embedding this seven-real-dimensional manifold in an abstract eight-real-dimensional manifold, which is then to be thought of as the desired "global twistor space." Global twistors are useful in treating the scattering of zero-rest-mass particles; see above and [44].

The major failing of such a global twistor space is that, while it can be given a symplectic structure (compare Equations XI.29), it cannot generally be given a meaningful complex structure to make it into a four-complex-dimensional complex manifold. The existence of a symplectic structure is related to the fact that a twist-free congruence of null geodesics remains twist-free when it passes through a region of conformal curvature, whereas the non-existence of a complex structure is related to the fact that an initially shearfree null congruence generally "picks up shear" in passing through a region of conformal curvature.

Because of this difficulty, still another type of curved space twistor concept is introduced, that of <u>hypersurface twistors</u>. These are constructed as follows. Consider a spacetime M and a hypersurface $\mathcal{S} \subsetneq M$ which is either spacelike or null and intersects each null geodesic of M exactly once. Take the complex extension $\mathbb{C}M$ of M and the complex extension $\mathbb{C}\mathcal{S} \subsetneq \mathbb{C}M$ of $\mathcal{S}$. If M is in fact Minkowski space $\mathbb{M}$, then we can consider a Hansen-Newman twistor surface S in the complex extension $\mathbb{C}\mathbb{M}$, with covariantly constant spinor $\pi_{A'}$ defined all over S. Then the intersection $S \cap \mathbb{C}\mathcal{S}$ of S (two-complex-dimensional) with $\mathbb{C}\mathcal{S}$ (three-complex-dimensional) will be a (one-complex-dimensional) <u>complex line ξ</u> lying in $\mathbb{C}\mathcal{S}$. The equation of such a curve is found to be (see [68])

$$n^{AB'}\pi_{B'}\pi^{A'}\nabla_{AA'}\pi_{C'} = 0, \tag{XI.60}$$

where $n^a = \sigma^a{}_{AB'}n^{AB'}$ is a type (1,0) normal vector to $\mathbb{C}\mathcal{S}$. The type (1,0) tangent vector $t^a = \sigma^a{}_{AA'}t^{AA'}$ to the curve ξ is given by

$$t^{AA'} = n^{AB'}\pi_{B'}\pi^{A'}. \tag{XI.61}$$

The situation is as illustrated in Figure XI.3 below, where numbers in parentheses refer to the complex dimensionality of the corresponding sets. When M is <u>not</u> Minkowski space, we retain Equation XI.60 and use it to define curves ξ in $\mathbb{C}\mathcal{S}$ with tangent vectors XI.61, which are determined by specifying $\pi_{A'}$ at one point of $\mathbb{C}\mathcal{S}$ and then propagating according to Equation XI.60; $\pi_{A'}$ is in fact covariantly constant along the curve so obtained. We call the pair $(\xi, \pi_{A'})$ a <u>hypersurface twistor</u> with respect to $\mathbb{C}\mathcal{S}$. (When M is Minkowski space, there is a unique correspondence between the hypersurface twistors and the twistor surfaces in $\mathbb{C}\mathbb{M}$.)

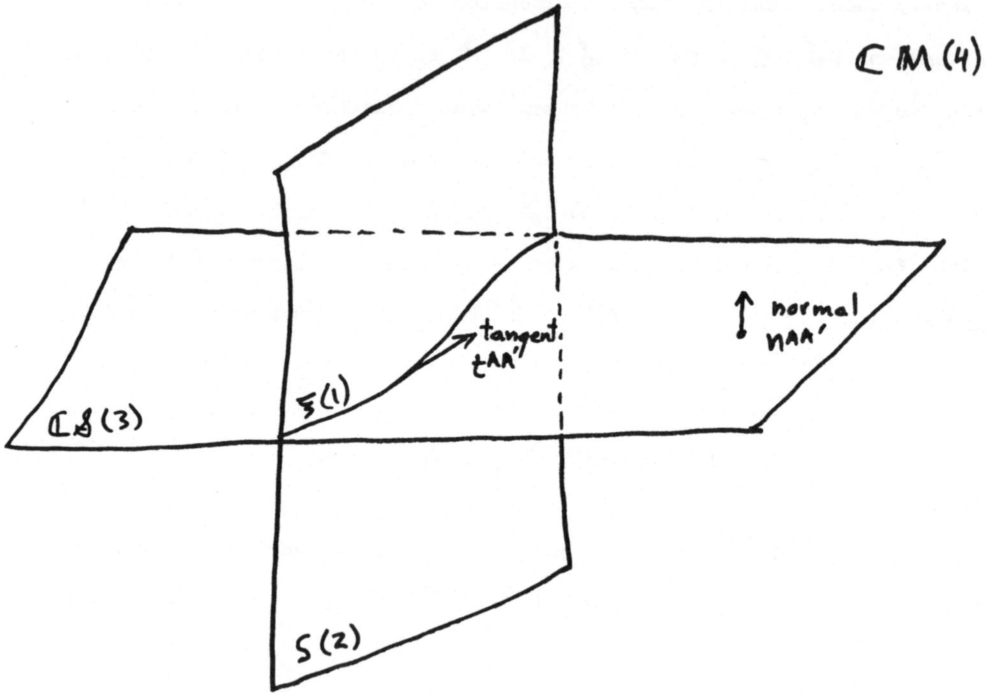

Figure XI.3

The set of all hypersurface twistors for a given $\mathcal{S}$ can be shown to form a four-complex-dimensional complex manifold, which we call hypersurface twistor space $\mathbb{T}(\mathcal{S})$. Thus we obtain the structure which was missing for global twistors (hypersurface twistor spaces also have well-defined symplectic structures as do global twistor spaces). This complex structure can be thought of as arising because the curves ξ are holomorphic with respect to the complex structure of the complex extension $\mathbb{C}M$.

Following Penrose [68], we can denote a hypersurface twistor by $Z^{\cdot} \longleftrightarrow (\xi, \pi_{A'})$. Complex scalar multiplication is defined by $\lambda Z^{\cdot} \longleftrightarrow (\xi, \lambda\pi_{A'})$, $\lambda \in \mathbb{C}$, and this leads to the concept of projective hypersurface twistor space $\mathbb{P}\mathbb{T}(\mathcal{S})$. A hypersurface twistor $Z^{\cdot}$ is said to be null if the corresponding curve ξ in $\mathbb{C}\mathcal{S}$ contains a real point $p \in \mathcal{S}$. It can be shown that ξ can contain at most one such real point. Such a null hypersurface twistor $Z^{\cdot}$ corresponds uniquely to a null global twistor in M : given a null $Z^{\cdot}$, go to the real point p of the corresponding curve, determine $\pi_{A'}$ at this point, take the null geodesic $\underline{Z}$ in M through p with tangent vector $\bar{\pi}_A \pi_{A'}$, and parallelly propagate $\pi_{A'}$ along $\underline{Z}$. Using this correspondence and the complex structure of $\mathbb{T}(\mathcal{S})$, a Kerr's Theorem for hypersurface twistors can be proven; the theorem generates congruences of null geodesics in M which have vanishing shear at their points of intersection with $\mathcal{S}$.

Dual hypersurface twistors can be defined by taking the "twiddled" versions of Equations XI.60 and XI.61:

$$n^{A'B}\, \tilde{\pi}_B\, \tilde{\pi}^A\, \nabla_{AA'}\, \tilde{\pi}_C = 0$$

$$t^{AA'} = n^{A'B}\, \tilde{\pi}_B\, \tilde{\pi}^A \, . \qquad \text{(XI.62)}$$

The resulting curves $\hat{\tilde{\xi}}$ in $\mathbb{C}\mathscr{S}$ with spinors $\hat{\tilde{\pi}}_A$ defined along them give the dual hypersurface twistors $\tilde{Z}. \longleftrightarrow (\hat{\tilde{\xi}}, \hat{\pi}_A)$. Given a hypersurface twistor $Z^{\cdot} \longleftrightarrow (\xi, \pi_{A'})$, the complex conjugate twistor is defined as $\bar{Z}. \longleftrightarrow (\bar{\xi}, \bar{\pi}_A)$, where $\bar{\xi}$ is the complex conjugate curve. The curves ξ and $\bar{\xi}$ intersect if and only if $Z^{\cdot}$ is null.

In general there are two problems with hypersurface twistors. First, in general no scalar product can be defined for hypersurface twistors. Secondly, in a general spacetime there is no canonical choice of a hypersurface $\mathscr{S}$, and when $\mathscr{S}$ is changed, the complex structure of $\Pi(\mathscr{S})$ "shifts" in the presence of conformal curvature. Both of these problems can be remedied in the case of asymptotically flat spacetimes by choosing $\mathscr{S}$ to be $\mathscr{I}^+$ or $\mathscr{I}^-$, the future or past null cone at infinity (Penrose in [11], [43]). The hypersurface twistors in these cases are called future or past asymptotic twistors.

We shall discuss future asymptotic twistors on $\mathscr{I}^+$. The situation for past asymptotic twistors is similar. $\mathscr{I}^+$ can be thought of as a three-real-dimensional manifold with topology $S^2 \times \mathbb{R}$. We shall assume that the asymptotically flat spacetime M is either a vacuum or an Einstein-Maxwell spacetime. Then $\mathscr{I}^+$ can be shown to have the following properties (see [61]). A singular metric, defined only up to conformal rescalings, exists on $\mathscr{I}^+$, and coordinates $u, \zeta, \bar{\zeta}$ can be found in which this metric takes the form

$$\Omega^2 d\ell^2 = 0 \cdot du^2 + \frac{1}{\frac{1}{4}(1+\zeta\bar{\zeta})^2} \, d\zeta \, d\bar{\zeta} \,. \qquad \text{(XI.63)}$$

The coordinates $u, \zeta, \bar{\zeta}$ together with a conformal frame for which $\Omega^2 = 1$ are said to comprise a Bondi frame. $\mathscr{I}^+$ possesses a second singular conformal metric, given by

$$\Omega^2\ ds^2 = du^2 + 0 \cdot d\zeta\ d\bar{\zeta} \quad . \qquad \text{(XI.64)}$$

Under a conformal rescaling, $d\ell^2$ and ds^2 are to pick up the same conformal factor, so that $ds/d\ell$ remains invariant. The singular metric ds^2 can be thought of as defining a preferred affine parameter u along the null generators of the null cone $\mathscr{I}^+$.

Let a spin frame for M be chosen in such a way that on $\mathscr{I}^+$, $\iota^A \bar{\iota}^{A'}$ is everywhere tangent to the generators of $\mathscr{I}^+$, i.e., $\iota_A \bar{\iota}_{A'} \propto du$. To construct (future) asymptotic twistors (hypersurface twistors for the hypersurface $\mathscr{I}^+$), we first take the complex extension $\mathbb{C}\mathscr{I}^+$ of $\mathscr{I}^+$. The complex normals $n^{AA'}$ to $\mathbb{C}\mathscr{I}^+$ will then be given by $n^{AA'} = \iota^A \tilde{\iota}^{A'}$. The hypersurface twistor equation XI.60 then becomes

$$\iota^A \tilde{\iota}^{B'} \pi_{B'} \pi^{A'} \nabla_{AA'} \pi_{C'} = 0. \qquad \text{(XI.65)}$$

It can be shown that the resulting curves are <u>geodesics</u> in $\mathbb{C}\mathscr{I}^+$, in the sense of Definition XI.8 with respect to the analytic continuation of the metric XI.63 inherited from the metric of M. With respect to the liberated coordinates $u, \zeta, \tilde{\zeta}$ for $\mathbb{C}\mathscr{I}^+$, these geodesics can be shown to lie entirely in $\tilde{\zeta}$ = constant surfaces. The dual asymptotic twistors are obtained from Equations XI.62 and are found to be geodesics in $\mathbb{C}\mathscr{I}^+$ which lie entirely in ζ = constant surfaces.

Since asymptotic twistors are a special case of hypersurface twistors, asymptotic twistor space has both a symplectic structure and a complex structure. An asymptotic Kerr Theorem can be proven which produces asymptotically shearfree null congruences for M.

Unlike general hypersurface twistors, a scalar product can be defined for asymptotic twistors (see[68]). When applied to an asymptotic twistor $Z^{\cdot}$ and its conjugate $\bar{Z}.$, this scalar product defines a norm $K = \langle Z^{\cdot}, \bar{Z}.\rangle$ which is a real-valued function on asymptotic twistor space $\mathcal{T}(\mathscr{I}^+)$. This real-valued function can be taken to be a Kähler scalar for $\mathcal{T}(\mathscr{I}^+)$, and hence it is seen that asymptotic twistor space $\mathcal{T}(\mathscr{I}^+)$ admits a Kählerian metric. The Kählerian structure of $\mathcal{T}(\mathscr{I}^+)$ contains information on the asymptotic curvature of M , by virtue of the way in which the scalar product is defined. (It is defined in terms of twistor transport of local twistors on $C\mathscr{I}^+$.)

Penrose has shown ([68], [64]) that by considering holomorphic curves in projective asymptotic twistor space, a four-complex-dimensional complex manifold of such curves can be constructed. This manifold is in fact identical to the "heavenly spacetime" associated with M , which is obtained by Newman in an entirely different way. We now turn our attention to these heavenly spacetimes.

Heavenly Spacetimes

Consider an asymptotically flat vacuum or Einstein-Maxwell spacetime M , so that a "future null cone at infinity" $\mathscr{I}^+$ exists, with the structure discussed above. There is a symmetry group associated with $\mathscr{I}^+$ which preserves the conformal metrid $d\ell^2$ of Equation XI.63 and the ratio $ds/d\ell$, with ds as in Equation XI.64. This is the BMS group (Bondi-Metzner-Sachs group; see [61] and Sachs in [14]). In a Bondi frame with the coordinates $u, \zeta, \bar{\zeta}$, a BMS transformation is given by

$$u' = K(u + \alpha(\zeta, \bar{\zeta}))$$
$$\zeta' = \frac{a\zeta + b}{c\zeta + d} \qquad \text{(XI.66)}$$

where $\alpha(\zeta, \bar{\zeta})$ is a regular real function, and

$$K = (1 + \zeta\bar{\zeta})\left[|a\zeta + b|^2 + |c\zeta + d|^2\right]^{-1} \qquad \text{(XI.67)}$$

and $\quad ad - bc = 1, \quad a,b,c,d \in \mathbb{C}$.

The BMS transformations for which $K = 1$,

$$u' = u + \alpha(\zeta, \zeta)$$
$$\zeta' = \zeta \qquad \text{(XI.68)}$$

form a subgroup of the BMS group called the supertranslation subgroup.

Of immediate interest to us are the two-surfaces on $\mathcal{J}^+$ defined by u = constant for some Bondi frame. We call such two-surfaces cuts on $\mathcal{J}^+$. Any Bondi frame on $\mathcal{J}^+$ determines a congruence of such cuts (one through each point of $\mathcal{J}^+$) corresponding to all possible values of u.

At any given point of a cut, the tangent space to the cut will be orthogonal to $du \equiv \iota_A \bar{\iota}_{A'} \sigma_a^{AA'} dx^a$, because the cut is defined by u = constant. Hence the tangent space to the cut is spanned by real vectors which are linear combinations of $m^a \leftrightarrow o^A \bar{\iota}^{A'}$ and $\bar{m}^a \leftrightarrow \iota^A \bar{o}^{A'}$, whereby o^A is determined up to a scaling. By normalizing according to $\iota^A o_A = 1$, we thus have a unique spinor o_A defined at every point of the cut. When $\mathcal{J}^+$ is thought of as the boundary of the spacetime M , we have a null vector $l_a \equiv \sigma_a^{AA'} o_A \bar{o}_{A'}$, defined at each point of the cut which points inward toward the interior of M . The situation is as depicted in Figure XI.4.

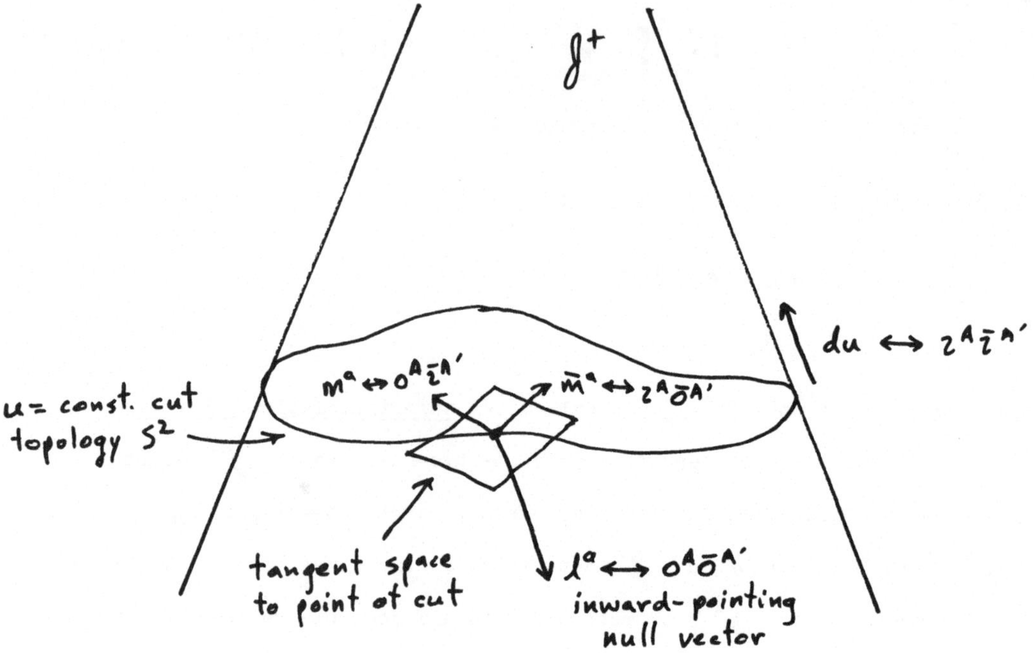

$\mathscr{I}^+$
$du \leftrightarrow \iota^A \bar{\iota}^{A'}$
$m^a \leftrightarrow o^A \bar{\iota}^{A'}$
$\bar{m}^a \leftrightarrow \iota^A \bar{o}^{A'}$
u = const. cut
topology S^2
tangent space
to point of cut
$l^a \leftrightarrow o^A \bar{o}^{A'}$
inward-pointing
null vector

Figure XI.4

We now note that $\sigma^{o} \equiv \ell_{a;b}\, m^{a}\, m^{b}$ depends only on information on the cut itself, hence is well-defined for a single cut. (σ^{o} has an appropriate scaling behavior under conformal rescalings of $\mathscr{I}^{+}$.) When we consider a whole congruence of u = constant cuts given by any Bondi frame, we see that a function $\sigma^{o}(u,\zeta,\bar{\zeta})$ is defined on $\mathscr{I}^{+}$ for each Bondi frame. If we parallelly propagate ℓ^{a} back into M , we have a tetrad defined on a neighborhood of $\mathscr{I}^{+}$ which gives the conformally rescaled metric on $M \cup \mathscr{I}^{+}$. Clearly, $\sigma^{o}(u,\zeta,\bar{\zeta})$ is just the shear of ℓ^{a} on $\mathscr{I}^{+}$ with respect to this conformally rescaled metric. Hence we call $\sigma^{o}(u,\zeta,\bar{\zeta})$ the shear of the cut u = constant.

If M is Minkowski space, it can be shown that the intersection of $\mathscr{I}^{+}$ with the null cone of any interior point gives a cut of $\mathscr{I}^{+}$ characterized by $\sigma^{o} = 0$. We call such a cut a <u>good cut</u>. Conversely, if we start with a good cut on $\mathscr{I}^{+}$ for Minkowski space and follow the rays defined by parallel propagation of ℓ^{a} (see Figure XI.4) back into the interior, we find that these rays focus at a point. When M is only asymptotically flat, we continue to call a cut with $\sigma^{o} = 0$ a good cut, but now when we follow the rays ℓ^{a} back into the interior, we find that they are "warped" by the curvature of M and do not focus. Hence a good cut for a general asymptotically flat M can be characterized only by saying that <u>viewed from $\mathscr{I}^{+}$</u> the associated rays <u>look as if</u> they would focus to an interior point.

The idea of Heaven, as developed by Newman, is then as follows. When M is Minkowski space, there is a one-to-one correspondence between interior points and good cuts on $\mathscr{I}^{+}$. When M is only asymptotically flat, we would like to identify good cuts with points in a fictitious space, the Heavenly spacetime $\mathcal{H}(M)$ associated with M . However, there is a serious problem: for a general M , there will be no good cuts at all. This problem is resolved by going to the complex extension $\mathbb{C}\mathcal{J}^+$ of $\mathcal{J}^+$ and considering complex cuts of $\mathbb{C}\mathcal{J}^+$. These are the two-complex-dimensional surfaces defined by $u = \text{constant}$, where $u, \zeta, \tilde{\zeta}$ are the liberated coordinates assoicated with a Bondi frame. (To get all of the complex Bondi frames, we must admit complex supertranslations $u' = u + \alpha(\zeta, \tilde{\zeta})$ where α is no longer required to be real.) As in the real cases, a shear $\sigma^{\circ}(u, \zeta, \tilde{\zeta})$ can be associated with these complex cuts, but now of course $\tilde{\sigma}^{\circ}$ will not be the complex conjugate of σ°. It turns out that when we ask for complex good cuts, i.e., those for which $\sigma^{\circ} = 0$, a four-complex-parameter family of solutions will exist for a general asymptotically flat M . This set of complex good cuts forms a four-complex-dimensional complex manifold $\mathcal{H}(M)$, the Heavenly spacetime associated with M .

Newman has shown [61] that an analytic complex Riemannian metric, in the sense of Definition XI.7, can be defined on $\mathcal{H}(M)$ in terms of integrals over the complex good cuts. It is found, quite amazingly, that this metric is Ricci-flat and right-conformally flat, i.e., $\Phi_{ABA'B'} = \Lambda = \tilde{\Psi}_{A'B'C'D'} = 0$. If we had chosen complex cuts for which $\tilde{\sigma}^{\circ}$ instead of σ° vanishes, a Ricci-flat and left-conformally flat metric could have been obtained, $\Phi_{ABA'B'} = \Lambda = \Psi_{ABCD} = 0$, so we should really refer to left- or right-heavenly spacetimes, corresponding to $\Psi_{ABCD}=0$ or $\tilde{\Psi}_{A'B'C'D'}=0$.

When the original spacetime M is stationary, real good cuts exist and the associated Heaven is just complex Minkowski space $\mathbb{C}M$. But for a general asymptotically flat vacuum or Einstein-Maxwell M , the complex metric of $\mathcal{H}(M)$ will be non-trivial, and there will exist no four-real-dimensional subspace on which the metric is real (because $\tilde{\Psi}_{A'B'C'D'}$ vanishes while Ψ_{ABCD} does not, or vice versa).

The connection between Heavenly spacetimes and asymptotic twistors can now be described. Consider a complex good cut of $\mathbb{C}\mathcal{J}^+$, with $\sigma^\circ = 0$. This cut will intersect a surface $\tilde{\zeta}$ = constant in a complex line. Then it is not hard to show that the condition $\sigma^\circ = 0$ is precisely the condition that this complex line be a complex geodesic on $\mathbb{C}\mathcal{J}^+$, and hence corresponds to an asymptotic twistor. As we intersect different $\tilde{\zeta}$ = constant surfaces with the complex good cut, we get a holomorphic curve's worth of asymptotic twistors, and as we have mentioned above, such curves in twistor space can be used to build up the points of an associated complex spacetime. Clearly, each curve is associated with the Heavenly point defined by the complex good cut in question, and thus the complex spacetime obtained is just the associated Heaven. For more details on Penrose's derivation of Heavenly spacetimes, see [64]. The situation is illustrated schematically in Figure XI.5.

We conclude this chapter with a discussion of the second rather remarkable connection between twistor and Heaven theory, and our work in the previous chapters. We start with a definition motivated by the discussion above:

<u>XI.17: DEFINITION:</u> A <u>left-Heavenly spacetime</u> is a four-complex-dimensional complex manifold with a complex analytic metric (as in Definition XI.7) such that

$$\Psi_{ABCD} = \Phi_{ABA'B'} = \Lambda = 0 \,. \qquad \text{(XI.69)}$$

A <u>right-Heavenly spacetime</u> is defined similarly with

$$\tilde{\Psi}_{A'B'C'D'} = \Phi_{ABA'B'} = \Lambda = 0 \,. \qquad \text{(XI.70)}$$

It is important to note that any four-complex-dimensional complex manifold $\mathbb{C}M$ can be thought of as the complex extension of any of its four-real-dimensional subspaces M. On such a subspace the analytic metric will reduce to a generally complex metric. Spinors for $\mathbb{C}M$ are then defined by taking the analytic continuation of van der Waerden symbols $\sigma^{a}{}_{AA'}$ appropriate for M. Then the curvature quantities Ψ_{ABCD}, $\widetilde{\Psi}_{A'B'C'D'}$, $\Phi_{ABC'D'}$, and Λ are defined in the obvious way.

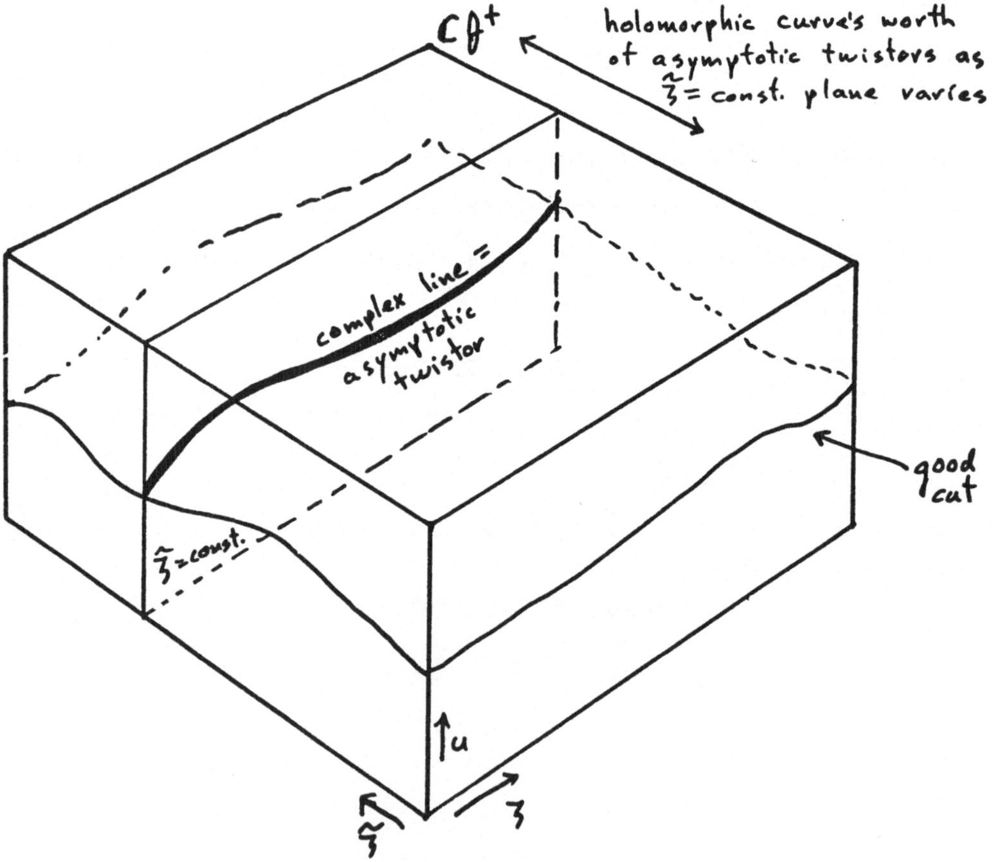

Figure XI.5

<u>XI.18: LEMMA</u>: Let $\mathbb{C}M$ be a four-complex-dimensional manifold with an analytic metric. Then $\mathbb{C}M$ admits a spinor dyad (o^A, ι^A) for which all of the "untwiddled" spin coefficients vanish if and only if $\mathbb{C}M$ is a left-Heavenly spacetime, and $\mathbb{C}M$ admits a dyad $(\tilde{o}^{A'}, \tilde{\iota}^{A'})$ for which all of the "twiddled" spin coefficients vanish if and only if $\mathbb{C}M$ is a right-Heavenly spacetime.

<u>Proof</u>: If a dyad exists for which $\alpha = \beta = \ldots = \tau = 0$, then the "untwiddled" NP Equations VII.20 can be easily shown to imply $\Psi_i = \Phi_{ij} = \Lambda = 0$, which is equivalent to $\Psi_{ABCD} = \Phi_{ABC'D'} = \Lambda = 0$. Conversely, if these curvature quantities vanish, Equations VII.82 imply that $\nabla_{[a} \nabla_{b]} \xi_A = 0$ for any spinor ξ_A. This means that parallel transport of unprimed spinors is integrable. Taking a spinor dyad (o^A, ι^A) at any point and parallel propagating it everywhere, we obtain a covariantly constant dyad field, for which $\alpha = \beta = \ldots = \tau = 0$. The second part of the lemma is proven similarly.

<u>XI.19: THEOREM</u>: A left-Heavenly spacetime is the complex extension of a vacuum left-Kähلerian spacetime, and a right-Heavenly spacetime is the complex extension of a vacuum right-Kählerian spacetime.

<u>Proof</u>: Let $\mathbb{C}M$ be a left-Heavenly spacetime. By the lemma above, a dyad (o^A, ι^A) exists for which all of the "untwiddled" spin coefficients vanish. Choose any dyad $(\tilde{o}^A, \tilde{\iota}^A)$; then the $J_a{}^b$ associated with these dyads satisfies $\kappa = \sigma = \nu = \lambda = \pi = \rho = \mu = \tau = 0$, hence is integrable and left-Kählerian. (More precisely, on any

four-real-dimensional subspace M of $\mathbb{C}$M , the metric is left-Kählerian.) Finally, the left-Heavenly spacetime is vacuum by Definition XI.17. The proof for right-Heavenly spacetimes is similar. Conversely, if $\mathbb{C}$M is vacuum and left-Kählerian, then we have Equations IX.49 and IX.50 holding, with $\Phi_{10} = \Phi_{11} = \Phi_{12} = \Lambda = 0$. But these latter conditions imply (see Equations IX.50) that $\Psi_2 = 0$; and since the other Ψ_i vanish already, $\mathbb{C}$M is left-Heavenly. The proof for vacuum right-Kählerian spacetimes is similar. (It is worthwhile stressing that this theorem is really an "if-and-only-if" proposition).

From our work in Chapter IX, we can say something about the form of a Heavenly spacetime:

XI.20: THEOREM: In a left- or right-Heavenly spacetime $\mathbb{C}$M , coordinates $u, \zeta, v, \tilde{\zeta}$ exist for which the analytic metric takes the form

$$ds^2 = 2\,A\,du\,dv + 2\,B\,d\zeta\,d\tilde{\zeta} + 2\,C\,du\,d\tilde{\zeta} + 2\,D\,d\zeta\,dv\,, \qquad \text{(XI.71)}$$

where

$$\begin{pmatrix} A & C \\ D & B \end{pmatrix} = \begin{pmatrix} K_{,uv} & K_{,u\tilde{\zeta}} \\ K_{,\zeta v} & K_{,\zeta\tilde{\zeta}} \end{pmatrix} \qquad \text{(XI.72)}$$

for some Kahler scalar $K(u, \zeta, v, \tilde{\zeta})$ subject to

$$\ln\Delta = \ln(AB-CD) = M(u,\zeta) + N(v,\tilde{\zeta}), \qquad \text{(XI.73)}$$

for some functions $M(u,\zeta)$ and $N(v,\tilde{\zeta})$.

Proof: Assume $\mathbb{C}M$ is left-Heavenly (if it is right-Heavenly, the same argument goes through with $J_a{}^b$ replaced by $\overline{J_a{}^b}$). Then by **Theorem** XI.19 it is the complex extension of a left-Kählerian spacetime, whence the results of Chapter IX give us at once Equations XI.71 and XI.72. Consulting Equations IX.50, the conditions $\Psi_{ABCD} = \Phi_{ABC'D'} = \Lambda = 0$ are seen to be equivalent to $\mathcal{D}\alpha = \delta\alpha = \mathcal{D}\gamma = \delta\gamma = 0$. From the definitions of $\mathcal{D}$ and δ (see Equations IX.50) these are equivalent to $\alpha_{,v} = \alpha_{,\tilde{\zeta}} = \gamma_{,v} = \gamma_{,\tilde{\zeta}} = 0$. From Equations IX.49 we have $\alpha = \frac{1}{2}(\ln\Delta)_{,\zeta}$ and $\gamma = -\frac{1}{2}(\ln\Delta)_{,u}$; so the Heavenly conditions become

$$(\ln\Delta)_{,uv} = (\ln\Delta)_{,u\tilde{\zeta}} = (\ln\Delta)_{,\zeta v} = (\ln\Delta)_{,\zeta\tilde{\zeta}} = 0 \,. \qquad \text{(XI.74)}$$

The general solution to these equations is easily seen to be Equation XI.73, where $M(u,\zeta)$ and $N(v,\tilde{\zeta})$ are arbitrary functions.

(NOTE: During the preparation of this work, two preprints have come to my attention in which some of our results have been obtained independently: J. F. Plebanski and S. Hacyan [70] have derived a theorem essentially equivalent to our Theorem XI.11, although the surfaces are not referred to as twistor surfaces. Also, Plebanski has examined [69] the field equations for Heavenly spacetimes, obtaining some results very similar to ours; in particular, his "key function" Ω is essentially our Kähler scalar K . Concerning our Theorem XI.20, Plebanski has obtained the stronger result $\Delta = 1$. Our result, which can be expressed as $\Delta = p(u,\zeta)\, Q(v,\tilde{\zeta})$ for some functions p and Q , is valid in any structure coordinate system. Since $\Delta = \det g_{\alpha\tilde{\beta}}$ is a scalar density, there are special structure coordinate systems in which $\Delta = 1$, in agreement with Plebanski. For example, one such structure coordinate system is given by $u' = a(u, \zeta)$, $\zeta' = \zeta$, $v' = b(v, \tilde{\zeta})$, $\tilde{\zeta}' = \tilde{\zeta}$, where a and b are functions such that $a_{,u} = \rho$ and $b_{,v} = Q$. I am grateful to J. Weinberg for pointing this out to me.)

In Figure XI.6 we summarize the increasing amount of structure obtained by applying the integrability condition, the left-Kählerian condition, and the vacuum or left-Heavenly condition on a spacetime.

$\kappa = 0$ $\sigma = \theta$ $\nu = 0$ $\lambda = 0$	Left-Hermitian spacetimes: including the type D vacuum and electrified spacetimes.
$\pi = 0$ $\rho = 0$ $\mu = 0$ $\tau = 0$	Left-Kählerian spacetimes: Kähler scalar and canonical form of metric given by Equations IX.45 and IX.46.
$\alpha = 0$ $\beta = 0$ $\gamma = 0$ $\epsilon = 0$	Vacuum left-Kählerian spacetime = left-Heavenly spacetime: canonical form of metric given by Equations XI.71, XI.72, and XI.73.

Figure XI.6

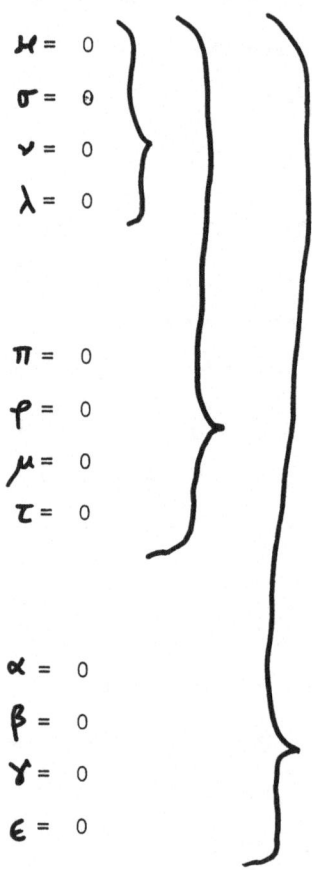

CHAPTER XII: DISCUSSION

We conclude by presenting in this chapter a brief discussion of possible directions for future research in complex manifold theory as applied to relativity.

As regards standard complex manifold theory, the most immediate application is to various aspects of twistor theory. The contour integral formulation of fields and scattering processes in twistor theory clearly brings in the homology theory of complex manifolds in a non-trivial way, when a choice between inequivalent contours over which to perform the integrations must be made. Also, in Penrose's construction of Heavenly spacetimes [64], the theory of deformations of a complex manifold [30] plays a central role. Related to this, it would be nice to have a clearer picture of the significance of the Kähler scalar for asymptotic twistor space, and its relation to the modified-Kähler scalar of the associated Heavenly spacetime.

There is another physical context in which standard complex manifold theory plays an important part. This is the theory of geometric quantization as developed by Kostant [60], Souriau [75], and others. In geometric quantization, a straightforward prescription for the passage from a classical physical system to the corresponding quantum system is given in terms of a symplectic manifold with a "polarization" (see e.g., Simms [74]) and with certain complex line bundles defined on it. A Kählerian manifold (in the genuine sense) is a particular example of a symplectic manifold with a polarization: the real, closed Kähler form is a symplectic form, and the decomposition of the complexified tangent space into type (1,0) and type (0,1) vectors define a polarization. Geometric quantization is important in its own right, and in addition, from our point of view, it would be very interesting to look for a connection between the geometric quantization associated with the Kähler polarization of twistor spaces, and the existing twistor quantization ideas.

Concerning our work on modified complex manifold theory, there are also many unanswered questions. In particular, the global properties of spacetimes admitting modified left-Hermitian and left-Kählerian structures have not been investigated here. For example, can embedding theorems analogous to those of standard complex manifold theory [30] be obtained? Also, what are the asymptotic properties of a Heavenly spacetime $\mathcal{H}(M)$? Is such a spacetime asymptotically flat in any useful sense? If so, can a heavenly spacetime $\mathcal{H}(\mathcal{H}(M))$ be constructed? A rather flamboyant conjecture along these lines would be that the right-Heaven associated with a left-Heavenly spacetime is always isometric to complex Minkowski space!

Another very important question is the treatment of non-analytic situations, both in twistor theory and in our modified complex analysis. Is the approximation of non-analytic functions by analytic ones the best procedure that can be hoped for? Can an analogue of the Newlander-Nirenberg theorem be obtained for our modified almost complex structures?

There are also some more specific questions suggested by our work. Can a general "complex transformation" algorithm be obtained which applies to a larger class of spacetimes? What is the connection, if any, between the four-complex-dimensional Hermitian spaces of the second algorithm of Chapter X, and Heavenly spacetimes? What useful generalizations of twistor surfaces can be made? Can the structure coordinates and Kähler scalar of a Heavenly spacetime be found in any direct way from data on $\mathscr{I}^+$? These remain exciting questions for the future.

APPENDIX A: PSEUDOGROUP STRUCTURES

A complex structure on a manifold is a covering by complex coordinate charts such that the transition functions for a change of coordinates are complex analytic functions. This idea of covering a manifold with certain coordinate charts which transform in a particular way can be generalized to the concept of "pseudogroup structures," which we shall now discuss. For details on these matters, Morrow and Kodaira [30] should be consulted; most of the following discussion is based on this reference.

Let X and Y be two topological spaces. A <u>local homeomorphism</u> f between X and Y is a homeomorphism of an open set $U \subseteq X$ onto an open set $f(U) \subseteq Y$. In a similar manner, a <u>local diffeomorphism</u> is defined. We shall be interested in the case $X = Y$, for which we speak of local homeomorphisms or local diffeomorphisms of X.

Now we let Θ be a domain of $\mathbb{R}^n$ or $\mathbb{C}^n$, with f, g diffeomorphisms of Θ. Then the <u>restriction</u> of f to $W \subseteq \Theta$, $f|_W$, the inverse of f, f^{-1}, and the composition $g \circ f$ are defined in the obvious manner.

Then we define a <u>pseudogroup</u> of transformations in Θ as a set Γ of local diffeomorphisms of Θ such that:

(i) $f \in \Gamma \Rightarrow f^{-1} \in \Gamma$;

(ii) $f, g \in \Gamma \Rightarrow g \circ f \in \Gamma$ when and where the latter is defined:

(iii) $f \in \Gamma \Rightarrow f|_W \in \Gamma$ for an open subset $W \subseteq \Theta$;

(iv) the identity diffeomorphism is in Γ ;

(v) if f is a local diffeomorphism of Θ, $\Theta = \bigcup U_j$, and $f|_{U_j} \in \Gamma$ for all j, then $f \in \Gamma$.

It is properties (iii) and (v) that distinguish a pseudogroup from a group in the ordinary sense. Note that associativity automatically holds for the composition of any mappings.

Now let Γ be a pseudogroup on Θ, and X a paracompact Hausdorff space. We define a set of <u>local Γ-coordinates</u> as a set $\{z_i, z_j, \ldots\}$ of local homeomorphisms z_i of X into Θ with the property that $z_i \circ z_j^{-1} \in \Gamma$ whenever this composition is defined. We define an equivalence relation for sets $\{z_i, z_j, \ldots\}$ and $\{w_i, w_j, \ldots\}$ of local Γ-coordinates by saying that $\{z_i, z_j, \ldots\}$ and $\{w_i, w_j, \ldots\}$ are <u>Γ-equivalent</u> if and only if $w_k \circ z^{-1} \in \Gamma$ whenever the composition is defined. Then a <u>Γ-structure</u> on X is an equivalence class of such sets of local Γ-coordinates on X. And finally, a <u>Γ-manifold</u> is a paracompact Hausdorff space X with an associated Γ-structure on X. We refer to a Γ-structure on a Γ-manifold as a <u>pseudogroup structure</u>.

Examples of pseudogroup structures include:

(i) complex structures, see Chapter III;

(ii) locally product structures, see Appendix B;

(iii) modified complex structures, see Chapter VIII;

(iv) symplectic structures, see Nelson [31];

(v) flat affine structures, see Morrow and Kodaira [30].

It is to be noticed that the "almost-structures" defined in the text are <u>not</u> in general induced by pseudogroup structures, unless certain "integrability conditions" are satisfied. The pseudogroup structures listed above have associated with them the following "almost-structures" and integrability conditions:

(i) almost complex structures with integrability conditions $N_{ab}{}^c = 0$, see Chapter V;

(ii) almost product structures with integrability conditions $N_{ab}{}^c = 0$, see Appendix B;

(iii) modified almost complex structures with integrability conditions $N_{ab}{}^{c} = 0$, see Chapter VIII;

(iv) almost symplectic structures with integrability condition the closure of a certain two-form, see Nelson [31];

(v) Riemannian structures, with integrability conditions $R^{a}{}_{bcd} = 0$.

When one pseudogroup structure is "contained" in another in the sense that every allowable chart of the former is an allowable chart of the latter, we have a "substructure." For example, a complex structure defines a unique differentiable structure which contains all of the real charts of the complex structure. We assume, in fact, that all of the structures we examine are contained in a unique differentiable structure.

APPENDIX B: LOCALLY PRODUCT STRUCTURES

There is a strong analogy between complex structures and the so-called locally product structures on manifolds, as we shall now see. The material in this appendix is based largely upon the presentation of Yano [54].

A locally product structure is a pseudogroup structure (see Appendix A) for which the allowed coordinate transformations are of the form:

$$w^{a'} = w^{a'}(w^b)$$

$$w^{x'} = w^{x'}(w^y) \quad ,$$

where a, b, a', b' range over values $1, \ldots, p$ and x, y, x', y' range over $p+1, \ldots, p+q = n$. A product manifold $M_1 \times M_2$ is an example of a manifold admitting a locally product structure.

A locally product manifold admits a tensor $J_i{}^j$ $(i, j = 1, \ldots, p, p+1, \ldots, n)$ with the property that $J_i{}^j J_j{}^k = +\delta_i{}^k$. In any structure coordinate system, $J_i{}^j$ has the numerical values $J_i{}^j = \mathrm{diag}(+1, \ldots, +1, -1, \ldots, -1)$, with p +1's occuring and $q = n - p$ -1's occurring. This inspires the definition of an almost product structure as a tensor $J_i{}^j$ satisfying $J_i{}^j J_j{}^k = +\delta_i{}^k$. The almost product structure is induced by a locally product structure if and only if it satisfies the integrability condition $N_{ij}{}^k = 0$, where $N_{ij}{}^k$ is the Nijenhuis tensor formed from $J_i{}^j$.

For a Riemannian manifold, it is desirable to have a compatibility condition relating the metric structure to the locally product structure (compare the Hermitian compatibility requirement for a complex manifold). We say that a locally product manifold with a Riemannian metric is a locally product Riemannian manifold if and only if the metric g_{ij} satisfies the compatibility requirement $J_i^{\ k} J_j^{\ \ell} g_{k\ell} = g_{ij}$. For a locally product Riemannian manifold, $J_{ij} = J_{ji}$, and the metric g_{ij} is "pure" in the structure coordinates, that is, $g_{ax} = g_{yb} = 0$, g_{ab}, $g_{xy} \neq 0$.

A locally product Riemannian manifold is said to be a <u>locally decomposable Riemannian manifold</u> if it satisfies the even stronger compatibility requirement $g_{ab} = g_{ab}(w^c)$ and $g_{xy} = g_{xy}(w^z)$. Then the only non-vanishing Christoffel symbols are given by $\Gamma_{ab}^{\ c}$ and $\Gamma_{xy}^{\ z}$. (This condition is the analogue of the Kählerian condition for an Hermitian manifold.) A necessary and sufficient condition for a locally product Riemannian manifold to be a locally decomposable Riemannian manifold is given by $\nabla_i J_j^{\ k} = \nabla_i J_{jk} = 0$.

APPENDIX C: COMPLEX VECTOR BUNDLES

The concept of a complex vector bundle is necessary for a precise definition of tensor fields on complex manifolds. (The introduction of spinor fields is accomplished by constructing a complex vector bundle with very special properties: see Chapter VII.) The facts we shall need are a subset of the more general theory of fibre bundles, concerning which Steenrod [76] and Trautman [77] may be consulted. Most of the material covered below may be found in Morrow and Kodaira [30] and Chern [8].

Let M be a differentiable manifold, which does not necessarily admit any complex structure. Let $\mathbb{C}^q$, with the usual topology, be thought of as a complex vector space of complex dimension q, and let the group $GL(q,\mathbb{C})$ act on $\mathbb{C}^q$ to the right. Thus we have, for example, $(\xi G)H = \xi(GH)$, where $\xi \in \mathbb{C}^q$ and $G,H \in GL(q,\mathbb{C})$; ξ is to be thought of as a $q \times 1$ row vector, and G,H as $q \times q$ matrices. Throughout this section, matrix indices will be suppressed, except when noted.

A <u>complex vector bundle</u> E over M, usually denoted more completely by $\pi:E \longrightarrow M$ is now defined to consist of:

(i) a topological space E, called the bundle, and

(ii) a continuous map π from E to M ; π is called the projection map, M is called the base space, and $\pi^{-1}(p)$ is called the fibre over $p \in M$.

In addition, the following properties are required:

(iii) for each point $p \in M$, there exists a neighborhood U and a homeomorphism ("chart")

$$\phi_U : U \times \mathbf{C}^q \longrightarrow \pi^{-1}(U)$$

with $\pi \circ \phi_U(p', \xi) = p'$ for all $p' \in U$, $\xi \in \mathbf{C}^q$,

and

(iv) in the intersection $U \cap V$ of two such charts, there exists a smooth map ("transition function")

$$G_{UV}: U \cap V \longrightarrow GL(q, \mathbf{C})$$

such that

$$\phi_U(p, \xi) = \phi_V(p, \xi') \qquad \xi G_{UV}(p) = \xi'.$$

We shall call q the <u>dimension</u> of the bundle; this will in general be different from the dimension of either E or M. If $q = 1$ we will call the bundle a <u>complex line bundle</u>.

Various slightly different definitions of complex vector bundles can be obtained by demanding stronger or weaker smoothness properties for the sets and maps in the above definition. Thus, for example, Morrow and Kodaira require that both the space E and the manifold M should be complex manifolds, and that the charts ϕ_U and the transition functions G_{UV} should be holomorphic maps. Under such stronger smoothness requirements, our results below still apply, <u>mutatis mutandis</u>.

From property (iv), there follow the compatibility conditions

$$\text{(i)} \quad \xi = \xi' \, G_{UV}^{-1}(p) = \xi' G_{VU}(p) \Rightarrow G_{UV}^{-1} = G_{VU}$$

$$\text{(ii)} \quad \xi = \xi'' \, G_{UW}(p) = (\xi' G_{UV}(p)) \, G_{VW}(p) \Rightarrow G_{UV} G_{VW} = G_{UW} \, .$$

Intuitively, it is clear that we may have two bundles which differ only by a "change of charts", for example, a change of charts induced by a change of coordinates on M . Clearly, we would like to identify two such bundles, and this is done by defining the concept of equivalence of bundles. Let $\pi : E \longrightarrow M$ and $\pi' : E' \longrightarrow M$ be two bundles of the same dimension q over the same base M. Suppose further that there exists an open cover $\{U, V, \ldots\}$ of M such that the charts and transition functions are given respectively by

$$\phi_U, \ \phi_V, \ldots \ ; \ G_{UV}, \ldots$$

and

$$\phi'_U, \ \phi'_V, \ldots ; \ G'_{UV}, \ldots \quad .$$

Then E and E' are said to be equivalent if for each patch U there exists a smooth map

$$T_U : U \longrightarrow GL(q, \mathbb{C})$$

such that for all $p \in U$,

$$\phi_U(p, \xi T_U) = \phi'_U(p, \xi) \ .$$

It is easy to show that this condition implies

$$G'_{UV} = T_U \, G_{UV} \, T_V^{-1} \ .$$

Accordingly, we shall now think of a bundle as being defined only up to equivalence classes $[G_{UV}]$ of transition functions.

New bundles may be constructed from given bundles in several ways. Let

$$\pi' : E' \longrightarrow M$$

and

$$\pi'' : E'' \longrightarrow M$$

be complex vector bundles over M with respective (equivalence classes of) transition functions $[G'_{UV}]$ and $[G''_{UV}]$. Then we can form the following bundles:

(i) the dual bundle of E' , denoted E'^{*} and defined by

$$[G'^{*}_{UV}] = [G'^{tr-1}_{UV}] \quad (\dim E'^{*} = \dim E');$$

(ii) the Whitney Sum (Direct Sum) bundle of E' and E'' , denoted $E' \oplus E''$ and defined by

$$[G_{UV}] = \left[\begin{pmatrix} G'_{UV} & 0 \\ 0 & G''_{UV} \end{pmatrix}\right] \qquad (\dim E' \oplus E'' = \dim E' + \dim E'').$$

(iii) the tensor product bundle of E' and E'' , denoted $E' \otimes E''$ and defined by

$$[G_{UV}] = [G'_{UV} \otimes G''_{UV}]$$

i.e., $$G^{(\alpha\lambda)}_{(\beta\mu)} = G'^{\alpha}_{\beta}\, G''^{\lambda}_{\mu} \qquad (\dim E' \otimes E'' = \dim E' \cdot \dim E'');$$

(iv) the <u>complex conjugate bundle</u> of E' ; denoted $\bar{E}'$ and defined by

$$[G_{UV}] = [\overline{G'_{UV}}] \qquad (\dim \overline{E'} = \dim E').$$

Next, suppose we have a bundle $\pi: E \longrightarrow M$ for which we can choose a covering $\{U,V,\ldots\}$ of M and charts ϕ_U, $\phi_V, \ldots$ such that

$$G_{UV} = \begin{pmatrix} A_{UV} & B_{UV} \\ 0 & C_{UV} \end{pmatrix},$$

where A_{UV} is an $m \times m$ matrix, B_{UV} an $m \times n$ matrix, and C_{UV} an $n \times n$ matrix, with $m + n = q$. Then the bundle defined by $[A_{UV}]$, of dimension m , is said to be a <u>subbundle</u> of E and the bundle $[C_{UV}]$ of dimension n is called the <u>quotient bundle</u> of E by $[A_{UV}]$ (see Morrow and Kodaira [30]).

Suppose now that M is a complex manifold. Then the following complex vector bundles are defined in terms of the complex structure $\{(U,z^{\alpha}), (U', z^{\alpha'}),\ldots\}$ of M:

(i) the holomorphic tangent bundle T_hM , defined by

$$[G_{UU'}] = \left[\left(\partial z^{\alpha'}/\partial z^{\beta}\right)\right] ;$$

(ii) the conjugate holomorphic tangent bundle $\overline{T_hM}$, defined by

$$[G_{UU'}] = \left[\overline{\left(\partial z^{\alpha'}/\partial z^{\beta}\right)}\right] = \left[\left(\partial \bar{z}^{\alpha'}/\partial \bar{z}^{\beta}\right)\right] ;$$

(iii) the complexified tangent bundle $\mathbb{C}TM = T_hM \oplus \overline{T_hM}$, defined by

$$[G_{UU'}] = \left[\begin{pmatrix} \partial z^{\alpha'}/\partial z^{\beta} & 0 \\ 0 & \partial \bar{z}^{\bar{\alpha}'}/\partial \bar{z}^{\bar{\beta}} \end{pmatrix}\right] .$$

Clearly, T_hM and $\overline{T_hM}$ are subbundles of $\mathbb{C}TM$, which have each other as their associated quotient bundles. (Also, the real tangent bundle TM is embedded smoothly and homomorphically in $\mathbb{C}TM$; see Atiyah [55].

We defined a (smooth) <u>section</u> over M or over $U \subseteq M$ of a bundle as a smooth map $f: M \longrightarrow E$ or $f: U \longrightarrow E$ such that

$$\pi(f(p)) = p \qquad \text{for all } p \in M \text{ or } U .$$

Then the complex tensor fields discussed informally in Chapter IV are just smooth sections of one of the <u>tensor product bundles</u>

$$\mathbb{C}TM \otimes \ldots \otimes \mathbb{C}TM \otimes \mathbb{C}TM^* \otimes \cdots \otimes \mathbb{C}TM^*$$
$$\overline{\mathbb{C}TM} \otimes \cdots \otimes \overline{\mathbb{C}TM} \otimes \overline{\mathbb{C}TM^*} \otimes \cdots \otimes \overline{\mathbb{C}TM^*}$$

Tensor fields of type (p,q) can also be defined rigorously in these terms. For example, a one-form ω of type (1,0) is a section of $\mathbb{C}TM^*$ which in addition can be thought of as a section of T_hM^* . (More precisely, ω is a section of $\mathbb{C}TM^*$ whose image lies in the image of the map $\psi: T_hM^* \longrightarrow \mathbb{C}TM^*$ which embeds T_hM^* in $\mathbb{C}TM^*$; see Atiyah [55].)

As a word of caution, it should be pointed out that in the literature the term "tangent bundle" is used inconsistently by different authors; the term may refer to either the real tangent bundle TM , the holomorphic tangent bundle T_hM , or the complexified tangent bundle $\mathbb{C}TM$. (Similarly for the term "cotangent bundle".) Often, careful investigation is required to ascertain which usage is in force.

We now briefly consider connections and curvature on a complex bundle. Let $\pi : E \longrightarrow M$ be a complex vector bundle, and let $\mathbb{C}TM^*$ be the complexified cotangent bundle ($= T_hM^* \oplus \overline{T_hM^*}$). Let $\Gamma(E)$ denote the set of sections of E and $\Gamma(E \otimes \mathbb{C}TM^*)$ the set of sections of $E \otimes \mathbb{C}TM^*$. (E.g., if E is a tensor product bundle, an element of $\Gamma(E)$ is a tensor field $S^{a\ldots b}{}_{c\ldots d}$ and an element of $\Gamma(E \otimes \mathbb{C}TM^*)$ is a tensor field $T^{a\ldots b}{}_{c\ldots de}$.) Then a <u>connection on E</u> is an operator (map) $D: \Gamma(E) \longrightarrow \Gamma(E \otimes \mathbb{C}TM^*)$ with the properties:

(i) $D(\gamma_1 + \gamma_2) = D(\gamma_1) + D(\gamma_2)$ (linearity)

and (ii) $D(f\gamma) = \gamma \otimes df + f\, D(\gamma)$,

where $\gamma_1, \gamma_2, \gamma \in \Gamma(E)$, $f \in \{\text{smooth functions from } M \text{ to } \mathbb{C}\}$, and "d" is the exterior derivative operator. (If γ is a section of E, then $f\gamma$ is the section defined by $f\gamma(p) \equiv f(p)\gamma(p)$, the complex number $f(p)$ times the vector $\gamma(p)$; similarly, $\gamma \otimes df$ and $fD\gamma$ are defined as sections of $E \otimes \mathbb{C}TM^*$ in the obvious way.)

For a "sufficiently small" open subset U of M, we can always find a <u>frame field</u> $e_1, e_2, \ldots, e_q$ over U, that is, a set of q sections of E over U such that $e_1(p), e_2(p), \ldots, e_q(p)$ are linearly independent vectors in the fiber over each point $p \in U$. Let e denote the $1 \times q$ column vector whose entries are e_i. Then the linearity of D allows us to write

$$De = \omega e ,$$

where De is a 1 x q column vector and ω is a q x q matrix of one-forms ($\omega_i^j \in$ CTM*) called the <u>connection matrix</u>. The connection matrix ω determines the connection D uniquely, because any section ξ of E (over some appropriate U) can be written uniquely as

$$\xi = \sum_i \xi^i e_i$$

where ξ^i are q functions. Then by the defining properties of the connection, we can write

$$D\xi = \sum_i \xi^i D(e_i) + \sum_i e_i \otimes d\xi^i$$

$$= \sum_{i,j} \xi^i \omega_i^j e_j + \sum_i e_i \otimes d\xi^i$$

where matrix indices have been written out explicitly.

Now let

$$e' = Ge$$

denote a change of frame, where G is a q x q non-singular matrix of smooth functions over U . Then it is easy to show that

$$De' = \omega' e' ,$$

where $$\omega' = dGG^{-1} + G\omega G^{-1}$$

is the connection matrix with respect to the new frame field.

We now define the <u>curvature matrix</u> Ω of the connection D as follows:

$$\Omega \equiv d\omega - \omega \wedge \omega$$

(or writing out the matrix indices,

$$\Omega_i^{\ k} = d\omega_i^{\ k} - \sum_j \omega_i^{\ j} \wedge \omega_j^{\ k} \text{).}$$

Ω is a $q \times q$ matrix of two-forms. A short calculation shows that under the change of frame $e' = G\,e$, we have

$$\Omega' = G \Omega G^{-1}.$$

Therefore, if $\Omega = 0$ for one frame, $\Omega' = 0$ for any other frame. A connection D for which $\Omega = 0$ is said to be a <u>flat connection</u>.

Finally, for any connection D , it is not hard to prove the following identity, known as the <u>Bianchi identity</u>:

$$d\Omega + \Omega \wedge \omega - \omega \wedge \Omega = 0$$

or in matrix components,

$$d\Omega_i^{\ k} + \sum_j \Omega_i^{\ j} \wedge \omega_j^{\ k} - \sum_j \omega_i^{\ j} \wedge \Omega_j^{\ k} = 0.$$

It is a straightforward matter to show that the "connections" $\gamma^a_{\ bc}$ and $\Gamma^a_{\ bc}$, defined in Chapter VI, in fact correspond to connections in the present sense for each of the tensor product bundles of an Hermitian manifold. (Also, the connections on the holomorphic tensor bundles which arise from $\gamma^\alpha_{\ \beta\gamma}$ are examples of the unique "admissible connections of type (1,0)" on Hermitian holomorphic vector bundles, which are discussed by Chern [8].)

BIBLIOGRAPHY

[1] Adler, A.; "The Second Fundamental Forms of S^6 and $\mathbf{P}^n(\mathbb{C})$"; Amer. J. Math. 91 (1969) 657.

[2] Aharonov, Y., and L. Susskind; "Observability of the Sign Change of Spinors Under 2 Rotations"; Phys. Rev. 158 (1967) 1237.

[3] Aronson, B., R. Lind, J. Messmer, and E. Newman; "A Note on Asymptotically Flat Spaces"; J. Math. Phys. 12 (1971) 2462.

[4] Bishop, R. L., and S. I. Goldberg; Tensor Analysis on Manifolds, MaxMillan, New York, 1968.

[5] Boyer, R. H., and R. W. Lindquist; "Maximal Analytic Extension of the Kerr Metric"; J. Math. Phys. 8 (1967) 265.

[6] Brans, C. H.; "Complex Structures and Representations of the Einstein Equations"; preprint.

[7] Chern, S. S.; "Several Complex Variables"; Bull. Amer. Math. Soc. 62 (1956) 101.

[8] Chern, S. S.; Complex Manifolds Without Potential Theory, van Nostrand, Princeton, New Jersey, 1967.

[9] Churchill, R.; Complex Variables and Applications, 2nd ed., McGraw-Hill, New York, 1960.

[10] Debever, R.; "La supr-energie en relativite generale"; Bull. Soc. Math. Belg. 10 (1958) 112.

[11] Debney, G. C., R. P. Kerr, and A. Schild; "Solutions of the Einstein and Einstein-Maxwell Equations"; J. Math. Phys. 10 (1969) 1842.

[12] Demianski, M.; "New Kerr-Like Space-Time"; Phys. Lett. 42A (1972) 157.

[13] Demianski, M., and E. T. Newman; "A Combined Kerr-NUT Solution of the Einstein Field Equations"; Bull. Acad. Polon. XIV (1966) 653.

[14] DeWitt, C., and B. S. DeWitt, eds.; Relativity, Groups, and Topology, Gordon and Breach, New York, 1964.

[15] Eckmann, B.; "Complex-Analytic Manifolds"; Proc. Intem. Congr. Math. Vol. II (1950) 420.

[16] Eisenhart, L. P.; Riemannian Geometry, Princeton University Press, Princeton, New Jersey, 1949.

[17] Frölicher, A.; "Zur Differentialgeometrie der komplexen Strukturen"; Math. Ann. 129 (1955) 50.

[18] Geroch, R.; "Spinor Structure of Space-Times in General Relativity I"; J. Math. Phys. 9 (1968) 1739.

[19] Geroch, R.; "Spinor Structure of Space-Times in General Relativity II"; J. Math. Phys. 11 (1970) 343.

[20] Geroch, R., A. Held, and R. Penrose; "A Space-Time Calculus Based on Pairs of Null Directions"; J. Math. Phys. 14 (1973) 874.

[21] Goldberg, J. N., and R. K. Sachs; "A Theorem on Petrov Types"; Acta Phys. Polon. 22 (1962) suppl., 13.

[22] Helgason, S.; Differential Geometry and Symmetric Spaces, Academic Press, New York, 1962.

[23] Hughston, L. P., and P. Sommers; "Spacetimes with Killing Tensors"; Commun. Math. Phys. 32 (1973) 147.

[24] Janis, A. I., and E. T. Newman; "Structure of Gravitational Sources"; J. Math. Phys. 6 (1965) 902.

[25] Kerr, R. P., and A. Schild; "Some Algebraically Degenerate Solutions of Einstein's Gravitational Field Equations"; Applications of Nonlinear Partial Differential Equations in Mathematical Physics, Proceedings of Symposia in Applied Mathematics; American Mathematical Society, Providence, R. I., Vol. XVII (1965) 199.

[26] Kinnersley, W. M.; Type D Gravitational Fields, Ph.D. thesis, California Institute of Technology (1969).

[27] Kinnersley, W. M.; Type D Vacuum Metrics"; J. Math. Phys. 10 (1969) 1195.

[28] Landau, L. D., and E. M. Lifshitz; The Classical Theory of Fields, 3d rev. English ed., Addison-Wesley, Reading, Massachusetts, and Pergamon, London, 1971.

[29] Misner, C.W., K. S. Thorne, and J. A. Wheeler; Gravitation, W. H. Freeman and Company, San Francisco, 1973.

[30] Morrow, J., and K. Kodaira; Complex Manifolds, Holt, Rinehart and Winston, New York, 1971.

[31] Nelson, E.; Tensor Analysis, Princeton University Press, Princeton, New Jersey, 1967.

[32] Newlander, A., and L. Nirenberg; "Complex Analytic Coordinates in Almost Complex Manifolds"; Ann. of Math. 65 (1957) 391.

[33] Newman, E. T.; "Maxwell's Equations and Complex Minkowski Space"; J. Math. Phys. 14 (1973) 102.

[34] Newman, E. T.; "Complex Coordinate Transformations and the Schwarzschild-Kerr Metrics"; preprint.

[35] Newman, E. T., E. Couch, K. Chinnapared, A. Exton, A. Prakash, and R. Torrence; "Metric of a Rotating, Charged Mass"; J. Math. Phys. 6 (1965) 918.

[36] Newman, E. T., and A. I. Janis; "Note on the Kerr Spinning-Particle Metric"; J. Math. Phys. 6 (1965) 915.

[37] Newman, E. T., and R. Penrose; "An Approach to Gravitational Radiation by a Method of Spin Coefficients"; J. Math. Phys. 3 (1962) 566.

[38] Nijenhuis, A.; "X_{n-1}-Forming Sets of Eigenvectors"; Proc. Kon. Ned. Akad. Amsterdam, 54 = Indagationes Math., 13 (1951) 200.

[39] Nomizu, K.; Fundamentals of Linear Algebra, McGraw-Hill, New York, 1966.

[40] Pennisi, L. L.; Elements of Complex Variables, Holt, Rinehart and Winston, New York, 1963.

[41] Penrose, R.; "A Spinor Approach to General Relativity"; Ann. Phys. 10 (1960) 171.

[42] Penrose, R.; "The Structure of Space-Time" in Battelle Rencontres in Mathematics and Physics: Seattle 1967, C. DeWitt, ed., W. A. Benjamin, Inc., New York, 1968.

[43] Penrose, R.; "Asymptotic Properties of Fields and Space-Times"; Phys. Rev. Lett. 10 (1963) 66.

[44] Penrose, R., and M. A. H. MacCallum; "Twistor Theory: An Approach to the Quantization of Fields and Spacetime"; Physics Reports 6C (1973) 242.

[45] Petrov, A. Z.; Sci. Not. Kazan State University 114 (1954) 55.

[46] Pirani, F. A. E.; "Invariant Formulation of Gravitational Radiation Theory"; Phys. Rev. 105 (1957) 1089.

[47] Sachs, R.; "Gravitational Waves in General Relativity VI. The Outgoing Radiation Condition"; Proc. R. Soc. London A264 (1961) 309.

[48] Schild, A.; "Lectures on General Relativity Theory"; Lectures in Applied Mathematics, Vol. 8, Relativity Theory and Astrophysics, American Mathematical Society, Providence, R. I.

[49] Sommers, P.; "Killing Tensors and Type {2,2} Spacetimes"; Ph.D. Dissertation, University of Texas at Austin (1973).

[50] Spivak, M.; A Comprehensive Introduction to Differential Geometry, Brandeis University, Waltham, Massachusetts, 1970.

[51] Talbot, C. J.; "Newman-Penrose Approach to Twisting Degenerate Metrics"; Commun. Math. Phys. 13 (1969) 45.

[52] Trautman, A.; "Analytic Solutions of Lorentz-Invariant Linear Equations"; Proc. Roy. Soc. A270 (1962) 326.

[53] Trautman, A., F. A. E. Pirani, and H. Bondi; Lectures on General Relativity, Brandeis 1964 Summer Institute on Theoretical Physics, Vol. 1, Prentice-Hall, Englewood Cliffs, N.J., 1965.

[54] Yano, K.; Differential Geometry on Complex and Almost Complex Spaces (Pergamon) MacMillan, New York, 1965.

[55] Atiyah, M. F.; K-Theory; W. A. Benjamin, Inc., New York, 1964.

[56] Cox, D., and E. J. Flaherty, Jr.; "A Conventional Proof of Kerr's Theorem"; to be published.

[57] Ehlers, J., and W. Kundt; "Exact Solutions of the Gravitational Field Equations; in Gravitation: an Introduction to Current Research, ed. L. Witten, John Wiley and Sons, Inc., New York, 1962.

[58] Ehresmann, C.; "Sur les Variétés Presque Complexes"; Proc. Int. Congr. Math. II (1950) 412.

[59] Hansen, R. O., and Ezra T. Newman; "A Complex Minkowski Space Approach to Twistors"; preprint.

[60] Kostant, B.; "Quantization and Unitary Representations"; in Lectures in Modern Analysis and Applications III, ed. C. T. Taam, Springer Lecture Notes in Mathematics 170 (1970).

[61] Newman, E. T., and R. Penrose; "H-Space - Its Definition and Properties", preliminary version.

[62] Newman, Ezra T., and Jeffrey Winicour; "A Curiosity Concerning Angular Momentum"; J. Math. Phys. 15 (1974) 1113.

[63] Newman, Ezra T., and Jeffrey Winicour; private correspondence.

[64] Penrose, R.; "Non-Linear Gravitons and Curved Twistor Theory"; talk delivered at Riddle of Gravitation Conference, Syracuse University, March 1975; to appear in GRG.

[65] Penrose, R.; "Relativistic Symmetry Groups"; preprint.

[66] Penrose, R.; "Twistor Algebra"; J. Math. Phys. 8 (1967) 345.

[67] Penrose, R.; "Twistor Quantisation and Curved Space-Time"; Int. J. Theor. Phys. 1 (1968) 61

[68] Penrose, R.; "Twistor Theory: Its Aims and Achievements"; in Quantum Gravity: An Oxford Symposium, eds. C. J. Isham, R. Penrose, and D. W. Sciama, Oxford University Press, 1975.

[69] Plebański, J. F.; "Heaven, Hell and Einstein Equations"; preprint.

[70] Plebański, J. F., and S. Hacyan; "Null Geodesic Surfaces and Goldberg-Sachs Theorem in Complex Riemannian Spaces"; preprint.

[71] Robinson, Ivor; "Null Electromagnetic Fields"; J. Math. Phys. 2 (1961) 290.

[72] Robinson, I., and A. Trautman; "Some Spherical Gravitational Waves in General Relativity"; Proc. Roy. Soc. A265 (1962) 463.

[73] Shutrick, H. B.; "Complex Extensions"; Quart. J. Math. Oxford (2) 9 (1958) 189.

[74] Simms, D. J.; "Geometric Quantization of Symplectic Manifolds"; preprint.

[75] Souriau, J.-M.; Structures des Systemes Dynamiques; Dunod, Paris, 1970.

[76] Steenrod, Norman; The Topology of Fibre Bundles; Princeton University Press, Princeton, New Jersey, 1951.

[77] Trautman, A.; "Fibre Bundles Associated with Space-Time"; Reports on Math. Phys. 1 (1970) 29.

[78] Buchdahl, H. A.; "On the Compatibility of Relativistic Wave Equations for Particles of Higher Spin in the Presence of a Gravitational Field"; Nuov. Cim. X (1958) 96.

[79] Penrose, Roger; "Solutions of the Zero-Rest-Mass Equations"; J. Math. Phys. 10 (1969) 38.

Lecture Notes in Physics

Bisher erschienen/Already published

Vol. 1: J. C. Erdmann, Wärmeleitung in Kristallen, theoretische Grundlagen und fortgeschrittene experimentelle Methoden. II, 283 Seiten. 1969.

Vol. 2: K. Hepp, Théorie de la renormalisation. III, 215 pages. 1969.

Vol. 3: A. Martin, Scattering Theory: Unitarity, Analyticity and Crossing. IV, 125 pages. 1969.

Vol. 4: G. Ludwig, Deutung des Begriffs „physikalische Theorie" und axiomatische Grundlegung der Hilbertraumstruktur der Quantenmechanik durch Hauptsätze des Messens. 1970. Vergriffen.

Vol. 5: M. Schaaf, The Reduction of the Product of Two Irreducible Unitary Representations of the Proper Orthochronous Quantum-mechanical Poincare Group. IV, 120 pages. 1970.

Vol. 6: Group Representations in Mathematics and Physics. Edited by V. Bargmann. V, 340 pages. 1970.

Vol. 7: R. Balescu, J. L. Lebowitz, I. Prigogine, P. Résibois, Z. W. Salsburg, Lectures in Statistical Physics. V, 181 pages. 1971.

Vol. 8· Proceedings of the Second International Conference on Numerical Methods in Fluid Dynamics. Edited by M. Holt. 1971. Out of print.

Vol. 9: D. W. Robinson, The Thermodynamic Pressure in Quantum Statistical Mechanics. V, 115 pages. 1971.

Vol. 10: J. M. Stewart, Non-Equilibrium-Relativistic Kinetic Theory. III, 113 pages. 1971.

Vol. 11: O. Steinmann, Perturbation Expansions in Axiomatic Field Theory. III, 126 pages. 1971.

Vol. 12: Statistical Models and Turbulence. Edited by C. Van Atta and M. Rosenblatt. Reprint of the First Edition. VIII, 492 pages. 1975.

Vol. 13: M. Ryan, Hamiltonian Cosmology. VII, 169 pages. 1972.

Vol. 14: Methods of Local and Global Differential Geometry in General Relativity. Edited by D. Farnsworth, J. Fink, J. Porter and A. Thompson. V, 188 pages.

Vol. 15: M. Fierz, Vorlesungen zur Entwicklungsgeschichte der Mechanik. V, 97 Seiten. 1972.

Vol. 16: H.-O. Georgii, Phasenübergang 1. Art bei Gittergasmodellen. IX, 167 Seiten. 1972.

Vol. 17: Strong Interaction Physics. Edited by W. Rühl and A. Vancura. V, 405 pages. 1973.

Vol. 18: Proceedings of the Third International Conference on Numerical Methods in Fluid Mechanics, Vol. I. Edited by H. Cabannes and R. Temam. VII, 186 pages. 1973.

Vol. 19: Proceedings of the Third International Conference on Nemerical Methods in Fluid Mechanics, Vol. II. Edited by H. Cabannes and R. Temam. VII, 275 pages. 1973.

Vol. 20: Statistical Mechanics and Mathematical Problems. Edited by A. Lenard. VIII, 247 pages. 1973.

Vol. 21: Optimization and Stability Problems in Continuum Mechanics. Edited by P. K. C. Wang. V, 94 pages. 1973.

Vol. 22: Proceedings of the Europhysics Study Conference on Intermediate Processes in Nuclear Reactions. Edited by N. Cindro, P. Kulišić and Th. Mayer-Kuckuk. XIV, 329 pages. 1973.

Vol. 23: Nuclear Structure Physics. Proceedings 1973. Edited by U. Smilansky, I. Talmi, and H. A. Weidenmüller. XII, 296 pages. 1973.

Vol. 24: R. F. Snipes, Statistical Mechanical Theory of the Electrolytic Transport of Non-electrolytes. V, 210 pages. 1973.

Vol. 25: Constructive Quantum Field Theory. The 1973 "Ettore Majorana" International School of Mathematical Physics. Edited by G. Velo and A. Wightman. III, 331 pages. 1973.

Vol. 26: A. Hubert, Theorie der Domänenwände in geordneten Medien. XII, 377 Seiten. 1974.

Vol. 27: R. K. Zeytounian, Notes sur les Ecoulements Rotationnels de Fluides Parfaits. XIII, 407 pages. 1974.

Vol. 28: Lectures in Statistical Physics. Edited by W. C. Schieve and J. S. Turner. V. 342 pages. 1974.

Vol. 29: Foundations of Quantum Mechanics and Ordered Linear Spaces. Advanced Study Institute, Marburg 1973. Edited by A. Hartkämper and H. Neumann. VI, 355 pages. 1974.

Vol. 30: Polarization Nuclear Physics. Proceedings 1973. Edited by D. Fick. IX, 292 pages. 1974.

Vol. 31: Transport Phenomena. Sitges International Schools of Statistical Mechanics, June 1974. Edited by G. Kirczenow and J. Marro. XIV, 517 pages. 1974.

Vol. 32: Particles, Quantum Fields and Statistical Mechanics. Proceedings 1973. Edited by M. Alexanian and A. Zepeda. V, 132 pages. 1975.

Vol. 33: Classical and Quantum Mechanical Aspects of Heavy Ion Collisions. Proceedings 1974. Edited by H. L. Harney, P. Braun-Munzinger, and C. K. Gelbke. VII, 311 pages. 1975.

Vol. 34: One-Dimensional Conductors. GPS Summer School Proceedings, 1974. Edited by H. G. Schuster. VII, 371 pages. 1975.

Vol. 35: Proceedings of the Fourth International Conference on Numerical Methods in Fluid Dynamics. June 24–28, 1974, University of Colorado. Edited by R. D. Richtmyer. V, 457 pages. 1975.

Vol. 36: R. Gatignol, Théorie Cinétique des Gaz à Répartition Discrète de Vitesses. II, 219 pages. 1975.

Vol. 37: Trends in Elementary Particle Theory. Proceedings 1974. Edited by H. Rollnik and K. Dietz. V, 472 pages. 1975.

Vol. 38: Dynamical Systems, Theory and Applications. Proceedings 1974. Edited by J. Moser. VI, 624 pages. 1975.

Vol. 39: International Symposium on Mathematical Problems in Theoretical Physics. Proceedings 1975. Edited by H. Araki. XII, 562 pages. 1975.

Vol. 40: Effective Interactions and Operators in Nuclei. Proceedings 1975. Edited by B. R. Barrett. XII, 339 pages. 1975.

Vol. 41: Progress in Numerical Fluid Dynamics. Proceedings 1974. Edited by H. J. Wirz. V, 471 pages. 1975.

Vol. 42: H II Regions and Related Topics. Proceedings 1975. Edited by D. Downes and T. L. Wilson. XII, 488 pages. 1975.

Vol. 43: Laser Spectroscopy. Proceedings 1975. Edited by S. Haroche, J. C. Pebay-Peyroula, T. W. Hänsch, and S. E. Harris. X, 466 pages. 1975.

Vol. 44: R. A. Breuer, Gravitational Perturbation Theory and Synchrotron Radiation. VI, 196 pages. 1975.

Vol. 45: Dynamical Concepts on Scaling Violation and the New Resonances in e^+e^- Annihilation. Edited by B. Humpert. VII, 248 pages. 1975.

Vol. 46: E. J. Flaherty, Hermitian and Kählerian Geometry in Relativity. VIII, 365 pages. 1976.

SPRINGER TRACTS IN MODERN PHYSICS

Ergebnisse der exakten Naturwissenschaften

Editor: G. Höhler

Associate Editor:
E.A. Niekisch

Editorial Board:
S. Flügge, J. Hamilton, F. Hund, H. Lehmann, G. Leibfried, W. Paul

Springer-Verlag
Berlin
Heidelberg
New York

Volume 66

30 figures. III, 173 pages. 1973
ISBN 3-540-06189-4

Quantum Statistics

in Optics and Solid-State Physics

R. Graham: Statistical Theory of Instabilities in Stationary Nonequilibrium Systems with Applications to Lasers and Nonlinear Optics.
F. Haake: Statistical Treatment of Open Systems by Generalized Master Equations.

Volume 67

III, 69 pages. 1973
ISBN 3-540-06216-5

S. Ferrara, R. Gatto, A. F. Grillo:

Conformal Algebra in Space-Time

and Operator Product Expansion

Introduction to the Conformal Group in Space-Time. Broken Conformal Symmetry. Restrictions from Conformal Covariance on Equal-Time Commutators. Manifestly Conformal Covariant Structure of Space-Time. Conformal Invariant Vacuum Expectation Values. Operator Products and Conformal Invariance on the Light-Cone. Consequences of Exact Conformal Symmetry on Operator Product Expansions. Conclusions and Outlook.

Volume 68

77 figures. 48 tables. III, 205 pages. 1973
ISBN 3-540-06341-2

Solid-State Physics

D. Schmid: Nuclear Magnetic Double Resonance — Principles and Applications in Solid-State Physics.
D. Bäuerle: Vibrational Spectra of Electron and Hydrogen Centers in Ionic Crystals.
J. Behringer: Factor Group Analysis Revisited and Unified.

Volume 69

13 figures. III, 121 pages. 1973
ISBN 3-540-06376-5

Astrophysics

G. Börner: On the Properties of Matter in Neutron Stars.
J. Stewart, M. Walker: Black Holes: the Outside Story.

Volume 70

II, 135 pages. 1974
ISBN 3-540-06630-6

Quantum Optics

G. S. Agarwal: Quantum Statistical Theories of Spontaneous Emission and their Relation to Other Approaches.

Volume 71

116 figures. III, 245 pages. 1974
ISBN 3-540-06641-1

Nuclear Physics

H. Überall: Study of Nuclear Structure by Muon Capture.
P. Singer: Emission of Particles Following Muon Capture in Intermediate and Heavy Nuclei.
J. S. Levinger: The Two and Three Body Problem.

Volume 72

32 figures. II, 145 pages. 1974
ISBN 3-540-06742-6

D. Langbein:

Theory of Van der Waals Attraction

Introduction. Pair Interactions. Multiplet Interactions. Macroscopic Particles. Retardation. Retarded Dispersion Energy. Schrödinger Formalism. Electrons and Photons.

Volume 73

110 figures. VI, 303 pages. 1975
ISBN 3-540-06943-7

Excitons at High Density

Editors: H. Haken, S. Nikitine
Biexcitons. Electron-Hole Droplets. Biexcitons and Droplets. Special Optical Properties of Excitons at High Density. Laser Action of Excitons. Excitonic Polaritons at Higher Densities.

Volume 74

75 figures. III, 153 pages. 1974
ISBN 3-540-06946-1

Solid-State Physics

G. Bauer: Determination of Electron Temperatures and of Hot Electron Distribution Functions in Semiconductors.
G. Borstel, H. J. Falge, A. Otto: Surface and Bulk Phonon-Polaritons Observed by Attenuated Total Reflection.

Selected Issues from

Lecture Notes in Mathematics

Vol. 345: Proceedings of a Conference on Operator Theory. Edited by P. A. Fillmore. VI, 228 pages. 1973.

Vol. 346: Fučík et al., Spectral Analysis of Nonlinear Operators. II, 287 pages. 1973.

Vol. 348: A. M. Mathai and R. K. Saxena, Generalized Hypergeometric Functions with Applications in Statistics and Physical Sciences. VII, 314 pages. 1973.

Vol. 358: F. L. Williams, Tensor Products of Principal Series Representations. VI, 132 pages. 1973.

Vol. 361: J. W. Schutz, Foundations of Special Relativity: Kinematic Axioms for Minkowski Space-Time XX, 314 pages. 1973.

Vol. 362: Proceedings of the Conference on Numerical Solution of Ordinary Differential Equations. Edited by D. G. Bettis. VIII, 490 pages 1974.

Vol. 363: Conference on the Numerical Solution of Differential Equations. Edited by G. A. Watson. IX, 221 pages. 1974.

Vol. 365: R. P. Gilbert, Constructive Methods for Elliptic Equations. VII, 397 pages. 1974.

Vol. 373: A. E. R. Woodcock and T. Poston, A Geometrical Study of the Elementary Catastrophes. V, 257 pages. 1974.

Vol. 374: S. Yamamuro, Differential Calculus in Topological Linear Spaces. IV, 179 pages. 1974.

Vol. 376: I. J. Good and D. B. Osteyee, Information, Weight of Evidence. The Singularity between Probability Measures and Signal Detection. XI, 156 pages. 1974.

Vol. 377: A. M. Fink, Allmost Periodic Differential Equations. VIII, 336 pages. 1974.

Vol. 384: Functional Analysis and Applications. Proceedings 1972. Edited by L. Nachbin. V, 270 pages. 1974.

Vol. 388: R. L. Lipsman, Group Representations. X, 166 pages. 1974.

Vol. 394: W. M. Patterson, 3rd, Iterative Methods for the Solution of a Linear Operator Equation in Hilbert Space – A Survey. III, 183 pages. 1974.

Vol. 395: Numerische Behandlung nichtlinearer Integrodifferential- und Differentialgleichungen. Tagung 1973. Herausgegeben von R. Ansorge und W. Törnig. VII, 313 Seiten. 1974.

Vol. 399: Functional Analysis and its Applications. Proceedings 1973. Edited by H. G. Garnir, K. R. Unni, and J. H. Williamson. II, 584 pages. 1974.

Vol. 408: J. Wermer, Potential Theory. VIII, 146 pages. 1974.

Vol. 409: Fonctions de Plusieurs Variables Complexes. Séminaire François Norguet 1970–1973. XIII, 612 pages. 1974.

Vol. 410: Séminaire Pierre Lelong (Analyse) Année 1972–1973. VI, 181 pages. 1974.

Vol. 413: M. Bruneau, Variation Totale d'une Fonction. XIV, 332 pages. 1974.

Vol. 415: Ordinary and Partial Differential Equations. Proceedings 1974. XVII, 447 pages. 1974.

Vol. 416: M. E. Taylor, Pseudo Differential Operators. IV, 155 pages. 1974.

Vol. 417: H. H. Keller, Differential Calculus in Locally Convex Spaces. XVI, 131 pages. 1974.

Vol. 419: Topics in Analysis. Proceedings 1970. Edited by O. E. Lehto, I. S. Louhivaara, and R. H. Nevanlinna. XIII, 392 pages. 1974.

Vol. 425: P. R. Chernoff and J. E. Marsden, Properties of Infinite Dimensional Hamiltonian Systems. IV, 160 pages. 1974.

Vol. 426: M. L. Silverstein, Symmetric Markov Processes. X, 287 pages. 1974.

Vol. 427: H. Omori, Infinite Dimensional Lie Transformation Groups. XII, 149 pages. 1974.

Vol. 430: Constructive and Computational Methods for Differential and Integral Equations. Proceedings 1974. Edited by D. L. Colton and R. P. Gilbert. VII, 476 pages. 1974.

Vol. 433: W. G. Faris, Self-Adjoint Operators. VII, 115 pages. 1975.

Vol. 434: P. Brenner, V. Thomée, and L. B. Wahlbin, Besov Spaces and Applications to Difference Methods for Initial Value Problems. II, 154 pages. 1975.

Vol. 440: R. K. Getoor, Markov Processes: Ray Processes and Right Processes. V, 118 pages. 1975.

Vol. 442: C. H. Wilcox, Scattering Theory for the d'Alembert Equation in Exterior Domains. III, 184 pages. 1975.

Vol. 446: Partial Differential Equations and Related Topics. Proceedings 1974. Edited by J. A. Goldstein. IV, 389 pages. 1975.

Vol. 448: Spectral Theory and Differential Equations. Proceedings 1974. Edited by W. N. Everitt. XII, 321 pages. 1975.

Vol. 449: Hyperfunctions and Theoretical Physics. Proceedings 1973. Edited by F. Pham. IV, 218 pages. 1975.

Vol. 458: P. Walters, Ergodic Theory – Introductory Lectures. VI, 198 pages. 1975.

Selected Issues from

Lecture Notes in Mathematics